Polymeric Drug Delivery Systems

DRUGS AND THE PHARMACEUTICAL SCIENCES
A Series of Textbooks and Monographs

Polymeric Drug Delivery Systems

Glen S. Kwon

University of Wisconsin
Madison, Wisconsin, U.S.A.

Taylor & Francis
Taylor & Francis Group

Boca Raton London New York Singapore

Published in 2005 by
Taylor & Francis Group
6000 Broken Sound Parkway NW, Suite 300
Boca Raton, FL 33487-2742

International Standard Book Number-10: 0-8247-2532-8 (Hardcover)
International Standard Book Number-13: 978-0-8247-2532-7 (Hardcover)

Library of Congress Cataloging-in-Publication Data

Catalog record is available from the Library of Congress

Taylor & Francis Group
is the Academic Division of T&F Informa plc.

Visit the Taylor & Francis Web site at
http://www.taylorandfrancis.com

Contents

Preface

Despite massive efforts in drug discovery fueled by combinatory chemistry, recombinant DNA technology, and high-throughput screening, surprisingly few molecules make it through the drug development process. While the reasons are debated, it is certain that many new chemical entities (NCEs) suffer from recurring problems that hinder development—low water solubility, instability, or inadequate pharmacokinetics. An estimated 43% of NCEs are poorly water-soluble. An entire conference, the 44th Annual Pharmaceutical Research and Development Conference, was devoted to the theme of strategies for formulating poorly water-soluble drugs. Chemical and physical stability problems of peptides and proteins are well documented. Pharmacokinetics is a major reason for failure during drug development. Many companies have abandoned "challenging" NCEs because of unsatisfactory drug delivery solutions yet these molecules have the potential to fill unmet needs in therapy and to serve as a basis for groundbreaking modalities such as gene therapy.[a]

---

[a] Davis SS, Illum L. Drug delivery systems for challenging molecules. Int J Pharm 1998; 176:1–8.

The purpose of this book is to survey efforts to adapt, modify, or tailor polymers for challenging molecules. As stated by Robert Langer, the needs of materials in drug delivery can be broken down into two categories—the creation of new materials and understanding better how to manipulate existing materials.[b] In either case, "you go to the unmet needs." These needs include reducing drug toxicity, increasing drug absorption, and improving drug release profiles. The goal is ultimately the spatial and temporal control of molecules or drug targeting. Meanwhile, the delivery of these challenging molecules by an array of polymers has captured the attention and imagination of scientists from diverse fields throughout the world. This book focuses on their efforts.

This book focuses on four classes of polymers in drug delivery—water-soluble polymers, hydrogels, biodegradable polymers, and assembled polymers. In this context, this book highlight efforts in the delivery of poorly absorbed molecules, poorly water-soluble or toxic molecules, peptide and proteins, and plasmid DNA. Well-established polymers with established biocompatibility such as the poly(α-hydroxy acids) and poly(ethylene glycols) have been included, emphasizing efforts to fill unmet needs in therapy. An attempt in this book has been made to highlight promising efforts in polymer synthesis, building structure-property correlations for drug delivery, and polymer assembles, which hold promise as nanoscopic carrier systems for a variety of biological molecules.

This book is written for graduate students, post-docs, and senior scientists in academia and the pharmaceutical and biotechnology industry. Efforts in drug delivery are extensive and interdisciplinary, and advances invariably will be made by collaboration, especially in the delivery of biological molecules—peptides, proteins, and plasmid DNA. This book is written for scientists in drug development occupied

---

[b] Henry CM. Materials scientists look for new materials to fulfill unmet needs. C&E News, 2002; 80(34):39–47.

with challenging molecules. This book may interest polymer scientists and engineers drawn to problems in drug delivery.

A unique strength of polymers is their versatility, which permits the creation of tailor-made materials to match the diversity of molecules in drug development, biotechnology, and therapy. An attempt has been made to reveal this versatility in this book. Certainly, not all the important developments in this rapidly evolving field of drug delivery have been included in this book. However, several of the acknowledged leaders have taken time away from their extremely busy lives to graciously contribute chapters and provide their perspectives on the field of drug delivery. In addition, several relatively young scientists have kindly given their time to describe their unique contributions in the field of drug delivery. I would like to thank them all for their considerable efforts toward the realization of this book.

Contributors

You Han Bae   Department of Pharmaceutics and Pharmaceutical Chemistry, University of Utah, Salt Lake City, Utah, U.S.A.

Younsoo Bae   Department of Materials Science and Engineering, Graduate School of Engineering, The University of Tokyo, Tokyo, Japan

Jeroen M. Bezemer   OctoPlus Technologies B.V., Lerden, The Netherlands

Henry Brem   Departments of Neurological Surgery and Oncology, The Johns Hopkins School of Medicine, Baltimore, Maryland, U.S.A.

Daan J. A. Crommelin   Department of Pharmaceutics, Utrecht Institute for Pharmaceutical Sciences (UIPS), Utrecht University, Utrecht, The Netherlands

Ruth Duncan   Centre for Polymer Therapeutics, Welsh School of Pharmacy, Cardiff University, Cardiff, U.K.

Yourong Fu   Departments of Pharmaceutics and Biomedical Engineering, Purdue University, West Lafayette, Indiana, U.S.A.

Hamidreza Ghandehari   Department of Pharmaceutical Sciences and Greenebaum Cancer Center, University of Maryland, Baltimore, Maryland, U.S.A.

Wim E. Hennink   Department of Pharmaceutics, Utrecht Institute for Pharmaceutical Sciences (UIPS), Utrecht University, Utrecht, Maryland, The Netherlands

Seong Hoon Jeong   Departments of Pharmaceutics and Biomedical Engineering, Purdue University, West Lafayette, Indiana, U.S.A.

Alexander V. Kabanov   College of Pharmacy, Department of Pharmaceutical Sciences, Nebraska Medical Center, Omaha, Nebraska, U.S.A.

Jichao Kang   Formulation Development Protein Design Labs, Fremont, California, U.S.A.

Kazunori Kataoka   Department of Materials Science and Engineering, Graduate School of Engineering, The University of Tokyo, Tokyo, Japan

Alex A. Khalessi   Department of Neurological Surgery, The Johns Hopkins School of Medicine, Baltimore, Maryland, U.S.A.

Akihiko Kikuchi   Institute of Advanced Biomedical Engineering and Science, Center of Excellence (COE) Program for the 21st Century, Tokyo Women's Medical University, and Core Research for Evolutional Science and Technology (CREST), Japan Science and Technology Agency, Tokyo, Japan

Sung Wan Kim   Center for Controlled Chemical Delivery (CCCD), Department of Pharmaceutics and Pharmaceutical Chemistry, University of Utah, Salt Lake City, Utah, U.S.A.

Glen S. Kwon Faculty of Pharmacy and Pharmaceutical Sciences, University of Alberta, Edmonton, Alberta, Canada

Young Min Kwon Center for Controlled Chemical Delivery (CCCD), Department of Pharmaceutics and Pharmaceutical Chemistry, University of Utah, Salt Lake City, Utah, U.S.A.

Robert S. Langer Department of Chemical Engineering and Division of Bioengineering, Massachusetts Institute of Technology, Cambridge, Massachusetts, U.S.A.

Anurag Maheshwari Center for Controlled Chemical Delivery, University of Utah, Salt Lake City, Utah, U.S.A.

Zaki Megeed Department of Pharmaceutical Sciences, University of Maryland, Baltimore, Maryland, U.S.A.

Kun Na Department of Pharmaceutics and Pharmaceutical Chemistry, University of Utah, Salt Lake City, Utah, U.S.A.

Yukio Nagasaki Department of Materials Science and Technology, Tokyo University of Science, Noda, Japan

Motoi Oishi Department of Materials Science and Technology, Tokyo University of Science, Noda, Japan

Teruo Okano Institute of Advanced Biomedical Engineering and Science, Center of Excellence (COE) Program for the 21st Century, Tokyo Women's Medical University, and Core Research for Evolutional Science and Technology (CREST), Japan Science and Technology Agency, Tokyo, Japan

Kinam Park Departments of Pharmaceutics and Biomedical Engineering, Purdue University, West Lafayette, Indiana, U.S.A.

John Samuel Faculty of Pharmacy and Pharmaceutical Sciences, University of Alberta, Edmonton, Alberta, Canada

Steven P. Schwendeman Department of Pharmaceutical Sciences, The University of Michigan, Ann Arbor, Michigan, U.S.A.

Cornelus F. van Nostrum   Department of Pharmaceutics, Utrecht Institute for Pharmaceutical Sciences (UIPS), Utrecht University, Utrecht, The Netherlands

Paul P. Wang   Department of Neurological Surgery, The Johns Hopkins School of Medicine, Baltimore, Maryland, U.S.A.

James W. Yockman   Center for Controlled Chemical Delivery, University of Utah, Salt Lake City, Utah, U.S.A.

Masayuki Yokoyama   Kanagawa Academy of Science and Technology, Takatsu-ku, Kawasaki-shi, Kanagawa-ken, Japan

Jian Zhu   College of Pharmacy, Department of Pharmaceutical Sciences, Nebraska Medical Center, Omaha, Nebraska, U.S.A.

1

N-(2-Hydroxypropyl) methacrylamide Copolymer Conjugates

RUTH DUNCAN

Centre for Polymer Therapeutics, Welsh School
of Pharmacy, Cardiff University,
Cardiff, U.K.

1. BACKGROUND

Improved therapies for the treatment of life-threatening and chronic debilitating disease are urgently needed. Many seek to achieve more effective chemotherapy through the use of nanosized drug carriers—now being called "nanomedicines." The technologies being explored include nanoparticles (1), polymer-coated liposomes (2), antibodies (3), and polymer therapeutics (4). As liposomes, antibodies (and immunoconjugates), and polyethyleneglycolylated-proteins (5) are all now

1

routinely used clinically, and an increasing number of polymer-drug conjugates have entered clinical development, the field of innovative drug delivery is at last coming of age. Whilst polymers play a pivotal role in most of these technologies, the phrase "polymer therapeutics" was specifically coined (6) to include polymeric drugs, polymer-drug conjugates, polymer-protein conjugates, polymeric micelles (to which drug is covalently bound), and those multi-component polyplexes being developed as non-viral vectors for intracytoplasmic delivery. The water-soluble N-(2-hydroxypropyl)methacrylamide (HPMA) conjugates described here are polymer therapeutics. From the industrial standpoint they are new chemical entities (NCEs) rather than conventional drug delivery systems or formulations which simply entrap, solubilize or control drug release without resorting to chemical conjugation. Since the 1970s an impressive literature has grown reporting the synthesis, characterization, solution, and biological properties of an ever-larger family of HPMA copolymer conjugates. These include polymer-drug conjugates (many also bearing targeting ligands), polymer-protein conjugates, water-soluble crosslinked polymer conjugates, "star-shaped" conjugates, and even vesicles entrapping HPMA copolymer conjugates (Fig. 1, Tables 1 and 2).

de Duve et al.'s realization that the endocytic pathway might be an ideal gateway for drug targeting ("lysosomotropic drug delivery") (7), and Ringsdorf's vision of the idealized polymer chemistry for drug conjugation (8,9), began in the 1970s to set the landscape for research in the field of polymer-drug conjugates. A vast number of hydrophilic polymers have since been explored as carriers [reviewed in (10–12)]. Usually they have molecular weight between 5000 and 100,000 g/mol, and include natural polymers (such as dextran, dextrin, and chitosan), pseudo-synthetic polymers [the polyaminoacids, e.g., poly-L-glutamic acid (PGA), poly-L-lysine] and both linear and hyperbranched synthetic polymers including most recently, dendrimers and dendronized polymers. A plethora of conjugates have now been synthesized, but few have been tested in vivo, and still fewer have reached clinical development. Only polyethyleneglycol (PEG) (13,14), PGA (15), polysaccharides

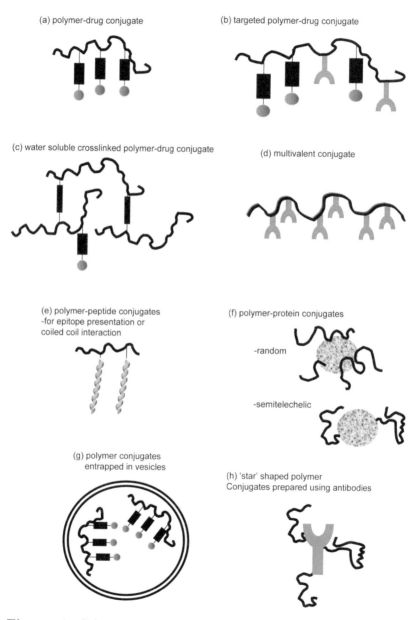

Figure 1 Schematic representation of the variety of HPMA copolymer conjugate architectures and compositions under study.

Table 1   HPMA Copolymer-Drug Conjugates

Drug bound	Use	Reference
Daunomycin	Anticancer	(59,60)
Doxorubicin	Anticancer	(20,61,62)
Methoxyporpholino-doxorubicin	Anticancer	(81)
Sacrolysin/melphalan	Anticancer	(82,83)
Paclitaxel	Anticancer	(76,84)
Camptothecin	Anticancer	(77,85)
Mesochlorin e6	Anticancer	(80,86,87)
5-Fluorouracil	Anticancer	(88,89)
Platinates	Anticancer	(90–92)
Emetine	Anticancer	(93)
9-Aminocamptothecin	Anticancer	(94)
Methotrexate	Anticancer	(95)
Aminoellipticine	Anticancer	(96)
Geldanamycin derivatives	Anticancer	(97,98)
Wortmannin	Anticancer	(99)
1,5-Diazaanthraquinone	Anticancer	(100)
Aminoglutethamide, doxorubicin	Anticancer	(101)
Ampicillin	Antibacterial	(102)
Cyclosporine A	Immunosuppression, colon inflammatory disease	(103,104)
Puromycin	Protein synthesis inhibition	(59)
L-*trans*-Epoxysuccinylleucyl amido (4-amino)butane (Ep459)	Cysteine protease inhibition	(105)
8-Aminoquinoline	Antileishmania	(106)

(16,17) and HPMA copolymer conjugates have been explored clinically. The first synthetic polymer-anticancer conjugate to enter Phase I trials did so in February 1994. This was an HPMA copolymer-doxorubicin conjugate (PK1, FCE28068). Since then, four HPMA anticancer conjugates, and two HPMA copolymer-derived gamma camera imaging agents have also been tested clinically [reviewed in (18–20)].

This chapter describes the evolution of HPMA copolymer-anticancer conjugates, their current clinical status, and reevaluates their proposed mechanism of action. The biological rationale for design, and methods used to synthesize, characterize, and evaluate HPMA copolymer-drug and HPMA

Table 2 HPMA Copolymer Conjugates Containing Proteins and Peptides

Proteins and peptides	Reference
Proteins	
Insulin	(31,138)
Transferrin	(139)
Basic fibroblast growth factor (bFGF)	(140)
Cathepsin B	(141)
β-Lactamase	(142)
8.1.1.1.1 Ribonuclease A	(143)
Antibodies	
Anti-Thy-1.2 antibody	(144,145)
Anti-transferrin receptor antibodies	(139,146)
Anti-BCL1 antibody	(147,158)
Anti-OV-TL16 antibody and its Fab′ fragment	(109,148)
Peptides	
MSH	(149)
8.1.1.1.2 VP2—human rhinovirus antigenic determinant	(138)
TAT—peptide to promote cellular uptake	(150)
EBV—peptide to promote targeting	(151)
Melittin—bioactive peptide to promote membrane damage	(152)

copolymer-protein conjugates are described. Prospects for use of HPMA copolymer conjugates in other applications are also briefly discussed.

2. HPMA COPOLYMER CONJUGATES—AN HISTORICAL PERSPECTIVE

Almost all the publications describing HPMA copolymer conjugates arise from just four research teams (with also collaborations between these laboratories): Kopecek and colleagues (Institute of Macromolecular Chemistry, Prague and subsequently the University of Utah); Ulbrich and colleagues (Institute of Macromolecular Chemistry, Prague); Rihova and colleagues (Institute of Microbiology, Prague) and my own group and colleagues (Department of Biological Sciences and the CRC Centre for Polymer Controlled Drug Delivery at Keele University, Farmitalia Carlo Erba/Pharmacia, Milan and most recently the Centre for Polymer Therapeutics at The London and Welsh Schools of Pharmacy). During the late 1960s, the Kopecek group synthesized HPMA homopolymer (pHPMA). It was studied as a potential plasma expander (21–23). This novel polymer was also viewed as a potential platform for drug delivery (24–26), and in the 1970s HPMA copolymers were prepared to contain libraries of peptidyl spacers chosen for cleavage by model enzymes including trypsin and chymotrypsin (27–29). The first HPMA copolymer-drug conjugates (30) and HPMA copolymer-protein conjugates (e.g., insulin) were also synthesized at this time (31,32). Subsequent evolution of the Kopecek program has been well documented (33–42).

At the same time we—Margaret Pratten, John Lloyd, and I—were conducting experiments to unravel the mysteries of the mechanism of pinocytosis. Polyvinylpyrolidone (PVP), divinylethermaleic anhydride (DIVEMA), and polycations (including poly-L-lysine, poly-L-α-ornithine and DEAE dextran) were chosen as model probes (43–48). Reports from de Duve, and a collaboration with Helmut Ringsdorf, inspired our search for effective lysosomotropic polymer-drug conjugates. One early goal was the identification of a polymer molecular weight that would ensure cellular internalization, and this was defined using molecular weight fractions of PVP (49). With the growing realization that an ideal polymer-drug conjugate would need a lysosomally degradable, peptidyl linker, Ringsdorf catalyzed the collaboration between the groups at Keele and Prague. In 1979, I began to study the ability of mixtures of lysosomal enzymes (tritosomes) to cleave peptidyl side-chains in HPMA copolymer conjugates using a library of HPMA copolymer conjugates. Initially we imagined that cathepsin D would be active against these substrates, but studies showed little or no sidechain degradation (50). This was not surprising, as later it was clear that longer (at least penta-peptidyl) sequences were required for cathepsin D activity (51). At this time studies investigating the endocytic uptake of HPMA copolymer conjugates clearly showed intracellular degradation to liberate the drug model *p*-nitroaniline (52). Re-investigation led to the discovery that it was the lysosomal thiol-dependent proteases that were crucial for conjugate degradation (53), and this observation prompted the design of more appropriate sequences, including Gly-Phe-Leu-Gly, for conjugation of anticancer agents (54). These studies also underlined the physiological importance of undertaking degradation experiments with mixtures of enzymes that the substrate is likely to encounter in vivo. Fundamental studies on the endocytic mechanism also underlined the need for prudent conjugate design. At high polymer-loading hydrophobic and cationic moieties caused substantial non-specific membrane binding of HPMA copolymer conjugates (55,56). This is a non-desirable feature if in vivo targeting is the goal. In addition, synthesis of HPMA copolymers bearing sugars (galactose and mannose) demonstrated that multivalent polymer conjugates could be designed to promote receptor-mediated targeting in vivo (57,58).

Throughout (at Keele University from 1975 to 1992), my research was supported by the UK Cancer Research Campaign. The ultimate goal was always to identify polymer anticancer-drug conjugates suitable for clinical evaluation. Emergence of the anthracyclines in the early 1980s led to our choice of daunomycin (59,60) and subsequently doxorubicin (61,62) as the most appropriate drug candidates for conjugate synthesis. The evolution of our research program has also been well documented (4,6,37,63–74).

It is important to emphasize that the close collaboration between Keele and Prague with pivotal contributions of Blanka Rihova and colleagues, and Karel Ulbrich and colleagues, brought together a unique interdisciplinary team that within five years designed two HPMA copolymer-doxorubicin conjugates. After patent filing in 1984 (75), conjugates were transferred into Phase I/II clinical trial as PK (Prague-Keele) 1 (FCE28068[a]) and PK2 (FCE28069[a]). The project was made possible by the UK Cancer Research Campaign Phase I/II Committee, Czechoslovak Academy of Sciences, and not least,

---

[a] It should be noted that in the text these compounds are described as either PK1 (FCE28068) or PK2 (FCE28069) to acknowledge the distinction between the product prepared at the laboratory scale (PK1, PK2) or industrially (FCE28068, FCE28069). The latter have a defined specification.

the Polymer Laboratories and Farmitalia Carlo Erba (became Pharmacia), Milan. Within Farmitalia Carlo Erba we established the protocols for product synthesis to Good Manufacturing Practice (GMP) and protocols for Phase I/II trials to Good Clinical Practice (GCP) guidelines. Although it took almost a decade to transfer first polymer conjugates from laboratory to clinic, these meticulous efforts gave credibility to this new class of therapeutics within the pharmaceutical industry, ensuring further industrial development of not only HPMA copolymer-drug conjugates, but also of the other polymer-drug conjugates that have since emerged.

Into the 1990s development of anticancer conjugates continued. Building on experience, the time for lead identification and preclinical development shortened (3–5 years) and it is now comparable with that seen for any other NCE. The taxanes and camptothecins emerged as new classes of anticancer agents, and colleagues in Pharmacia developed second-generation HPMA copolymer conjugates of paclitaxel (76) and camptothecin (77). On my return to academia we designed a family of HPMA copolymer platinates (78) that, following license of technology to Access Pharmaceuticals Inc., have also entered clinical trial. New concepts for polymeric anticancer combinations, including polymer-enzyme polymer-drug therapy (PDEPT) (79) and, from the Kopecek team, HPMA anticancer conjugates containing photoactivatable drugs (80) have also emerged. Readers are directed to the reviews cited in this section for a full list of primary studies conducted from the 1970s to the 1990s. The remaining sections of this chapter focus on the current state of the art and discuss the future prospects for use of HPMA copolymer conjugates.

3. SYNTHESIS AND CHARACTERIZATION OF HPMA COPOLYMER CONJUGATES

3.1. Polymer-Drug Conjugates

Many HPMA copolymer conjugates have been described. They contain drugs (Table 1), proteins, antibodies, and targeting ligands (Table 2). Although many subtle changes in synthetic chemistry have evolved in an attempt to prepare better-defined structures, most conjugates are still prepared by polymer analogous reaction (33,37). Free radical precipitation co-polymerization of HPMA monomer and methacryolylated (MA) peptides terminating in reactive p-nitrophenoxy groups (ONp) gives a polymeric intermediate to which drugs, drug analogs or targeting ligands containing an aliphatic amine can be subsequently bound. This is undertaken either directly by aminolysis, or after further polymer side-chain modification to generate the required functionality. This procedure (Fig. 2) allows synthesis of a library of conjugates using a common intermediate of defined molecular weight characteristics, and it has been used to generate the HPMA anticancer conjugates that have progressed to clinical trial. Drugs (e.g., doxorubicin) or targeting ligands (e.g., galactosamine) containing an aliphatic $-NH_2$ are bound to a peptidyl linker via an amide bond (Fig. 3a,b). A monomer feed ratio of HPMA:MA-peptide-ONp (95:5) was used to prepare PK1 (Fig. 3a), and this was changed to 90:10, to allow addition of both doxorubicin and galactosamine, in PK2 (Fig. 3b). Increasing the content of pendant side-chains can also afford a higher drug loading. Clinically used gamma camera imaging agents have been prepared by *ter*-polymerization to

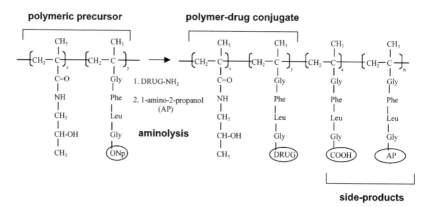

Figure 2 Synthesis of HPMA copolymer conjugates via polymer analogous reaction.

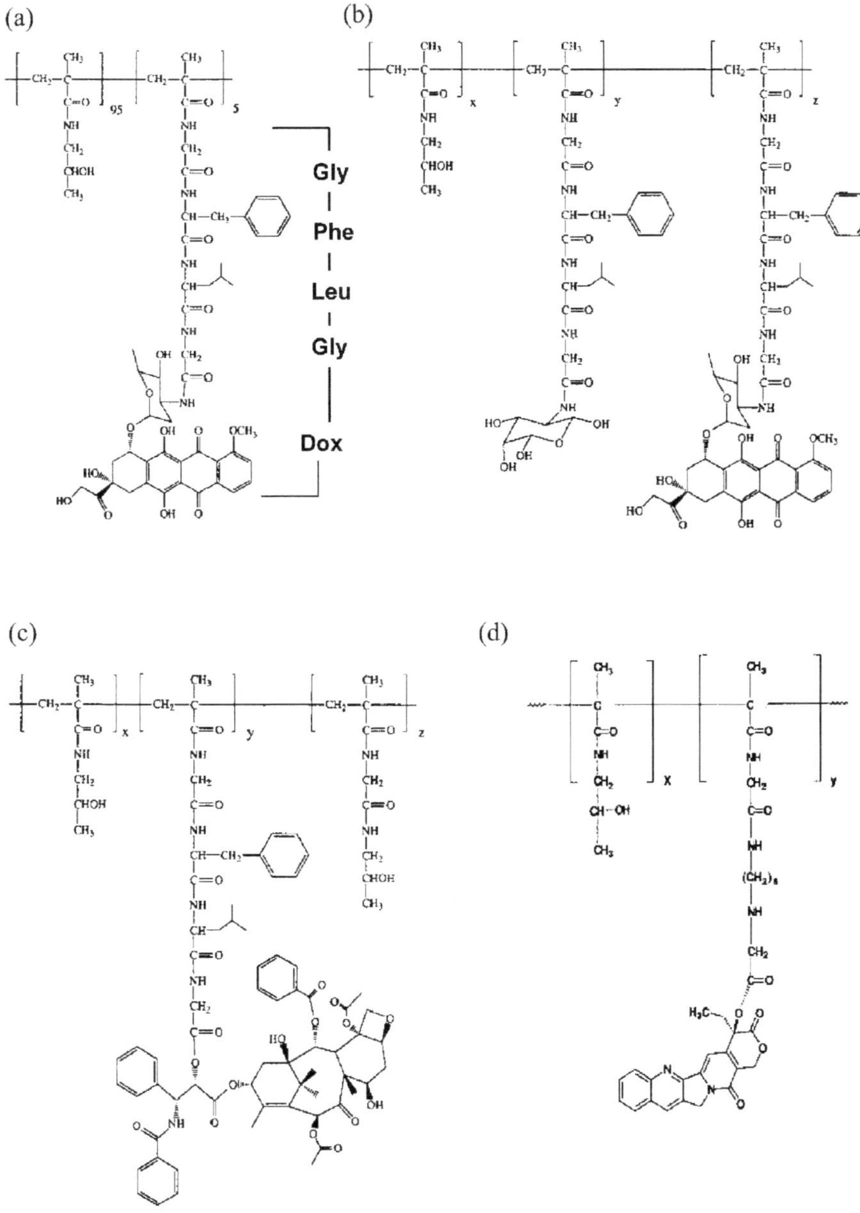

(a)   (b)

(c)   (d)

Figure 3 HPMA copolymer-anticancer conjugates. Panel (a): HPMA copolymer-doxorubicin (PK1, FCE28068); panel (b): HPMA copolymer-doxorubicin-galactosamine (PK2, FCE28069); panel (c): HPMA copolymer-paclitaxel; panel (d): HPMA copolymer-camptothecin.

include MA-tyrosinamide as a third comonomer ($\sim 1\,\text{mol}\%$), which can subsequently be radiolabeled.

There are multiple synthetic routes to the creation of an HPMA copolymer bearing, for example, a pendant tetrapeptide side-chain bearing drug (Fig. 4a) (37). In some cases, synthesis of peptidyl prodrugs before polymer conjugation has become the preferred route. For example, glycine derivatives of paclitaxel via the 2' position and camptothecin modified at the C-20 α-hydroxy group (Fig. 3c,d) were conjugated to the polymer. This can be advantageous as it enables conjugation of well-characterized drug derivatives to shorter side-chains, e.g., MA-glycine-ONp (MAG) (76,77).

Conjugates can also be prepared by co-polymerization (or *ter*-polymerization) of HPMA, MA-drug and/or MA-targeting ligand monomers (Fig. 4). MA-*N*-(4-aminobenzenesulfonyl) N'-0-butylurea (107), MA-D-galactose (108) and MA-PEG-Fab'OV-TL have been introduced into HPMA copolymer conjugates in this way (109). Use of novel comonomers does, however, require careful optimization of the polymerization kinetics, on a case-by-case basis, to ensure adequate control of the molecular weight characteristics, comonomer incorporation and conversion (polymerization yield) at different loadings. The fact that free radical precipitation polymerization of HPMA and MA-Gly-Phe-Leu-Gly-ONp (5–10 mol%) reproducibly gives a polymeric intermediate of $\text{Mw} \sim 20{,}000\text{--}25{,}000\,\text{g/mol}$ and relatively low polydispersity (1.2–1.5), which additionally has established clinical safety, continues to make it an attractive starting point for drug conjugation.

Nevertheless, synthesis via the polymer analogous reaction route does have limitations. Quantitation of side-products including –COOH terminating peptidyl side-chains (hydrolysis of the active ester) and 2-hydroxypropylamide terminating units (from the termination reaction) is difficult (Fig. 2). Additionally, removal of free drug can be troublesome especially when potent anticancer agents such as anthracyclines are used. Recently, Godwin et al. (110) prepared HPMA copolymer conjugates via a homopolymer intermediate synthesized by atom-transfer free radical polymerization (ATRP) using

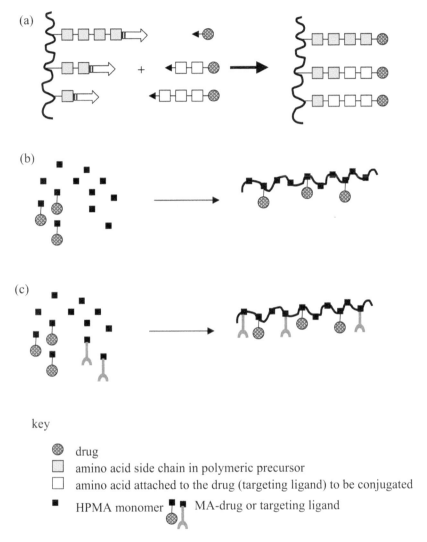

Figure 4 Synthetic routes to generate HPMA copolymer-linker-drug-targeting residue conjugates. Panel (a): routes to the synthesis of a tetrapeptide linker via aminolysis of an HPMA copolymer precursor; panel (b): preparation of an HPMA copolymer conjugate by co-polymerization of HPMA and MA-drug-containing monomers panel (c): *ter*-polymerization of HPMA, MA-drug-bearing and MA-ligand-bearing co-monomers.

N-methacryloxysuccinimide as monomer with a copper catalyst. The resultant polymeric intermediates had an Mn of 10,000–40,000 g/mol and narrow polydispersity (Mw/Mn = 1.1–1.2). Reaction with model amines verified the ability to generate HPMA homopolymer (using 1-amino-2-propanol) and also conjugates containing pendant side-chains.

A recognized limitation of HPMA as a polymeric carrier is the lack of biodegradability of the –C–C– polymer chain. This excludes safe parenteral administration of polymers > 40,000 g/mol and prohibits chronic use of its conjugates. Ingenious methods have been sought to increase "apparent" molecular weight whilst using individual polymer chains of lower molecular weight. Water-soluble HPMA copolymers crosslinked via biodegradable peptidyl sequences can be prepared (33,111–113) (Fig. 1c), and used to generate longer circulating carriers for anthracyclines (114). Another option, pioneered by Ulbrich and colleagues, has been the systematic design of a family of "star-shaped" conjugates (Fig. 1h). An antibody serves as the central "core" that is conjugated to semi-telechelic HPMA copolymers prepared by radical precipitation copolymerization in the presence of 3-sulfanylpropionic acid as a chain terminating agent subsequently reacted with N-succinimide (115).

3.2. Polymer-Drug Linkers

Biological and clinical studies have underlined the fundamental importance of the polymer-drug linker. When a conjugate is administered parenterally its stability in the circulation and ability to release drug at an appropriate rate on arrival in the target tissue is a principal determinant of therapeutic index. A variety of chemistries can be used to create pendant linkers (Fig. 5) (10). Lysosomally degradable peptides ensure site-specific intracellular activation, but many drug candidates do not have a convenient aliphatic –NH$_2$ for this purpose.

5-Fluoruracil has been linked to HPMA copolymer bearing peptide side-chains via α-glycine (88), and hydrolytically labile terminal ester bonds were used to prepare conjugates of paclitaxel and camptothecin (76,77,116). The endosomal/

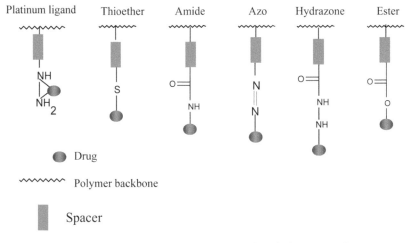

Figure 5   Linking chemistries used to bind drugs and targeting residues to HPMA copolymer conjugates.

lysosomal proton pump lowers intravesicle pH to 4.0–5.5 (lysosomes) and 5.6–6.5 (endosomes). As extracellular pH is ~7.4, pH-sensitivity can be used to accelerate drug release intracellularly. Ulbrich and colleagues have exploited pH-sensitive hydrazone or *cis*-aconityl linkers to explore this strategy for anthracycline conjugation (117–120). To create a library of HPMA copolymer-platinates we used peptidyl spacers (Gly-Gly or Gly-Phe-Leu-Gly) terminating in –COOH, ethylenediamine (en), malonate (mal), or aspartate as platinum ligands. Cisplatin-like (en) and carboplatin-like (mal) polymeric platinates were thus prepared. The conjugates based on the -Gly-Phe-Leu-Gly-en-Pt linker only released active platinum species after side-chain cleavage (90,91) (Fig. 6). The other platinum chelates, including the malonate derivative AP5280, released platinum species by hydrolysis (Fig. 7). Use of pendant azo linkers was pioneered by Kopecek and colleagues, with the aim of achieving site-specific colonic cleavage by bacterial azo-reductases (121,122). Other conjugation options have been used, for example, HPMA copolymers bearing pendant side-chains terminating in maleimide have been used to bind antibody Fab′ fragments (109) and peptides (e.g., TAT peptide) (123) via thioether linkages.

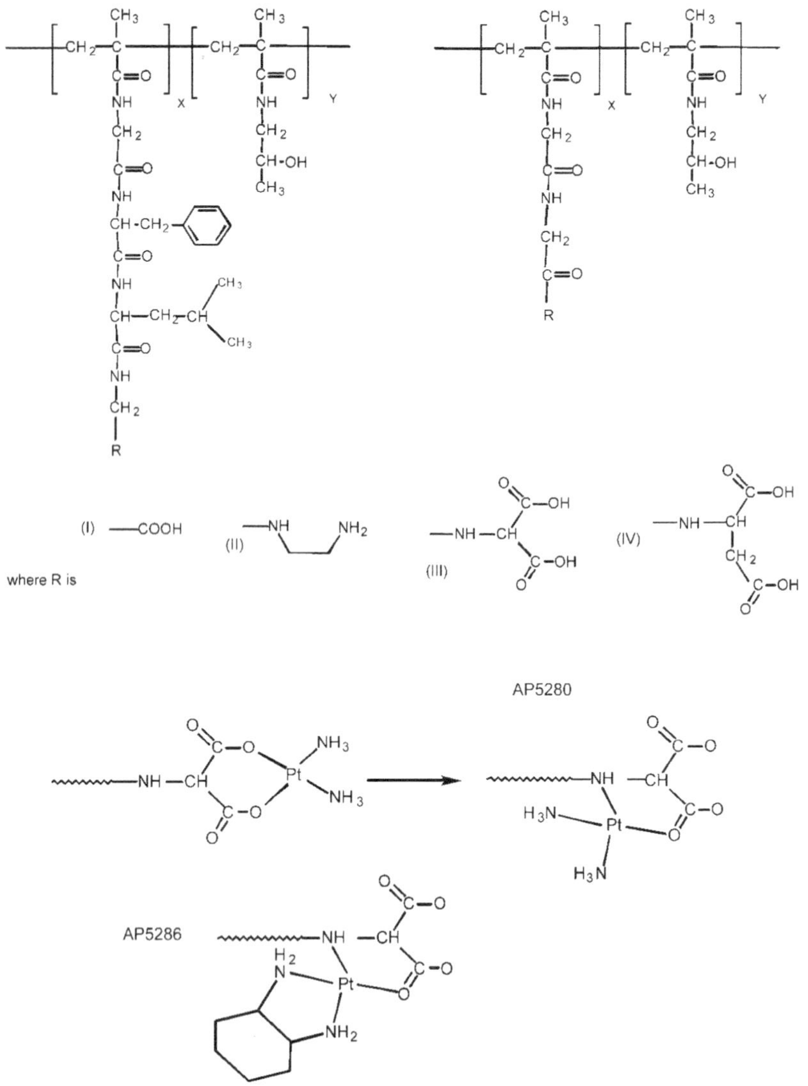

Figure 6   HPMA copolymer-platinates.

Whichever linkage chemistry is used, an important feature to note is the clear influence of drug loading on conjugate conformation in solution. This in turn governs drug release rate. Solution conformation determines rates

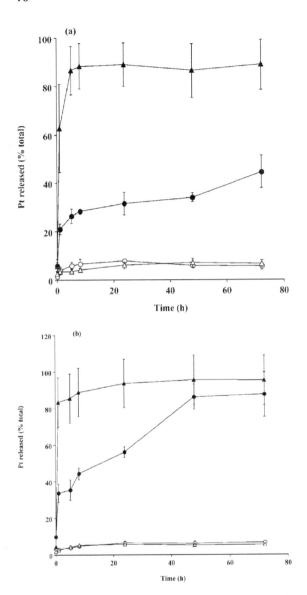

Figure 7 Release of platinates from HPMA copolymer-platinates in vitro. Panel (a): release in phosphate buffered saline (pH 7.4); panel (b): release in citrate phosphate buffer (pH 5.5). HPMA copolymer-Gly-Phe-Leu-Gly-en-Pt (△—△); HPMA copolymer-Gly-Gly-en-Pt (○—○); HPMA copolymer-Gly-Phe-Leu-Gly-COO-Pt (▲—▲); HPMA copolymer-Gly-Gly-COO-Pt (●—●). (From Ref. 90.)

of both hydrolytic and enzymatic degradation. This phenomenon was again evident in recent experiments with HPMA copoly-mer-aminoellipticine conjugates (Fig. 8). High loading with hydrophobic drugs can reduce the rate of pro-drug activation with a consequent reduction in antitumor activity.

3.3. Characterization and Formulation of Polymer-Drug Conjugates

Transfer of HPMA copolymer-anticancer conjugates into clinical trial required careful characterization of polymer-drug conjugates to Regulatory Authority Standards. Colleagues

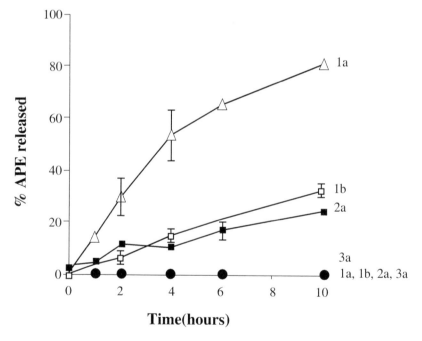

Figure 8   Release of aminoellipticine (APE) from HPMA copolymer conjugates during incubation with mixtures of isolated lysosomal enzymes (tritosomes). The results shown represent HPMA copolymer-Gly-Phe-Leu-Gly(5 mol%)-APE (1a,1b); HPMA copolymer-Gly-Phe-Leu-Gly(10 mol%)-APE (2a) and HPMA copolymer-Gly-Gly(5 mol%)-APE (3a). The numbers in bold represent the same samples but incubated without tritosomes. (From Ref. 96.)

at Farmitalia Carlo Erba/Pharmacia in Milan undertook this pioneering work. High field (600 MHz) 2D NOESY and TOCSY ^{1}H NMR were used to define the identity of the HPMA copolymer-doxorubicin conjugates FCE28068 and FCE28069 (124). Proton assignment, by comparison with reference compounds, enabled visualization of the connection between doxorubicin and the polymer chain, the D and L isomeric forms of Phe (inadvertently introduced by the synthetic chemistry used), and the alternative side-chains terminating in –COOH and 2-hydroxyproylamide. Such high field NMR also reveals contaminating solvents, free 2-propanolamine, free doxorubicin, and in the case of FCE28069, free galactosamine with great sensitivity < 0.1%. Using NMR the bound galactosamine present in FCE28069 was found to be present in the four isomeric α- and β-pyranose and furanose forms with the α-pyranose form predominating (124).

For clinical use, precise definition of molecular weight characteristics and batch-to-batch reproducibility is essential. Commercially available gel permeation chromatography (GPC) columns and compound-specific universal calibrations derived from narrow molecular weight fractions of the polymer-drug conjugate have been used for this purpose (125,126). In this context, 13 fractions of the HPMA copolymer-paclitaxel conjugate (PNU166945) and 6 fractions of the HPMA homopolymer were characterized by size-exclusion chromatography, viscometry, and light scattering to give the molar mass distribution, intrinsic viscosity, and size dimensions (127). Whereas the UV extinction coefficient is often used to estimate total drug in a conjugate, inaccuracies arising from changes in the extinction coefficient during drug conjugation can give a dangerous measure of clinical dose. Specialized HPLC techniques were used to determine total and free drug content of conjugates (128,129) and also to quantitate total and free galactose in FCE28069 (129). For FCE28068 the assay had a sensitivity of 0.01–0.02% and recovery of free doxorubicin of 97.7% (129).

HPMA copolymer-drug conjugation greatly increases the solubility of hydrophobic drugs, e.g., doxorubicin solubility increases > 10-fold in FCE28068, but despite this high solubility ($\sim$5%) the dissolution rate of FCE28068 and FCE28069 in aqueous solution is relatively slow ($>$ 30 min). For clinical use, faster dissolving, filter sterilized, freeze dried formulations were developed containing polysorbate 80 as a dissolution enhancer, lactose, and a small amount of ethanol (130). The resulting lyophilized cake could easily be reconstituted with water for injection or NaCl and dissolution occurred then in $\sim$2 min. To ensure safety, absence of particulates in pharmaceutical formulations for injection is essential, so considerable care is needed as HPMA copolymer-drug conjugates containing hydrophobic side-chains tend to form unimolecular micelles and/or multi-molecular aggregates in aqueous solution (131–134). Recent experiments using small angle neutron scattering (SANS) have shown subtle differences in the coil structure of PK1 and PK2 (135) that would account for the differences in their maximum tolerated dose (MTD) seen in patients (discussed later).

3.4. Polymer-Peptide and Polymer-Protein Conjugates

The concept of protein modification by PEGylation is now well established clinically (5,136,137), and is used to increase protein solubility and stability, reduce immunogenicity, prevent rapid renal clearance of small proteins, and prevent receptor-mediated clearance by cells of the reticuloendothelial system. Many HPMA copolymer-antibody, -protein, and-peptide conjugates have also been synthesized (Table 2). In the main, protein incorporation has been used to promote cell-specific drug targeting, but conjugates of biologically active proteins, enzymes and coiled-coil peptide domains have also been reported.

   Historically conjugates were prepared by non-specific aminolysis using HPMA copolymer-ONp resulting in random binding to $-NH_2$ groups in the protein or peptide, but this does lead to very heterogeneous products. Synthetic routes to more uniform products are now being used. HPMA copolymer-OV-TL-TL16 conjugates were prepared by: (i) random aminolysis; (ii) reaction with aldehyde groups arising from antibody carbohydrate oxidation (155), and also most recently; (iii) using HPMA copolymer-maleimido intermediates to create thioether linkages to the antibody Fab' (109). Comparison of antigen recognition has shown that site-specific conjugation of Fab' results in highest retention of antigen-binding activity. As mentioned above, Ulbrich and colleagues were the first to create semitelechelic HPMA copolymers (156). This tool has been useful to synthesize more homogeneous enzyme (chymotrypsin and bovine ribonuclease A) (143,156), and antibody (B1 monoclonal) conjugates (157,158). Using semitelechelic HPMA copolymer-doxorubicin intermediates, it was possible to synthesize star-shaped immunoconjugates in which 40–45% of the $-NH_2$ groups in the antibody were modified (157,158). By elimination of random antibody crosslinking, the star product had a lower molecular weight (300,000–350,000 g/mol) compared to conjugates prepared by the classical route (1,300,000 g/mol).

Peptides have been incorporated into HPMA conjugates to promote receptor-mediated targeting. Melanocyte stimulating hormone (MSH) has been used to target melanoma (149), and an Epstein-Barr virus (EBV) peptide (EDPGFFNVE) to target T and B cell lymphoma (151). MSH was coupled via classical non-specific aminolysis; nonetheless HPMA copolymer-MSH retained biological activity, as demonstrated by the ability of conjugates to stimulate both tyrosinase activity and melanin production. Comparison of HPMA copolymer-doxorubicin conjugates with and without MSH in the subcutaneous (sc) B16F10 murine melanoma model in vivo confirmed greater antitumor activity (two-fold) of the targeted conjugate (149). The HPMA copolymer-EDPGFFNVE conjugate, also prepared by aminolysis, showed binding to B lymphocytes (Raji B cells) and T cells (CCRF-CEM and CCRF-HSB-2). Binding to T cells was approximately two-fold higher; it increased with increasing peptide substitution per chain (multivalency), and also when a longer spacer (Gly-Phe-Leu-Gly) was used (151,153). Polymer-protein conjugates present significant challenges for validated characterization. GPC is often used to estimate conjugate molecular weight but the elution profiles are difficult to interpret. Standardized GMP manufacture and quality control for clinical trial will require very careful validation.

4. BIOLOGICAL RATIONALE FOR DESIGN AND POTENTIAL APPLICATIONS

The factors important for polymer-drug conjugate design are now well established (4,18). The polymeric carrier must be "biocompatible," a term defined as "the ability of a material to perform, with an appropriate host response in a *specific application*" (159). No polymer can be designated "biocompatible" unless qualified by information on its route, and frequency of administration. Polymers for parenteral use must not elicit a cellular or humoral (IgG or IgM) immune response. They must be readily eliminated, and preferably be biodegradable. A polymeric carrier requires sufficient drug-carrying capacity, tailored on a case-by-case basis to drug potency, and targeting efficiency. (HPMA copolymers are well equipped in this respect due to their multifunctional nature.) The conjugate should protect its payload from premature metabolism/inactivation during transit, and display the whole-body pharmacokinetic pertinent to proposed use. For intravenous (iv) injection and disease-specific targeting, the conjugate must avoid clearance by the reticuloendothelial system (RES). Specific applications may require transport across biological barriers, e.g., the gastrointestinal tract or blood brain barrier (BBB). In all cases, conjugation seeks to concentrate the drug in the target tissue, while minimizing access to potential sites of toxicity. At the cellular level those conjugates designed to deliver via a specific intracellular compartment (for lysosomotropic or endosomotropic delivery) must be able to enter the target cell and display intracellular trafficking appropriate to the proposed function (Fig. 9). Finally, the conjugate must be able to deliver the drug, protein or oligonucleotide at an appropriate rate without premature release (inactivation) in transit.

Those anticancer conjugates successful in clinical studies were systematically designed with careful optimization of

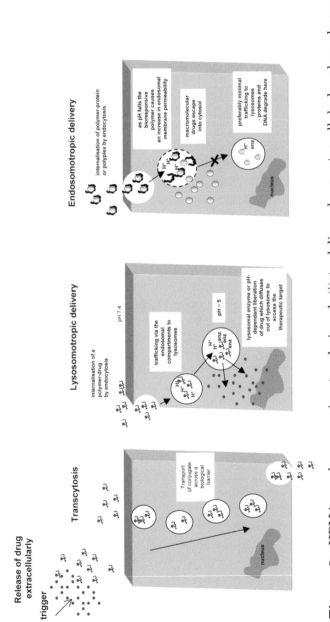

Figure 9 HPMA copolymer conjugates designed: (i) to deliver drug extracellularly when cleaved by an external trigger (e.g., azo linkers), (ii) to transfer drugs across a biological barrier (e.g., oral peptide vaccine delivery), (iii) for lysosomotropic delivery of anticancer conjugates, or (iv) to deliver macromolecular drugs into the cytosol via the endosomotropic route.

each of these features. For other therapeutic indications conjugates must be tailored on a case-by-case basis.

4.1. Biocompatibility

Biocompatibility of pHPMA was initially studied in the context of its use as a plasma expander. [^{14}C]pHPMA showed renal elimination (of lower molecular weight polymers), and lack of mitogenicity, hematotoxicity, and immunogenicity (160,161). pHPMA had a maximum tolerated dose (MTD) in rabbits of >30 g/kg. The pioneering research of Rihova and colleagues defined the principal immunological issues to be considered when constructing any polymer-drug conjugate. Studies exploring HPMA copolymer conjugates helped set these basic principles (162,163). HPMA copolymers bearing pendant peptide side-chains are only weakly immunogenic, acting as thymus-independent antigens. The antibodies raised are against both the peptide side-chain and the HPMA polymer backbone. Like PEG, when bound to proteins, HPMA copolymers significantly reduce protein immunogenicity (164–167). Copolymers are neither mitogenic, nor cytotoxic, and do not activate complement (168). Whereas higher molecular weight polymer chains accumulate in the RES and skin over time (169,170), lower molecular weight fractions of ^{125}I-labeled HPMA copolymers do not show organ tropism and are readily eliminated via the kidneys (169).

Following administration of HPMA copolymer conjugates, the polymeric carrier and the bound drug can be considered as the two primary metabolites. Documentation of the fate of each component is important, but it is imperative to understand the toxicological profile of the conjugate as a whole, i.e., the drug substance to be given. HPMA copolymer-anticancer conjugates have typically been significantly less toxic than free drugs in animals (171). After a single iv injection to C57black mice, PK1 had an MTD of ~95 mg/kg doxorubicin-equivalent. This can be compared to an MTD of ~20 mg/kg for doxorubicin in this strain (62). In MF1 mice, the MTD for PK1 was 45 mg/kg doxorubicin-equivalent

(172). Evaluation of immunogenicity showed that neither PK1 nor PK2 elicited antibody production (173); they did not cause bone marrow toxicity at the doses used. Both conjugates displayed significantly less cardiotoxicity (a known side-effect of anthracyclines) than free doxorubicin in a Wistar rat model (174,175).

Preclinical toxicology of FCE28068 is the only study so far conducted to "good laboratory practice" (GLP) guidelines (171). The compound was administered as a single dose to MF1 mice (22.5 or 45 mg/kg) or as multiple doses to MF1 mice or Wistar rats weekly for 5 consecutive weeks (12.0 or 22.5 mg/kg, mice; or 3 and 5 mg/kg, rats). Hematological changes were observed shortly after treatment, and in the single dose study alanine- and aspartate-aminotransferase levels were elevated at higher doses. Liver damage was seen only in rat tissue during histological examination. Other histological changes induced included thymic and testicular atrophy, bone marrow depletion, gastrointestinal tract changes and, in the multiple dose study, an increase in nuclear size in the proximal tubules of the kidney (although no changes in urine biochemistry were seen). Recovery from these effects in rats was seen at 59 days. These toxicities are generally typical of anthracyclines, and the study recommended an FCE28068 safe starting dose for Phase I clinical trials of $20 \, \text{mg/m}^2$ doxorubicin-equivalent.

4.2. HPMA Copolymer-Anticancer Conjugates

More than 80% of publications describing HPMA copolymer conjugates refer to *anticancer conjugates*. Although driven by the desire to create improved chemotherapy, this indication also allowed progress to early clinical evaluation. The HPMA copolymer conjugate PK2 (Fig. 3b) realized the ambition of a lysosomotropic conjugate able to target via receptor-mediated endocytosis. It contains -Gly-Phe-Leu-Gly-doxorubicin pendant side-chains, stable in the circulation (176) and degraded intracellularly by the lysosomal thiol-dependent proteases, particularly cathepsin B, to liberate drug over a 24–48 h period. The degradation rate is similar in vitro (177), in vivo

(178,179), and apparently also in man (20). Although intellectually attractive, tumor targeting by receptor-mediated endocytosis has since been difficult to realize in vivo. PK2 is still the only "targeted" conjugate to enter clinical trial (180), and even so PK2 localizes to both normal hepatocytes and hepatocellular carcinoma via the asialoglycoprotein receptor (ASGR) (57,58,181). This is organ-specific, rather than tumor-specific, targeting. The magnitude of PK2 targeting in vivo (% dose) is also markedly dose-dependent due to ASGR receptor saturation (179). Matching conjugate dose and receptor number/saturation is a challenge clinically, particularly as receptor number may be up- or down-regulated in the presence of ligand. A PK2 gamma camera imaging analog-enabled non-invasive monitoring of conjugate fate in patients helped to guide the clinical protocol for dose escalation (182–184).

4.2.1. Passive and Active Tumor Targeting

As more than ten anticancer conjugates have now been transferred to clinical trial and all except PK2 are without a biorecognition motif, the mechanism of HPMA copolymer tumor targeting is worthy of specific discussion. In the 1980s it became clear that conjugates exhibit significant passive tumor targeting by virtue of the enhanced permeability and retention (EPR) effect (185,186). This phenomenon (Fig. 10) arises due to the hyperpermeability of tumor vasculature allowing preferential extravasation of circulating macromolecules into tumor tissue. Absence of tumor lymphatic drainage further promotes conjugate retention. EPR-mediated targeting was first observed using HPMA copolymer-daunomycin in a Walker sarcoma model (187), and later confirmed for PK1 (188) and HPMA copolymer platinates in a B16F10 melanoma model (90,91) (Fig. 11). Since then, numerous studies using different in vivo tumor models and various HPMA copolymer conjugates have verified the importance of this phenomenon (188–190).

Both polymer- and tumor-related characteristics govern the extent of EPR-mediated targeting. Using HPMA copolymer

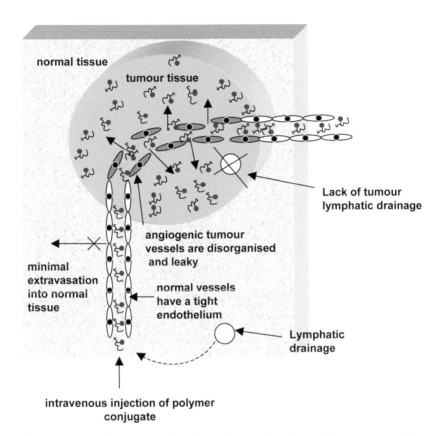

Figure 10 Schematic showing the mechanism of tumor targeting by the EPR effect.

molecular weight fractions in the range 10,000–800,000 g/mol as probes (191,192), we found that tumor uptake had broad size tolerance. In the human colon xenograft LS174T, murine B16F10, and rat sarcoma 180 models, accumulation was independent of HPMA copolymer size. Conjugate plasma

Figure 11 (*Facing page*) HPMA copolymer-anticancer conjugates demonstrate improved targeting of chemotherapy to an sc B16F10 solid tumor models in vivo. The data shown represent the time-dependent accumulation of (a) PK1 and free doxorubicin and (b) HPMA copolymer-platinate. (From Refs. 90,188.)

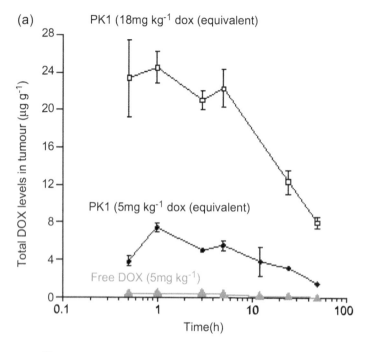

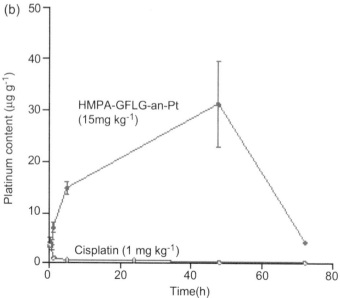

Figure 11 (*Caption on facing page*)

concentration was the main driving force controlling the extent of tumor capture (191,192). Visualization of tumor accumulation using the window chamber and a P22 carcinosarcoma model showed that HPMA copolymer conjugate extravasation begins within seconds of iv injection and homogenous intratumoral distribution is seen within minutes (193).

The magnitude of EPR-mediated tumor uptake determines conjugate antitumor activity in vivo and this, in turn, governs the predictive usefulness of tumor models. A large panel of murine and xenograft models was used to study the effect of tumor type on targeting, and tumor size on the extent of EPR-mediated targeting (194). The dye Evans blue (which binds to albumin and is a known marker of vascular permeability), and PK1 were used as probes. Approximately 50% of the tumor models studied showed tumor size-dependent targeting of PK1, with smaller tumors (<200 mg) displaying highest uptake in terms of percentage dose per gram tumor (194) (Fig. 12). The other tumor models showed tumor size-independent localization. After 1 h the levels of PK1 detected in tumor tissue varied between 2 and 18% dose/g tumor, a 10-fold variation depending on tumor type.

Importantly, intratumoral free doxorubicin (liberated from PK1) varied to a much greater extent (~200-fold) (194), suggesting a greater variation in activating enzyme level than EPR-mediated targeting. Comparison of three doxorubicin drug-carriers in a standardized B16F10 murine melanoma model showed highest accumulation for the liposomal formulation Doxil® (up to 13.2% dose/g tumor), followed by PK1 (7.8% dose/g tumor) and then a polyamidoamine (PAMAM) dendrimer-doxorubicin conjugate (2% dose/g tumor). Again, tumor levels showed a direct correlation with the plasma half-life of each carrier (195). It is noteworthy that the clinical dose of PK1 is >4-fold higher than can be safely administered for Doxil®.

The increased uptake of PK1 in smaller tumors (expressed as % dose/g) (194) (Fig. 12b), and the observation that maximum PK1 extravasation occurs in the tumor periphery where angiogenesis is taking place (196), posed the question, "Would vascular morphology, vascular permeability,

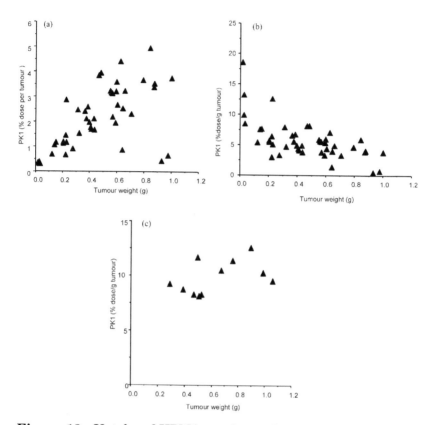

Figure 12 Uptake of HPMA copolymer-doxorubicin (PK1) at 1 h in tumors of different sizes. Panel (a) shows tumor accumulation in sc B16F10 tumors in terms of % dose per tumor. As the tumors increased in size, uptake increased. Panel (b) shows B16F10 tumor uptake in terms of % dose per gram of tumor. As size increases uptake in these terms reduces. Panel (c) shows uptake (% dose per gram of tumor) in MAC 15A colon xenografts of different sizes. In this case non-size dependence of uptake is seen. (From Ref. 194.)

and/or cathepsin B activity be rate-limiting in determining PK1 efficacy?". Experiments using the murine colon tumor models MAC15A and MAC26 showed that PK1 (40 mg/kg doxorubicin equivalent) was significantly more effective than doxorubicin (10 mg/kg) against MAC15A tumors, but not against MAC26 tumors (197). Although similar levels of PK1 were detected in both tumors, the peak tumor levels of released doxorubicin were seven-fold greater in the responsive MAC15A. Therefore, the differences in tumor response were best correlated with cathepsin B activity (197).

There is continued hope that receptor-mediated targeting will further augment the targeting advantage that the EPR effect brings. To this end, the search continues to find ligands, effective at tumor targeting in vivo, and in particular antibody conjugates are being studied (42,198,199). Many tumors exhibit high levels of the transferrin receptor. HPMA copolymer-anthracycline conjugates containing transferrin, or anti-transferrin receptor antibody (B3/25 or CD71) have been described (139,200). HPMA copolymer-B3/25 conjugates were internalized by human fibroblasts in vitro nine-fold faster than conjugates without (200). Conjugates containing doxorubicin and anti-CD71antibody have been studied in a B-cell lymphoma model 38C13 in vitro and in vivo. In vitro the anti-CD71 conjugate had greater cytotoxicity (nine-fold) than PK1, and was 4-fold more active than a transferrin conjugate (201). The anti-CD71 conjugate was also most effective in suppressing the growth of established 38C13 tumors in vivo. Both the transferrin and antibody conjugate were more active compared to PK1. The mechanism responsible for these differences is not completely clear. Both the antibody and transferrin conjugates have a higher molecular weight than PK1; hence plasma pharmacokinetics and potentially EPR-mediated targeting will be different. In addition it has been shown that these different types of conjugates release drugs at different rates.

Anti-thy-1.2 antibody-targeted HPMA copolymer conjugates have also been explored as a means of targeting the photosensitizer mesochlorin e_6 (Mce$_6$) and doxorubicin (155,202,203). Binding of antibody via oxidized carbohydrate give a 2-fold better antigen binding. The HPMA copolymer-Mce$_6$ conjugates were phototoxic towards primary mouse splenocytes and T lymphocytes, and such conjugates appear to act via singlet oxygen generation in endosomal or lysosomal vesicles, causing localized membrane damage (203,204). Kopecek and colleagues have systematically studied the targeting of the OA-3 antigen of ovarian cancer using conjugates containing OV-TL 16 antibody or its Fab' fragments (109,155,202,204,205,227). Conjugates containing doxorubicin or Mce$_6$, and antibody showed increased activity against OVCAR-3 xenografts (204). In addition, Fab'-targeted conjugates of Mce$_6$ were more cytotoxic towards OVCAR-3 cells in vitro (IC$_{50}$ = 2.6 µM) than non-targeted Mce$_6$ conjugate (IC$_{50}$ = 230 µM) or free Mce$_6$(IC$_{50}$ = 7.9 µM) (205). Recently, HPMA copolymer-doxorubicin conjugates containing RGD-terminating side-chains, designed to target tumor endothelial cells have been described (206). Enhanced uptake by ECV304 cells was demonstrated in vitro, but experiments to confirm in vivo selectivity are awaited.

4.2.2. Mechanism of Action of HPMA Anticancer Conjugates

Reappraisal of the current understanding of the mechanism of action of HPMA copolymer-anticancer conjugates is important given the growing database of in vitro, in vivo and clinical data (Fig. 13). Data interpretation is often complex, and in vivo, many factors (Fig. 13) will act in concert to produce the antitumor effect observed. Nevertheless, the success of second-generation conjugates will rely on design to capitalize on the mechanism of action of the conjugate rather than solely on the mechanism of action of the parent drug.

Change in Drug Pharmacokinetics Following Polymer Conjugation (Fig. 13a)

Unequivocally drug conjugation dramatically alters biodistribution. This has been proven in vivo and clinically. After iv administration the conjugate is initially retained in the vascular compartment so that drug $t_{1/2\alpha}$ is increased (20,178,188). The levels of free drug detected in plasma are very low (>100–1000 times less than seen for the conjugated drug) (20). Preclinical rodent pharmacokinetic studies and clinical pharmacokinetics correlate well (20,188). The plasma half-life of HPMA copolymer-anticancer conjugates is typically 1–6 h and elimination occurs predominantly via the

(a) Whole organism level

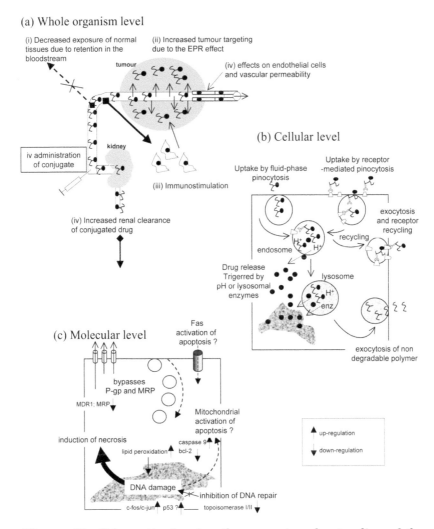

Figure 13 Schematic showing the current understanding of the mechanism of action of HPMA copolymer-anticancer conjugates at the: (a) whole organism level, (b) the cellular level, and (c) the molecular level. The exact molecular mechanisms leading to cytostatic and cytotoxic action are not conclusively proven (see discussion in the text), so suggested molecular mechanisms are illustrated and the possible up- and down-regulation of molecular markers indicated.

kidney with >50% of conjugated drug excreted within 24 h. The only exception is PK2. Following liver targeting, hepato-biliary elimination plays a major role.

The biodistribution of HPMA copolymer conjugates reduces drug access to potential sites of toxicity, including the heart and bone marrow. This, together with the enhanced elimination of inactive, conjugated drug via the kidney (when the polymer-drug linker is stable), explains the significant reduction in toxicity of anticancer conjugates like PK1 and PK2 in vivo and in man (20,172). EPR-mediated targeting can significantly increase the tumor drug concentration area under the curve (AUC)>70-fold in animal models (90,187,188). Improved antitumor activity of conjugates compared to free drug relies on EPR-mediated targeting with an appropriate rate of drug liberation (activation) at the target site. Mechanisms of clinical resistance show a maximum of 5–10-fold decrease in sensitivity, so EPR mediated targeting could theoretically provide high enough localized drug concentrations to overcome both inherent and acquired resistance simply by the pharmacokinetic modification. Gamma camera imaging has shown some evidence to support EPR-mediated tumor localization in patients (20,180). However, further studies are needed in the clinical setting to better understand the clinical significance of the EPR effect in human tumors of different tissue origin, the extent of targeting at different stages of tumor development (primary, metastatic, post-surgery, etc.), and not least the extent of targeting that can be achieved using different dosing protocols and drug and/or radiotherapy combinations.

Changes in Drug Intracellular Trafficking
Following Polymer Conjugation (Fig. 13b)

Small molecule drugs usually enter cells rapidly (within minutes) by passage across the plasma membrane. HPMA copolymer conjugates are taken into cells much more slowly by endocytosis (52,206,207) and this makes comparative in vitro screening of activity almost meaningless (18,67). Endocytic internalization of conjugates has been verified using a variety of cell lines, together with ^{125}I-labeled probes,

HPLC assay of drug, and both epifluorescence and confocal microscopy (128,205,207–210). Conjugates lacking a targeting ligand are usually internalized slowly as a solute by fluid-phase pinocytosis, but in some cell types hydrophobic conjugates interact with the plasma membrane leading additionally to non-specific adsorptive uptake. Non-specific binding seems to vary according to the cell line used, and probably relates to membrane morphology (with or without microvilli) and the biochemistry of the glycocalyx.

Addition of targeting ligands such as galactose, transferrin, MSH, -GRGD, and antibodies to conjugates promotes receptor-mediated internalization (128,149,200,206–209). Subsequent intracellular trafficking is governed primarily by the specific receptor-ligand, but the conjugate physicochemical nature is also important. For example, galactose-containing HPMA copolymer conjugates are directed to lysosomes. Subcellular fractionation showed time-dependant liberation of [^{3}H]daunomycin from an HPMA copolymer-[^{3}H]daunomycin-galactose conjugate and redistribution into the nucleus (207). Similar events can be visualized using fluorescence microscopy (208), but care must be taken with interpretation of these images. Artifacts arise due to cell fixation, changes in fluorescence of polymer-bound and free drug, and the issues of pH- and concentration-dependent fluorescence quenching. When using transferrin as a targeting ligand, conjugate is also directed toward the transferrin receptor recycling pathway (200). This has implications for linker design, as any pH-sensitive linker must be able to off-load its drug payload during the rapid (5–10 min) intracellular transit. Peptidyl linkers designed for lysosomal degradation are unsuitable for use in transferrin-conjugates; illustrated by the decrease in the in vivo antitumor activity of HPMA copolymer-Gly-Phe-Leu-Gly-daunomycin-transferrin conjugates that are directed away the lysosomal compartment (139).

Jensen et al. (211) recently suggested that a small fraction of HPMA copolymer conjugate initially entrapped within lysosomes could escape into the cytosol and then traffic to the nucleus. The process was promoted by addition of nuclear localization sequences and/or use of photoactivatable conjugates to enhance endosomal and/or lysosomal escape (211). If quantitative transfer can be achieved, this might provide a novel way to encourage conjugate-mediated cytosolic delivery of macromolecular drugs. Irrespective of the ultimate intracellular localization, concern has been expressed that the lack of polymer main-chain biodegradability might lead to harmful accumulation. Lysosomal sequestration could theoretically create an artificial "lysosomal storage disease" syndrome (212). No evidence to date suggests that HPMA copolymer conjugates do cause lysosomal dysfunction (for example, by enzyme inhibition), lead to increased lysosomal membrane permeability (58), or promote increased cytosolic transfer of polymer anthracyclines (207). However, as conjugates with different physicochemical form are devised, perhaps even for chronic use, careful experimentation must discount this possibility.

Can Conjugate Interaction with the Plasma Membrane Initiate Tumor Cell Killing (Fig. 13c)?

This is still a contentious question. When Rihova and colleagues studied the cellular fate of HPMA copolymer-Gly-Phe-Leu-Gly-doxorubicin (and related anti-Thy-1.2 and anti-CD71 conjugates) by EL-4 T-cell lymphoma, SW620 colorectal carcinoma and OVCAR-3 using laser scanning confocal microscopy and fluorescent microscopy, and HPLC (207), they were unable to detect free doxorubicin in the nucleus (up to 72 h). To explain the mechanism of cell killing it was suggested that, at least in part, toxicity might be due to a direct effect of conjugated drug at the level of the plasma membrane. This is plausible as hydrophobic conjugates do interact with the surface of some cell types in vitro. In vivo experiments however argue against this theory. While non-degradable conjugates sometimes demonstrate in vitro cytotoxicity (this is likely due to contaminating free drug not the conjugate per se), in vivo experiments have shown that drug liberation is a prerequisite for activity. Non-degradable conjugates such as HPMA copolymer-Gly-melphalan (83), HPMA copolymer-Gly-Gly-doxorubicin and -Gly-Gly-daunorubicin

(59,61) and HPMA copolymer-Gly-Gly-en-Pt (90) are all completely inactive in vivo. On the contrary, the corresponding conjugates containing degradable polymer-drug linkers demonstrate greater antitumor activity in vivo than seen for the parent drug.

Anticancer Action Due to Immunostimulation (Fig. 13a)

Growing evidence supports the immunostimulatory role of HPMA copolymer-anticancer conjugates. Rihova postulated that the early antitumor activity in vivo occurs via cytotoxic or cytostatic action, but that secondary immunostimulatory action of circulating low levels of conjugate supplement this effect (213–215). This hypothesis is supported by the following observations: (i) Pre-treatment of animals with immunosuppressive agents (such as doxorubicin and cyclosporine A) accelerates the growth of subsequently implanted tumors whereas pre-treatment with HPMA copolymer-doxorubicin does not. (ii) An increase in circulating natural killer (NK) cell numbers and anticancer antibodies are seen in animals treated with conjugate. (iii) Moreover, increased NK and lymphokine-activated killer (LAK) cells have been seen in breast cancer patients treated with HPMA copolymer-Dox-IgG (215). Prolonged low dose chemotherapy is well known to be immunostimulatory (reviewed in 216), so that immunostimulatory action of HPMA copolymer conjugates would be consistent, prolonged liberation of low concentrations of drug from the conjugate. In our laboratory, Flanagan et al. (217) showed that tumors taken from Walker W256 sarcoma-bearing rats 4 days after treatment with HPMA copolymer-melphalan conjugates contained significantly higher numbers of macrophages, B-cells and T-cells than those from both untreated controls and animals treated with melphalan alone. Pre-immunization of mice with HPMA copolymer-doxorubicin before implantation of L1210 tumor did not impair subsequent antitumor response to the conjugate (218). The observation was taken as evidence that the conjugate did not induce neutralizing antibodies, but this could also be consistent with immunostimulation.

*Do HPMA Anticancer Conjugates Act at the
Level of Tumor Vasculature?*

It is generally assumed that anticancer conjugates act, after EPR-mediated targeting, at the level of the tumor cell, and not at the level of the tumor vascular endothelium. The P22 carcinosarcoma window chamber model was initially standardized to document the activity of the anti-vascular agent combretastatin A4 phosphate—a drug that causes vascular shutdown within 1 h (219). The fact that PK1 had no acute effect of tumor blood flow in this model confirmed a mechanism different from that of combretastatin A4 (193). Comparison of in vivo effects of free doxorubicin and PK1 in a sc human ovarian xenograft A2470/AD model (a doxorubicin resistant cell line), however, showed that whereas free doxorubicin causes VEGF gene over-expression, surprisingly PK1 causes a down-regulation of expression and a reduction in vascular permeability. This was puzzling as the conjugate was more active than free drug in this model (220). This is clearly an important observation as the gateway for conjugate delivery is via permeable vasculature and shutdown would prohibit subsequent doses of conjugate entering (220,221). Clinical imaging before, and after, treatment would be the best way to resolve this question.

Recent studies have described the first HPMA copolymer antiangiogenic conjugates (222). They were synthesized to contain *O*-(chloractyl-carbamoyl) fumagillol (TNP470). This antiangiogenic agent has undergone clinical evaluation, and although active it showed unacceptable clinical neurotoxicity (222). TNP-470 conjugation enhanced and prolonged the angiogenic activity of the drug in several in vivo models and abolished neurotoxicity. It was again shown that drug release was essential for activity (222).

*Molecular Mechanisms of Anticancer Activity
(Fig. 13c)*

As the molecular mechanisms of tumor cell cycle regulation, control of apoptosis and necrosis, DNA repair, and drug resistance become better understood, it is interesting to investigate whether polymer-drug conjugation changes its fundamental mechanism of action at the molecular level. These experiments are technically difficult to design and interpret, as conjugate and free drugs have such different cellular pharmacokinetics. Moreover, all conjugates contain 1.0–0.01% free drug. These traces can contribute significantly to the in vitro activity (18,67). Minko and Kopecek and colleagues have undertaken a number of studies aimed at trying to understand the effect of HPMA copolymer conjugates on gene expression in normal and resistant ovarian carcinoma cells (reviewed in (42,189,223–230). The effects of conjugates on activation of signaling pathways involved in apoptosis, necrosis, and DNA repair, and expression of the multidrug resistance efflux pumps MDR1 and MRP were investigated.

Inherent and acquired drug resistance, particularly multi-drug resistance (MDR), is one of the most important reasons for the failure of cancer chemotherapy. Intellectually, lysosomotropic delivery provides an ideal opportunity to bypass plasma membrane-localized efflux pumps and facilitate direct intra-cytosolic drug delivery (Fig. 13c). It has been reasoned that conjugates using this routing should circumvent p-glycoprotein-mediated anthracycline resistance (128), but direct in vivo evidence to support the hypothesis has been hard to find (128). In vitro and in vivo, resistant tumors grow at different rates, and moreover in vivo passive tumor targeting may itself lead to drug concentrations high enough to overcome the plethora of resistance mechanisms. Whereas PK1 was not exceptionally active against resistant P388R (murine leukemia) (128), the parent P388 tumor regressed completely following administration of PK1. Resistant Lovo (human colon) tumors also did not respond to PK1 in vivo (Duncan, unpublished). However, antitumor activity seen in an epirubicin-resistant breast cancer patient (20) at low doxorubicin dose $(80 \, mg/m^2)$ established the clinical potential of drug conjugation. PK1 activity has also been seen in a doxorubicin-resistant ovarian carcinoma xenograft (A2780/AD), the magnitude of the effect being equivalent to that observed against the parental tumor A2780 (189). On balance current evidence supports the likelihood that HPMA copolymer conjugates can overcome MDR, at least in part, by cellular or molecular mechanisms. This remains to be proven clinically.

A2780/AD cells display a 40-fold resistance to free doxorubicin in vitro. In contrast, Minko and colleagues showed that the PK1 IC$_{50}$ only 20% higher than seen in the parental A2780 cells (223). Whereas acute exposure to free doxorubicin induced MDR1 gene expression, PK1 did not. Even during chronic exposure, PK1 failed to induce MDR1 gene expression (225). Similarly, when St'astny et al. examined the in vitro activity of PK1 and HPMA copolymer-doxorubicin conjugates targeted with antibodies or transferrin in various normal and MDR cell lines (including P388-MDR) (231), they found increased conjugate cytotoxicity in the resistant cell lines. This was dependent on the targeting ligand used and extent of internalization of the conjugates, and interestingly, combination of conjugates with sensitizers, such as cyclosporin A, almost completely restored the sensitivity of the resistant cell lines (231).

When ovarian carcinoma cells were incubated with free doxorubicin up-regulation of genes encoding the efflux pumps (MDR1, MRP), genes implicated in metabolism, detoxification and genes involved in DNA repair were seen. In contrast, incubation with PK1 led to down-regulation of the MRP, HSP-70, GST-p, BUDP, topoisomerase-II α,β and tyrosine kinase 1 genes (reviewed in 42,189). In these experiments it was concluded that whereas PK1 and free doxorubicin both activate the apoptosis signaling pathways (conjugate activates apoptosis more strongly), only PK1 suppressed protective mechanisms. PK1 appeared to activate p53, the c-fos (in A2780 cells) or c-jun (in A2780/AD cells) signaling pathways, increased lipid peroxidation and displayed increased DNA damage compared to free doxorubicin (227). Later more detailed experiments studying the time- and concentration-dependence of PK1 and doxorubicin action in the same ovarian carcinoma cell lines reached a different conclusion suggesting that PK1 causes cell death primarily by necrosis (229). The proposal that HPMA copolymer conjugates increase the expression of Fas after 36–48 h in A2780 and that cell death is at least in part due to activation of the Fas-receptor pathway (232) complicates this story.

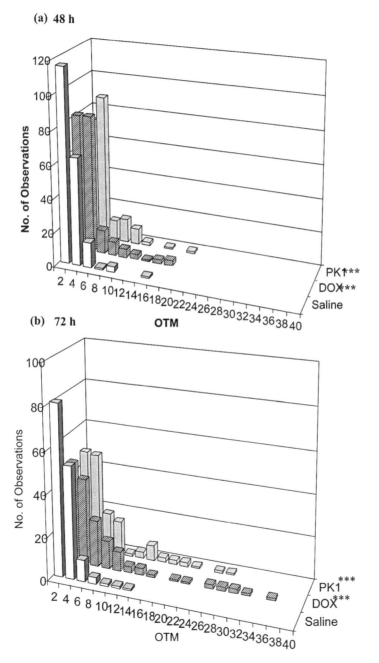

Figure 14 (*Caption on facing page*)

Our studies using B16F10 tumors in vitro and in vivo also concluded PK1-induced cell killing was by necrosis (233). No apoptosis was detected using YO-PRO®-1 or Alexa Fluor 488® annexin V in B16F10 cells incubated with PK1 (10 or 150 µg/mL /mL doxorubicin-equivalent) or free DOX (10 µg/mL) in vitro. When using the Comet assay, there was no significant difference in DNA damage in vivo caused by PK1 or doxorubicin when care was taken with tissue processing (Fig. 14) (233). Tumor freezing prior to assay can itself cause DNA damage.

Using a number of different cell lines, including OVCAR-3, Rihova and colleagues (209,234) also found that unlike free doxorubicin treatment (which was a strong enhancer p53 expression) HPMA copolymer conjugates never increased the expression of p53. Although a number of interesting observations are emerging, conclusions are conflicting. It is too early to ascribe any particular mechanism of action to HPMA copolymer conjugate antitumor activity. In vitro experiments are complicated by the traces of free drug in the conjugate, cellular pharmacokinetics of the conjugate compared to free drug which will vary form one cell type to another (endocytic uptake rates and rate of lysosomal activation). Moreover, methods used for cell and tissue processing in these experiments can introduce artifacts demanding a vast array of controls before definitive conclusions can be drawn.

4.2.3. The Next Generation of HPMA Copolymer-Based Anticancer Therapy

Based on the premise that the EPR effect will aid access of polymeric anticancer conjugates to tumor tissue in humans,

---

Figure 14 (*Facing page*) DNA damage determined using the COMET assay in B16F10 tumors after treatment of mice with PK1 (15 mg/kg), doxorubicin (7.5 mg/kg) or saline as a control. Damage is measured as the olive tail moment (OTM) and the data represent $n = 3$. p value ***<0.0001 relative to saline control. Panel (a) shows DNA damage at 48 h and panel (b) shows DNA damage at 72 h. Even at earlier time points no significant difference was seen in damage caused by free doxorubicin and PK1. (From Ref. 233.)

acting as a gateway for both passive and, in future, receptor-mediated targeting, a number of strategies are being explored as second-generation therapies. These include novel anticancer conjugates for lysosomotropic delivery, conjugates that might promote cytosolic access of peptides, proteins and oligonucleotides, conjugates that act extracellularly at the level of the plasma membrane, and combination therapies designed to release drug in the tumor interstitium. As an alternative, conjugates designed to deliver antiangiogenic and target tumor vasculature have recently been described.

Novel Anticancer Conjugates for Lysosomotropic Delivery

Having established the clinical profile of HPMA copolymer-anticancer conjugates using traditionally used cancer chemotherapy (anthracyclines, platinates, taxanes, and camptothecins), recently synthesized conjugates have evolved to contain novel antitumor agents. These have been either compounds that have failed in early clinical development due to unaccepted toxicity [e.g., ellipticines (96) and TNP-470 (222)], or interesting novel natural product antitumor agents [e.g., geldanamycin derivatives (97,98), 1,5-diazaanthraquinones (100), wortmannin (99)]. In all cases, HPMA copolymers of traditional structure (molecular weight characteristics and a Gly-Phe-Leu-Gly linker) have been used as a platform with the terminal linker dependent on compound chemistry. With the exception of the HPMA copolymer-TNP-470 conjugate, which shows remarkable in vivo therapeutic index (222), more in vivo data are required to verify whether any of these new conjugates might warrant clinical evaluation.

Combination Chemotherapy for Lysosomotropic Delivery

In the clinic, to maximize therapeutic index, cancer chemotherapy is routinely administered as drug combinations. HPMA copolymer technology has been moving in this direction too. Kopecek and colleagues developed a chemotherapy-phototherapy combination using a combination of HPMA copolymer-Mce_6 and HPMA copolymer-doxorubicin. In vivo treatment of neuroblastoma and human ovarian carcinoma xenografts showed an antitumor activity of the combination that was greater than seen for each element alone (86,87). Photodynamic therapy is clinically limited by the depth of laser penetration, and thus cannot be used to treat disseminated metastatic disease. Nevertheless the added benefit of site-specific "targeting" that laser activation brings can be put to good use in other situations such as treatment of superficial disease, compartment-localized disease (e.g., ovarian cancer) and post-surgical removal of a tumor to eradicate any residual tumor cells. Combination of the immunomodulators β-glucan and AM-2 with HPMA copolymer-doxorubicin conjugates targeted with anti-ELA lymphoma antibody conjugates led to enhanced antitumor activity in vivo (235). This observation, and the enhanced activity seen when conjugates were combined with the p-glycoprotein inhibitor cyclosporine A (231), show the potential for exploration of combinations in future clinical studies.

HPMA copolymers provide an ideal platform from which to deliver more than one drug to a tumor as an optimized drug combination therapy. Conjugates containing both doxorubicin and melphalan pendant on the same polymer chain were already synthesized in the 1980s (unpublished). Recently, the first conjugates for use in the treatment of hormone-responsive cancer (breast and prostate) have been described in the form of HPMA copolymer-aminoglutethimide (Fig. 15) (101). A combination conjugate, HPMA copolymer-aminoglutethimide-doxorubicin was less hemolytic compared to free drug and also showed greater cytotoxicity in vitro against MCF-7 cells than the individual HPMA copolymer-doxorubicin or HPMA copolymer-aminoglutethimide conjugates alone. Interestingly, addition of these two drugs to the same polymer chain influences their drug release rate, so careful optimization of the linker is needed to maximize therapeutic index in vivo.

Combination Chemotherapy for Intratumoral Drug Delivery

Tumor cells will be resistant to polymer-drug conjugate therapy designed for lysosomotropic delivery if: (i) endocytic

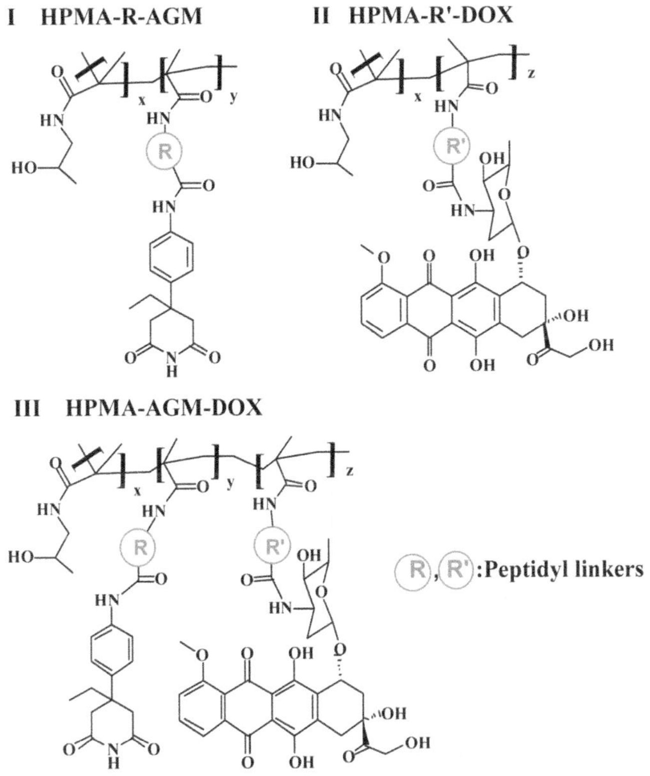

I **HPMA-R-AGM**   II **HPMA-R'-DOX**

III **HPMA-AGM-DOX**

$\textcircled{R},\textcircled{R'}$:**Peptidyl linkers**

Figure 15 Structure of HPMA copolymer conjugates containing doxorubicin, aminoglutethimide and a combination with both drugs pendant to the same chain. Peptidyl linkers -R- comprising Gly-Gly or Gly-Phe-Leu-Gly have been used. (From Ref. 101.)

internalization is inadequate, or (ii) levels of activating enzyme are too low. To circumvent these problems we devised an alternative approach called "polymer directed enzyme pro-drug therapy" (PDEPT) (79,141,142). This two-step approach uses a combination of a polymeric pro-drug and polymer-enzyme conjugate able to generate cytotoxic drug selectively within the tumor, but extracellularly (Fig. 16a). To achieve proof of concept HPMA copolymer-Gly-Phe-Leu-Gly-doxorubi-cin (PK1) was selected as the model pro-drug, and HPMA copolymer-cathepsin B as the activating enzyme conjugate.

(a) Step I. Administration of polymer - drug

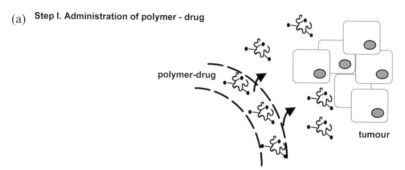

Step II. Administration of polymer- enzyme
when polymer - drug is no longer in the circulation

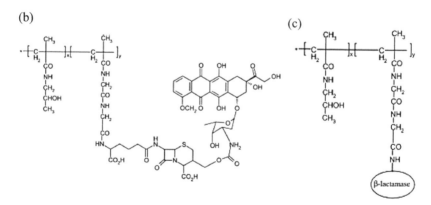

(b)   (c)

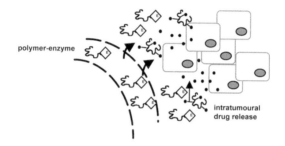

Figure 16 Polymer directed enzyme prodrug therapy (PDEPT). Panel (a) is a schematic showing the two-step approach taken to realize PDEPT; panels (b) and (c) show a combination that utilizes HPMA-copolymer-glycine-glycine-cephalosporin-doxorubicin as the macromolecular prodrug and HPMA copolymer β-lactamase conjugate as the activating enzyme. (From Refs. 141,142.)

In vivo,[125]I-labeled HPMA copolymer-cathepsin B and PK1 showed tumor targeting by the EPR effect in a sc B16F10 model. Moreover, when PK1 (10 mg/kg doxorubicin-equivalent) iv injection was followed after 5 h by the cathepsin B conjugate, there was a rapid increase in doxorubicin release within the tumor tissue (3.6-fold increase in the AUC compared to that seen for PK1 alone). This confirmed the ability of triggered drug release in the extracellular space (141). Subsequently, a non-mammalian enzyme combination, HPMA-copolymer-Gly-Gly-cephalosporin-doxorubicin (HPMA copolymer-C-Dox) and HPMA copolymer-Gly-Gly-β-lactamase were synthesized (Fig. 16b,c) (142). Again the two-step administration led to release of free Dox in vivo, and this PDEPT combination caused a significant decrease in tumor growth, in a B16F10 tumor model that was non-responsive to free Dox and HPMA copolymer-C-Dox. As the PDEPT combination displayed no general toxicity at the doses used and did not lead to an increase in free doxorubicin concentration in normal tissues further development of this concept is warranted (141,142).

Conjugates Designed to Act at the Plasma Membrane

Design of HPMA copolymer conjugates that are able to act at the level of the plasma membrane is an attractive alternative. With this objective in mind we synthesized melittin conjugates containing up to $38.9 \pm 2.5\%$ w/w peptide (152,154). A conjugate with medium loading and Gly-Gly spacer showed optimal activity, with reduced hemolytic behavior relative to free melittin, but maintained in vitro cytotoxic activity against B16F10 cells. The MTD of HPMA copoly mer-MLT was four-fold greater than that of free MLT and body distribution showed three- to four- fold increased circulation time and improved tumor targeting (~two-fold) of the conjugate after 4 h.

Anticancer Protein Conjugates

HPMA copolymer-bovine pancreatic ribonuclease (RNase A) conjugates containing one RNase A and two polymer chains retained activity in vitro compared to RNase A, although the conjugate seemed to be internalized more slowly than free enzyme (143). In vivo the conjugates showed enhanced activity after iv and intraperitoneal (ip) administration in mice bearing human xenografts (236). Tumor histology showed potent cytotoxicity (143), and confirmed the potential of this novel antitumor conjugate.

The Tumor Vasculature as a Target

The first HPMA copolymer-antiangiogenic conjugate (HPMA copolymer-Gly-Phe-Leu-Gly-TNP470) displayed selective accumulation in tumor tissue due to the EPR effect and improved antiangiogenic activity in both in vivo tumor models (A258 human melanoma and Lewis Lung carcinoma) and a hepatectomy model. Drug conjugation also prevented TNP470 crossing the BBB and thus eliminating the neurotoxicity associated with its clinical use (222). This very promising approach is worthy of further development. In fact we hope that a combination of HPMA copolymer conjugates containing RGD motifs to promote integrin $\alpha_v\beta_3$ receptor-mediated targeting (206) with anti-endothelial chemotherapy might enhance the effects seen further. In situations where tumor is well vascularized but vasculature permeability is poor, this strategy might be essential.

It is noteworthy that there have been hints from clinical studies (activity in lung cancer patients previously treated with radiotherapy) that a combination of radiotherapy with polymer conjugates leads to enhanced tumor targeting and improved antitumor activity. Radiotherapy, directly or indirectly, may induce an increase in vascular permeability, so exploration of clinical protocols combining HPMA copolymer conjugates with radiotherapy is an important option to explore.

4.3. Conjugates for Intracellular Delivery of Oligonucleotides and Proteins

Many synthetic non-viral vectors are under development with the hope of promoting intracellular delivery of macromolecular

drugs including genes. Usually the polymers used are polycations, or polymers that demonstrate pH-dependent conformational change to facilitate endosomal escape. Whereas
hydrophilic, charge-neutral HPMA copolymers would not
seem an obvious choice for use as non-viral vectors, recent

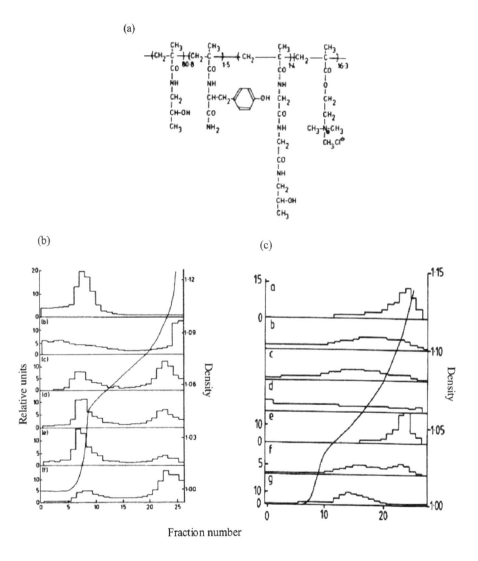

Figure 14 (*Caption on facing page*)

observations suggest that HPMA copolymer conjugates can escape endosomes and lysosomes albeit in low amounts (211). HPMA copolymer-antisense conjugates prepared from sulfhydryl-terminating antisense oligonucleotides linked to HPMA copolymers via a disulfide linkage were reportedly efficiently taken up by cells in vitro (237). Furthermore, HPMA copolymers synthesized to contain a synthetic peptide TAT sequence (H-Gly-Arg-(Lys)$_2$-(Arg)$_2$-Gln-(Arg)$_2$-Gly-Tyr-Lys-Cys-OH) labeled with FITC (1 TAT peptide per polymer chain) appear to show rapid, energy-independent cytosolic access and nuclear delivery of doxorubicin (123).

Early studies synthesized cationic HPMA copolymer conjugates containing tri-methyloxyethylammonium chloride a pendant side-chains (56) these we strongly membrane binding and subcellular fractionation studies showed that these cationic HPMA copolymer conjugates remained plasma membrane-associated for much longer than galactose-containing HPMA copolymers (58) (Fig. 17). Presumably this was due to non-specific membrane adsorption to non-internalizing membrane elements.

A series of experiments conducted over the last decade (238–246) by Seymour, Ulbrich, and Oupicky and colleagues has systematically tried to optimize the use of HPMA copolymer-containing non-viral vectors for oligonucleotide (including genes and antisense) delivery. Diblock copolymers of

---

Figure 17 *(Facing page)* Comparison of the subcellular distribution of HPMA copolymer-galactosamine (58) and cationic HPMA copolymers containing pendant side-chains terminating in oxyethytrimethylammonium chloride (56), structure in panel (a). The profiles shown in panels (b) and (c) show rat liver fractionation using a percoll gradient at various times after iv administration of the conjugates. Panel (b) shows the distribution of ^{125}I-labeled cationic HPMA copolymers at -e- 10 min, -d- 20 min and -c- 60 min. Aryl sulfatase distribution (lysosomes) is shown in -f-. Panel (c) shows the distribution of ^{125}I-labeled HPMA copolymer-galactosamine in this case, -g- 10 min, -f- 20 min and -c- 60 min. Arylsulfatase distribution (lysosomes) is shown in -a- and 5'-nucleotidase (plasma membrane) in -c-.

p(HPMA) and 2-(trimethylammonio) ethyl methacrylate (TMAEMCl) formed interpolyelectrolyte complexes (IPECS) with DNA. The most effective compositions had a cationic block of lower molecular weight than the pHPMA block (239). These pHPMA-pTMAEM block copolymers complexed to bcl-2 antisense produced small nanoparticles (~36 nm). Unfortunately their lack of stability precluded in vivo systemic use (243). To develop longer circulating vectors, poly-lysine was grafted with short pHPMA side chains (241) and coated complexes were prepared by covalent attachment of semi-telechelic pHPMA to the surface of preformed poly-lysine/DNA IPECS (244). Despite improved properties in vitro, these complexes also lacked stability in vivo. An alternative approach covalently conjugated HPMA copolymer-Gly-Phe-Leu-Gly-ONp precursors (i.e., multiple points of attachment) to the surface of the poly-lysine IPEC (245). The resultant complexes were more stable, and showed prolonged circulation times in vivo (246), but lacked the ability to transfect in vitro unless transferrin was added to promote cellular uptake (245). Recent studies have explored nano-sized IPECs prepared by polyethyleinimine-SH complexation of DNA subsequently coated with HPMA copolymers containing pendant 2-pyridyldisulfanyl or maleimide groups, the aim being to form degradable disulfide (reduced on addition of dithiothreitol) linkages or stable thioether linkages. In vitro studies confirmed that the degradable constructs had a 40- to 100-fold higher transfection efficiency (247).

5. CLINICAL STATUS OF HPMA COPOLYMER CONJUGATES

5.1. HPMA Copolymer-Doxorubicin (PK1, FCE28068)

Standard Phase I entry criteria were used in GCP controlled trials, although patients known to have brain metastases were excluded due to concern that brain vascular permeability might lead to neurotoxicity. Appreciating that conjugate pharmacokinetics could lead to unexpected anthracycline toxicities, hepatic and renal function was also closely monitored (20). FCE28068 was administered as a short infusion (4.16 mL/ min and drug concentration 2 mg/mL doxorubicin-equivalent) every 3 weeks and during dose escalation the infusion time was gradually extended. Dose escalation was made from the start- ing dose at $20 \, mg/m^2$ (doxorubicin-equivalent) to the MTD at $320 \, mg/m^2$ (doxorubicin-equivalent) (20). This MTD can be compared with a normal clinical dose for doxorubicin of $\sim 80 \, mg/m^2$. No polymer-related toxicity (or immunogenicity) was observed, and the dose-limiting toxicities (DLT) were typical of the anthracyclines, for example febrile neutropenia and mucositis. Alopecia was not seen at doses $<180 \, mg/m^2$ and nausea was mild without the need for anti-emetics until doses of $\geq 240 \, mg/m^2$. Anthracycline cardiotoxicity was absent even though individual cumulative FCE28068 doses of up to $1680 \, mg/m^2$ (doxorubicin-equivalent) were given. It should be noted that these cumulative doses represent $>20 \, g/m^2$ of HPMA copolymer and are considerably higher than the maxi- mum allowable cumulative dose of doxorubicin $(550 \, mg/m^2)$.

Two partial and two minor responses were seen in the study, in non-small cell lung cancer (NSCLC), colorectal cancer and anthracycline-resistant breast cancer at $80 \, mg/m^2$ (doxorubicin-equivalent). Activity in these chemotherapy refractory patients at relatively low conjugate dose is consis- tent with targeting by the EPR effect. Clinical pharmacoki- netics and gamma camera imaging showed a distribution profile similar to that seen in the preclinical studies, i.e., prolonged plasma circulation, absence of liver accumulation, and significant renal elimination (50–75% over 24 h) over time. FCE28068 had a $t_{1/2\alpha} = 1.8 \, h$ and $t_{1/2\beta} = 93 \, h$ (20,248). There was no evidence of dose dependency of pharmacoki- netics. Although imaging with a ^{131}I-labeled FCE28068 analog generally showed poor resolution, uptake was seen in known tumor sites in 6 of the 21 patients studied. The con- jugate was particularly visible in a head and neck primary tumor where levels of radioactivity were 2.2% dose at 2–3 h, 1.3% dose at 24 h, and 0.5% dose after 8 days. Phase II trials were conducted in colon, NSCLC and breast cancer patients using an FCE28068 dose of $280 \, mg/m^2$ again administered using a once-in-3-weeks schedule. Although no activity was seen in colorectal cancer patients, partial responses were again seen in breast and NSCLC patients refractory to conventional chemotherapy (249).

5.2. HPMA Copolymer-Doxorubicin-Galactosamine (PK2, FCE28069)

Phase I evaluation of FCE28069 was conducted in 31 patients, of which 23 had primary hepatocellular carcinoma (179). Conjugate was given iv every 3 weeks, initially at an infusion rate of 4.16 mL/min (2 mg/mL doxorubicin-equivalent). Due to pain the rate of infusion was reduced to 2 mL/min (1.0 mg/mL solution). Six patients were given FCE28069 by 24 h infusion to see if this would improve targeting efficiency by minimizing ASGR saturation. The starting dose was again $20 \, mg/m^2$ (doxorubicin-equivalent). FCE28069 displayed a significantly lower MTD ($160 \, mg/m^2$) than seen for FCE28068, but the DLT was typical of anthracyclines; myelosuppression and mucositis (179).

Two patients with primary hepatocellular carcinoma had measurable partial responses lasting >26 and >47 months respectively (179); a third showed reduction in tumor volume and 11 had stable disease. FCE28069 did not show dose-dependency in plasma pharmacokinetics with <0.1% of the drug in plasma present as free doxorubicin. In this case urinary excretion at 24 h was only 5% and SPECT gamma camera imaging showed FCE28069 liver levels of 15–20% dose at 24 h (179,183). The majority of conjugate was present in normal liver (after 24 h, 16.9% dose) with lower accumulations within hepatic tumor (3.2% dose). However, it was estimated that this hepatoma-associated drug was still 12- to 50-fold higher than could be achieved with administration of free doxorubicin.

5.3. HPMA Copolymer-Paclitaxel (PNU166945)

In Phase I clinical trials PNU166945 was administered by a 1 h infusion every 3 weeks (84). A starting dose of $80 \, mg/m^2$ (paclitaxel-equivalent) was used and the highest PNU166945 dose administered was $196 \, \mathrm{mg/m^2}$ (paclitaxel-equivalent). The toxicity observed was consistent with commonly observed paclitaxel toxicities: flu-like symptoms, mild nausea and vomiting, mild hematological toxicity, and neuropathy. Neurotoxicity grade 2 occurred in two patients at a dose of $140 \, \mathrm{mg/m^2}$ (although grade 1 was pre-existing on their entry) and one patient at $196 \, \mathrm{mg/m^2}$ had grade 3 neuropathy after the fourth cycle. Although no DLTs were reported dose escalation was discontinued prematurely due to concerns of potential clinical neurotoxicity.

In this small patient cohort antitumor activity was also observed. A paclitaxel-refractory breast cancer patient showed remission of skin metastases after two courses at $100 \, \mathrm{mg/m^2}$ (paclitaxel-equivalent). Two other patients had stable disease at a dose of $140 \, \mathrm{mg/m^2}$. Plasma pharmacokinetics were measured over 48 h using HPLC and was linear with dose both for PNU166945 and the released paclitaxel. The conjugate had a $t_{1/2} \sim 6.5 \, \mathrm{h}$ and its volume of distribution indicated plasma circulation. Free paclitaxel released from the conjugate had a $t_{1/2} \sim 1.2 \, \mathrm{h}$ and free drug levels were low, with $\sim 1\%$ of the paclitaxel present in plasma as conjugate (84).

5.4. HPMA Copolymer-Camptothecin (MAG-CPT; PNU 166148)

PNU166148 (MAG-CPT) containing the Gly-C6-Gly-camptothecin pendant side-chain was selected for Phase I clinical studies. Two dosing schedules were studied during Phase I evaluation. In the first study, 62 patients were entered starting at a dose of $30 \, \mathrm{mg/m^2}$ (camptothecin-equivalent) and conjugate was given as an iv infusion as given 30 min every 28 days (21). Dose escalation progressed to an MTD of $240 \, \mathrm{mg/m^2}$ with $200 \, \mathrm{mg/m^2}$ being the recommended dose for further studies. At $240 \, \mathrm{mg/m^2}$ the DLTs included grade 4 neutropenia, thrombocytopenia, and grade 3 diarrhea. Severe and unpredictable cystitis was also seen.

In another Phase I study (250) MAG-CPT was administered as a 30 min infusion on 3 consecutive days every 4 weeks. The starting dose was $17\,mg/m^2/day$ and this was escalated to $130\,mg/m^2/day$, i.e., total dose per cycle= $390\,mg/m^2$. Hematological toxicity was rare, but cumulative bladder toxicity was dose-limiting at doses of $68\,mg/m^2$ or greater. No objective clinical responses were seen in either of these trials; however, one patient with renal cell carcinoma had tumor shrinkage and a colon patient had stable disease for 62 days. A validated assay was developed to measure bound and free drug in clinical samples (251). No dose-dependency of plasma clearance of either MAG-CPT or the released drug was seen (51,52). Plasma levels of free camptothecin were 100 times lower than conjugated drug. The $t_{1/2\alpha}$ was similar to that seen for other HPMA copolymer conjugates and urinary excretion of the MAG-CPT was again high (66% at 24 h). There was no significant difference in terminal half-life ($\sim$8–10 days) of free and polymer bound drug and the conjugate was still detectable in urine after 4 weeks (51,52). It is probable that the observed bladder toxicity relates to the altered biodistribution of the camptothecin conjugate and lack of stability of the polymer-drug linker during renal elimination.

To seek clinical evidence to support the concept of EPR-mediated targeting, 10 colorectal cancer patients were given a single dose of MAG-CPT ($60\,mg/m^2$ camptothecin-equivalent) at 24 h, 3 days, or 7 days prior to surgery (252). At surgery plasma, tumor, and adjacent normal tissue samples were collected and analyzed for bound and free camptothecin. The concentrations of free and bound drug in plasma, tumor, and normal tissue achieved equilibrium by 24 h after dosing. However, while the conjugate achieved similar levels in tumor and normal tissue at 24 h, free drug levels were lower in tumor than in normal tissue indicating a lack of preferential accumulation by the EPR effect. These results were in contrast to previous data from animal tumor xenograft studies (116).

5.5. HPMA Copolymer-Platinates
(AP5280; AP5346)

First Phase I studies were undertaken with an HPMA copolymer-Gly-Phe-Leu-Gly-malonato-platinate (a carboplatin analog) (AP5280; Fig. 6) (253–255). Twenty-nine patients reported were given AP5280 at doses from 90 to $2300\,mg/m^2$ (Pt-equivalent) by short infusion (1 h; 500 mL) (254). The MTD was 2300 and dose limiting toxicities were uncontrollable vomiting (not possible to relieve) and grade 2 thrombocytopenia. A Phase II program is being planned.

More recently, HPMA copolymer-platinates containing a 1,2-diaminocyclohexyl (DACH) analog have been described. AP5286 contains the DACH analog bound to an HPMA copolymer-Gly-Phe-Leu-Gly-malonato (Fig. 6), and AP5346 is an HPMA copolymer-Gly-Gly-Gly-malonato-DACH (255,256). In in vivo studies using sc B16F10 tumors, AP5346 produced higher levels of tumor Pt-DNA complexes when compared to oxaliplatin, and showed superior ability to inhibit tumor growth. This compound is currently in Phase I clinical trials designed to establish its MTD and pharmacokinetics in patients with solid tumors.

5.6. HPMA Copolymer-Antibody-Doxorubicin Conjugates

Rihova and Ulbrich and colleagues have reported preliminary clinical experiments in six patients with refractory disease (angiosarcoma and breast carcinoma) (214,257). HPMA copolymer-Gly-Phe-Leu-Gly-doxorubicin (or epirubicin)-human immunoglobulin (HuIg) has been synthesized on a case-by-case basis for patient treatment. The HuIg used was either prepared was autologous IgG from sera by precipitation with 40% ammonium sulfate, or was a commercially available allogenic human γ-globulin. The primary aims of these preliminary studies were evaluation of the immunomodulatory effects of these conjugates. Disease progression was monitored and a large number of biochemical and immunological parameters were assessed. Although it is difficult to assess objectively the data obtained in the context of GCP guidelines, it is interesting to note that in some patients, antitumor effects were seen; the conjugate did not seem to induce anti-Ig antibodies and increased the levels of CD16$^+$56 and CD4$^+$ cells in peripheral blood, and the activation of NK and LAK cells supported the suggestion that HPMA copolymer-doxorubicin conjugates can be immunostimulatory.

6. OTHER APPLICATIONS

Few other applications of HPMA copolymer conjugates have been explored in depth, and none has yet been tested clinically. Rihova and colleagues (258–260) studied HPMA copolymer-cyclosporine A conjugates. Cyclosporine A is routinely used as an immunosuppressive agent to treat transplantation patients and also autoimmune diseases. Its use is however limited by nephrotoxicity and interference with normal thymocyte differentiation. HPMA copolymer conjugates containing cyclosporine A, with or without antibodies (anti-thymocyte serum, anti-human CD3, and anti-mouse CD4), overcame the cyclosporine A-induced abnormal thymocyte differentiation (260). Moreover, the severe kidney and thymus lesions induced by free drug were absent when conjugate was administered to rats (259).

6.1. Conjugates for Oral Delivery

Oral delivery is a preferred route for all drug administration. Nonetheless many formulations still require optimization in respect of (i) site-specific drug delivery within the gastrointestinal (GI) tract, (ii) prolongation of GI transit time to improve and prolong oral bioavailability, and (iii) the ability to protect and assist the transfer of macromolecular drugs, particularly proteins and peptides, across the GI barrier. Cartlidge et al. (113) and Bridges et al. (261,262) discovered that HPMA copolymers with pendant sugars have the potential to promote GI lectin binding. Since then, systematic investigations by Kopecek, Rihova, Rubenstein and colleagues have looked carefully at HPMA copolymer conjugate features pertinent to use in advanced oral delivery systems. Conjugates bearing pendant side-chains terminating in sugars, initially galactose (113), fucose (261,262) (selected as colonic bacteria are known to interact with the GI mucosa via a fucose-mediated ligand), and cationic residues

(261,262) were bioadhesive in vitro. Although strongly bioadhesive in vitro, in vivo cationic HPMA copolymers (56) appeared to induce mucin precipitation leading to more rapid elimination in the feces than observed for unmodified HPMA copolymers. They were therefore abandoned as oral bioadhesives (unpublished).

Designing appropriate polymer-drug linkers for orally administered conjugates is another challenge. Peptidyl sequences tailored for site-specific cleavage by brush border enzymes, i.e., at the enterocyte surface (263), were explored, and also azo linkers for colon-specific, e.g., of anti-inflammatory agents like aminosalicylic acid (264).

HPMA copolymer-sugar/lectin conjugates have been advanced. HPMA copolymer-fucose conjugates showed enhanced binding to rat and guinea pig everted colonic sacs and enterocytes in vitro and in vivo (265–267), but the magnitude of the binding effect was quite variable. The targeting potential of HPMA copolymer-doxorubicin conjugates containing galactosamine, peanut agglutinin, wheat germ agglutinin or anti-Thy-1.2 antibody was studied using primary (HT29 and SW 480) and metastatic (SW 620) human colon carcinoma cells and also an SW 620 line transfected with Thy-1.2 (SW 620T) (268). The wheat germ agglutinin- and anti-Thy-1.2-containing conjugates displayed greatest cytotoxicity in vitro, but the conjugates of peanut agglutinin were more selective. Kopecek and colleagues (269–271) also prepared HPMA copolymer-peanut agglutinin, -wheat germ agglutinin, and anti-Thomsen-Friedenreich antigen antibody conjugates in an attempt to localize more specifically to diseased or pre-cancerous lesions. Lectin-conjugation did not change the lectin binding affinity. The differences in normal vs. diseased tissue localization (peanut agglutinin conjugates do not localize to normal healthy tissue) might provide targeting opportunities for anti-inflammatory drugs, e.g., cyclosporine A to ulcerative colitis, and anticancer agents; e.g., aminocamptothecin to colon cancer (94). However, the fact that the colonic bacterial population influences the mucosal glycoprotein composition (target) makes the in vivo models for efficacy evaluation and clinical trials complicated to design

(271). Recently, the interaction of multivalent galactose-bearing HPMA copolymer conjugates with human colon carcinoma cells suggested an alternative opportunity to localize such conjugates to human colon cancer via the oral route (272).

We examined HPMA copolymer-peptide conjugates as potential oral vaccines (138). A human rhinovirus peptide VP2 (VKAETRLNPDLQPTETSQDVANAIVC) was covalently linked to an HPMA-Gly-Gly-ONp polymeric precursor. Conjugation improved peptide stability during incubation with brush border membrane and luminal peptidases transport but studies on GI transport and oral immunization failed to show conjugate advantage.

HPMA copolymer conjugates, as would be predicted, show minimal ability to cross the BBB although transfer in rats was elevated using an osmotic infusion protocol (that has been safely applied clinically) to transiently open the BBB endothelium (273). Polymer transfer was influenced by molecular weight, charge and hydrophobicity. Using a cortical slice model coupled with optical imaging it was recently shown that HPMA copolymer-FITC or Texas red narrow molecular weight labeled fractions diffuse in the brain extracellular space with the same tortuosity as small molecules (274). This is consistent with observations in the tumor interstitium (193).

Improved anti-infective and antiparasitic agents are a major clinical need, but to date few (102,275) HPMA copolymer conjugates have been designed for this purpose. Nan et al. (275) recently described the first HPMA copolymer-antileishmanial agent. It contains 8-aminoquinoline (NPC1161) and N-acetylmannosamine targeting residues chosen to locate the conjugate to macrophages which harbor the parasite. In vitro the HPMA copolymer-NPC1161 conjugates with at least 5 mol% N-acetylmannosamine were significantly more active than non-targeted conjugates, and in vivo also they were more effective.

FITC-labeled HPMA copolymer conjugates containing aldendronate and aspartic acid peptide have recently been synthesized (276). Their ability to target bone was examined in vitro using a hydroxyapatite model and in vivo in mice, and preliminary experiments suggest it would be interesting to quantitate targeting efficiency in vivo (276).

7. RELATED TECHNOLOGIES

For completeness, it is important to briefly review the extended family of HPMA copolymer conjugate compositions now available. In most cases these technologies were developed to prolong plasma circulation time and thus increase passive tumor targeting by the EPR effect.

7.1. Water-Soluble Crosslinked Conjugates

Water-soluble crosslinked HPMA copolymer-doxorubicin conjugates (Mw 100,000–1,000,000g/mol) were synthesized using N^2N^5-bis(N-methyacryloyl-Gly-Phe-Leu-Gly-doxorubicin) and a new crosslinking agent N^2N^5-bis(N-methyacryloyl-Gly-Phe-Leu-Gly) ornithine methylester (140,277). In this case lysosomal enzymes can cleave both the pendant side-chain and crosslinking sequences. In vivo studies using OVCAR-3 ovarian cancer xenografts confirmed the increased plasma circulation half-life of the higher molecular weight conjugates, compared to PK1, and increased tumor targeting did correlate with increased antitumor activity.

7.2. HPMA Copolymer-Coated Nanoparticles

Polymeric nanoparticles and nanospheres can be prepared to have a high drug carrying capacity, but unfortunately recognition by the RES often leads to premature liver and spleen clearance resulting in poor tumor targeting. In an attempt to overcome this limitation, nanospheres based on methylmethacrylate, maleic anhydride, and methacrylic acid were covalently coated with semitelechelic pHPMA. The resultant nanospheres displayed decreased protein adsorption and with increasing pHPMA molecular weight (Mn 5000–15,000 g/mol) they displayed long-circulating properties and decreased liver uptake (278).

7.3. Encapsulation of HPMA Copolymer Conjugates Within Vesicles

We decided that a convenient way to modulate the pharmacokinetics of HPMA copolymer conjugates and also to build in a
two-step process for triggered drug liberation would be
entrapment (PK1 was used as an example) within vesicles
(279,280). Niosomes were prepared using a variety of compositions. Whereas passive association of PK1 with preformed
vesicles was low (3–4%), dehydration (freeze drying) followed
by rehydration of the formulation increased the entrapment
to 40–60%. The mean size of $C_{16}G_2$ niosomes was 235 nm,
and transmission electron microscopy revealed an electron-
dense core as evidence of intravesicular concentration of
PK1. Degradation experiments showed slow release of conjugate from the vesicle and only once liberated enzymatically
triggered drug release. Preliminary biodistribution studies
showed that these particular formulations require further
optimization if they are to avoid RES capture.

8. FUTURE PROSPECTS

Water-soluble polymer-protein and polymer-drug conjugates
are becoming an accepted addition to the repertoire of
clinically useful chemotherapy. Transfer of HPMA copolymer-anticancer conjugates into clinical trials verified the
practicality of reproducible manufacture of HPMA copolymers to a predefined GMP specification. The data acquired
from >250 patients have also established the safety (including
lack of immunogenicity) of this novel polymer when administered parenterally as a short course of treatment (6–10 doses).
Whilst the anthracycline conjugates PK1 and PK2 displayed
promising antitumor activity, sadly they have never been
subjected to the intensive clinical development program
needed to define optimum dose and dosing schedule. Their
true potential is still as yet unknown. Failure of the paclitaxel
and camptothecin conjugates underlines the need for
conjugate rational design with consideration of whole body

pharmacokinetics and linker stability. Second generation anticancer conjugates containing novel natural products, therapy for hormone-dependant cancers and also polymer-drug combinations are now progressing towards clinical trial. With the aid of HPMA copolymer-imaging agents [for both gamma camera imaging and positron emission tomography (PET) imaging] it will be possible to assess the clinical significance of both EPR-mediated and also receptor-mediated targeting.

The proven clinical safety of the HPMA copolymer per se supports further development of HPMA copolymer-drug conjugates for treatment of other life-threatening diseases in circumstances where a short course of treatment will bring therapeutic benefit. Lack of biodegradability of the polymer main-chain precludes chronic parenteral administration although repeated oral dosing would be a practical option. The scene is now set to progress to better defined HPMA copolymer conjugates (control of molecular weight, structure, and defined 3D architecture), to conjugates with more imaginative polymer-drug linkers and to conjugates that will address modern challenges in macromolecular delivery and promotion of tissue engineering and repair.

ACKNOWLEDGMENTS

Few have the opportunity to see a concept through from the laboratory to clinic. I was lucky, and would like to express my gratitude to all at Keele, Prague and in Milan that made the journey so enjoyable. Also I would like to acknowledge all colleagues across the globe, in industry and academia, whose technical expertise and vision helped to bring the first synthetic polymeric anticancer conjugate into clinical trial. Throughout this sometimes challenging, interdisciplinary project, there were several folks who where always there to help, who always gave encouragement. Some of them are no longer with us. This is just the beginning. Although the field of polymer therapeutics is undoubtedly now well established there is still a job to finish.

REFERENCES

1. Brigger I, Dubernet C, Couvreur P. Nanoparticles in cancer therapy and diagnosis. Adv Drug Del Rev 2002; 54:631–651.

2. Allen TM. Ligand-targeted therapeutics in anticancer therapy. Nature Rev Drug Discov 2002; 2:750–763.

3. Brekke OH, Sandlie I. Therapeutic antibodies for human diseases at the dawn of the twenty-first century. Nature Rev Drug Discov 2003; 2:52–62.

4. Duncan R. The dawning era of polymer therapeutics. Nature Rev Discov 2003; 2:347–360.

5. Harris JM, Chess RB. Effect of pegylation on pharmaceuticals. Nature Rev Drug Discov 2003; 2:214–221.

6. Duncan R, Dimitrijevic S, Evagorou E. The role of polymer conjugates in the diagnosis and treatment of cancer. STP Pharma 1996; 6:237–263.

7. de Duve C, de Barsy T, Poole B, Trouet A, Tulkens P, van-Hoof F. Lysosomotropic agents. Biochem Pharmacol 1974; 23:2495–2531.

8. Ringsdorf H. Structure and properties of pharmacologically active polymers. J Polym Sci Polymer Symp 1975; 51:135–153.

9. Gros L, Ringsdorf H, Schupp H. Polymeric antitumour agents on a molecular and cellular level. Angew Chemie Int Ed Eng 1981; 20:301–323.

10. Brocchini S, Duncan R. Pendent drugs: release from polymers. In: Mathiowitz E, ed. Encyclopaedia of Controlled Drug Delivery. New York: John Wiley & Sons, 1999:786–816.

11. Kim KH, Hirano T, Ohashi S. Anticancer polymeric prodrugs. In:Salomone JC, ed. Polymer Materials Encyclopedia. Boca Raton: CRC Press, 1996:272–285.

12. Sezaki H, Takakura Y, Hashida M. Soluble macromolecular carriers for delivery of antitumour agents. Adv Drug Del Rev 1989; 3:247–266.

13. Greenwald RB, Choe YH, McGuire, Conover CD. Effective drug delivery by PEGylated drug conjugates. Adv Drug Del Rev 2003; 55:217–250.

14. Greenwald RB. Drug delivery systems: anticancer prodrugs and their polymeric conjugates. Expert Opin Therap Patents 1997; 7:601–609.

15. Garfield D. New form of paclitaxel shows promise. Lancet Oncol 2001; 2:192.

16. Danauser-Reidl S, Hausmann E, Schick H, Bender R, Dietzfelbinger H, Rastetter J, Hanauske A-R. Phase-I clinical and pharmacokinetic trial of dextran conjugated doxorubicin (AD-70, DOX-OXD). Invest New Drugs 1993; 11:187–195.

17. Inoue K, Kumazawa E, Kuga H, Susaki H, Masubuchi N, Kajimura T. CM-dextran-polyalchol-camptothecinconjugate. In: Maeda H, Kabanov A, Kataoka K, Okano T, eds. Polymer Drugs in the Clinical Stage: Advantages and Prospects. New York: Kluwer Academic/Plenum Publishers, 2003:145–177.

18. Duncan R. Polymer-drug conjugates. In: Budman D, Calvert H, Rowinsky E, eds. Handbook of Anticancer Drug Development, Philadelphia: Lippincott, Williams & Wilkins, 2003: 239–260.

19. Vasey P, Duncan R, Twelves C, Kaye S, Strolin-Benedetti M, Cassidy J. Clinical and pharmacokinetic phase 1 study of PK1(HPMA) copolymer doxorubicin. Ann Oncol 1996; 7:97.

20. Vasey P, Kaye SB, Morrison R, Twelves C, Wilson P, Duncan R, Thomson AH, Murray LS, Hilditch TE, Murray T, Burtles S, Fraier D, Frigerio E, Cassidy J. Phase I clinical and pharmacokinetic study of PK1 [N-(2-hydroxypropyl)metha- crylamide copolymer doxorubicin]: first member of a new class of chemotherapeutic agents-drug-polymer conjugates. Clin Cancer Res 1999; 5:83–94.

21. Kopecek J, Bazilova H. Poly[N-(hydroxypropyl)methecryla- mide]-I. Radical polymerisation and copolymerisation. Eur Polymer J 1973; 9:7–14.

22. Kopecek J, Sprincl L, Lim D. New types of synthetic infusion solutions. I. Investigation of the effect of solutions of some hydrophilic polymers on blood. J Biomed Mater Res 1973; 7:179–191.

23. Sprincl L, Exner J, Sterba O, Kopecek J. New types of syn- thetic infusion solutions. III. Elimination and retention of

poly[N-(2-hydroxypropyl)methacrylamide] in a test organism.
J Biomed Mater Res 1976; 10:953–963.

24. Kopecek J. Reactive copolymers of N-(2-hydroxypropyl)-methacrylamide with N-methacryloylated derivatives of L-leucine and L-phenylalanine. Makromol Chem 1977; 178: 2169–2183.

25. Drobnik J, Kopecek J, Labsky J, Rejmanova P, Exner J, Saudek V, Kalal J. Enzymatic cleavage of side-chains of synthetic water-soluble polymers. Makromol Chem 1976; 177: 2833–2848.

26. Rejmanova P, Labsky J, Kopecek J. Aminolyses of monomeric and polymeric 4-nitrophenyl esters of N-methacryloylamino acids. Makromol Chem 1977; 178:2159–2168.

27. Kopecek J, Rejmanova P, Chytry V. Polymers containing enzymatically degradable bonds. 1. Chymotrypsin catalyzed hydrolysis of p-nitroanilides of phenylalanine and tyrosine attached to sidechains of copolymers of N-(2-hydroxypropyl)-methacrylamide. Makromol Chem 1981; 182:799–809.

28. Ulbrich K, Strohalm J, Kopecek J. Polymers containing enzymatically degradable bonds. 3. Poly[N-(2-hydroxypropyl)-methacrylamide] chains connected by oligopeptide sequences cleavable by trypsin. Makromol Chem 1981; 182: 1917–1928.

29. Ulbrich K, Zacharieva EI, Obereigner B, Kopecek J. Polymers containing enzymatically degradable bonds. 5. Hydrophilic polymers degradable by papain. Biomaterials. 1980; 1: 199–204.

30. Ringsdorf H, Schmidt B, Ulbrich K. Bis(2-chloroethyl)amine bound to copolymers of N-(2-hydroxypropyl)methacrylamide and methacryloylated oligopeptides via biodegradable bonds. Macromol Chem Phys 1987; 188:257–264.

31. Chytry V, Vrana A, Kopecek J. Synthesis and activity of a polymer which contains insulin covalently bound on a copolymer of N-(2-hydroxypropyl)methacrylamide and N-methacryloylglycylglycine 4-nitrophenyl ester. Makromol Chem 1978; 179:329–336.

32. Chytry V, Kopecek J, Sikk P, Sinijarv R, Aaviksaar A. A convenient model system for the study of the influence of water-soluble polymer carrier on the interaction between proteins. Makromol Chem Rapid Commun 1982; 3:11–15.

33. Kopecek J. Biodegradation of polymers for biomedical use. In: Benoit H, Rempp P, eds. IUPAC Macromolecules. Oxford: Pergamon Press, 1982:305–320.

34. Kopecek J, Rejmanova P. Enzymatically degradable bonds in synthetic polymers. In: Bruck SD, ed. Controlled Drug Delivery. Boca Raton: CRC Press, 1983:81–124.

35. Kopecek J. Controlled biodegradability of polymers-key to drug delivery systems. Biomaterials 1984; 5:19–25.

36. Rihova B, Kopecek J. Biological properties of targetable poly[N- (2-hydroxypropyl)methacrylamide]-antibody conjugates. J Control Rel 1985; 2:289–310.

37. Duncan R, Kopecek J. Soluble synthetic polymers as potential drug carriers. Adv Polymer Sci 1984; 57:51–101.

38. Kopecek J, Rejmanova P, Duncan R, Lloyd JB. Controlled release of drugs model from N-(2-hydroxypropyl)methacrylamide copolymers. Annals New York Acad Sci 1985; 446: 93–104.

39. Kopecek J. The potential of water-soluble polymeric carriers in targeted and site-specific drug delivery. J Cont Rel 1990; 11:279–290.

40. Krinick NL, Kopecek J. Soluble polymers as targetable drug carriers. In: Juliano RL, ed. Targeted Drug Delivery, Handbook of Experimental Pharmacology. Berlin: Springer, 1991: 105–179.

41. Putnam D, Kopecek J. Polymer conjugates with anticancer activity. Adv Polym Sci 1995; 122:55–123.

42. Kopecek J, Kopeckova P, Minko T, Lu Z-R. HPMA copolymer-anticancer drug conjugates: design, activity and mechanism of action. Eur J Pharm Biopharm 2000; 50:61–81.

43. Duncan R, Pratten MK. Membrane economics in endocytic systems. J Theor Biol 1977; 66:727–735.

44. Pratten MK, Duncan R, Lloyd JB. A comparative study of the effects of polyamino acids and dextran derivatives on pinocytosis in the rat yolk sac and the rat peritoneal macrophage. Biochim Biophys Acta 1978; 540:455–462.

45. Duncan R, Lloyd JB. Pinocytosis in the rat visceral yolk sac. Effects of temperature, metabolic inhibitors and some other modifiers. Biochim Biophys Acta 1978; 544:647–655.

46. Duncan R, Pratten MK, Lloyd JB. Mechanism of polycation stimulation of pinocytosis. Biochim Biophys Acta 1979; 587:463–475.

47. Pratten MK, Duncan R, Lloyd JB. Adsorptive and passive pinocytic uptake. In: Ockleford CJ, Whyte A, eds. Coated Vesicles. Cambridge: Cambridge University Press, 1980:179–218.

48. Duncan R, Pratten MK. Pinocytosis: mechanism and regulation. In: Dean RT, Jessup W, eds. Mononuclear Phagocytes: Physiology and Pathology. Amsterdam: Elsevier Biomedical Press, 1985:27–51.

49. Duncan R, Pratten MK, Cable HC, Ringsdorf H, Lloyd JB. Effect of molecular size of ^{125}I-labelled poly(vinylpyrrolidone) on its pinocytosis by rat visceral yolk sacs and rat peritoneal macrophages. Biochem J 1980; 196:49–55.

50. Duncan R, Lloyd JB, Kopecek J. Degradation of side chains of N-(2-hydroxypropyl)methacrylamide copolymers by lysosomal enzymes. Biochem Biophys Res Comm 1980; 94:284–290.

51. Duncan R, Lloyd JB. Biological evaluation of soluble synthetic polymers as drug carriers. In: Anderson JM, Kim SW, eds. Recent Advances in Drug Delivery Systems. NewYork: Plenum Press, 1984:9–22.

52. Duncan R, Rejmanova P, Kopecek J, Lloyd JB. Pinocytic uptake and intracellular degradation of N-(2-hydroxypropyl)-methacrylamide copolymers. A potential drug delivery system. Biochim Biophys Acta 1981; 678:143–150.

53. Duncan R, Cable HC, Lloyd JB, Rejmanova P, Kopecek J. Degradation of side-chains of N-(2-hydroxypropyl)methacrylamide copolymers by lysosomal thiol-proteinases. Biosci Reps 1983; 2:1041–1046.

54. Duncan R, Cable HC, Lloyd JB, Rejmanova P, Kopecek J. Polymers containing enzymatically degradable bonds.7. Design of oligopeptide side chains in poly [N-(2-hydroxypropyl)methacrylamide] copolymers to promote efficient degradation by lysosomal enzymes. Makromol Chem 1984; 184:1997–2008.

55. Duncan R, Cable HC, Rejmanova P, Kopecek J, Lloyd JB. Tyrosinamide residues enhance pinocytic capture of N-(2-hydroxypropyl) methacrylamide copolymers. Biochim Biophys Acta 1984; 799:1–8.

56. McCormick LA, Seymour LCW, Duncan R, Kopecek J. Interaction of a cationic N-(2-hydroxypropyl)methacrylamide copolymer with rat visceral yolk sacs culture in vitro and rat liver in vivo. J Bioact Compat Polym 1986; 1:4–19.

57. Duncan R, Kopecek J, Rejmanova P, Lloyd JB. Targeting of N-(2-hydroxypropyl)methacrylamide copolymers to liver by incorporation of galactose residues. Biochim Biophys Acta 1983; 755:518–521.

58. Duncan R, Seymour LCW, Scarlett L, Lloyd JB, Rejmanova P, Kopecek J. Fate of N-(2-hydroxypropyl)methacrylamide copolymers with pendant galactosamine residues after intravenous administration to rats. Biochim Biophys Acta 1986; 880:62–71.

59. Duncan R, Kopeckova-Rejmanova P, Strohalm J, Hume I, Cable HC, Pohl J, Lloyd JB, Kopecek J. Anticancer agents coupled to N-(2-hydroxypropyl) methacrylamide copolymers. 1. Evaluation of daunomycin and puromycin conjugates in vitro. Brit J Cancer 1987; 55:165–174.

60. Duncan R, Kopeckova P, Strohalm J, Hume I, Lloyd JB, Kopecek J. Anticancer agents coupled to N-(2-hydroxypropyl)-methacrylamide copolymers. II. Evaluation of daunomycin conjugates in vivo against L1210 leukaemia. Brit J Cancer 1988; 57:147–156.

61. Duncan R, Hume IC, Kopeckova P, Ulbrich K, Strohalm J, Kopecek J. Anticancer agents coupled to N-(2-hydroxypropyl)methacrylamide copolymers. 3. Evaluation of adriamycin conjugates against mouse leukaemia L1210 in vivo. J Cont Rel 1989; 10:51–63.

62. Duncan R, Seymour L, O'Hare K, Flanagan P, Wedge S, Ulbrich K, Strohalm J, Subr V, Spreafico F, Grandi M, Ripamonti M, Farao M, Suarato A. Preclinical evaluation of polymer-bound doxorubicin. J Cont Rel 1992; 19:331–346.

63. Duncan R, Lloyd JB. Biological evaluation of soluble synthetic polymers as drug carriers. In: Anderson JM, Kim SW, eds. Recent Advances in Drug Delivery Systems. New York: Plenum Press, 1984:9–22.

64. Duncan R. Biological effects of soluble synthetic polymers as drug carriers. CRC Critical Rev Therap Drug Carrier Sys 1985; 1(4):281–310.

65. Duncan R. Lysosomal degradation of polymers used as drug carriers. CRC Critical Rev Biocompat 1986; 2(2):127–146.

66. Duncan R. Selective endocytosis of macromolecular drug carriers. In: Robinson JR, Lee VH, eds. Sustained and Controlled Drug Delivery Systems. New York: Marcel Dekker, 1987:581–621.

67. Duncan R. Drug-polymer conjugates: potential for improved chemotherapy. Anti-Cancer Drugs 1992; 3:175–210.

68. Duncan R, Spreafico F. Polymer conjugates: pharmacokinetic considerations for design and development. Clin Pharmacokinetics 1994; 27:290–306.

69. Duncan R. Drug-polymer conjugates: reduction in anti-cancer drug toxicity. In: Zeller WJ, Eisenbrand G, Hellman K, eds. Reduction of Anticancer Drug Toxicity. Pharamacologic, Biologic, Immunologic and Gene Therapeutic Approaches. Heidelberg: Karger Press, Contrib Oncol, 1995:48: 170–180.

70. Duncan R. Polymer therapeutics for tumour specific delivery. Chem Ind 1997; 7:262–264.

71. Duncan R. Polymer conjugates for tumour targeting and intracytoplasmic delivery The EPR effect as a common gateway? Pharm Sci Technol Today 1999; 2(11):441–449.

72. Duncan R. Polymer therapeutics into the 21st century. In: Park K, Mrsny R (eds), Drug Delivery in the 21st Century. New York: ACS Books, 2000:350–363.

73. Duncan R. Polymer-anticancer drug conjugates. In: Muzykantov V, Torchilin V, eds. Biomedical Aspects of Drug Targeting. New York: Kluwer Academic Pubs, 2003:193–209.

74. Thanou M, Duncan R. Polymer-protein and polymer-drug conjugates in cancer therapy. Current Opin Invest Drugs 2003; 4(6):701–709.

75. Kopecek J, Rejmanova P, Strohalm J, Ulbrich K, Rihova B, Chytry V, Duncan R, Lloyd JB. Synthetic polymeric drugs. UK Patent Appl Jan 4th 1985: 8,500,209; US Patent 1991: 5,037,883.

76. Mongelli N, Pesenti E, Suarato A, Biasoli G. Polymer-bound paclitaxel derivatives. US Patent 1994: 5,362,831.

77. Angelucci F, Mongelli N, Pesenti E, Biasoli G, Suarato A. Polymer-bound paclitaxel derivatives. US Patent 1996: 5,569,720.

78. Duncan R, Evagorou EG, Buckley RG, Gianasi E. Polymer-platinum compounds. US Patent 1999: 5,965,118.

79. Duncan R, Satchi-Fainaro R. Pharmaceutical compositions containing antibody-enzyme conjugates in combination with prodrugs. US Patent 2002: 6,372,205.

80. Kopecek J, Krinick NL. Drug delivery system for the simultaneous delivery of drugs activatable by enzymes and light. US Patent 1993: 5,258,453.

81. Angelucci F, Ballinari D, Farao M. Synthesis and biological evaluation of polymer-bound 3'-deamino-3'[2(S)-methoxy-4-morpholinyl] doxorubicin. Proceedings of the XIII International Symposium on Medicinal Chemistry, Paris 1994.

82. Ulbrich K, Zacharieva EI, Kopecek J, Hume IC, Duncan R. Polymer-bound derivatives of sarcsolysin and their antitumour activity against mouse and human leukaemia in vitro. Makromol Chem 1987; 188:2497–2509.

83. Duncan R, Hume I, Yardley H, Flanagan P, Ulbrich K, Subr V, Strohalm J. Macromolecular prodrugs for use in targeted cancer chemotherapy: melphalan covalently coupled to N-(2-hydroxypropyl)methacrylamide copolymers. J Cont Rel 1991; 16:121–136.

84. Meerum Terwogt JM, ten Bokkel Huinink WW, Schellens JHM, Schot M, Mandjes IAM, Zurlo MG, Rocchetti M, Rosing H, Koopman FJ, Beijnen JH. Phase I clinical and pharmacokinetic study of PNU166945, a novel water soluble polymer-conjugated prodrug of paclitaxel. Anti-Cancer Drugs 2001; 12:315–323.

85. de Bono JS, Bissett D, Twelves C, Main M, Muirhead F, Robson L, Fraier D, Magne M, Porro MG, Speed W, Cassidy J. Phase I pharmacokinetic study of MAG-CPT (PNU 166148) a polymeric derivative of camptothecin. Proc Am Soc Clin Oncol 2000:771.

86. Krinick NL, Sun Y, Joyner D, Spikes JD, Straight RC, Kopecek J. A polymeric drug delivery system for the simultaneous delivery of drugs activatable by enzymes and/or light. J Biomater Sci Polym Ed 1994; 5:303–324.

87. Peterson CM, Lu JM, Sun Y, Peterson CA, Shiah JG, Straight RC, Kopecek J. Combination chemotherapy and photodynamic therapy with N-(2-hydroxypropyl)methacrylamide copolymer bound anticancer drugs inhibit human ovarian carcinoma heterotransplanted in nude mice. Cancer Res 1996; 56:3980–3985.

88. Putnam D, Kopecek J. Enantioselective release of 5-fluorouracil from N-(2-Hydroxypropyl)methacrylamide-based copolymers via lysosomal enzymes. Bioconj Chem 1995; 6:483–492.

89. Putnam D, Shiah J, Kopecek J. Intracellularly biorecognizable derivatives of 5-fluorouracil. Biochem Pharmacol 1996; 52:957–962.

90. Gianasi E, Wasil M, Evagorou EG, Keddle A, Wilson G, Duncan R. HPMA copolymer platinates as novel antitumor agents: in vitro properties, pharmacokinetics and antitumour activity in vivo. Eur J Cancer 1999; 35:994–1002.

91. Gianasi E, Buckley RG, Latigo J, Wasil M, Duncan R. HPMA copolymers platinates containing dicarboxylato ligands. J Drug Target 2002; 10:549–556.

92. Lin X, Zhang Q, Rice Jr, Stewart DR, Nowotnik DP, Howell SB. Improved targeting of platinum chemotherapeutics: the antitumour activity of HPMA copolymer platinum agent

AP5280 in murine tumour models. Eur J Cancer 2004; 40:291–297.

93. Dimitrijevic S, Duncan R. Synthesis and characterization of N-(2-hydroxypropyl)-methacrylamide (HPMA) copolymer-ementine conjugates. J Bioact Compat Polymers 1998; 13:165–178.

94. Sakuma S, Lu Z-R, Kopeckova P, Kopecek J. Biorecognisable HPMA copolymer-drug conjugates for colon-specific delivery of 9-aminocamptothecin. J Cont Rel 2001; 75:365–379.

95. Subr V, Strohalm J, Hirano T, Ito Y, Ulbrich K. Poly[N-(-2-hydroxypropyl)methacrylamide] conjugates of methotrexate. J Cont Rel 1997; 49:123–132.

96. Searle F, Gac-Breton S, Keane R, Dimitrijevic S, Brocchini S, Duncan R. *N*-(2-hydroxypropyl)methacrylamide copolymer-6-(3-aminopropyl)-ellipticine conjugates synthesis characterisation preliminary in vitro and in vivo studies. Bioconj Chem 2001; 12:711–718.

97. Kasuya Y, Lu Z-R, Kopeckova P, Tabibi SE, Kopecek J. Influence of the structure of drug moieties on the in vitro efficacy of HPMA copolymer-geldanamycin derivative conjugates. Pharm Res 2002; 19(2):115–123.

98. Nishiyama N, Nori A, Malugin A, Kasuya Y, Kopeckova P, Kopecek J. Free and N-(2-hydroxypropyl)methacrylamide copolymer-bound geldanamycin derivative induce different stress responses in A2780 human ovarian carcinoma cells. Cancer Res 2003; 63(22):7876–7882.

99. Varticovski L, Lu Z-R, Mitchell K, de Aos I, Kopecek J. Water-soluble HPMA copolymer-wortmannin conjugate retains phosphoinositide 3-kinase inhibitory activity in vitro and in vivo. J Cont Rel 2001; 74:275–281.

100. Vicent MJ, Manzanaro S, de la Fuente JA, Duncan R. HPMA copolymer-1, 5-diazaanthraquinone conjugates as novel anticancer therapeutics. J Drug Target, (2004) in press.

101. Vicent MJ, Greco F, Barrow D, Nicholson R, Duncan R. Using polymer conjugates to optimise combination therapy for treatment of hormone-dependent cancer. Proceedings of the

International Symposium Controlled Release of Biactive Materials 2003: 30:487–488.

102. Solovsjj MV, Ulbrich K, Kopecek J. N-(2-hydroxypropyl)-methacrylamide copolymers with antimicrobial activity. Biomaterials 1983; 4:44–48.

103. Rihova B, Jegorov A, Strohalm J, Matha V, Rossmann P, Fornusek L, Ulbrich K. Antibody-targeted cyclosporin A. J Cont Rel 1992; 19:25–40.

104. Rihova B, Strohalm J, Plocova D, Subr V, Srogl J, Jelinkova M, Sirova M, Ulbrich K. Cytotoxic and cytostatic effect of anti-Thy1.2 targeted doxorubicin and cyclosporin A. J Cont Rel 1996; 40:303–319.

105. Subr V, Duncan R, Hanada K, Cable HC, Kopecek J. A lyso-somotropic polymeric inhibitor of cysteine proteinases. J Cont Rel 2004; 94:115–127.

106. Nan A, Croft SL, Yardley V, Ghandehari H. NPC1161 macrophage targeted delivery of N-(2-hydroxypropyl) methacrylamide (HPMA) copolymer-antileishmanial drug conjugates. J Cont Rel 2004; 94:115–127.

107. Obereigner B, Buresova M, Vrana A, Kopecek J. Preparation of polymerizable derivatives of N-(4-aminobenzenesulfonyl)-N-butylurea. J Polym Sci Polym Symp 1979; 66:41–52.

108. Chytry V, Kopecek J, Leibnitz E, O'Hare K, Scarlett L, Duncan R. Copolymers of 6–0-methacryloyl-D-galactose and N-(2 hydroxypropyl) methacrylamide: Targeting to liver after intravenous administration to rats. New Polym Mater 1987; 1:21–28.

109. Lu ZR, Kopeckova P, Kopecek J. Polymerizable Fab' antibody fragments for targeting of anticancer drugs. Nat. Biotechnol 1999; 17:1101–1104.

110. Godwin A, Hartenstein M, Müller AHE, Brocchini S. Narrow molecular weight distribution precursors for polymer-drug conjugates. Angew Chem 2001; 113:614–617.

111. Cartlidge SA, Duncan R, Lloyd J, Rejmanova P, Kopecek J. Soluble crosslinked N-(2-hydroxypropyl)methacrylamide copolymers as potential drug carriers. 1. Pinocytosis by rat visceral yolk sacs and rat intestinal cultured in vitro. Effect

of molecular weight on uptake and intracellular degradation. J Cont Rel 1986; 3:55–66.

112. Cartlidge SA, Duncan R, Lloyd JB, Kopeckova-Rejmanova P, Kopecek J. Soluble, crosslinked N-(2-hydroxypropyl)-methacrylamide copolymers as potential drug carriers. 2. Effect of molecular weight on blood clearance and body distribution in the rat after intraveneous administration. Distribution of unfractionated copolymer after intraperitoneal, subcutaneous or oral administration. J Cont Rel 1987; 4:253–264.

113. Cartlidge S, Duncan R, Lloyd J, Kopeckova-Rejmanova P, Kopecek J. Soluble crosslinked N-(2-hydroxypropyl)metha-crylamide copolymers as potential drug carriers. 2. Effect of molecular weight on blood clearance and body distribution in the rat intravenous administration. Distribution of unfrac-tionated copolymer after intraperitoneal subcutaneous and oral administration. J Cont Rel 1986; 4:253–264.

114. Shiah J-G, Dvorak M, Kopeckova P, Sun Y, Peterson CM, Kopecek J. Therapeutic efficacy of long-circulating HPMA copolymer-doxorubicin conjugates. Eur J Cancer 2000; 37:131–139.

115. Ulbrich K, Subr V, Strohalm J, Plocova D, Jelinkova M, Rihova B. Polymeric drugs based on conjugates of synthetic and natural macromolecules. 1. Synthesis and physico-chemical characterisation. J Cont Rel 2000; 64:63–69.

116. Caiolfa VR, Zamal M, Fiorini A, Frigerio E, Pellizzoni C, d'Argy R, Ghiglieri A, Castelli MG, Farao M, Pesenti E, Gigli M, Angelucci, Suarato A. Polymer-bound camptothecin: Initial biodistribution and antitumour activity studies. J Cont Rel 2000; 65:105–119.

117. Etrych T, Chytil P, Jelinkova M, Rihova B, Ulbrich K. Synth-esis of HPMA copolymers containing doxorubicin bound via a hydrazone linkage. Effect of spacer on drug release and in vitro cytotoxicity. Macromol Biosci 2002; 2:43–52.

118. Ulbrich K, Etrych T, Chytil P, Jelinkova M, Rihova B. HPMA copolymers with pH-controlled release of doxorubicin. In vitro cytotoxicity and in vivo antitumour activity. J Cont Rel 2003; 87:33–47.

119. Rihova B, Etrych T, Pechar M, Jelinkova M, St'asny M, Hovorka O, Kovar M, Ulbrich K. Doxorubicin bound to HPMA copolymer carrier through hydrazone bond is effective also in a cancer cell line with a limited content of lysosomes. J Cont Rel 2001; 74:225–232.

120. Choi WM, Kopeckova P, Minko T, Kopecek J. Synthesis of HPMA copolymer containing adriamycin bound via an acid-labile spacer and its activity toward human ovarian carcinoma cells. J Bioact Comp Polym 1999; 14:447–556.

121. Kopecek J, Kopeckova P. N-(2-hydroxypropyl)methacrylamide copolymers for colon-specific drug delivery. In: Friend DR, ed. Oral, Colon-Specific Drug Delivery. Boca Raton: CRC Press, 1992:189–211.

122. Yeh PY, Kopeckova P, Kopecek J. Degradability of hydrogels containing azoaromatic crosslinks. Macromol Chem Phys 1995; 196:2183–2202.

123. Nori A, Jensen KD, Tijerina, Kopeckova P, Kopecek J. Tat-conjugates synthetic macromolecules facilitate cytoplasmic drug delivery to human ovarian carcinoma cells. Bioconj Chem 2003; 14:44–50.

124. Pinciroli V, Rizzo V, Angelucci F, Tato M, Vigevani A. Characterisation of two polymer-drug conjugates by ^{1}H-NMR: FCE 28068 and FCE 28069.. Magn Reson Chem 1997; 35:2–8.

125. Mendichi R, Rizzo V, Gigli M, Giacometti Schieroni A. Molecular characterisation of polymeric antitumour drug carriers by size exclusion chromatography and universal calibration. J Liq Chrom Rel Technol 1996; 19:1591–1605.

126. Mendichi R, Rizzo V, Gigli M, Razzano G, Angelucci F, Schieroni A. Molar mass distribution of a polymer-drug conjugate containing the antitumor drug paclitaxel by size exclusion chromatography and universal calibration. J Liq Chromatogr Relat Technol 1998; 21:1295–1309.

127. Mendichi R, Rizzo V, Gigli M, Schieroni AG. Dilute-solution properties of a polymeric antitumor drug carrier by size-exclusion chromatography, viscometry, and light scattering. J Appl Polym Sci 1998; 70:329–338.

128. Wedge S. Mechanism of action of polymer anthracyclines: potential to overcome multidrug resistance. Ph.D. Thesis, Keele University, UK, 1990.

129. Configliacchi E, Razzano G, Rizzo V, Vigevani A. HPLC methods for the determination of bound and free doxorubicin and of bound and free galactosamine in methacrylamide polymer-drug conjugates. J Pharm Biomed Anal 1996; 15: 123–129.

130. Cavallo R, Adami M, Magrini R, Colombo G. PK1/PK2 lyophilised formulations. HPMA doxorubicin conjugates with potentially improved antitumour activity. Proceedings of the First World Meeting on APGI/APV, Budapest, 1995:843.

131. Bohdanecky M, Bazilova H, Kopecek J. Poly[N-(2-hydroxypropyl) methacrylamide]. II. Hydrodynamic properties of diluted polymer solutions. Eur Polym J 1974; 10:405–410.

132. Ulbrich K, Konak C, Tuzar Z, Kopecek J. Solution properties of drug carriers based on poly[N-(2-hydroxypropyl)-methacrylamide] containing biodegradable bonds. Makromol Chem 1987; 188:1261–1272.

133. Uchegbu IF, Ringsdorf H, Duncan R. The lower critical solution temperature of doxorubicin polymer conjugates. Proceedings of the International Symposium on Controlled Release of Biactive Materials 1996; 23:791–792.

134. Shiah JG, Konak C, Spikes JD, Kopecek J. Solution and photoproperties of N-(2-hydroxypropyl)methacrylamide copolymer-meso-chlorin e6 conjugates. J Phys Chem 1997; 101: 6803–6809.

135. Vicent MJ, Paul A, Griffiths PC, Duncan R. Using small angle neutron scattering (SANS) to evaluate the conformation of PK1 and PK2: a potential explanation for their clinical behaviour. Proceedings of the Sixth International Symposium on Polymer Therapeutics: Laboratory to Clinical Practice, Cardiff, UK, 7–9th January 2004.

136. Monfardini C, Veronese FM. Stabilization of substances in circulation. Bioconj Chem 1998; 9:418–450.

137. Veronese FM, Harris JM (eds). Introduction and overview of peptide and protein pegylation. Adv Drug Del Rev 2002; 54:453–609.

138. Morgan SM, Subr V, Ulbrich K, Woodley JF, Duncan R. Evaluation of *N*-(2-hydroxypropyl)methacrylamide copolymer-peptide conjugates as potential oral vaccines. Studies on their degradation by isolated rat small intestinal peptidases and their uptake by adult rat small intestine in vitro. Int J Pharm 1966; 128:99–111.

139. Flanagan PA, Duncan R, Subr V, Ulbrich K, Kopeckova P, Kopecek J. Evaluation of protein-N-(2-hydroxypropyl)methacrylamide copolymer conjugates as targetable drug-carriers. 2. Body distribution of conjugates containing transferrin, antitransferrin receptor antibody or anti-Thy 1.2 antibody and effectiveness of transferrin-containing daunomycin conjugates against mouse L1210 leukaemia in vivo. J Cont Rel 1992; 18:25–38.

140. Sunnasee K. The evaluation of polymer-peptide and polymer-protein conjugates for targeted melanoma chemotherapy. Ph.D. Thesis, Keele University, UK, 1994.

141. Satchi R, Connors TA, Duncan R. PDEPT: Polymer directed enzyme prodrug therapy. I. HPMA copolymer-cathepsin B and PK1 as a model combination. Brit J Cancer 2001; 85:1070–1076.

142. Satchi-Fainao R, Hailu H, Davies JW, Summerford C, Duncan R. PDEPT: Polymer directed enzyme prodrug therapy. II. HPMA copolymer-β-lactamase and HPMA copolymer-C-Dox as a model combination. Bioconj Chem 2003; 14: 797–804.

143. Pouckova P, Zadinova M, Hlouskova D, Strohalm J, Plocova D, Spunda M, Olejar T, Zitko M, Matousek J, Ulbrich K, Soucek J. Polymer-conjugated bovine pancreatic and seminal ribonucleases inhibit growth of human tumors in nude mice. J Cont Rel 2004; 95:83–92.

144. Krinick NL, Rihova B, Ulbrich K, Strohalm J, Kopecek J. Targetable photoactivatable drugs. 2. Synthesis of N-(2-hydroxypropyl)methacrylamide copolymer-anti-thy-1.2 antibody chlorin e6 conjugates and a preliminary study of their

photodynamic effect on mouse splenopcytes in vitro. Makromol Chem 1990; 191:839–856.

145. Jelinkova M, Strohalm J, Plocova D, Subr V, Stastny M, Ulbrich K, Rihova B. Targeting of human and mouse T-lymphocytes by monoclonal antibody-HPMA copolymer-doxorubicin conjugates directed against different T-cell surface antigens. J Cont Rel 1998; 52:253–270.

146. Rihova B, Jelinkova M, Plundrova D, Kovar M, Novak M, Strohalm J, Pechar M, Ulbrich K. Mechanisms of action of polymeric drugs. Proceedings of the Third International Symposium on Polymer Therapeutics: From Laboratory to Clinical Practice. January 1998, London, UK, p. 9.

147. Kovar M, Mrkvan T, Strohalm J, Etrych T, Ulbrich K, Stastny M, Rihova B. HPMA copolymer-bound doxorubicin targeted to tumor-specific antigen of BCL1 mouse B cell leukemia. J Cont Rel 2003; 92:315–330.

148. Lu Z-R, Shiah JG, Kopeckova P, Kopecek J. Polymerizable Fab' antibody fragment targeted photodynamic cancer therapy in nude mice. STP Pharma Sci 2003; 13:69–75.

149. O'Hare KB, Duncan R, Strohalm J, Ulbrich K, Kopeckova P. Polymeric drug carriers containing doxorubicin and melanocyte stimulating hormone: in vitro and in vivo evaluation against murine melanoma. J Drug Target 1993; 1:217–229.

150. Nori A, Jensen KD, Tijerina M, Kopeckova P, Kopecek J. Subcellular trafficking of HPMA copolymer-TAT conjugates in human ovarian carcinoma cells. J Cont Rel 2003; 28: 1189–1210.

151. Omelyanenko V, Kopeckova P, Prakash RK, Ebert CD, Kopecek J. Biorecognition of HPMA copolymer-adriamycin conjugates by lymphocytes mediated by synthetic receptor binding epitopes. Pharm Res 1999; 16:1010–1019.

152. Musila R, Duncan R. Synthesis and evaluation of *N*-(2-hydroxypropyl)methacrylamide (HPMA) copolymer-melittin; a potential novel anticancer approach. Proceedings of the Fourth International Symposium on Polymer Therapeutics: From Laboratory to Clinical Practice, London, UK, January 2000, p. 78.

153. Tang AJ, Kopeckova P, Kopecek J. Binding and cytotoxicity of HPMA copolymer conjugates to lymphocytes mediated by receptor-binding epitopes. Pharm Res 2003; 20:360–367.

154. Musila R, Duncan R. Synthesis and evaluation of HPMA copolymer-melittin as a potential anticancer agent—in vivo studies in the mouse. J Pharm Pharmacol 2000; 52 (supplement):51.

155. Omelyanenko V, Kopeckova P, Gentry C, Shiah JG, Kopecek J. HPMA copolymer-anticancer drug-OV-TL16 antibody conjugates. 1. Influence of the method of synthesis on the binding affinity to OVCAR-3 ovarian carcinoma cells in vitro. J Drug Target 1996; 3:357–373.

156. Oupicky D, Ulbrich K, Rihova, B. Conjugates of semitelechelic poly[N-(2-hydroxypropyl)methacrylamide]with enzymes for protein delivery. J Bioact Biocomp Polymers 1999; 14: 213–230.

157. Jelinkova M, Strohalm J, Etrych T, Ulbrich K, Rihova B. Starlike vs. classic macromolecular prodrugs: Two different antibody-targeted HPMA copolymers of doxorubicin studied in vitro and in vivo as potential anticancer drugs. Pharm Res 2003; 20:1558–1564.

158. Kovar M, Strohalm J, Etrych T, Ulbrich K, Rihova B. Star structure of antibody-targeted HPMA copolymer-bound doxorubicin. A novel type of polymeric conjugate for targeted drug delivery with potent antitumour effect. Bioconj Chem 2002; 13:206–215.

159. Williams DF (ed.). Progress in biomedical engineering. Definitions in biomaterials. Proceedings of a Consensus Conference of the European Society for Materials. Amsterdam: Elsevier, 1987.

160. Sprincl L, Exner J, Sterba O, Kopecek J. New types of synthetic infusion solutions. III. Elimination and retention of poly[N-(2-hydroxypropyl)methacrylamide] in a test organism. J Biomed Mater Res 1976; 10:953–963.

161. Sprincl L, Vacik J, Kopecek J, Lim D. Biological tolerance of poly(N-substituted methacrylamides). J Biomed Mater Res 1971; 5:197–205.

162. Rihova B, Riha I. Immunological problems of polymer-bound drugs. CRC Crit Rev Therapeut Drug Carrier Syst 1984; 1:311–374.

163. Rihova B. Biocompatibility of biomaterials: haematocompatibility, immunocompatibility, and biocompatibility of solid polymeric materials and soluble targetable polymeric carriers. Adv Drug Del Rev 1996; 21:157–176.

164. Rihova B, Ulbrich K, Kopecek J, Mancal P. Immunogenicity of *N*-(2-hydroxypropyl)methacrylamide copolymers —potential hapten or drug carriers. Folia Microbiol 1983; 28:217–297.

165. Rihova B, Kopecek J, Ulbrich K, Pospisil J, Mancal P. Effect of the chemical structure of N-(2-hydroxypropyl)methacrylamide copolymers on their ability to induce antibody formation in inbred strains of mice. Biomaterials 1984; 5:143–148.

166. Rihova B, Kopecek J, Ulbrich K, Chytry V. Immunogenicity of N-(2-hydroxypropyl)methacrylamide copolymers. Makromol Chem 1985; 9:13–24.

167. Flanagan PA, Duncan R, Rihova B, Subr V, Kopecek J. Immunogenicity of protein-N-(2-hydroxypropyl)methacrylamide copolymer conjugates measured in A/J and B10 mice. J Bioact Compat Polym 1990; 5:151–166.

168. Simeckova J, Plocova D, Rihova B, Kopecek J. Activity of complement in the presence of N-(2-hydroxypropyl)methacrylamide copolymers. J Bioact Compat Polym 1986; 1:20–31.

169. Seymour L, Duncan R, Strohalm J, Kopecek J. Effect of molecular weight (Mw) of N-(2-hydroxypropyl)methacrylamide copolymers on body distributions and rate of excretion after subcutaneous, intraperitoneal and intravenous administration to rats. J Biomed Mater Res 1987; 21:1341–1358.

170. Goddard P, Williamson I, Brown J, Hutchinson L, Nicholls J, Petrak K. Soluble polymeric carriers for drug delivery—Part 4: Tissue autoradiography and following intravenous administration. J Bioact Compat Polym 1991; 6:4–24.

171. Duncan R. Reduction of anticancer drug toxicity. In: Zeller WJ, Eisenbrand G, Hellman K, eds. Pharmacologic, Biologic, Immunologic and Gene Therapeutic Approaches. Contrib Oncol 1995; 48:170–180.

172. Duncan R, Coatsworth JK, Burtles S. Preclinical toxicology of a novel polymeric antitumour agent: HPMA copolymer-doxorubicin (PK1). Human Exptl Toxicol 1998; 17:93–104.

173. Rihova B, Ulbrich K, Strohalm J, Vetvicka V, Bilej M, Duncan R, Kopecek J. Biocompatibility of N-(2-hydroxypropyl)-methacrylamide copolymers containing adriamycin. Immunogenicity, effect of haematopoietic stem cells in bone marrow in vivo and effect on mouse splenocytes and human peripheral blood lymphocytes in vitro. Biomaterials 1989; 10:335–342.

174. Yeung TK, Hopewell JW, Simmonds RH, Seymour LW, Duncan R, Bellini O, Grandi M, Spreafico F, Strohalm J, Ulbrich K. Reduced cardiotoxicity of doxorubicin given in the form of N-(2-hydroxypropyl)methacrylamide conjugates: an experimental study in the rat. Cancer Chemotherap Pharmacol 1991; 29:105–111.

175. Hopewell JW, Duncan R, Wilding D, Chakrabarti K. Modification of doxorubicin induced cardiotoxicity. I. Preclinical studies of cardiotoxicity of PK2 : A novel polymeric antitumour agent. Human Exptl Toxicol 2001; 20:461–470.

176. Rejmanova P, Kopecek J, Duncan R, Lloyd JB. Stability in rat plasma and serum of lysosomally degradable oligopeptide sequences in N-(2-hydroxypropyl)methacrylamide copolymers. Biomaterials 1985; 6:45–48.

177. Subr V, Strohalm J, Ulbrich K, Duncan R, Hume I. Polymers containing enzymatically degradable bonds. XII. Release of daunomycin and adriamycin from poly[N-(2-hydroxypropyl)-methacrylamide] copolymers. J Cont Rel 1992; 18:123–132.

178. Seymour L, Ulbrich K, Strohalm J, Kopecek J, Duncan R. Pharmacokinetics of polymer-bound adriamycin. Biochem Pharmacol 1990; 39:1125–1131.

179. Seymour L, Ulbrich K, Wedge S, Hume I, Strohalm J, Duncan R. N-(2-hydroxypropyl)methacrylamide copolymers targeted to the hepatocyte galactose-receptor: pharmacokinetics in DBA-2 mice. Brit J Cancer 1991; 63:859–866.

180. Seymour LW, Ferry DR, Anderson D, Hesslewood S, Julyan PJ, Payner R, Doran J, Young AM, Burtles S, Kerr DJ. Hepatic drug targeting: phase I evaluation of polymer bound doxorubicin. J Clin Oncol 2002; 20:1668–1676.

181. O'Hare K, Hume I, Scarlett L, Duncan R. Evaluation of anticancer agents coupled to N-(2-hydroxypropyl)methacrylamide copolymers. Effect of galactose incorporation on interaction with hepatoma in vitro. Hepatology 1989; 10:207–214.

182. Pimm MV, Perkins AC, Duncan R, Ulbrich K. Targeting of N-(2-hydroxypropyl)methacrylamide copolymer-doxorubicin conjugate to the hepatocyte galactose-receptor in mice: visualisation and quantification by gamma scintigraphy as a basis for clinical targeting studies. J Drug Cancer 1993; 1:125–131.

183. Pimm M, Perkins AC, Strohalm J, Ulbrich K, Duncan R. Gamma scintigraphy of the biodistribution of ^{123}I-labelled N-(2-hydroxypropyl)methacrylamide copolymer-doxorubicin conjugates in mice with transplanted melanoma and mammary carcinoma. J Drug Target 1996; 3:375–383.

184. Julyan PJ, Seymour LW, Ferry DR, Daryani S, Boivin CM, Doran J, David M, Anderson D, Christodolou C, Young AM, Hesselwood S, Kerr DJ. Preliminary clinical study of the distribution of HPMA copolymers bearing doxorubicin and galactosamine. J Cont Rel 1999; 57:281–290.

185. Matsumura Y, Maeda H. A new concept for macromolecular therapeutics in cancer chemotherapy; mechanism of tumoritropic accumulation of proteins and the antitumour agent SMANCS. Cancer Res 1986; 6:6387–6392.

186. Maeda H, Matsumura Y. Tumoritropic and lymphotropic principles of macromolecular drugs. CRC Crit Rev Ther Drug Carrier Sys 1989; 6:193–210.

187. Cassidy J, Duncan R, Morrison G, Strohalm J, Plocova D, Kopecek J, Kaye S. Activity of N-(2-hydroxypropyl)methacrylamide copolymers containing daunomycin against a rat tumour model. Biochem Pharmacol 1989; 38:875–879.

188. Seymour LW, Ulbrich K, Steyger PS, Brereton M, Subr V, Strohalm J, Duncan R. Tumour tropism and anti-cancer efficacy of polymer-based doxorubicin prodrugs in the treatment of subcutaneous murine B16F10 melanoma. Brit J Cancer 1994; 70:636–641.

189. Minko T, Kopeckova P, Kopecek J. Efficacy of chemotherapeutic action of HPMA copolymer-bound doxorubicin in a

solid tumor model of ovarian carcinoma. Int J Cancer 2000; 86:108–117.

190. Shiah J-G, Dvorak M, Kopeckova P, Sun Y, Peterson CM, Kopecek J. Biodistribution and antitumour efficacy of long-circulating N-(2-hydroxypropyl)methacrylamide copolymer-doxorubicin conjugates in nude mice. Eur J Cancer 2001; 37:131–139.

191. Seymour LW, Miyamoto Y, Brereton M, Subr V, Strohalm J, Duncan R. Influence of molecular size on passive tumour-accumulation of soluble macromolecular drug carriers. Eur J Cancer 1995; 5:766–770.

192. Noguchi Y, Wu J, Duncan R, Strohalm J, Ulbrich K, Akaike T, Maeda H. Early phase tumor accumulation of macromolecules: a great difference in clearance rate between tumor and normal tissues. Jpn J Cancer Res 1998; 89:307–314.

193. Keane RS, Wilson J, Vojnovic B, Tozer GM, Duncan R. Extravasation of PK1/FCE28068 in a P22 carcinosarcoma tumour: Preliminary studies using the rat window chamber model. Proceedings of the Fourth International Symposium on Polymer Therapeutics: From Laboratory to Clinical Practice 2000:76.

194. Sat YN, Burger AM, Fiebig HH. Comparison of vascular permeability and enzymatic activation of the polymeric prodrug HPMA copolymer-doxorubicin (PK1) in human tumour xenografts. Proceedings of the Annual Meeting of the American Association. Cancer Res 1999; 90:419.

195. Sat YN, Malik N, Turton JT, Duncan R. Tumour targeting by the EPR effect: comparison of three drug delivery system containing doxorubicin. Proceedings of the International Symposium on Controlled Release of Bioactive Materials 1999; 26:44–45.

196. Steyger PS, Baban DF, Brereton M, Ulbrich K, Seymour LW. Intratumoural distribution as a determinant of tumor responsiveness to therapy using polymer-based macromolecular prodrugs. J Cont Rel 1996; 39:35–46.

197. Loadman PM, Bibby MC, Double JA, Al-Shakhaa WM, Duncan R. Pharmacokinetics of PK1 and doxorubicin in

experimental colon tumour models with differing responses to PK1. Clin Cancer Res 1999; 5:3682–3688.

198. Rihova B. Antibody-targeted polymer-bound drugs. Folia Microbiol 1995; 40:367–384.

199. Rihova B. Receptor-mediated targeted drug or toxin delivery. Adv Drug Del Rev 1998; 29:273–289.

200. Flanagan, PA, Kopeckova P, Kopecek J, Duncan R. Evaluation of antibody-N-(2-hydroxypropyl)methacrylamide copolymer conjugates as targetable drug-carriers. 1. Binding, pinocytic uptake and intracellular distribution of anti-transferrin receptor antibody-conjugates. Biochim Biophys Acta 1989; 993:83–91.

201. Kovar M, Strohalm J, Ulbrich K, Rihova B. In vitro and in vivo effect of HPMA copolymer-bound doxorubicin targeted to transferrin receptor of B-cell lymphoma 38C13. J Drug Target 2002; 10:23–30.

202. Omelyanenko V, Gentry C, Kopeckova P, Kopecek J. HPMA copolymer-anticancer drug-OV-TL16 antibody conjugates. 2. Processing in epithelial ovarian carcinoma cells in vitro. Int J Cancer 1998; 75:600–608.

203. Rihova B, Krinik NL, Kopecek J. Targeted photoactivatable drugs. 3. In vitro efficiacy of polymer-bound chlorin-e6 toward human hepatocarcinoma cell line (PLC/PRF/5) targetd with galactosamine and to mouse splenocytes targeted with ant-thy-1.2 antibodies. J Cont Rel 1993; 25:71–87.

204. Tijerina M, Kopeckova P, Kopecek J. Mechanisms of cytotoxicity in human ovarian carcinoma cells exposed to free Mce(6) or HPMA copolymer-Mce(6) conjugates. Photochem Photobiol 2003; 77:645–652.

205. Shiah J-G, Sun Y, Kopeckova P, Peterson CM, Straight RC, Kopecek J. Combination chemotherapy and photodynamic therapy of targetable N-(2-hydroxypropyl)methacrylamide copolymer-doxorubicin/mesochlorin e_6-OV-TL16 antibody immunoconjugates. J Cont Rel 2001; 74:249–253.

206. Wan K-W, Vicent MJ, Duncan R. Targeting endothelial cells using HPMA copolymer-doxorubicin-RGD conjugates. Pro-

ceedings of the International Symposium on Controlled Release of Bioactive Materials 2003; 30:491–492.

207. Wedge SR, Duncan R, Kopeckova P. Comparison of the liver subcellular distribution of free daunomycin and that bound to galactosamine targeted N-(2-hydroxypropyl) methacrylamide copolymers following intravenous administration in the rat. Brit J Cancer 1991; 63:546–549.

208. Omelyanenko V, Kopeckova P, Gentry C, Kopecek J. Targetable HPMA copolymer-adriamycin conjugates. Recognition, internalization, and subcellular fate. J Cont Rel 1998; 53:25–37.

209. Hovorka O, St'astny M, Etrych T, Subr V, Strohalm J, Ulbrich K, Rihova B. Differences in the intracellular fate of free and polymer-bound doxorubicin. J Cont Rel 2002; 80:101–117.

210. Seib FP, Hann A, Jones AT, Duncan R. Establishing a quantitative subcellular fractionation method in B16F10 cells: Use to monitor the intracellular trafficking of polymer therapeutics. Proceedings of the International Symposium on Controlled Release of Bioactive Materials 2003; 30:519–520.

211. Jensen KD, Nori A, Tijerina P, Kopeckova P, Kopecek J. Cytoplasmic delivery and nuclear targeting of synthetic macromolecules. J Cont Rel 2003; 87:89–105.

212. Lloyd JB. The lysosome/endosome membrane: A barrier to polymer-based drug delivery?. Macromol Symp 2001; 172: 29–34.

213. Rihova B, Strohalm J, Kubackova K, Jelinkova M, Hovorka O, Kovar M, Plocova D, Sirova M, St'astny M, Rozprimova L, Ulbrich K. Acquired and specific immunological mechanisms co-responsible for efficacy of polymer-bound drugs. J Cont Rel 2002; 78:97–114.

214. Rihova B, Strohalm J, Prausova J, Kubackova K, Jelinkova M, Rozprimova L, Sirova M, Plocova D, Etrych T, Subr V, Mrkvan T, Kovar M, Ulbrich K. Cytostatic and immunomobilizing activities of polymer-bound drugs: experimental and first clinical data. J Cont Rel 2003; 91:1–16.

215. Rihova B, Strohalm J, Kubackova K, Jelinkova M, Rozprimova L, Sirova M, Plocova D, Mrkvan T, Kovar M, Pokorna

J, Etrych T, Ulbrich K. Drug-HPMA-HuIg conjugates effective against human solid cancer. In: Maeda H, Kabanov A, Kataoka K, Okano T, eds. Polymer Drugs in the Clinical Stage: Advantages and Prospects Advances in Experimental Medicine and Biology. New York: Kluwer Academic/Plenum blishers, 2003:159:125–143.

216. Spreafico F, Vecchi F, Coletta F, Mantovani A. Cancer chemotherapies as immunomodulators. Seminars Immunopathol 1985; 8:361–374.

217. Flanagan PA, Rihova B, Kopecek J, Duncan R. Immunogenicity of N-(2-hydroxypropyl)methacrylamide copolymer drug conjugates. Harden Conference on "Cellular Barriers and Targeting" Wye, UK 1989.

218. Flanagan PA, Strohalm J, Ulbrich K, Duncan R. Effect of pre-immunisation on the activity of polymer-doxorubicin against murine L1210 leukaemia. J Cont Rel 1993; 26:221–228.

219. Tozer GM, Prise VE, Wilson J, Locke RJ, Vojnovic B, Stratford MRL. Combretastatin A-4 phosphate as a tumor vascular-targeting agent: Early effects in tumors and normal tissues. Cancer Res 1999; 59:1626–1634.

220. Minko T, Kopeckova P, Pozharov V, Jensen K, Kopecek J. The influence of cytotoxicity of macromolecules and of VEFG gene modulated vascular permeability on the enhanced permeability and retention effect in resistant solid tumor. Pharm Res 2000; 17:505–514.

221. Duncan R. Polymer conjugates for tumour targeting and intracytoplasmic delivery. The EPR effect as a common gateway? Pharm Sci Technol Today 1999; 2(11):441–449.

222. Satchi-Fainaro R, Puder M, Davies JW, Tran HT, Sampson DA, Greene AK, Corfas G, Folkman J. Targeting angiogenesis with a conjugate of HPMA copolymer and TNP-470. Nature Med 2004; 10:225–261.

223. Minko T, Kopeckova P, Pozharov V, Kopecek J. HPMA copolymer bound adriamycin overcomes MDR1 gene encoded resistance in a human ovarian carcinoma cell line. J Cont Rel 1998; 54:223–233.

224. Minko T, Kopeckova P, Kopecek J. Comparison of the anticancer effect of free and HPMA copolymer-bound adriamycin in human ovarian carcinoma cells. Pharm Res 1999; 16:986–996.

225. Minko T, Kopeckova P, Kopecek J. Chronic exposure to HPMA copolymer-bound adriamycin does not induce multidrug resistance in a human ovarian carcinoma cell line. J Cont Rel 1999; 59:133–148.

226. Tijerina M, Flowers KD, Kopeckova P, Kopecek J. Chronic exposure of human ovarian carcinoma cells to free or HPMA copolymer-bound mesochlorin e_6 does not induce p-glycoprotein-mediated multidrug resistance. Biomaterials 2000; 21:2203–2210.

227. Kunath K, Kopeckova P, Minko T, Kopecek J. HPMA copolymer-anticancer drug-OV-TL16 antibody conjugates. 3. The effect of free and polymer-bound adriamycin on the expression of some genes in the OVCAR-3 human ovarian carcinoma cell line. Eur J Pharm Biopharm 2000; 49:11–15.

228. Minko T, Kopeckova P, Kopecek J. Preliminary evaluation of caspaes-dependent apoptosis signaling pathways of free and HPMA copolymer-bound doxorubicin in human ovarian carcinoma cells. J Cont Rel 2001; 71:227–237.

229. Demoy M, Minko T, Kopeckova P, Kopecek J. Time- and concentration-dependent apoptosis and necrosis induced by free and HPMA copolymer-bound doxorubicin in human ovarian carcinoma cells. J Cont Rel 2000; 69:185–196.

230. Minko T, Kopeckova P, Kopecek. Mechanisms of anticancer action of HPMA copolymer-bound doxorubicin. Macromol Symp 2001; 172:35–37.

231. St'astny M, Strohalm J, Plocova D, Ulbrich K, Rihova B. A possibility to overcome P-glycoprotein (PGP)-mediated multidrug resistance by antibody-targeted drugs conjugated to N-(2-hydroxypropyl)methacrylamide (HPMA) copolymer carrier. Eur J Cancer 1999; 35:459–466.

232. Malugin A, Kopeckova P, Kopecek J. HPMA copolymer-bound doxorubicin induces apoptosis in human ovarian cancer cells by a Fas-independent pathway. J Cont Rel 2003; 91:254.

233. Musila R, Quarcoo N, Kortenkamp A, Duncan R. In vivo assessment of time-dependent DNA damage induced by HPMA copolymer-doxorubicin (PK1) using the alkaline single cell electrophoresis (Comet) assay. Proceedings of the International Symposium on Controlled Release of Bioactive Materials 2001; 28:7131–7132.

234. Kovar L, Ulbrich K, Kovar M, Strohalm J, Stastny M, Etrych T, Rihova B. The effect of HPMA copolymer-bound doxorubicin conjugates on the expression of genes involved in apoptosis signaling. J Cont Rel 2003; 91:247–248.

235. Rihova B, Jelinkova M, Strohalm J, Subr M, Plocova D, Hovorka O, Novak M, Plundrova D, Germano Y, Ulbrich K. Polymeric drugs based on conjugates of synthetic and natural macromolecules. II. Anti-cancer activity of antibody or (Fab')₂-targeted conjugates and combined therapies with immunomodulators. J Cont Rel 2000; 64:241–261.

236. Soucek J, Pouckova P, Strohalm J, Plocova D, Hlouskova D, Zadinova M, Ulbrich K. Poly[*N*-(2-hydroxypropyl)methacrylamide] conjugates of bovine pancreatic ribonuclease (RNase A) inhibit growth of human melanoma in nude mice. J Drug Target 2002; 10:175–183.

237. Wang L, Kristensen J, Ruffner DE. Delivery of antisense oligonucleotides using HPMA polymer: synthesis of a thiol polymer and its conjugation to water-soluble molecules. Bioconj Chem 1998; 9:749–757.

238. Toncheva V, Wolfert MA, Dash PR, Oupicky D, Ulbrich K, Seymour LW, Schacht EH. Novel vectors for gene delivery formed by self-assembly of DNA with poly(L-lysine) grafted with hydrophilic polymers. Biochim Biophys Acta 1998; 1380:354–368.

239. Oupicky D, Konak C, Ulbrich K. Preparation of DNA complexes with diblock copolymers of poly[N-(2-hydroxypropyl)-methacrylamide] and polycations. Mat Sci Eng 1999; 7:59–65.

240. Oupicky D, Konak C, Ulbrich K. DNA complexes with block and graft copolymers of N-(2-hydroxypropyl)methacrylamide and 2-(trimethylammonio)ethyl methacrylate. J Biomat Sci Polym Ed 1999; 10:573–590.

241. Konak C, Mrkvickova L, Nazarova O, Ulbrich K, Seymour LW. Formation of DNA complexes with diblock copolymers of poly[N-(2-hydroxypropyl)methacrylamide] and polycations. Supramolecular Sci 1998; 5:67–74.

242. Dautzenberg H, Konak C, Reschel T, Zintchenko A, Ulbrich K. Cationic graft copolymers as carriers for delivery of antisense-oligonucleotides. Macromol Biosci 2003; 3:425–435.

243. Read ML, Dash P, Clark A, Howard KA, Oupicky D, Toncheva V, Alpar HO, Schacht EH, Ulbrich K, Seymour LW. Physicochemical and biological characterisation of an antisense oligonucleotide targeted against bcl-2 mRNA complexed with cationic-hydrophilic copolymers. Eur J Pharm Sci 2000; 10:169–177.

244. Oupicky D, Howard KA, Konak C, Dash P, Ulbrich K, Seymour LW. Steric stabilisation of poly-L-lysine/DNA complexes by the covalent attachment of semitelechelic poly[N-(2-hydroxypropyl)methacrtylamide]. Bioconj Chem 2000; 11:492–501.

245. Read ML, Dash PR, Read ML, Fischer K, Howard KA, Wolfert M, Oupicky D, Subr V, Strohalm J, Ulbrich K, Seymour LW. Decreased binding of proteins and cells of polymeric gene delivery vectors surface modified with a multivalent hydrophilic polymer and retargeted through attachemnt of transferrin. J Biol Chem 2000; 275:3793–3802.

246. Oupicky D, Ogris M, Howard KA, Dash PR, Ulbrich K, Seymour LW. Importance of lateral and steric stabilisation of polyelectrolyte gene delivery vectors for extended systemic circulation. Molecul Therap 2002; 5:463–472.

247. Carlisle RC, Etrych T, Briggs SS, Preece JA, Ulbrich K, Seymour LW. Polymer-coated polyethylenimine/DNA complexes designed for triggered activation by intracellular reduction. J Gene Med 2004; 6:337–344.

248. Thompson AH, Vasey PA, Murray LS, Cassidy J, Fraier D, Frigerio E, Twelves C. Population pharmacokinetics in phase I drug development: a phase I study of PK1 in patients with solid tumors. Brit J Cancer 1999; 81:99–107.

249. Cassidy J. PK1: Results of Phase I studies. Proceedings of the Fifth International Symposium on Polymer Therapeutics: From Laboratory to Clinical Practice, Cardiff, UK, 2000, p. 20.

250. Schoemaker NE, van Kesteren C, Rosing H, Jansen, Swart M, Lieverst J, Fraier D, Breda M, Pellizzoni C, Spinelli R, Grazia PM, Beijnen JH, Schellens JH, Bokkel Huinink WW. A phase I clinical and pharmacokinetic study of MAG-CPT, a water soluble polymer conjugate of camptothecin. Brit J Cancer 2002; 87:608–614.

251. Schoemaker NE, Frigerio E, Fraier D, Schellens JHM, Rosing H, Jansen S, Beijnen J. High-performance liquid chromatographic analysis for the determination of a novel polymer-bound camptothecin derivative (MAG-camptothecin) and free camptothecin in human plasma. J Chromatogr 2001; 763:173–183.

252. Sarapa N, Britto MR, Speed W, Jannuzzo MG, Breda M, James C, Porro MG, Rocchetti M, Nygren P. Targeted delivery and preferential uptake in solid cancer of MAG-CPT, a polymer bound prodrug of camptothecin—a trial in patients undergoing surgery for colorectal carcinoma. Cancer Chemother Pharmacol 2003; 52:424–430.

253. Rice JR, Stewart DR, Safaei R, Howell SB, Nowotnik D. Preclinical development of water-soluble polymer-platinum chemotherapeutics AP5280 and AP5286. Proceedings of the Fifth International Symposium on Polymer Therapeutics: From Laboratory to Clinical Practice, Cardiff, UK, 2002, p. 54.

254. Nowotnik DP. Preclinical development and Phase I results with HPMA copolymer platinates. Proceedings of the Sixth International Symposium on Polymer Therapeutics: From Laboratory to Clinical Practice, Cardiff, UK, January 7–9, 2004.

255. Bouma M, Nuijen B, Stewart DR, Shannon KF, St. John JV, Rice JR, Harms R, Jansen BAJ, van Zutphen S, Reedijk J, Bult A, Beijnen JH. Pharmaceutical quality control of the investigational polymer-conjugated platinum anti-cancer agent AP 5280, PDA. J Pharm Sci Technol 2003; 57:198–207.

256. Rice JR, Stewart DR, Nowotnik DP. Enhanced antitumour activity of a new polymer-linked DACH-platinum complex. Proc Am Assoc Cancer Res 2002.

257. Rihova B. Antibody-targeted copolymer-bound anthracycline antibiotics. Drugs Future 2003; 28:1189–1210.

258. Rihova B, Kopeckova P, Strohalm J, Rossmann P, Vetvicka V, Kopecek J. Antibody directed affinity therapy applied to the immune system: in vivo effectiveness and limited toxicity of daunomycin conjugates to HPMA copolymers and targeting antibody. Clin Immunol Immunopathol 1988; 46:100–114.

259. Rossmann P, Rihova B, Strohalm J, Ulbrich K. Morphology of rat kidney and thymus after native and antibody-coupled cyclosporin A application (reduced toxicity of targeted drug). Folia Microbiol 1997; 42:277–287.

260. St'astny M, Ulbrich K, Strohalm J, Rossmann P, Rihova B. Abnormal differentiation of thymocytes induced by free cyclosporine is avoided when cyclosporine bound to N-(2-hydroxypropyl)-methacrylamide copolymer is used. Transplantation 1997; 63:1818–1827.

261. Bridges JF, Woodley JF, Duncan R, Kopecek J. Soluble N-(2-hydroxypropyl)methacrylamide copolymers as a potential oral, controlled-release drug delivery system. I. Bioadhesion to the rat intestine in vitro. Int J Pharm 1988; 44:213–223.

262. Bridges JF, Woodley JF, Duncan R, Kopecek J. In vitro and in vivo evaluation of N-(2-hydroxypropyl)methacrylamide copolymers as a potential oral drug delivery system. Proceedings of the International Symposium on Controlled Release of Bioactive Materials 1988; 14:14–15.

263. Kopeckova P, Longer MA, Woodley JF, Duncan R, Kopecek J. Release of p-nitroaniline from oligopeptide side-chains attached to N-(2-hydroxypropyl) methacrylamide copolymers during incubation with rat intestinal brush border enzymes. Makromol Chem Rapid Comm 1991; 12:101–106.

264. Grim Y, Kopecek J. Bioadhesive water-soluble polymeric carriers for site-specific oral drug delivery. Synthesis, characterisation and relesae of 5-aminosalicylic acid by Streptococcus faecium in vitro. New Polym Mater 1991; 3:49–59.

265. Rathi RC, Kopeckova P, Rihova B, Kopecek J. N-(2-Hydroxy-propyl) methacrylamide copolymers containing pendent saccharide moieties. Synthesis and bioadhesive properties. J. Polym Sci A 1991; 29:1895–1991.

266. Rihova B, Rathi RC, Kopeckova P, Kopecek J. *In vitro* bioadhesion of carbohydrate-containing N-(2-hydroxypropyl)-methacrylamide copolymers to the GI tract of guinea pigs. Int J Pharm 1992; 87:105–116.

267. Rathi RC, Kopeckova P, Kopecek J. Biorecognition of sugar containing N-(2-hydroxypropyl)methacrylamide copolymers by immobilized lectin. Macromol Chem Phys 1997; 198: 1165–1180.

268. Rihova B, Jelinkova M, Strohalm J, St'astny M, Hovorka O, Plocova D, Kovar M, Draberova L, Ulbrich K. Antiproliferative effects of a lectin and anti-thy-1.2 antibody targeted HPMA copolymer-bound doxorubicin on primary and metastatic human colorectal carcinoma and on human colorectal carcinoma transfected with mouse thy-1.2 gene. Bioconj Chem 2000; 11:664–673.

269. Wroblewski S, Berenson M, Kopecekova P, Kopecek J. Biorecognition of HPMA copolymer lectin conjugates as an indicator of differentiation of cell surface glycoproteins in development, maturation, and diseases of human and rodent gastrointestinal tissues. J Biomed Mater Res 2000; 5:329–342.

270. Wroblewski S, Berenson M, Kopecekova P, Kopecek J. Potential of lectin-N-(2-hydroxypropyl)methacrylamide copolymer-drug conjugates for the treatment of pre-cancerous conditions. J Cont Rel 2001; 74:283–293.

271. Wroblewski S, Rihova B, Rossmann, Hudcovicz T, Rehakova Z, Kopecekova P, Kopecek J. The influence of colonic microbiota on HPMA copolymer lectin conjugate binding in rodent intestine. J Drug Target 2001; 9:85–94.

272. David A, Kopeckova P, Minko T, Rubinstein A, Kopecek J. Design of a multivalent galactoside ligand for selective targeting of HPMA copolymer-doxorubicin conjugates to human colon cancer cells. Eur J Cancer 2004; 40:148–157.

273. Armstrong BK, Smith Q, Rapoport SI, Strohalm J, Kopecek J, Duncan R. Osmotic opening of the blood-brain barrier per-

meability to *N*-(2-hydroxypropyl)methacrylamide copolymers. Effect of M_w charge and hydrophobicity. J Cont Rel 1989; 10:27–35.

274. Prokopova-Kubinova S, Vargova L, Tao L, Subr V, Sykova E, Nicholson C. Poly[N-(2-hydroxypropyl)methacrylamide] polymers diffuse in brain extracellular space with the same totruosity as small molecules. Biophys J 2001; 80:542–548.

275. Nan A, Croft SL, Yardley V, Ghanderhari H. Targetable water-soluble polymer-drug conjugates for the treatment of visceral leishmaniasis. J Cont Rel 2003; 94:115–127.

276. Wang D, Miller S, Sima M, Kopekova P, Kopecek J. Synthesis and evaluation of water-soluble polymeric bone-targeted drug delivery systems. Bioconj Chem 2003; 14:853–859.

277. Dvorak M, Kopeckova P, Kopecek J. High-molecular weight HPMA copolymer-adriamycin conjugates. J Cont Rel 1999; 60:321–332.

278. Kamei S, Kopecek J. Prolonged blood circulation in rats of nanspheres surface modified with semi-telechelic poly [N-(2-hydroxypropyl)methacrylamide]. Pharm Res 1995; 12: 663–668.

279. Gianasi E, Cociancich F, Uchegbu IF, Florence AT, Duncan R. Pharmaceutical and biological characterisation of a doxorubicin-polymer conjugate (PK1) entrapped in sorbitan monostearate Span 60 niosomes. Int J Pharm 1997; 148:139–148.

280. Uchegbu IF, Duncan R. Niosomes containing N-(2-hydroxypropyl) methacrylamide-doxorubicin (PK1): effect of method of preparation and choice of surfactant on niosome characteristics and a preliminary study of body distribution. Int J Pharm 1997; 155:7–17.

2

Functional PEG for Drug Delivery

MOTOI OISHI and YUKIO NAGASAKI
Department of Materials Science and
Technology, Tokyo University of
Science, Noda, Japan

KAZUNORI KATAOKA
Department of Materials Science and
Engineering, Graduate School of
Engineering, The University of
Tokyo, Tokyo, Japan

1. INTRODUCTION

Poly(ethylene glycol) (PEG) is a polymer possessing unique physicochemical properties such as high water solubility and flexibility. From these characteristics it is known that PEG is bioinert, and it has attracted increasing attention in the sphere of biomedical applications. Since Abchowski et al. (1) reported that end-functionalized PEG can modify biologically active proteins, an interaction that caused not only lowered immunogenicity but also an alteration of pharmacokinetics, PEG modification chemistry has become increasingly important in the field of protein drugs. The modification of the protein by PEG is the so-called *"PEGylation,"*

93

and the term has been utilized from the beginning of the 1990s (2,3). As can be seen in Figure 1, the number of publications on PEGylation chemistry is increasing significantly every year. Thus, the merits of PEGylation chemistry have been becoming clear to the research community.

The conventional PEG that is utilized for PEGylation chemistry is semi-telechelic PEG and, especially, the commercially available methoxy-ended PEG (4,5). A particular form— the commercially available methoxy-PEG-OH, in which the hydroxyl group was converted to another functional group— is used for the modification of proteins. The beard-like PEG chains that are conjugated to a protein molecule are believed to prevent recognition by a group of scavenger cells, namely the reticuloendothelial system (RES) that is located at the liver, the spleen, and the lungs. In addition, the increase in

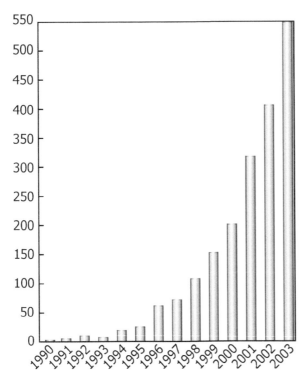

Figure 1   Number of publications on PEGylation chemistry.

the molecular mass of the protein by PEGylation reduces clearance by renal excretion, resulting in prolonged blood circulation. Thus, the stealth effect of the PEG chain conjugated to biologically active molecules is fairly effective in vivo. Indeed, several types of PEGylated protein drugs such as adenosine deaminase, asparaginase, and interleukin 2 have been successfully applied in clinical applications (6–8). Now, PEGylation chemistry is extending to other in vivo systems such as drug carriers. For instance, the PEGylated liposome improved the performance significantly as compared to that of the liposome without PEG chains on the surface. Therefore, PEGylation chemistry is improving rapidly in the field of bioconjugation and drug delivery systems (9–11).

PEGylation is usually carried out via the covalent linkage between the end group of the PEG chain and biologically active molecules such as proteins, as mentioned above. The other end group of the PEG chain, usually the methoxy-group, should be inert. In other words, it is not possible to use the other end group for further linking to ligand molecules. In order to improve PEGylation chemistry, we have been focusing on the synthesis of heterotelechelic PEG, which denotes PEG possessing a functional group at one end and another functional group at the other chain end, selectively and quantitatively. Several types of heterotelechelic PEGs, including end-functionalized block copolymers, have been prepared so far. In this chapter, the synthesis, characteristics, and application of these compounds are described.

2. SYNTHESIS OF HETEROTELECHELIC POLY(ETHYLENE GLYCOL)S

There are several reports on the preparation of heterotelechelic PEG (12–14). Most of them, however, started from commercially available PEGs possessing hydroxyl groups at both ends. In particular, one of the hydroxyl groups at the PEG ends was converted to another functional group first; then, the other hydroxyl group was converted to one of the

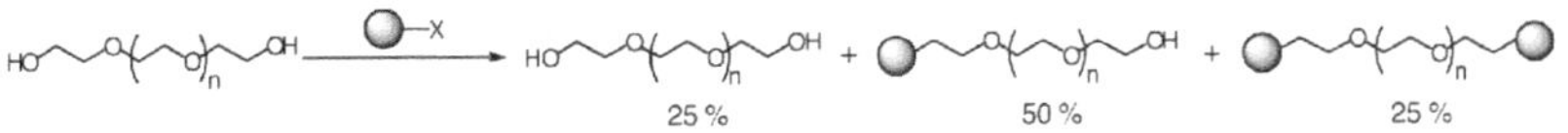

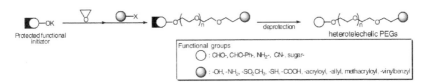

Scheme 1 Preparation of heterotelechelic PEGs from commercially available PEG possessing hydroxyl groups at both ends.

other functional groups. These complex consecutive steps not only make the overall process troublesome but also cause several problems. Because the reactivity of hydroxyl groups at both PEG ends is the same, the resulting PEG must be a mixture of three types of PEG homologues in the first step as shown in Sch. 1. In this case, the mono-reacted PEG should be separated from the mixture. It is well known that the separation of polymeric compounds is very difficult and, thus, leads to low yield and low purity.

In order to obtain heterotelechelic PEGs that give rise to both high yield and high purity, we have applied a functional initiator to a living anionic ring-opening polymerization of ethylene oxide in order to design new routes (15–18). In this case, the initiator was introduced to one of the ends of the obtained PEG quantitatively as shown in Sch. 2. Several types of initiators were examined for the effective polymerization of ethylene oxide. One of the most important and convenient compounds we prepared so far is an acetal-ended PEG. The acetal group at the PEG chain end can be easily converted to an aldehyde group by acid treatment (19). The aldehyde group is known to rapidly react with primary amine, generating a Schiff's base intermediate (–CH=N–), which can be converted to a secondary amino group (–CH$_2$–NH–) by the addition of a reducing agent. Also, the aldehyde group is

Scheme 2 Novel synthetic approach to heterotelechelic PEGs using the functional initiators.

stable under the neutral media. Thus, the aldehyde-ended heterotelechelic PEG can be utilized as the ligand installation via the reductive amination reaction.

As a typical example, a synthesis of α-acetal-ω-hydroxyl-PEG (acetal-PEG-OH) was carried out as follows: To 1 mmol of potassium 3,3-diethoxypropanolate in dry THF under argon atmosphere, 130 mmol of ethylene oxide was added via a cooled syringe. The mixture was stirred magnetically and allowed to react for 48 h at room temperature. The resulting polymer was precipitated into ether, and the precipitate was filtered. Finally the collected sample was freeze-dried with benzene. Figure 2 shows the size exclusion chromatogram (SEC) of the obtained polymer. The molecular weight could be controlled by the monomer/initiator ratio, retaining a low molecular weight distribution factor. Figure 3 shows the ^{13}C NMR spectra of the obtained polymer before and after the acid treatment. The signals based on the end acetal group can be observed at 15, 34, 62, and 101 ppm before the acid treatment. The acetal signals disappeared completely after the acid treatment. Instead, a new signal based on the carbonyl carbon at 202 ppm appeared after the acid treatment, indicating that the acetal group at the PEG chain end can be converted completely to the aldehyde group. Thus, α-aldehyde-ω-hydroxy-PEG (CHO-PEG-OH) was easily obtained. Further

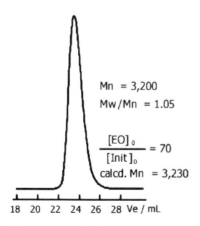

Figure 2   SEC chromatogram of the acetal-PEG-OH.

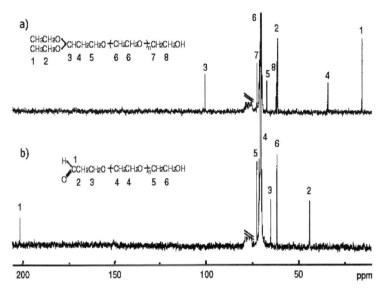

Figure 3 ^{13}C NMR spectra of (a) acetal-PEG-OH and (b) aldehyde-PEG-OH.

functionalization of the hydroxy group at the ω-end was been attained in several ways. One typical example was methane sulfonation, which took place after the polymerization was attained. After the monomer was completely consumed in the polymerization of ethylene oxide, methanesulfonyl chloride was added to the reaction mixture to convert the mixture from the potassium alkoxide chain end to the methane sulfonate chain end (Sch. 3). For the amination of the ω-end group, acetal-PEG-SO$_2$CH$_3$ was dissolved in a 25% ammonia aqueous solution. The ω-end group (methane sulfonate group) was quantitatively converted to primary amine by this reaction.

For the O-ethyldithiocarbonate-ended heterotelechelic PEG, the methane sulfonate end group was reacted with potassium O-ethyldithiocarbonate. Figure 4 shows both the ^{13}C NMR spectra of the acetal-PEG possessing methane sulfonate (acetal-PEG-SO$_2$CH$_3$) and the O-ethyldithiocarbonate end groups at the ω-end of the PEG chain [acetal-PEG-S(C=S)Oet]. The O-ethyldithiocarbonate end group can be

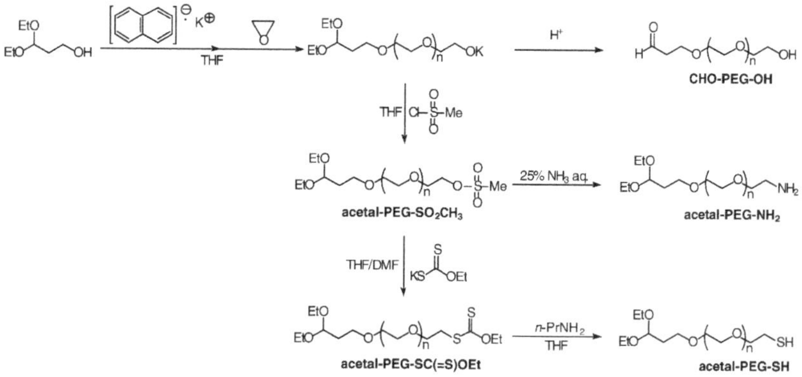

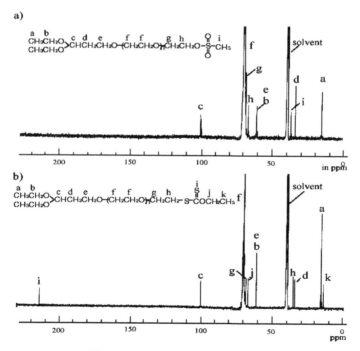

Scheme 3 Synthesis of acetal-ended heterotelechelic PEGs.

Figure 4 ^{13}C NMR spectra of (a) acetal-PEG-SO$_2$CH$_3$ and (b) acetal-PEG-SC(=S)OCH$_2$CH$_3$.

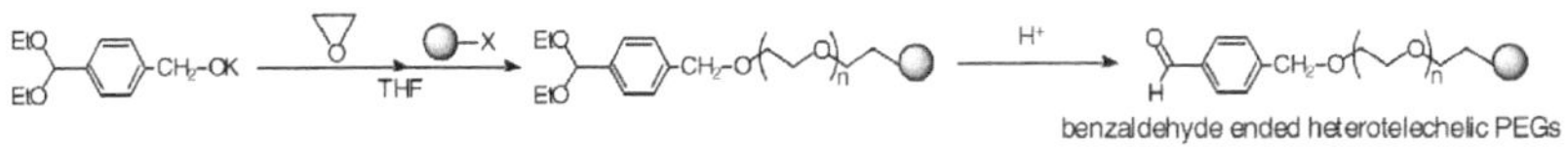

benzaldehyde ended heterotelechelic PEGs

Scheme 4 Synthesis of benzaldehyde-ended heterotelechelic PEGs initiated by potassium 4-(diethoxymethyl)benzylalkoxide.

easily converted to the mercapto group by the addition of propylamine (20). Thus, acetal-PEG-NH$_2$ and acetal-PEG-SH can be synthesized quantitatively. These heterotelechelic PEGs are very useful not only for protein modification but also for surface modification.

We have examined several initiators other than the acetal group for the polymerizations of ethylene oxide. Potassium 4-(diethoxymethyl)benzylalkoxide can be utilized as the functional initiator for the polymerization of ethylene oxide to produce various benzaldehyde-ended heterotelechelic PEGs (Sch. 4) (21). Owing to the lack of the α-protons, the benzaldehyde-ended heterotelechelic PEGs do not occur in the dimerization between the aldehyde-PEG ends via the aldol reaction, which is a major side reaction utilization of aldehyde-ended PEGs; therefore, the potassium 4-(diethoxymethyl)benzylalkoxide initiator system is more useful for the obtaining of the heterotelechelic PEGs than the potassium 3,3-diethoxy-1-propanolate initiator system. The polymerization by the potassium amide of 2,2,5,5-tetramethylcyclopentadisilazane formed a PEG with the organosilyl-protected amino group at the α-chain end, as shown in Sch. 5. The silyl group can be easily deprotected by the weak acid treatment to produce an α-primary-amino-ended PEG-OH (NH$_2$-PEG-OH). Protected monosaccharide can be also utilized as an initiator so that sugar-ended heterotelechelic PEGs can

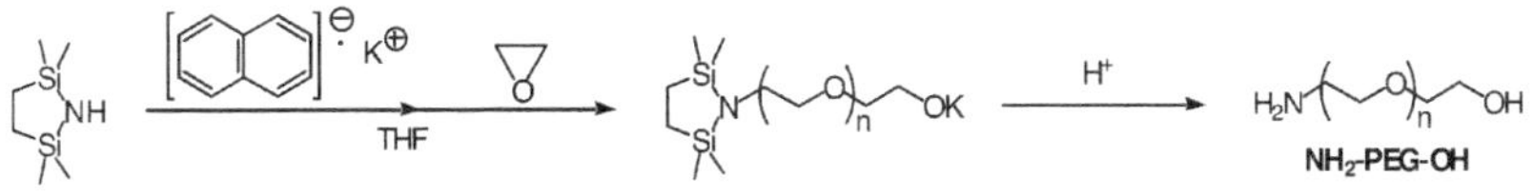

Scheme 5 Synthesis of primary amine-ended heterotelechelic PEG initiated by potassium amide of 2,2,5,5-tetramethylcyclopentadisilazane.

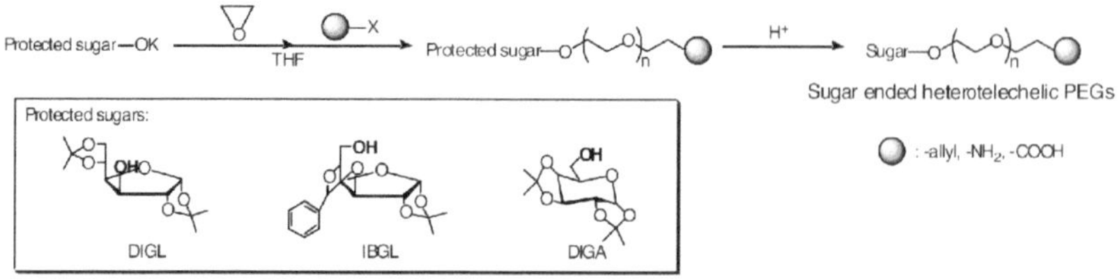

Scheme 6 Synthesis of sugar-ended heterotelechelic PEGs initiated by protected monosaccharide.

be obtained. Four hydroxyl groups in the monosaccharide were regioselectively protected via acetal linkage such as 1,2,5,6-di-O-isopropylidene-D-glucofuranose (DIGL), 1,2,3, 4-di-O-isopropylidene-D-galactopyranose (DIGA), and 1,2-O-isopropylidene-3,5-O-benzylidene-D-glucofuranose (IBGL) (Sch. 6) (22,23). The remaining hydroxyl group can be used as an initiator for the polymerization of ethylene oxide. The resulting PEGs possess the corresponding sugar moiety at the α-chain end and a hydroxyl group at the ω-chain end. The ω-chain end could be converted to several functional groups such as allyl group, amino group, and carboxylic acid group with high yield and high purity. Such heterotelechelic PEGs possessing a monosaccharide residue at the α-end are promising as one of the tools for bioconjugate chemistry.

3. SYNTHESIS OF AN END-FUNCTIONALIZED BLOCK COPOLYMER POSSESSING A PEG SEGMENT

Block copolymers, consisting partly of PEG, have become attractive in the field of drug delivery systems (9,24–26). For example, the PEG/poly(lactide) block copolymer composed of hydrophilic and hydrophobic segments forms core-shell type polymeric micelles in aqueous media with a 10–100 nmr particle size, and this block copolymer can be utilized as a nanospherical drug carrier (Sch. 7). Diverse drugs with a hydrophobic nature can be loaded with high efficacy into the core of polymeric micelles, allowing drugs to be solubilized in aqueous

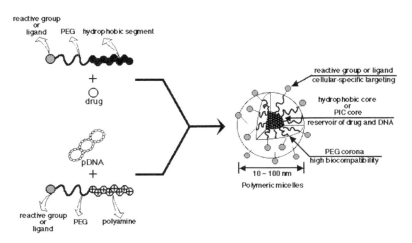

Scheme 7   Schematic illustration of polymeric micelles.

media. Indeed, polymeric micelles loaded with the anticancer drug, doxorubicin, were shown to stably circulate in the blood stream for a prolonged time period, eventually accumulating in the solid tumor by the enhanced permeable and retention effect (EPR) (27,28). Recently, we reported that PEG/polyamine block copolymers were found to form polyion complex (PIC) micelles with negatively charged DNA through electrostatic interaction, and PIC micelles have a size of less than 100 nm (Sch. 7) (29–38). The PIC micelles exhibited excellent solubility in aqueous media, tolerance against enzymatic degradation, and minimal interaction with the cell membrane and the serum compartment due to the steric stabilization of the high-density PEG corona surrounding the PIC core. Thus, PIC micelles have attracted much attention as nonviral vectors for gene therapy.

   However, most of the polymeric micelles, including PIC micelles, that have been prepared so far possess no reactive group on the surface. If reactive groups can be introduced on the surface, any kind of ligand molecule can be installed on the surface of the micelle. For the preparation of reactive polymeric micelles, it is important to synthesize block copolymers possessing a functional group at the PEG chain end. By using heterotelechelic PEGs thus prepared, new types of block copolymers possessing a functional group at the PEG chain end can be designed. In this section, the new molecular design of the three types of PEG end-functionalized block copolymers is described.

3.1. Acetal-PEG-Poly(D,L-lactide) Block Copolymer (39,40)

In the field of drug delivery systems, PEG-poly(D,L-lactide) (PEG-PLA) block copolymers (41–43) have been extensively studied because of the following reasons: (i) the potassium alcholate species of the PEG chain end can initiate the ring-opening polymerization of the D,L-lactide (LA) monomer without any side reactions (44); and (ii) PLA is biodegradable and nontoxic (45). After the preparation of acetal-PEG-OK in THF, a certain amount of LA was added to obtain the acetal-PEG-PLA block copolymer, as shown in Sch. 8. Figure 5 shows the SEC chromatograms and ^{1}H NMR spectrum of the obtained block copolymer. As can be seen in the SEC chromatograms, the molecular weight of the block copolymer increased, retaining a narrow molecular weight distribution (MWD) as well as no remaining prepolymer peak, indicating the high efficiency of block copolymerization. In addition to the signals based on both the PEG and PLA segments on the ^{1}H NMR spectrum, the end acetal protons were clearly observed at 4.7 ppm. Thus, the PEG-PLA (hydrophilic-hydrophobic) block copolymer possessing an acetal group at the PEG chain end was quantitatively obtained.

3.2 Acetal-PEG-Poly(2-*N,N*-dimethylaminoethyl Methacrylate) Block Copolymer (46)

Recently, we have found that a methacrylic ester possessing an amino group at the β-position of the ester moiety can be

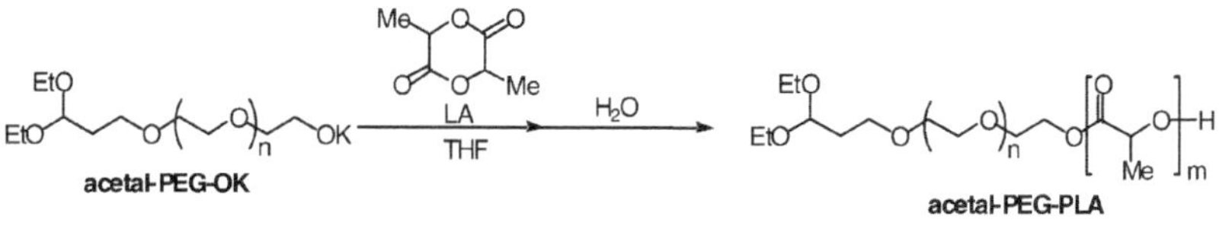

Scheme 8 Synthesis of acetal-PEG-PLA block copolymer.

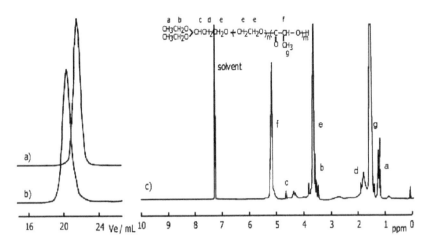

Figure 5 SEC chromatograms of (a) acetal-PEG-OH, (b) acetal-PEG-PLA block copolymer, and (c) ^{1}H NMR spectrum of the acetal-PEG-PLA block copolymer.

polymerized by a simple alkoxide species such as potassium ethoxide (47). We proposed the increased reactivity of an alkoxide initiator that was generated by a chelation which was located between the monomer molecule and the initiator. This polymerization system can be applicable to a new block copolymer synthesis (48). In the same way as mentioned above, the acetal-PEG-OK was prepared in THF, followed by the addition of a certain amount of 2-N,N-dimethylaminoethyl methacrylate (AMA) to the reaction mixture, which was stirred for several minutes (Sch. 9) (49). Figure 6 shows the SEC chromatograms and the ^{1}H NMR spectrum of the obtained block copolymer. By the addition of an AMA monomer to the polymerization system after all of the ethylene oxide was consumed, block polymerization proceeded, although a small

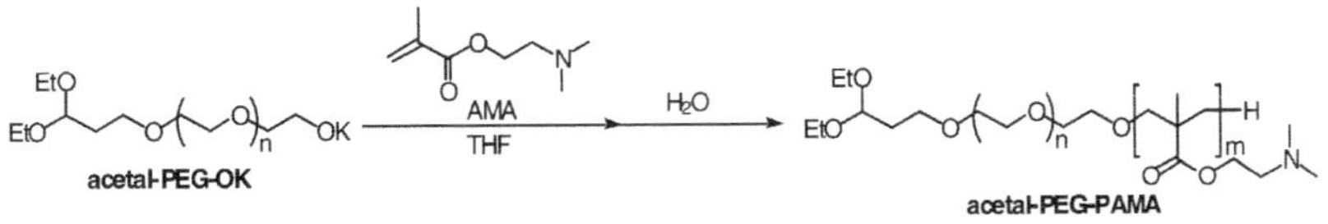

Scheme 9 Synthesis of acetal-PEG-PAMA block copolymer.

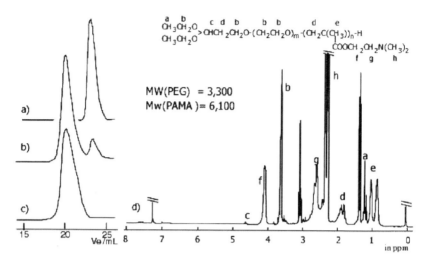

Figure 6 SEC chromatograms of (a) acetal-PEG-OH, (b) crude acetal-PEG-PAMA block copolymer, (c) acetal-PEG-PAMA block copolymer purified by ion exchange resin, and (d) ^{1}H NMR spectrum of the purified acetal-PEG-PAMA block copolymer.

amount of the prepolymer remained, as shown in Fig. 6b. It should be noted that the MWD of the obtained block copolymer was still low, indicating that the possible ester exchange reaction was negligible. After the treatment with ion exchange resin, the remaining prepolymer was removed completely, as shown in Fig. 6c. Along with the signals based on both the PEG and PAMA segments, the end acetal protons were observed on the ^{1}H NMR spectrum (Fig. 6d), indicating that, through the use of heterotelechelic PEG as a macroinitiator, a new end-functionalized PEG/polyamine block copolymer can be synthesized. In addition, this new end-functionalized PEG/polyamine block copolymer (acetal-PEG-PAMA) and DNA will spontaneously form the PIC micelle that can be utilized as the targetable nonviral gene vector.

3.3. Acetal-PEG/Linear-Poly(ethylenimine) Block Copolymer (50)

Linear-poly(ethylenimine) (L-PEI) has often been used for gene delivery as DNA/L-PEI complexes due to its high transfection efficiency, because PEI—by virtue of its "buffer" or "proton-sponge" effect—is known to overcome the critical barrier of endosomal escape (51). The acetal-PEG-SO$_2$CH$_3$ can be also utilized as the macroinitiator for the cationic ring-opening polymerization of 2-methyl-2-oxazoline (Oz) so that the block copolymer acetal-PEG/POz (Sch. 10) can be obtained, and this process is important because sulfonate derivatives can initiate the cationic polymerization of several oxazoline monomers (52). The cationic ring-opening polymerization of the Oz monomer in the presence of acetal-PEG-SO$_2$CH$_3$ was carried out in nitromethane at 70°C for 70 h. Figure 7 shows the SEC chromatograms and the ^{1}H NMR spectrum of the obtained block copolymer. As can be seen in the SEC chromatograms, the molecular weight of the block copolymer increased, retaining narrow MWD as well as no remaining prepolymer peak, indicating the high efficiency of the block copolymerization. In addition to the signals based on both the PEG and the POz segments on the ^{1}H NMR spectrum, the end acetal protons were clearly observed at 1.1 and 1.9 ppm. Thus, the PEG/POz block copolymer possessing an acetal group at the PEG chain end was quantitatively obtained. The hydrolysis of acetal-PEG/POz using NaOH in ethylene glycol/ethanol (1:1) produced an acetal-PEG/L-PEI, of which the ^{1}H NMR spectrum is shown in Fig. 8. The methyl protons of the acetyl group of the POz segment at 2.1 ppm disappeared completely after hydrolysis, and the shift of both the methylene protons adjacent to the nitrogen atom from 3.4 to 2.6 ppm and the end acetal protons at 1.1 and

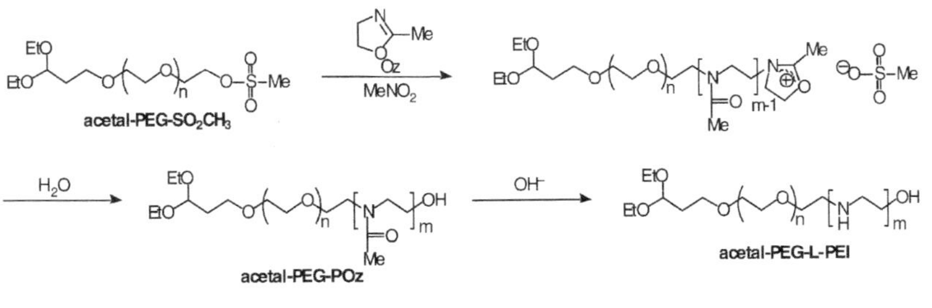

Scheme 10   Synthesis of acetal-PEG-L-PEI block copolymer.

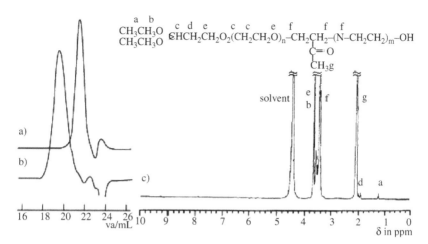

Figure 7 SEC chromatograms of (a) acetal-PEG-SO$_2$CH$_3$, (b) acetal-PEG-POz block copolymer, and (c) ^{1}H NMR spectrum of acetal-PEG-POz block copolymer.

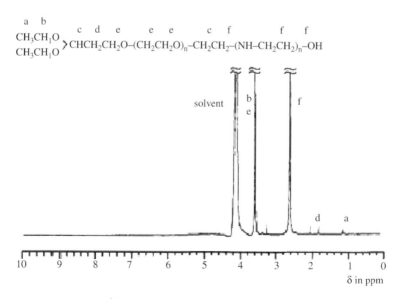

Figure 8 The ^{1}H NMR spectrum of the acetal-PEG-L-PEI block copolymer.

1.8 ppm were also observed, indicating that the acetal-PEG/L-PEI was successfully synthesized. PIC micelles composed of the obtained acetal-PEG/L-PEI block copolymer and DNA may also have a potential utility as a targetable nonviral gene vector.

4. BIOCONJUGATION USING HETEROTELECHELIC PEG

Biologically active molecules such as proteins and oligodeoxy-nucleotides (ODNs) have attracted much attention as a class of therapeutic agents in recent years. Nevertheless, the therapeutic value of the proteins and ODNs under in vivo conditions has not been fully proven to be effective, owing to several obstacles including nonspecific interaction with plasma proteins, low stability against enzymatic degradation, and preferential liver and renal clearance. Achievement of these desired improvements is being addressed in many different ways. Most notable is the introduction of the PEG chain to biologically active molecules (PEG conjugate). In fact, it is well known that PEG is a unique molecule that, when covalently bonded to biologically active molecules, increases their solubility in aqueous solution, decreases the immunogenic effect, and extends the body lifetime. Therefore, numerous reports on the preparation of PEG conjugates have been reported (1–3).

However, most of the PEG conjugates including the PEG-protein conjugate and the PEG-ODN conjugate prepared so far possess no ligand molecule at the PEG end (6–8). Therefore, a way to install the ligand molecules at the PEG chain end is still required for the achievement of cellular-specific targeting. If reactive groups can be introduced to the PEG chain end, any kind of ligand molecule can be installed at the PEG chain end of biologically active molecule-PEG conjugates. By using the heterotelechelic PEGs thus prepared, new types of biologically active molecule-PEG conjugates possessing a reactive group or a ligand molecule at the PEG chain can be designed. In this section, a new PEG end-functionalized PEG-protein conjugate and the PEG-ODN conjugate are described.

4.1. Modification of Protein by Heterotelechelic PEG

Our interest is focused on the Superoxide dismutase (SOD) enzyme as the therapeutic protein, because the biological function of SOD is known to play a very important role in preventing oxidative damage of DNA by the reactive oxygen species (53,54). To achieve the cellular-specific targeting of SOD, we synthesized heterobifunctional PEGs having a lactose group at one end and a carboxylic acid group at the other end (lac-PEG-COOH) by the reductive amination reaction of α-aldehyde-ω-carboxylic acid-PEG (CHO-PEG-COOH) with *p*-aminophenyl-β-D-lactopyranoside as shown in Sch. 11. The SODs conjugated to PEG (lac-PEG-SOD and acetal-PEG-SOD conjugates) were prepared using acetal-PEG-COOH and lac-PEG-COOH, which were thus synthesized via an active ester method in the presence of water-soluble carbodiimide. The PEG modification number was found to be 8.3 PEG molecules per SOD, as determined by the TNBS method.

Using FITC-labeled acetal-PEG-SOD and lac-PEG-SOD conjugates thus synthesized, cellular-specific uptake was evaluated by flow cytometry. For in vitro evaluation, HL-60

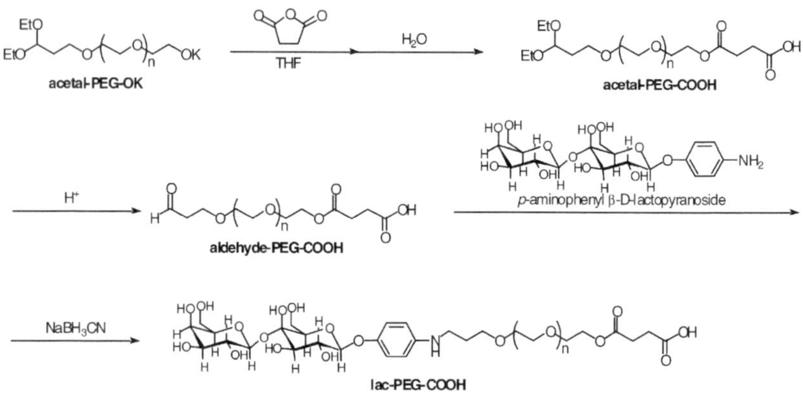

Scheme 11 Synthesis of lactose-PEG-SOD conjugate.

a)   b)

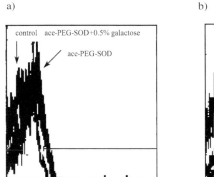 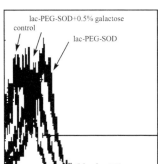

Figure 9 Flow cytometry diagrams of HL-60 cells treated with (a)
an acetal-PEG-SOD conjugate and (b) a lac-PEG-SOD conjugate.

cells (human promyelocytic leukemia cells), which are known
to have an abundance of asialoglycoprotein (ASGP) receptors
on the cell surface (55), were used. The sample concentration
was 250 μg/mL and was added to the cells, and then incubated
at 37°C for 3 h. Cellular uptake observations are shown in
Figure 9. Compared to the control, both the acetal-PEG-SOD
and the lac-PEG-SOD conjugates showed apparent shifts of
the cell distribution peak to high fluorescence intensity. When
0.5% of galactose was added as an inhibitor, the peak of the
acetal-PEG-SOD conjugate stayed in the same position as that
of galactose free. On the contrary, the peak of the lac-PEG-
SOD conjugate shifted back to lower fluorescence intensity.
These results suggest that, on the one hand, the acetal-PEG-
SOD conjugate was taken up via endocytosis and, on the other
hand, the lac-PEG-SOD conjugate was taken up effectively via
the ASGP receptor-mediated endocytosis on the surface of the
HL-60 cells. By using heterotelechelic PEGs, a targetable
protein-PEG conjugate can be successfully synthesized.

4.2. Modification of Oligodeoxynucleotide by Heterotelechelic PEG (56)

Heterotelechelic PEG can be also utilized for conjugation with
ODN. The synthetic route of the acetal-PEG conjugated to

antisense ODN through an acid-labile linkage (β-thiopropio-nate linkage) is shown in Sch. 12. The key issue of the introduction of an acid-labile linkage between the PEG and ODN segments is to release the active (free) antisense ODN in response to the endosomal pH (6.0~5.0) (57) leading to efficient interaction of antisense ODN with the target mRNA in the cytoplasm. To introduce both the reactive group and the acid-labile linkage in the PEG-ODN conjugate, the heterobifunctional PEG possessing an acetal group at the α-end and an acrylate group at the ω-end (acetal-PEG-acrylate) was synthesized. As mentioned above, this synthesis was based on the anionic ring-opening polymerization of ethylene oxide and involved the use of, first, the potassium 3,3-diethoxy-1-propanolate initiator system, and, then, 3.0 equivalents of acryloyl chloride in the presence of triethylamine so that the acetal-PEG-acrylate could be obtained. To obtain the acetal-PEG-ODN conjugate, a Michael reaction of the 3'-thiol-modified ODN (5'-AGG ACA GGT TCA GTG GAT C-$CH_2CH_2CH_2SH$-3', dopamine D2 receptor antisense sequence 19-mer) with excess acetal-PEG-acrylate was carried out in water (pH 8.0) at room temperature for 6 h. There was no remaining 3'-thiol-modified ODN after a 6 h reaction, as confirmed by SEC, suggesting that the Michael reaction quantitatively proceeded. Figure 10 shows the ^{1}H NMR spectrum of the acetal-PEG-ODN conjugate in D_2O with assignments. Along with the signals based on both the PEG segment and the ODN segment, the end acetal protons were observed at δ 1.1, 2.5, and 4.7 ppm. From the integral ratio of the ethylene

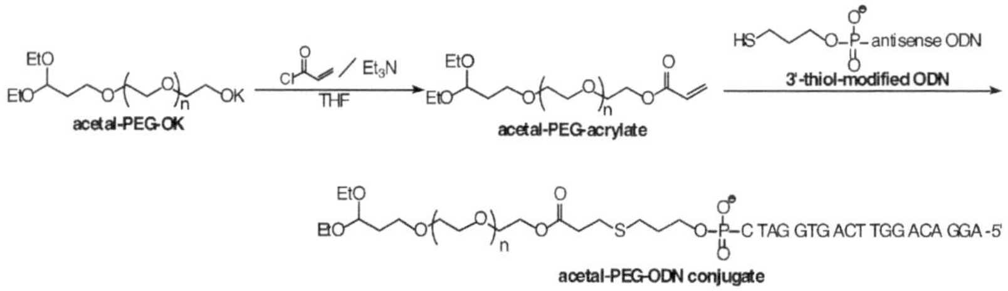

Scheme 12 Synthesis of pH-responsive acetal-PEG-ODN conjugate.

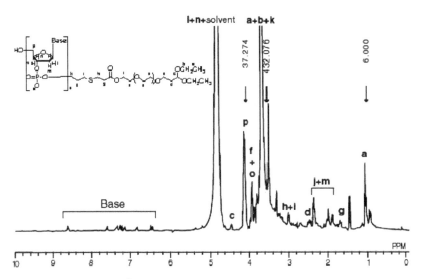

Figure 10 The ^{1}H NMR spectrum of the acetal-PEG-ODN conjugate.

protons on the PEG-backbone (3.7 ppm $-OCH_2CH_2-$) and methylene protons of ODN (4.1 ppm, $5'-CH_2-$), the unit number of deoxynucleotides in the ODN segment was calculated to be 18.6 bases, which is in good accordance with that of the starting $3'$-thiol-modified ODN (19-bases).

Hydrolysis of the acid-labile linkage (β-thiopropionate linkage) of the acetal-PEG-ODN conjugate was carried out at pH 5.5 and 7.4, so that the effect of the environmental pH on the stability of the conjugate could be estimated. Aliquots of the reaction mixture at 37°C were taken at appropriate time intervals, and were subjected to an SEC analysis. Figure 11 shows the cleavage reaction profiles of the PEG-ODN conjugate as a function of incubation time. The remaining amount of the intact acetal-PEG/ODN conjugate after incubation for 24 h at pH 5.5 was found to be 3% of the initial value, whereas almost no cleavage of the acid-labile linkage was observed at pH 7.4 even after incubation for 24 h. The end-functionalized antisense ODN-PEG conjugate through an acid-labile linkage (β-thiopropionate linkage) can be successfully synthesized by a Michael reaction of

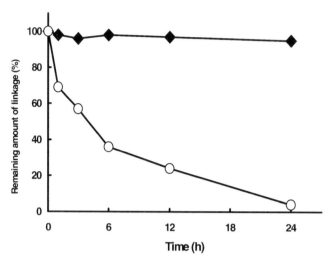

Figure 11 Stability of the acid-labile linkage in an acetal-PEG-ODN conjugate at pH 7.4 (◆) and 5.5 (○) at 37°C. Cleavage of the linkage is expressed as percentage of remaining amount of linkage over time based on SEC data.

3′-thiol-modified ODN with a heterotelechelic PEG. Underlying this process is an aim to develop a novel ODN delivery system.

5. FUNCTIONALIZED POLYMERIC MICELLES BY A SELF-ASSEMBLING OF END-FUNCTIONALIZED PEG BLOCK COPOLYMERS

As described in Sec. 2, polymeric micelles that are formed through either hydrophilic-hydrophobic interaction of PEG-based block copolymers or electrostatic interaction of PEG-polycation block copolymers with negatively charged macromolecules have attracted considerable attention in the field of drug delivery and gene delivery systems, owing to their excellent biocompatibility, long blood circulation time, and nontoxicity. In this section, the preparation and the characterization of end-functionalized polymeric micelles composed of end-functionalized PEG block copolymer copolymers is described.

5.1. Aldehyde-PEG-PLA Core-Shell Type Polymeric Micelles (39,40)

It is known that, when exposed to aqueous media, amphiphilic block copolymers with a suitable hydrophilic-hydrophobic balance form micelle structures (24–26). In our case, after the block polymer was dissolved in *N,N*-dimethylacetamide (DMAc), the solution was dialyzed against water. Dynamic light scattering (DLS) measurements made possible an angular dependent analysis, from which it was found that the acetal-PEG-PLA ($DP_{PEG} = 52$, $DP_{PLA} = 56$) micelle has a spherical shape and a size of ca. 30 nm. Figure 12 shows the representative data for the gamma-distribution of the obtained polymeric micelle. The acetal-PEG-PLA micelles thus obtained possess a unimodal distribution in the histogram analysis.

The conversion of the acetal end group into an aldehyde end group was conducted after the micelle formation. Hydrochloric acid was used to adjust the acetal-PEG-PLA (52/56) micelle solution to pH = 2. After a predicted period, the

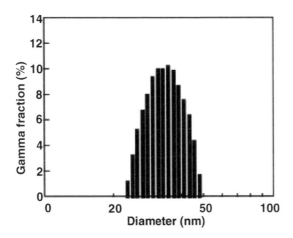

Figure 12   Gamma-distribution of the acetal-PEG-PLA polymeric micelle obtained from histogram analysis of DLS (detection angle, 90°; temperature, 25°C).

reaction was quenched by neutralization with $NaOH_{aq}$ and the polymeric micelle was purified by dialysis. The conversion reaction of the acetal group into the aldehyde group was monitored by means of ^1H NMR spectroscopy. Figure 13 shows the ^1H NMR spectra of acetal-PEG-PLA (52/56) in $CDCl_3$ and D_2O after the hydrolysis reaction. As can be seen in both of the figures, the end-aldehyde proton appeared at 9.8 ppm, whereas the acetal methine proton disappeared at about 4.6 ppm. Protons based on the PLA segment almost disappeared when the spectrum was monitored in D_2O. This disappearance indicates that the PLA segments form a solid core in aqueous media, the solidity, in turn, causing a broadening effect due to the restricted mobility in NMR spectroscopy. Therefore, it is reasonable to consider that the aldehyde-PEG-PLA micelle possesses a core-shell structure having aldehyde groups on its surface. The extent of the

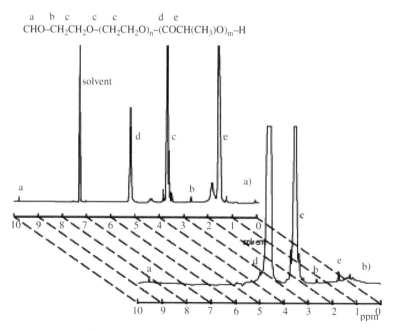

Figure 13 ^1H NMR spectra of the acetal-PEG-PLA (52/56) in (a) $CDCl_3$ and (b) D_2O after the acid treatment.

conversion from the acetal group to the aldehyde group was determined by the ^{1}H NMR spectrum. More than 80% of the acetal group was converted to the aldehyde group by the 4 h reaction. This reactive polymeric micelle will be utilized as the precursor for the targetable drug carrier.

5.2. Targetable PIC Micelle Composed of Lactose-PEG/PAMA Block Copolymer and pDNA (58)

As stated above, the PEG-polyamine block copolymer, through electrostatic interaction, forms a PIC micelle with negatively charged DNA. A lactosylated PEG/PAMA block copolymer was obtained from the reductive amination reaction of the aldehyde-PEG/PAMA block copolymer with *p*-aminophenyl-β-D-lactopyranoside, and this step, in turn, initiated the introduction of the lactose group (ligand) to the acetal-PEG/PAMA block copolymer. Figure 14 shows the ^{1}H NMR spectrum of the lactose-PEG/PAMA block copolymer with assignment. The protons of the acetal (4.6 ppm) and aldehyde

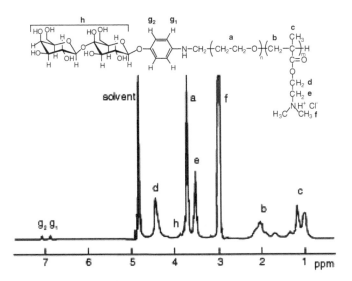

Figure 14 The ^{1}H NMR spectrum of the lactose-PEG-PAMA block copolymer.

groups (9.8 ppm) completely disappeared in conjunction with the appearance of the protons assignable to the aromatic residue of the p-aminophenyl-β-D-lactopyranoside. From the integral ratio between the aromatic protons and PEG-backbone protons, the degree of the functionality of the lactose moiety was determined to be 28.5%.

Lactose-installed PIC micelles were prepared by the simple and direct mixing of pDNA (pGL3 plasmid DNA encoding firefly luciferase) with lactose-PEG/PAMA block copolymer solutions at various charge ratios (cationic charge/anionic charge; N/P ratio). All the PIC micelles prepared at various N/P ratios had a unimodal distribution (μ_2/Γ^2=0.16–0.22) with an average diameter of approximately 100 nm, as revealed by DLS measurements. Figure 15 shows representative data for the gamma-distribution of the obtained PIC micelles at $N/P = 6.25$. The small size of the PIC micelles compared to the dimension of free pDNA strongly suggests the compaction of complexed pDNA, which forms the collapsed core of the micelles. In addition, the PIC micelles exhibited

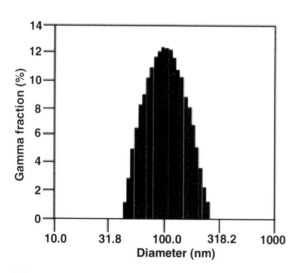

Figure 15 Gamma-distribution of the PIC micelle composed of a lactose-PEG-PAMA block copolymer and pDNA at $N/P = 6.25$ obtained from a histogram analysis of DLS (detection angle, 90°; temperature, 25°C).

not only lower absolute values of zeta-potential compared to the reported zeta-potential values of the polyplex that was composed of both simple polycations (without a PEG segment) (59) and pDNA but also stability against deoxyribonuclease (DNase I) digestion. This is reasonably assumed to be due to the formation of a corona surrounding the PIC core of the micelles.

To estimate the transfection ability of the lactose-installed PIC micelles against cultured HepG2 cells (human hepatoma cells) possessing an abundance of ASGP receptors on the cell surface (60), a transfection study was carried out in the presence of 5% fetal bovine serum (FBS). PIC micelles without the lactose moiety (acetal-PIC micelle) and lipofect AMINE/pDNA complexes (lipoplex) were used as controls. HepG2 cells were co-incubated with lactose-installed PIC micelles ($N/P = 6.25$), acetal-PIC micelles ($N/P = 6.25$), or lipoplex ($N/P = 5.0$) for varying periods of time so that the time-dependent gene transfection could be observed, as shown in Figure 16. After approximately 1 h of an induction period, both the lactose-installed and the acetal-PIC micelles exhibited a time-dependent increase in transfection efficiency that was greater than that of the lipoplex. Interestingly, the lactose-installed PIC micelles achieved significantly higher transfection efficiency than the acetal-PIC micelles, suggesting that interaction between the lactose moieties of the PIC micelles and the ASGP receptors on the HepG2 cells (ASGP receptor-mediated endocytosis) may play a role in this phenomenon. In order to confirm the ASGP receptor-mediated endocytosis mechanism, a competitive assay using asialofetuin (ASF) was performed for both the lactose-installed and the acetal-PIC micelles. ASF, which is a natural ligand for the ASGP receptor, should serve as the inhibitor of the lactose-installed PIC micelle, unless the ASGP receptor-mediated mechanism doses not play a substantial role therein. The results are shown in Figure 17. The transfection efficiency of the lactose-installed PIC micelles was higher than that of acetal-PIC micelles in the absence of ASF. In contrast, the presence of excess ASF (4 mg/mL) resulted in a significant decrease in transfection efficiency of the

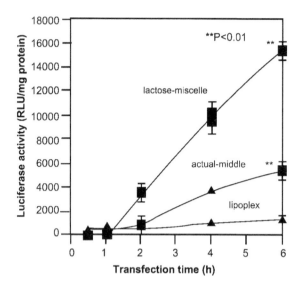

Figure 16 Effect of transfection time on gene expression. HepG2 cells were transfected with acetal- and lactose-PIC micelles at $N/P = 6.25$ in a medium (DMEM + 5% FBS) containing 100 μM of hydroxychloroquine (HCQ). Transfection with lipoplexe was done in the same medium except in the absence of HCQ. Data are the mean ± S.E.M.; $n = 4$.

lactose-installed PIC micelles ($P < 0.01$). On the contrary, the transfection efficiency of the acetal-PIC micelles was not affected by the presence of ASF up to 4 mg/mL, indicating that ASF in the medium has a negligible effect on the cellular uptake of the acetal-PIC micelle. Thus, it can be concluded that an appreciable fraction of the lactose-installed PIC micelles is taken up into HepG2 cells through an ASGP receptor-mediated endocytosis mechanism, yet a fluid-phase endocytosis pathway may concomitantly take part in the maintenance of as much transfection efficiency as the acetal-PIC micelles, even in the presence of excess ASF (4 mg/mL).

The installation of lactose (ligand) moieties increases the gene transfer efficiency of the PIC micelles composed of PEG/PAMA block copolymers and pDNA against HepG2 cells through the contribution of the receptor-mediated endocytosis

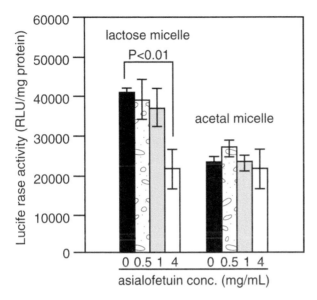

Figure 17 Inhibitory effect of ASF on gene transfer to HepG2 cells. HepG2 cells were transfected with acetal- and lactose-PIC micelles at $N/P = 6.25$ in a medium (DMEM + 5% FBS) containing 100 μM of HCQ. Transfection time was fixed at 6 h. Data are the mean ± S.E.M.; $n = 4$.

mechanism, all of which indicates that the polymeric micelles are a promising contribution to cellular targetable gene delivery systems.

5.3. pH-Responsive PIC Micelle Composed of Acetal-PEG/ODN Conjugate and Poly(ethylenimine) (56)

Because the acetal-PEG-ODN conjugate is one of the homologs of PEG/polyanion block copolymers, it is expected that the conjugate will, based on the electrostatic interaction, mix with the appropriate polycation and, thus, form a PIC micelle. Linear-PEI was chosen as the counter polycation because it has often been used for the gene delivery of DNA/L-PEI complexes and displays high transfection efficiency, which is due to the "buffer" or "proton-sponge" effect (51). So that the PIC micelle would be formed, the acetal-PEG-ODN conjugate was mixed with commercially available L-PEI that featured an equal unit molar ratio of the phosphate group in the acetal-PEG-ODN conjugate and the amino group in L-PEI ($N/P = 1$). The PIC micelle had a hydrodynamic diameter of 102.5 nm with a relatively narrow polydispersity index (μ_2/Γ^2) of 0.096, and showed a DLS-measured unimodal distribution at 37°C, as shown in Figure 18. The prepared solution showed no precipitation even after one-week storage at room temperature.

The stability of ODN against nuclease is an important issue for in vivo application. Both the acetal-PEG-ODN conjugate and its PIC micelle with L-PEI were incubated with deoxyribonuclease (DNase I) for appropriate time intervals (5, 30, 60, and 120 min), and the digestion of ODN was quantified by means of an SEC analysis. So that an exact evaluation of the stability against DNase I attack of ODN in PIC micelles could be obtained, excess dextran sulfate (counter polyanion) was added to the PIC micelles solution. This step

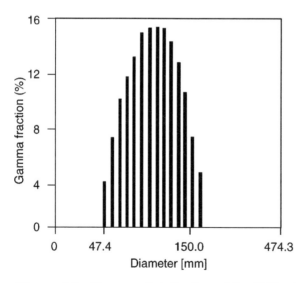

Figure 18 Gamma-distribution of the PIC micelle composed of an acetal-PEG-ODN conjugate and L-PEI at $N/P = 1$ obtained from a histogram analysis of DLS (detection angle, 90°; temperature, 37°C).

released the acetal-PEG-ODN conjugate from the PIC micelle prior to the SEC analysis. Figure 19 shows that the DNase I digestion profiles of the acetal-PEG-ODN conjugate and its PIC micelle with L-PEI are a function of incubation time. More than 50% of the acetal-PEG-ODN conjugate was digested in 120 min, whereas at least 90% of the acetal-PEG-ODN conjugate in the PIC micelle still remained intact. The improved stabilization against DNase I of the acetal-PEG-ODN conjugate in the micelle apparently results when the compartmentalization of the ODN segment that occurs in the core of the PIC micelle, which is surrounded by a highly dense PEG corona layer, effectively masks the ODN segment, thus, thwarting any enzyme attack. The stability of PIC micelle in 10% FBS was also investigated by means of an SEC analysis, as seen in Figure 20. Even in the presence of 10% FBS, no change in the retention time and the peak

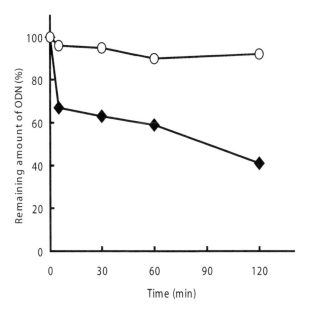

Figure 19 DNase I digestion of acetal-PEG-ODN conjugate in free form (◆) and in PIC micelle (○). The remaining amount of acetal-PEG-ODN conjugate was established from the peak area ratio in SEC.

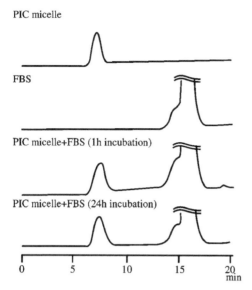

Figure 20 SEC profiles for a PIC micelle in the presence of 10% FBS.

intensity of the PIC micelle was observed for 24 h, suggesting minimal interaction between the PIC micelle and FBS. It may be reasonable to assume that the PIC micelle composed of an acetal-PEG-ODN conjugate and L-PEI might achieve an improved half-life in blood owing to the effect of the PEG corona surrounding the PIC core. These characteristics regarding the PIC micelle's entrapping of the pH-responsive PEG-ODN conjugate suggest that the PIC micelle possesses great utility as a part of a new ODN delivery system.

6. CONCLUSION

We have described here a novel approach to both the precise synthesis of the heterotelechelic PEGs and their application to drug delivery and gene delivery systems. Use of potassium 3,3-diethoxy-1-propananolate as an initiator of the polymerization of ethylene oxide not only produced various heterotelechelic PEGs but also end-functionalized PEG block copolymers such as acetal-PEG-PLA, acetal-PEG-PAMA, and acetal-PEG-L-PEI with well-defined structures, high yield, and high purity. In addition to the preparation of the end-functionalized PEG block copolymers, heterotelechelic PEGs were found to be useful for the conjugation of biologically active macromolecules so that both the lactosylated PEG-SOD conjugate and the end-functionalized PEG-ODN conjugate could be obtained. The end-functionalized block copolymers and end-functionalized PEG-ODN conjugate thus prepared, formed functionalized polymeric micelles that exhibited cellular-specific uptake, tolerated enzymatic degradation, and interacted minimally with serum components owing to the steric stabilization of the highly dense end-functionalized PEG corona surrounding the micelle core. Therefore, the synthesis of block polymers and bioconjugation using the heterotelechelic PEGs offers researchers in the field a promising approach to targetable drugs, proteins, and gene delivery systems.

REFERENCES

1. Abuchowski A, van Es T, Palczuk NC, Davis FF. J Biological Chem 1997; 252:3578.

2. Katre NV. J Immunology 1990; 144:209.

3. Goodson RJ, Katre NV. Bio/Technology 1992; 8:343.

4. Harris JM, ed. Poly(ethylene glycol) Chemistry: Biotechnical and Biological Applications. New York: Plenum Press, 1992; and references therein.

5. Kataoka K, Harada A, Nagasaki Y. Adv Drug Delivery Rev 2001; 47:113.

6. Deguchi Y, Kurihara A, Pardridge EM. Bioconjugate Chem 1999; 10:32.

7. Greenwald RB, Yang K, Zhao H, Conover CD, Lee S, Filpula D. Bioconjugate Chem 2003; 14:395.

8. Harris JM, Zalipsky S, eds. Polyethylene Glycol: Chemistry and Biological Applications, ACS Symp. Ser. 680. Washington DC: American Chemical Society, 1997.

9. Fuertges F, Abuchowski A. J Contr Rel 1990; 11:139.

10. Park YK, Abuchowski A, Davis S, Davis F. Anticancer Res 1981; 1:373.

11. Ramachandran R, Katzenstein D, Winters MA, Kundu SK, Merigan TC. J Infect Dis 1996; 173:1005.

12. Zalipsky S. Bioconjugate Chem 1995; 6:150; and references therein.

13. Kaiser K, Marek M, Haselgrubler T, Schindler H, Gruber H. J Bioconjugate Chem 1997; 8:545.

14. Bettinger T, Remy JS, Erbacher P, Behr JP. Bioconjugate Chem 1998; 9:842.

15. Kim YJ, Nagasaki Y, Kataoka K, Kato M, Yokoyama M, Okano T, Sakurai Y. Polym Bull 1994; 33:1.

16. Cammas S, Nagasaki Y, Kataoka K. Bioconjugate Chem 1995; 6:702.

17. Nagasaki Y, Kutsuna T, Iijima M, Kato M, Kataoka K. Bioconjugate Chem 1995; 6:231.

18. Nagasaki Y, Iijima M, Kato M, Kataoka K. Bioconjugate Chem 1995; 6:702.

19. Greene TW. Protecting Groups in Organic Synthesis. Wiley-Interscience, 1991.

20. Akiyama Y, Otsuka H, Nagasaki Y, Kato M, Kataoka K. Bioconjugate Chem 2000; 11:947.

21. Akiyama Y, Nagasaki Y, Kataoka K. Bioconjugate Chem 2003; 15:424.

22. Nakamura T, Nagasaki Y, Kato M, Kataoka K. Bioconjugate Chem 1998; 8:300.

23. Yasugi K, Nakamura T, Nagasaki Y, Kato M, Kataoka K. Macromolecules 1999; 32:8024.

24. Gref R, Domb A, Quellec P, Blunk T, Müller RH, Verbavatz JM, Labger R. Adv Drug Delivery Rev 1995; 16:215.

25. Kwon GS, Kataoka K. Adv Drug Delivery Rev 1995; 16:295.

26. Otsuka H, Nagasaki Y, Kataoka K. Adv Drug Delivery Rev 2003; 55:403.

27. Matsumura Y, Maeda H. Cancer Res 1986; 46:6387.

28. Maeda H, Seymour LW, Miyamoto Y. Bioconjugate Chem 1992; 3:351.

29. Kataoka K, Togawa H, Harada A, Yasugi K, Matsumoto T, Katayose S. Macromolecules 1996; 29:8556.

30. Katayose S, Kataoka K. Bioconjugate Chem 1997; 8:702.

31. Katayose S, Kataoka K. J Pharm Sci 1998; 87:160.

32. Harada A, Togawa H, Kataoka K. Eur J Pharm Sci 2001; 13:35.

33. Harada-Shiba M, Yamauchi K, Harada A, Takamisawa I, Shimokado K, Kataoka K. Gene Therapy 2002; 9:407.

34. Kakizawa Y, Harada A, Kataoka K. Biomacromolecules 2002; 2:491.

35. Itaka K, Yamauchi K, Harada A, Nakamura K, Kawaguchi H, Kataoka K. Biomacromolecules 2002; 3:841.

36. Kakizawa Y, Kataoka K. Langmuir 2002; 18:4539.

37. Itaka K, Yamauchi K, Harada A, Nakamura K, Kawaguchi H, Kataoka K. Biomaterials 2003; 24:4495.

38. Miyata K, Kakizawa Y, Nishiyama N, Harada A, Yamasaki Y, Koyama H, Kataoka K. J Am Chem Soc 2004; 126:2355.

39. Nagasaki Y, Kada T, Scholz C, Iijima M, Kato M, Kataoka K. Macromolecules 1998; 31:1473.

40. Iijima M, Nagasaki Y, Okada T, Kato M, Kataoka K. Macromolecules 1999; 32:1140.

41. Scholz C, Iijima M, Nagasaki Y, Kataoka K. Macromolecules 1995; 28:7295.

42. Yasugi K, Nagasaki Y, Kato M, Kataoka K. J Contr Rel 1999; 62:89.

43. Yamamoto Y, Nagasaki Y, Kato M, Kataoka K. Colloid Surf B 1999; 16:135.

44. Inoue S, Aida T. Anionic ring opening polymerization: copolymerization. In: Allen SG, Bevington JC, eds. Comprehensive

Polymer Science: The Synthesis, Characterization, Reactions and Applications of Polymers. Vol. 3. Elmsford, NY: Pergamon Press, 1989:553.

45. Kimura Y. Biocompatible polymers. In: Tsuruta T, Hayashi T, Kataoka K, Ishihara K, Kimura Y, eds. Biomedical Applications of Polymeric Materials. Boca Raton, FL: CRC Press, 1993:164.

46. Kataoka K, Harada A, Wakebayashi D, Nagasaki Y. Macromolecules 1999; 32:6892.

47. Iijima M, Nagasaki Y, Kato M, Kataoka K. Polymer 1997; 38:1197.

48. Nagasaki Y. Recent Res Dev Macromol Res 1997; 2:11.

49. Nagasaki Y, Sato Y, Kato M. Macromol Rapid Commun 1997; 18:827.

50. Akiyama Y, Harada A, Nagasaki Y, Kataoka K. Macromolecules 2000; 33:5841.

51. Boussif O, Lezoualc'h F, Zanta MA, Mergny MD, Scherman D, Demeneix B, Behr JP. Proc Natl Acad Sci U.S.A 1995; 92:7297.

52. Saegusa T, Chujo Y. Makromol Chem, Macromol Symp 1991; 51:1.

53. Oberley LW, Buettner GR. Cancer Res 1979; 39:1141.

54. Farmer KJ, Sohal RS. Free Radical Biol Med 1989; 7:23.

55. Paietta E, Gallagher R, Wiernik PH, Stockert R. J Cancer Res 1988; 48:280.

56. Oishi M, Sasaki S, Nagasaki Y, Kataoka K. Biomacromolecules 2003; 4:1426.

57. Mukherjee S, Ghosh RN, Maxfield FR. Physiol Rev 1997; 77:759.

58. Wakebayashi D, Nishiyama N, Yamasaki Y, Itaka K, Kanayama N, Harada A, Nagasaki Y, Kataoka K. J Contr Rel 2004; 95:653.

59. Cherng JY, Wetering P van de, Talsma H, Crommelin DJA, Hennink WE. Pharm Res 1996; 13:1038.

60. Schwartz AL, Fridovich SE, Knowles BB, Lodish HF. J Biol Chem 1981; 256:8878.

3

pH-Sensitive Polymers
for Drug Delivery

KUN NA and YOU HAN BAE
Department of Pharmaceutics and
Pharmaceutical Chemistry, University of Utah,
Salt Lake City, Utah, U.S.A.

1. INTRODUCTION

In the last two decades, the development of polymers which change their structures and properties in response to environmental stimuli such as pH, temperature, and light has attracted a great deal of attention (1–3). Such polymers have been called "smart polymers," "intelligent polymers," "stimulus-sensitive polymers," or "responsive polymers." They have been used in many applications, ranging from bioactive agent delivery to separation (4,5). Various delivery systems based on the smart polymers have been proposed because of their unique potential in the modulation of drug release and targeting functionality (6–8).

A pH-sensitive polymer in drug delivery applications is different from other stimuli-responsive polymers that require external stimuli such as temperature and light. The pH-sensitive polymers can take advantage of variations (whether they occur naturally or under pathological conditions) of pH in the body.

In the discussion of new trends in the field of drug delivery, this chapter will summarize recent research on drug carriers that use pH-sensitive polymers.

2. pH VARIATION IN THE BODY

The most pronounced pH variation in the human body occurs in the gastrointestinal tract (GIT). The GIT is a continuous tube that runs from the mouth to the anus. The interdigestive migration of a drug (or a dosage form) is governed by GI motility, wherein the drug is exposed to different pHs at different time periods, as summarized in Table 1 (9). The stomach has an acidic environment (a pH of 1–2 in a fasting condition and pH 4 during digestion) induced by hydrochloric acid from the gastric mucosa. The acidic pH in the stomach increases up to a pH of 5.5 at the duodenum, in which the acidic chime mixes with the bicarbonate from pancreatic juices. The pH then increases progressively from the duodenum to the small intestines (a pH of 6–7) and reaches a pH of 7–8 in the distal ileum. After the ileocecal junction, the pH falls sharply to 5.6 owing to the presence of short-chain fatty acids and then climbs up to neutrality during transit through the colon (a pH of above 7–7.5) because of free fatty-acid absorption (10).

A blood pH of 7.40 corresponds to 40 nM $[H^+]$, which is the mean of a normal range of 7.36–7.44 on a pH scale, or 44–36 nM $[H^+]$. If the blood pH falls below 7.36 ($[H^+] >$ 44 nM), the condition is called acidemia. On the other hand, if the pH rises above 7.44 ($[H^+] < 36$ nM), the condition is called alkalemia. When it is in the pH range of 7.20–7.50, for every change of 0.01 pH unit, there is a change of

Table 1 Physiological Parameters of the GIT Related to Dug Delivery

GIT segment	Approximate residence time	Approximate pH of segment	Principal catabolic activities
Oral cavity	Seconds to minutes	6.5	Polysaccharidases
Esophagus	Seconds	—	—
Stomach	0.2–2.0 h	1.0–2.0	Proteases, lipases
Duodenum	30–40 min	4.0–5.5	Polysaccharidases; oligosaccharidases, proteases, peptidases, lipases
Jejunum	1.5–2.0 h	5.5–7.0	Oligosaccharidases
Ileum	—	7.0–8.0	Oligosaccharidases
Colon and rectum	13–68 h	7.0–7.5	Broad spectrum of bacterial enzymes (glycosidases, azoreductase, polysaccharidases)

approximately $1\,nM$ $[H^+]$ in the opposite direction (11). The extracellular pH of most tissues and organs is in equilibrium with the blood pH; however, it is seen that certain diseased states lead to acidosis.

Extracellular pH (pH_e) in most tumors are more acidic ($pH_e < 7.0$) than in normal tissues (12). Although there is a distribution range of in vivo pH_e among human patients with a variety of solid tumors (adenocarcinoma, squamous cell carcinoma, soft tissue sarcoma, and malignant melanoma) that are located at accessible areas of the body (limbs, neck, or chest wall), the mean pH value was reported to be 7.0 with a full range being from 5.7 to 7.8 (13). The variation mainly depends on the tumor's histology and volume. Recent measurements of pH_e by noninvasive technology involving magnetic resonance spectroscopy (^{19}F, ^{31}P, or ^1H probes) in the human tumor xenografts in animals showed and thus further verified a consistently low pH_e. All measurements (involving either invasive or noninvasive methods) of the pH_e of human and animal solid tumors showed that more than 80% of all measured values fell below a pH of 7.2 (14,15). The high rate of the glycolysis of tumor cells under either aerobic or anaerobic conditions has been thought to be a major cause of low pH_e. Through both mitochondrial oxidative phosphorylation and glycolysis, the tumor cells synthesize ATP. Glycolysis produces 2 mol of lactic acid ($pK_a = 3.9$) and 2 mol of ATP from 1 mol of glucose with the hydrolysis of ATP-producing protons. Despite the high production rate of protons in tumor cells, their cytosolic pH remains alkaline, a condition that is favorable for the glycolysis. The mechanism by which protons are exported from a cell remains unknown. However, the inadequate blood supply, poor lymphatic drainage, and high interstitial pressure in tumor tissues have all been implicated in a low pH_e (16). Yet, it is of interest to note that the glycolysis-deficient variant cells (lacking lactate dehydrogenase) producing negligible quantities of lactic acid are still present in acidic environments (17). In interpreting this finding, researchers have suggested that the acidity is a phenotype of tumor cells rather than a consequence of cellular metabolic events. The acidic environment may benefit the cancer cells because it promotes invasiveness (metastasis) by destroying the extracellular matrix of the surrounding normal tissues.

Although the cytosolic pH corresponding to most cells' intracellular levels is approximately 7.2, the pH in the endosomes and the Golgi is slightly acidified and ranges from a pH of 6.6 to 6.0. Hence, the lysosomes are more acidic (pH 5.0) (18–22). This acidification is generated by the work of vacuolar-type (V) ATPases, which are a family of ATP-dependent proton pumps. They produce two to three protons per ATP hydrolyzed (23). As for endosome, the acidification plays an important role in intracellular transport by endocytosis (24,25). During the receptor-mediated endocytosis, a low pH in early endosomes triggers a dissociation of internalized ligand-receptor complexes, which allows the recycling of receptors to the plasma membrane.

The pH of intracellular compartments is influenced by various transport proteins. The pH of endosomes and of clathrin-coated vesicles depends on the activity of a chloride channel that dissipates the membrane potential generated by the V-ATPase. A defect in the *chloride channel protein-5* gene leads to defective endosomal acidification in the kidney (26). Interestingly, disruption of the *chloride channel protein-7* (*CLC-7*) gene blocks the acid secretion of osteoclasts, an event that leads to osteopetrosis (27). In addition, the passive proton conductance of a membrane is an important determinant of the steady-state pH. For example, the activity of a Zn^{2+}- sensitive proton channel is of importance in controlling the lumenal pH of the Golgi (28).

Intracellular pH in tumor cells is different from normal cells. The pH of a tumor cell is considerably more acidic than that of a normal cell. The cytosol of a tumor cell has a pH of about 6.7, and the lumen of recycling endosomes and the trans-Golgi network has a pH of 6.7. This demonstrates a significant lack of acidification in the organelles of tumor cells (29). Interestingly, the pH gradient of multidrug-resistant (MDR) tumor cells strongly resembles that of normal cells. For example, in cases of MDR MCF-7 cells (a human breast cancer cell) and K-562 cells (a human erythroleukemic cell line), the cytosolic pH is neutral (7.2), endosomes and Golgi become acidified (6.0), and the lysosomes turn very acidic (4.9) (30).

3. pH-SENSITIVE POLYMERS

The pH-sensitive polymers containing acidic groups (e.g., carboxylic and sulfonic acids) or basic amino groups can accept or release protons in response to changes in environmental pH. Figure 1 illustrates the structural examples of anionic and cationic polyelectrolytes and their pH-sensitive ionization values. Poly(acrylic acid) (PAA) becomes ionized at a high pH, whereas poly(N,N'-diethylaminoethyl methacrylate) (PDEAEM) becomes ionized at a low pH. As shown in Figure 1, cationic polyelectrolytes, such as PDEAEM, are more soluble at a low pH. If they are cross-linked, they swell at this pH owing to ionization. On the other hand, polyanions, such as PAA, swell at a high pH (31).

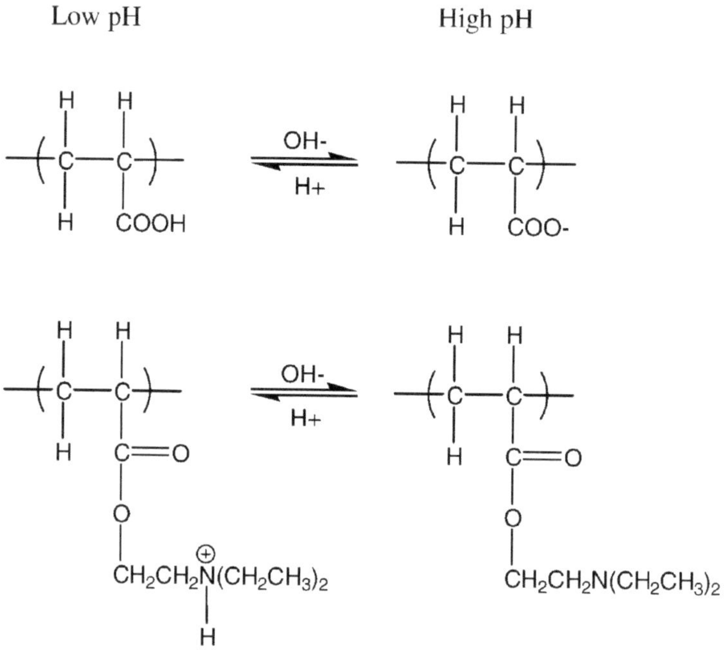

Low pH High pH

Figure 1 pH-Dependent ionization of polyelectrolytes. Poly (acrylic acid) (top) and poly(N',N'-diethylaminoethyl methacrylate) (bottom). (Adapted from Ref. 31.)

In general, the pH-sensitive polymers manifest their sensitivity to changes in pH, as soluble-insoluble phase transition, i.e., swelling-shrinking changes, or as conformational changes. These properties are dependent on the degree of the ionization of the ionizable groups in the polymers; this degree is related to the pK values (pK_a or pK_b) of monomers and the environmental local pH. Thus, the selection of an ionizable monomer is a fundamental parameter for the control of pH-dependent properties; in the meantime, the sensitivity is further influenced by the nature of ionizable groups, the polymer composition, the ionic strength, and the hydrophobicity of the polymer backbone. Conformational changes were also induced by altering the pH (32,33).

4. ORAL DELIVERY

Oral intake is considered to be one of the most convenient routes for administering drugs to patients. Due to the pH variation in the GIT, pH-sensitive polymers have been historically utilized as an enteric coating material. Enteric-coated products featuring pH-sensitive polymers include tablets, capsules, and pellets and are designed to keep an active substance intact in the stomach and then to release it to the upper intestine. In addition, the coat reduces the side effects such as foul taste and severe gastric irritation. Similar coating materials, but with a different pH sensitivity, have been applied for colon-targeted drug delivery. This special application is due to a pH change in the ileocecal junction. The polymers that are designed for colon targeting are able to withstand the low pH levels of the stomach and proximal part of the small intestine and then disintegrate at the neutral or slightly alkaline pH of the terminal ileum and preferably at the ileocecal junction (10). A process such as this distributes the drug throughout the large intestine and improves drug availability to the colon. In addition, the polymer is used for peptide/protein oral delivery systems. The pH-dependent solubility or swelling allows for the possibility that polymers can be engineered to conserve the stability of proteins at a low pH in the stomach and intestine and then release these proteins at a higher pH.

4.1. Enteric Coating

pH-Sensitive polymers that are used for enteric coating are anionic polymethacrylates [methacrylic acids copolymerized with methylmethacrylate or ethyl acrylate (Eudragit®)], cellulose-based polymers (e.g., cellulose acetate phthalate; Aquateric®), or polyvinyl derivatives [e.g., poly(vinyl acetate phthalate) (PVAP); Coateric®]. The threshold pH of acidic polymers that are commonly employed for enteric coating is summarized in Table 2 (34).

A general chemical structure of Eudragit is presented in Figure 2. Its properties are manipulated by copolymer composition and the ionizable group. For instance, Eudragit L100

Table 2 Threshold pH of Commonly Used Polymers for Enteric Coating

Polymer	Threshold pH
Eudragit L100	6.0
Eudragit S100	7.0
Eudragit L-30D	5.6
Eudragit FS 30D	6.8
Eudragit L100-55	5.5
Polyvinylacetate phthalate	5.0
Hydroxypropylmethylcellulose phthalate	4.5–4.8
Hydroxypropylmethylcellulose phthalate 50	5.2
Hydroxypropylmethylcellulose phthalate 55	5.4
Cellulose acetate trimelliate	4.8
Cellulose acetate phthalate	5.0

Source: Ref. 34.

and S100 (Fig. 2) are copolymers of methacrylic acid and methylmethacrylate. The ratio of carboxyl to ester groups is approximately 1:1 in Eudragit L100 and 1:2 in Eudragit S100. The disintegration rate and drug release behavior of a coated tablet is controlled through various combinations of different methacrylic acid copolymers. A capsule coated with

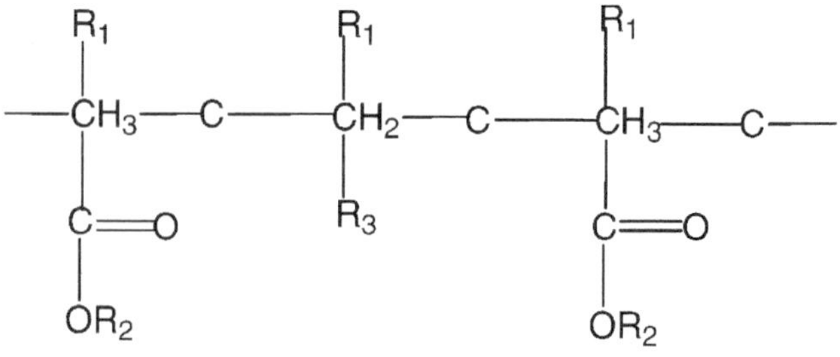

$R_1 = CH_3$; H
$R_2 = CH_3$, CH_3CH_2
$R_3 =$ COOH (Eudragit® L and S) or
$R_3 = COOCH_2CH_2N(CH_3)_3Cl$ (Eudragit® RL and RS)

Figure 2 Chemical structures of various formulations of Eudragit. (Adapted from Ref. 34)

Eudragit L-30 showed gastro resistance for 3 h at a pH of 1.2
(34). The mixtures of Eudragit L100-55 and Eudragit S100
with weight ratios of 1:0, 4:1, 3:2, 1:1, 2:3, 1:4, 1:5, and 0:1
were investigated insofar as they pertained to tablet-coating
material for site-specific oral drug delivery (35). In the formulations, the dissolution rates decreased with the introduction
of Eudragit S100. The effect of a buffer media pH on the
release profiles of the three selected formulations (1:4, 1:5,
and 1:0) is shown in Figure 3. The release profiles were signif-

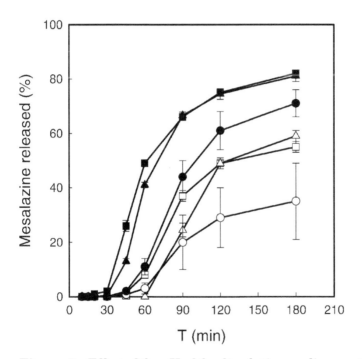

Figure 3 Effect of the pH of the dissolution medium on the release
of mesalazine tablets tested in pH 6.5 (open symbols) and 7.0 (closed
symbols) mixed phosphate buffers. The mesalazine core tablets were
coated with Eudragit L100-55 and Eudragit S100 combinations of
1:4 (squares), 1:5 (triangles), and 0:1 (circles) ratios (w/w) and tested
in 0.1 N HCl for 120 min prior to the buffer stage. The dissolution
test was performed using United States Pharmacopoeia method B
for extended-release tablets. The vertical bars indicate standard
errors of the means ($n = 6$). (Adapted from Ref. 35.)

icantly faster at a pH of 7.0 than at a pH of 6.5. Recently, Eudragit microspheres that were prepared according to an emulsion-solvent evaporation method have been used for the delivery of anti-inflammatory agents such as indomethacin, ibuprofen, and ketoprofen (36–38). The formulations showed suppression of drug release in the acidic medium. Table 3 summarizes various applications of Eudragit.

Polymers with ionizable phthalic acid groups dissolve faster at a lower pH than polymers with acrylic or methacrylic acid groups (Table 2). The simple coacervation of hydroxypropyl methylcellulose phthalate (HPMCP) with the addition of a 20% sodium sulfate solution was studied. Coacervation is the method of choice for the production of pharmaceutical preparations having a high active ingredient content and a smaller particle size of core materials used. With an increase in pH, the electrolyte (sodium sulfate) amount required to

Table 3 Various Applications of Eudragit

Enteric coatings	Eudragit L100-55 or L 30 D-55—delivery to the duodenum (pH > 5.0). Exact pH-controlled drug release can be adjusted by a combination of polymers and coating thickness Eudragit L100—delivery to the jejunum (pH > 6.0) Eudragit FS 30 D—delivery to the colon (pH > 7.0) Eudragit S 100—delivery to the intestine (pH 6.0–7.5), site-specific drug delivery can be achieved by combining with Eudragit L types
Sustained-release coatings	Eudragit RL and RS—release profile determined by the ratio of RL to RS polymers, and the film thickness is applied
Taste masking	Eudragit RD 100, p1H independent, fast disintegrating film Eudragit E 100 and EPO, pH-dependent cationic polymer, soluble in gastric fluid up to pH 5.0, swellable and permeable above pH 5.0
Rapidly disintegrating films	Eudragit RD 100

Source: From Eudragit Technology Bulletin, Rohm America

induce the phase separation of HPMCP increased from 5.1% (w/w; pH 4.75) to 7.7% (w/w; pH 6.75). Thus, no coacervate was formed below a pH of 5.25, since the polymer separated immediately as a precipitated form. Above a pH of 5.25, HPMCP was primarily salted out as a coacervate, which then was transformed to a precipitate upon the addition of more electrolytes (sodium sulfate). The charge density increased from 1.66 mEq/g at a pH of 4.3–2.21 mEq/g at a pH of 7.0. Considering that the coacervate phases occurred only in a pH above 5.25, approximately 95% of the phthalyl carboxyl groups had to be ionized for the coacervate formation (39). Coated films from an aqueous dispersion of Aquateric, which is a spray-dried cellulose acetate phthalate (CAP) powder composed of CAP, Pluronic F-68, Myvacet 9-45, and Polysorbate 60, showed no superior acid resistance for theophylline beads (40). Additionally, CAP was used with a subcoating material such as aqueous amylopectin for more acid resistance. The effect of an aqueous amylopectin subcoating on the acidic resistance and the dissolution behavior of CAP was studied by using riboflavin sodium phosphate. According to this study, it seems that the amylopectin subcoating prevented the influx of the dissolution medium into the pellet core owing to the higher viscosity and the branched molecular structure of amylopectin, thus decreasing the premature dissolution and release of the drug from the enteric-coated pellets in 0.1 N HCl solutions (41). Poly(vinyl acetate phthalate) (Coateric, threshold pH 5.0) is a commercial product that has been available since the late 1960s. It is prepared by the esterification of partially hydrolyzed poly(vinyl acetate) with phthalic anhydride. It is currently used as an ingredient in coating systems for marketed pharmaceutical oral solid dosage forms with approximately 10–78 mg of the polymer being applied per dosage form (42). It is used in prescription drugs and in over-the-counter (OTC) drugs, such as enteric-coated aspirin.

4.2. Colon-Targeted Drug Delivery

Colon targeting is extremely beneficial when applied to the treatment of colon diseases. In high local concentrations, colon-targeted drugs can remedy diseases such as ulcerative colitis, carcinomas, and infections. In addition, site-specific absorption in the colonic regions offers excellent source of treatment. It targets maladies with diurnal rhythms such as asthma, arthritis, and inflammation (43–45).

Eudragit produces several advantages in colon-specific drug delivery owing to its superior pH sensitivity. A study was conducted using Eudragit on drug delivery systems. The drugs used for the study were prednisolone (46), quinolones (47), sulfasalazine (48), cyclosporine (49), beclomethasone dipropionate (50), and naproxen (51). Recently, a microsphere prepared from a mixture of different Eudragit has been used to prolong the duration of anticancer drug release. This was done to avoid over-dose and its side effects. Six doses of 5-fluorouracil (5-FU) that were injected into each of the patients at above 500 mg/sq m/week caused the patients to experience side effects such as diarrhea and toxicity (52). At the same time, microspheres from Eudragit P-4135F showed a pH-sensitive drug-release pattern, exhibiting a dissolution threshold pH of slightly above 7.2. Even though the microsphere was found to release up to 35% of the drug at a pH of 6.8 within 6 h, a nearly 100% release within 30 min occurred at a pH of 7.4. This release behavior caused side effects. For the prolongation of the drug-release time, the microspheres from a mixture of polymers of P-4135F and RS100 (the mixing ratios of P-4135F and RS100 are 9:1, 8:2, and 7:3 wt ratios) were used as a 5-FU delivery carrier (53). 5-Fluorouracil was slowly released from the mixed microspheres when tested in buffer systems at pHs of 1.2 and 6.8, maintaining at least 65–70% of the initial drug load after 6 h of incubation. An extended release was observed at a pH of 7.4, which delivered about 100% of the incorporated drug in 100 min. Generally, slower release is found with larger particles because of prolonged diffusion and transport distances in the matrix. The mixture ratio of Eudragit RS100 and Eudragit P-4135F did not affect the particle size of the microspheres, drug loading efficiency, and the prolongation of the drug release.

Ashford and colleagues (54) showed that pH-sensitive polymers (Eudragit) are not effective for colon-targeted drug delivery systems owing to the systems' poor site specificity. The in vivo disintegration of Eudragit was extremely variable in both time (5.0–15 h) and position. The disintegration sites varied from the ileum to the splenic flexure and showed a lack of site specificity. This problem can be partially resolved by multiple-unit dosage forms prepared from the combination of different coating processes. Markus and associates (55) developed a multi-unit dosage form containing 5-amino salicylic acid (5-ASA) for the treatment of ulcerative colitis. Pellets [lactose (20%) and 5-ASA (80%)] were prepared by a granulation and spheronization process and then coated with a pH-sensitive polymer (Eudragit FS 30D; Table 2) to achieve a site-specific drug release close to the ileocecal valve. The 5-ASA release from the pellets coated with Eudragit FS 30D at a pH of below 6.8 was not observed. The release at a pH of 7.2 increased by 20% per hour with the threshold pH of Eudragit FS 30D at 6.8. However, at a pH of 7.5, the release was completed within 2 h. This means that drug-release behavior can be controlled through the use of a multi-unit dosage form.

Another multi-unit dosage form was constituted from drug-loaded cellulose acetate butyrate (CAB) microspheres coated with an enteric polymer (Eudragit S). Both CAB cores and pH-sensitive capsules (Eudragit S) were prepared according to an emulsion-solvent evaporation technique in an oil phase. Being potentially applicable to the local treatment of intestinal disorders, the most interesting drugs, Ondansetron and Budesonide, were efficiently microencapsulated in CAB microspheres at different polymer concentrations (6 and 8%) (56). These CAB cores (about 60 and 110 μm in size) were then encapsulated with Eudragit S, resulting in a multinucleated structure (between 200 and 400 μm in size). In in vitro drug release studies, the dosage showed that no drug was released below a pH of 7. When the microspheres were exposed to a pH above 7.0, CAB microspheres efficiently controlled the release of Budesonide.

CODES^{TM} is a unique colon-specific drug delivery technology that was designed to reduce the lack of site specificity (57). The design of CODES exploited the advantages of

certain polysaccharides that are degraded only by bacteria
available in the colon. These bacteria produce organic acids
that subsequently lower the pH. Usually, a CODES is coupled
with a pH-sensitive polymer coating. Because the degradation
of polysaccharides occurs only in the colon, this system exhibits the potential to achieve the consistent and reliable delivery of drugs to the colon. One typical configuration of CODES
consists of a core tablet covered with three layers of polymer
coats. The first coating (next to the core tablet) is an acid-soluble polymer (Eudragit E; granules, pH dependent, soluble in
gastric fluid up to 5.0, swellable, and permeable above a pH
of 5.0), and the second coating is enteric with an HPMC barrier layer to prevent any possible interaction between the
oppositely charged polymers (an anionic polymer for the
enteric coating and a basic polymer for the acid soluble
layers). The core tablet is comprised of one or more polysaccharides and other desirable excipients. During its transit
through the GIT, CODES remains intact in the stomach
owing to the protection provided by the enteric coating. When
the drug enters the small intestine, the enteric and barrier
coating dissolves where the pH is above 6. Because Eudragit
E starts to dissolve at a pH ≤5, the inner Eudragit E coating
becomes only slightly permeable and swellable in the small
intestines. Upon entry into the colon, the polysaccharide
inside the core tablet dissolves and diffuses through the coating. The bacteria enzymatically degrade the polysaccharide
into organic acids. Consequently, the reduced local pH in
the system facilitates the dissolution of the acid-soluble
coating and subsequent drug release.

    Chitosan (CS) has often been employed as a coating
material for the colon-targeted delivery of drugs because of
its pH sensitivity, its complete digestion by the colonic bacteria, and its low toxicity (58,59). Chitosan is a cationic polysaccharide; the pK_a of the NH_3^+ group of CS in water is 6.2,
whereas that of D-glucosamine HCl, a monomer of CS, is
found to be 7.8. Thus, it becomes slightly soluble at or below
a pH of 6.0 and turns completely soluble at a pH of 4.5 (61).
A CS-based multiparticulate dosage form was prepared for
effective colon targeting combining specific biodegradability

and pH-dependent release behavior (58). To begin with, a CS microcore [with diclofenac sodium (DS)] was prepared, according to a spray-drying method (size, 1.8–2.9 μm), and the drug was efficiently entrapped. Then, with an oil-in-oil solvent evaporation method (the ratio of Eudragit to the CS core is 5:1), the CS core was encapsulated within Eudragit L100 or Eudragit S100 so that a multireservoir system is formed. In turn, they formed microspheres, the sizes of which were 152 μm for the L100 or 223 μm for the S100. Even though CS dissolves very fast at a pH of 1.2, DS release from CS microcores was suppressed by the coating effect of Eudragit. However, when reaching the Eudragit pH solubility, a continuous release for a longer time period (8–12 h) was achieved. This result may have been induced by the deprotonization of CS at a neutral pH.

Also, a CS dispersed system (CDS), which is composed of an active ingredient reservoir and an outer drug release-regulating layer dispersing CS powder in a hydrophobic polymer (Fig. 4), was investigated in relation to colon-specific drug delivery (59). An aminoalkyl methacrylate copolymer RS (Eudragit RS) was selected as a hydrophobic polymer, since it had the ability to be sparsely dissolved in the acidic medium in which CS is freely soluble. The release rate was controlled by changes in the thickness of the coating layer. The resulting enteric-coated CDS capsules reached the large intestine within 1–3 h after oral administration and, in beagle dogs,

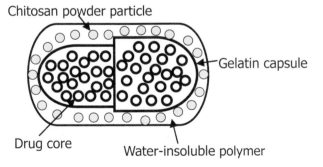

Figure 4 Fundamental structure of chitosan dispersed system (CDS) capsule. (Adapted from Ref. 59.)

degraded at the colon. The authors insisted that, for colon-specific drug delivery, an additional outer enteric coating was necessary for prevention of the drug's release from CDS capsules into the stomach because the dispersed CS in the layer was able to dissolve easily under acidic conditions.

4.3. Protein/Peptide Oral Drug Delivery

There are many limitations to the oral delivery of therapeutic proteins. These limitations result from such barriers as enzymatic degradation in the GIT, low epithelial permeability, and protein instability. In order to improve the bioavailability of a protein or peptide drug, a delivery system must increase the residence time, maintain drug stability until the drug reaches its absorption site, and enhance the drug's permeation into the systemic circulation. Numerous attempts have been made to deliver proteins orally and have involved the usage of pH-sensitive enteric-coated capsules (61,62), the capsules being protected in liposomes (63), the capsules having cyclodextrin (64), the capsules in erythrocytes (65), and in conjunction with absorption promoters (66), such as sodium salicylate, sodium glycocholate, and sodium cholate.

In early attempts to deliver protein orally, a number of different formulations of gelatin capsules that were coated with polyacrylic acids (PAA) and loaded with insulin were studied in rats. When delivered orally, the formulation, which was made of Eudragit and contained 16 IU of insulin, showed a statistically significant reduction of blood glucose concentration (67). When compared to the blood glucose reduction obtained with an intraperitoneal injection of 4 IU of insulin, these oral formulations demonstrated a pharmacological bioavailability of 9.3–12.7%. The result was encouraging, and many other researchers in the field subsequently applied similar strategies to the oral delivery of insulin. When Eudragit L100 and S100 capsules were used, the insulin release from the S100 capsules at a pH of 7.0 was slower than that from the L100 capsules. As a result, the L100 capsules released almost all of loaded the insulin into the stomach, duodenum, and jejunum. The S100 capsules were protective at a low pH and released more insulin in the ileal portions of the small intestines. The S100 capsules reduced a modest 10% of the glucose concentration in the blood. The glucose reduction was significantly enhanced through the use of a protease inhibitor, or specifically by aprotinin (68–70). Eudragit L100 or S100 capsules with absorption promoters showed a significant ($p < 0.01$) increase of the hypoglycemic effect over the 5 h period of the experiments (66,71).

Kim and coworkers have investigated the potential use of polymeric beads as an oral peptide drug delivery carrier involving pH/temperature-sensitive polymers that are composed of *N*-isopropylacrylamide (NIPAAm) (temperature-sensitive), butyl methacrylate (BMA), and acrylic acid (AA) (pH-sensitive). The pH- and temperature-dependent solubility of the polymers made the use of organic solvents unnecessary for the protein drug in the polymeric beads. The polymers prevented gastric degradation in the stomach while providing a controlled release of a peptide drug (72–76). The beads were formed by precipitation of the polymers at the interface of cold aqueous solution droplets, the warm oil phase, and the subsequent drying process. The bioactivities of the insulin that was released from the beads at a pH of 7.4 and was recovered from the beads which were kept in rats' stomachs for 5 h were well preserved (72). The loading of insulin from an aqueous medium into the beads was achieved by preparing a 7% or 10% (w/v) polymer solution with 0.2% (w/v) insulin concentration at a low pH and below the lower critical solution temperature (LCST) of the polymer (pH 2.0 and 4°C) (74). The release profile from different MW polymeric beads was studied at 37°C. The beads were placed for 2 h at a pH of 2.0 and then at a pH of 7.4 (Fig. 5) (76). A small fraction of the insulin (less than 10%) was released from all the polymer beads at a pH of 2.0 with the high MW polymeric beads showing minimum release. The polymers at a pH of 2.0 and placed under 37°C did not swell or dissolve. Rather, mainly surface-bound drug was released under these conditions. The beads that were at a pH of 7.4 and under 37°C swelled or eroded depending on the MW of the polymer, and all the polymers exhibited a release of drugs. The low MW

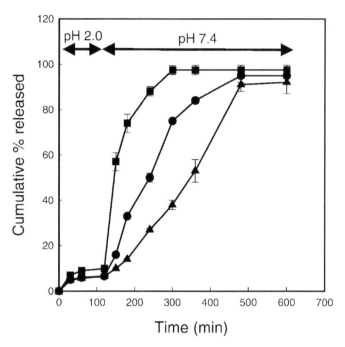

Figure 5   Modulating insulin release profile from pH/temperature-sensitive beads made of terpolymers of N-isopropylacrylamide (NIPAAm)/butyl methacrylate (BMA)/acrylic acid (AA) feed mol ratio 85/5/10 and increasing MW. Beads were placed at pH 2.0 and 37°C for 2 h and then at pH 7.4 and 37°C for the remainder of the release studies ($n = 6$). (Adapted from Ref. 76.)

polymeric beads released most of the insulin within 2 h; the intermediate MW polymeric beads released most of the insulin within 4 h; and the high MW polymeric beads released insulin over a period of 8 h. The polymer was also used for calcitonin delivery. Additionally, they reported that as the AA content increased from 0 to 10 mol%, the loading efficiency and the stability of calcitonin improved significantly (73).

With regard to protein oral delivery, an experiment was conducted on a pH-sensitive hydrogel polymer composed of poly[methacrylic acid-graft-poly(ethylene glycol)] (P(MAA-g-EG)) (77–80). The P(MAA-g-PEG) nanosphere system demonstrated low cytotoxicity and was capable of opening the tight junctions between epithelial cells. Thus, this significantly reduced the transepithelial electrical resistance using the chelation capability of the polymer (Ca^{2+}-binding properties) (81). Insulin was then loaded into the polymeric microspheres and was administered orally to healthy and diabetic Wistar rats. The gels that were exposed to acidity of the stomach did not swell owing to the formation of the intermolecular polymer complexes. The insulin did not leak out of the gel and did not go through a process of proteolytic degradation. In the areas of the intestine that had basic and neutral pH levels, the complexes became dissociated. The gel experienced rapid swelling and eventually released the insulin that was in the gels. Within 2 h of the administering of the insulin-containing polymers, strong dose-dependent hypoglycemic effects were observed in both the healthy and diabetic rats. According to this study, such effects lasted up to 8 h after the administration (80).

Chitosan capsules coated with HPMCP were also tested for insulin oral delivery (82). Using male Wistar rats, insulin-containing capsules were administered orally to the rats, each of which received, through polyethylene tubing, a total dose of 20 IU into their stomachs. The hypoglycemic effect did not start until 6 h after the administration. This was when the capsules had reached the colon. The bioavailability of the insulin from the CS formulation was 5.73% as compared to the intravenous injection. It was observed that certain absorption enhancers like sodium glycocholate increased the absorption of insulin in the large intestine (66).

5. PARENTERAL DRUG CARRIERS

5.1. Tumor Extracellular pH (pH$_e$) Targeting

The acidic tumor pH$_e$ prompted researchers to design pH-sensitive targeting systems that targeted these tumors. In general, fast-release kinetics from drug carriers leads to premature losses of agents causing systemic side effects. They fail to concentrate their agents at drug acting sites. Conversely, there is a high probability that low-release kinetics, which also has the potential to compromise drug efficacy at tumor sites, will give rise to multidrug resistance in initially sensitive tumor cells. This concern has provoked a strong desire in the field of pH-sensitive research to create "switching carriers" that change from slow release kinetics to rapid kinetics upon reaching target sites. This can be done during the drug's circulating phase.

Initially, tumor pH-induced drug release as a new mode of drug delivery involved pH-sensitive liposomes that incorporated the carboxylic group. However, owing to both the poor stability and responsibility to the range of tumor acidity (pH 6.5–7.2), the carriers were limited in their utility for tumor pH_e targeting. This finding necessitated the development of a polymeric nanocarrier that would be truly responsive to tumor pH_e. For this purpose, recent reports have described two nanocarrier systems that operate according to different mechanisms: (1) triggered release and improved cell interaction by aggregation/shrinking (using weak acid) at tumor pH_e (Fig. 6a) and (2) triggered release by carrier destabilization at tumor pH_e (Fig. 6b).

5.1.1. Carrier Aggregation/Shrinking System

Bae's group reported new pH-sensitive polymers based on sulfonamide (SD) derivatives (83–86). They revealed that polymeric systems bearing SD units (such as linear polymers, networks, oligomers, or single separate units conjugated to existing polymers) were, with the first-order-like transitions, capable of responding to pH changes as small as 0.2 pH units.

Sulfonamide is a generic name for the derivatives of *para*-aminobenzene sulfonamide (sulfanilamide). It was the first effective antibacterial agent employed systemically in the treatment of bacterial infections for human beings (87,88). Sulfonamide is a weak acid because the hydrogen atom of amide nitrogen (N^1) can be readily ionized so that a proton can be liberated in solution (Fig. 7a) (89). The oxygen atoms of the sulfonyl group (SO_2), having high electronegativity, strongly attract electrons from the sulfur atom. The resulting electron deficiency of the sulfur atom draws the

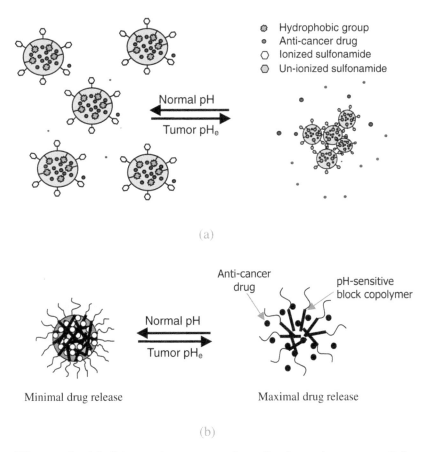

Hydrophobic group
Anti-cancer drug
Ionized sulfonamide
Un-ionized sulfonamide

Normal pH
Tumor pH$_e$

(a)

Anti-cancer drug

pH-sensitive block copolymer

Normal pH
Tumor pH$_e$

Minimal drug release

Maximal drug release

(b)

Figure 6 (a) Schematic presentation of polymeric nanoparticles with weak acid (sulfonamide groups). The polymeric nanoparticles show good stability at pH 7.4 but are aggregated/shrunken near the tumor where the pH is below tumor pH$_e$. (b) A schematic concept of anticancer drug release triggered by the change of hydrophobic core solubility in the polymeric micelle.

electrons of the N–H bond to the nitrogen atom due to the ionization (Fig. 7b). The pK_a of a SD comes under the substantial influence of the substituted R groups on N^1 nitrogen and varies from 3 to 11.

In approaching the design of new pH-sensitive polymers, researchers have modified selected SDs over a range of pK_a to polymerizable monomers. The pK_a of the monomers, of the

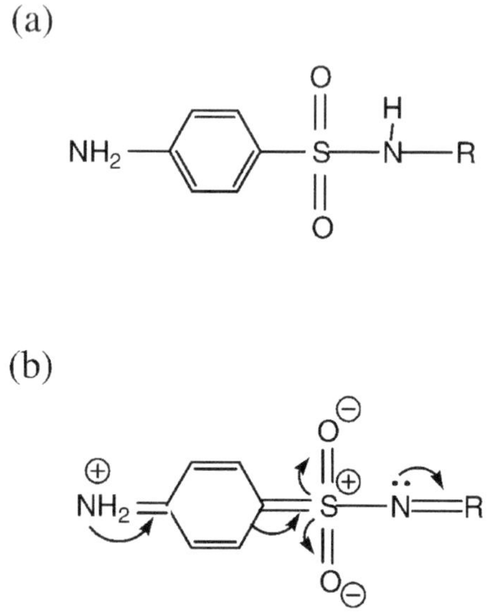

(a)

(b)

Figure 7 Sulfonamide (a) and resonance forms of ionized sulfonamide (b). (Adapted from Ref. 83.)

homopolymers, and of the copolymers with *N,N*-dimethyla-
crylamide was examined, as was the pH-induced phase transition behavior (particularly the solubility). The pK_a of both
the SD monomers and the polymers at 25°C was slightly
higher than the pK_a corresponding to the SDs. The solubility
transition of each homopolymer in aqueous solutions occurred
at a degree of ionization of 85–90% (83).

The solubility transition, which was observed by means
of light transmittance on the copolymers, occurred in a narrow pH range (0.2 pH units), and this transition pH shifted
to a higher pH region as the SD content in the copolymer
increased (Fig. 8) (83). This kind of polymer precipitation
occurs because of the relative balance of the overall hydrophi-
licity/hydrophobicity along the polymer chain. Because the
un-ionized form of the SD units is hydrophobic, the copolymer
with a higher content of SD units requires a higher degree of
ionization for solubilization. Hence, a transition occurs at a
higher pH. A series of new copolymers of the SD monomer
and *N,N*-dimethylacrylamide demonstrated a reversible

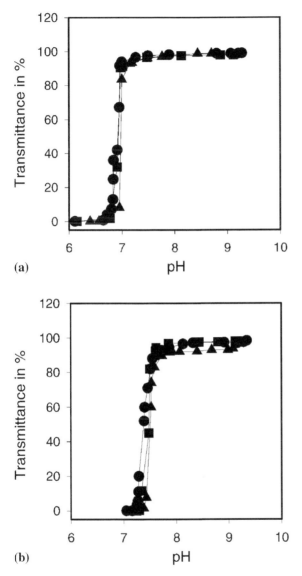

Figure 8 Transmittance of poly(methacryloyl sulfamethazine monomer-co-*N*,*N*-dimethyl acrylamide) as pH values of the solution at 37°C decrease. (a) 20 mol% sulfamethazine, (b) 30 mol% sulfamethazine (●, 0.5 wt%; ■, 1 wt%; ▲, 2 wt% solution). (Adapted from Ref. 83.)

first-order-like transition in solubility by pH. In principle, by varying the type of SD and the copolymer composition, polymers with SD were able to target a narrow range of pH for transition.

To test the pH sensitivity of the SD units in a polymeric micelle system, a pH-sensitive polymeric micelle was prepared in which the pH-sensitive SD oligomer was attached to the end of the PEG block of the ordinary diblock copolymer micelles. The polymeric micelles were composed of block copolymers of poly(L-lactide) (PLLA)/poly(ethylene glycol) (PEG)/polysulfonamide (PSD; M_w: 3000). Polysulfonamide was synthesized through the radical polymerization of a vinylated SD (SD monomer; SDM), which was itself based on the use of 2-aminoethanethiol as a chain transfer agent. Polysulfonamide was coupled to a carboxylic PEG ($M_w = 2000$)/PLLA ($M_w = 3500$) block copolymer (90). The polymeric micelles, which were prepared in accordance with a diafiltration method, showed a unimodal size distribution that was smaller than 60 nm. The PLLA/PEG-PSD micelles aggregated below a pH of 7.0 owing to the de-ionized SDM on the surface, whereas the control micelles turned out to be stable. Figure 9 shows that the polymeric micelles prepared in this study responded sharply to the change in pH around the physiological condition.

In another test, a sulfonamide was used to decorate an existing hydrophobically modified polysaccharide, pullulan acetate (PA). This technique endowed pullulan with pH sensitivity owing to the polysaccharide's conjugation to sulfonamide. The pullulan derivative was synthesized by conjugating a sulfonamide, sulfadimethoxine, to succinylated PA. The final structure of the resulting polymer conjugate is shown in Figure 10. The polymers formed self-assembled PA/sulfadimethoxine hydrogel nanoparticles in an aqueous medium. Confirmation of these polymers was obtained through the use of fluorometry and field emission-scanning electron microscopy. The examination of the pH-dependent behavior of the nanoparticles was based on the measuring of transmittance, particle size, and zeta potential (91). Doxorubicin (DOX) was loaded and released from the nanoparticles

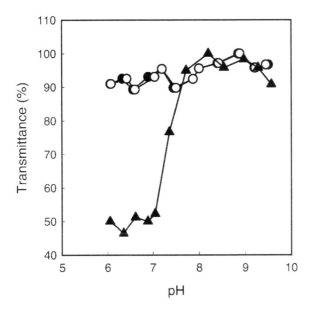

Figure 9 Changes of relative transmittance decreasing with pH values of the solution. PLLA/PEG (•), PLLA/PEG-SD (○), and PLLA/PEG-PSD (▼) block copolymer. (Adapted from Ref. 90.)

at various pHs. The mean diameter of all of the PA/sulfadimethoxine nanoparticles that were tested was <70 nm, with a unimodal size distribution.

As a result, the nanoparticles showed stability at a pH of 7.4 but shrank and became aggregated below a pH of 7.0. Figure 11, which features a comparison of in vitro cytotoxicity to the MCF-7 cells at two pHs (7.4 and 6.8), reveals that there was no noticeable difference in the cytotoxicity of the PA and the PA/sulfadimethoxine nanoparticles at the DOX concentration range of 1–10,000 ng/mL at a pH of 7.4 (Fig. 11a). When the cells were at a pH of 6.8, the DOX-loaded PA/sulfadimethoxine nanoparticles showed a significant increase in cytotoxicity. The DOX-loaded PA nanoparticles (a control sample), on the other hand, showed no pH effect on cytotoxicity (Fig. 11b). The pronounced cytotoxicity of PA/sulfadimethoxine nanoparticles at pH 6.8 may be partly attributed to the increased DOX release (Fig. 12).

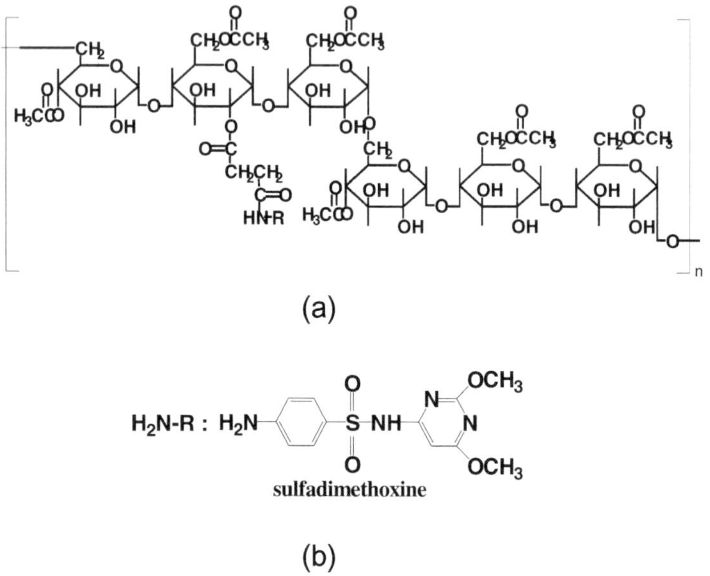

(a)

sulfadimethoxine

(b)

Figure 10 Chemical structures of (a) pullulan acetate (PA) and (b) sulfadimethoxine (SDM). (Adapted from Ref. 92.)

Although the amount of DOX added to the cell-culture medium was the same for both the free drug and the loaded drug, the release pattern of DOX from the nanoparticles followed the first-order kinetics. The total percentage of the DOX that was released after 24 h turned out to be less than 70% at a pH of 6.8. At a pH of 7.4, it was about 30%. The blank nanoparticles of PA and PA/sulfadimethoxine had no cytotoxicity and indicated that the drug-free nanoparticles did not influence the viability of MCF-7 cells. The cytotoxicity of the polymer by itself was measured and, at a polymer concentration of < 0.5 g/L, was found to be not significant ($>90\%$ viability). Therefore, if we consider that the free drug was added at the outset to the cell culture, the enhanced cytotoxicity of DOX-loaded PA/sulfadimethoxine nanoparticles does not rely entirely on the drug release rate, a finding that suggests the existence of other factors underlying the observed cytotoxicity.

The effect of pH on the interaction between PA/sulfadimethoxine nanoparticles and MCF-7 cells is presented

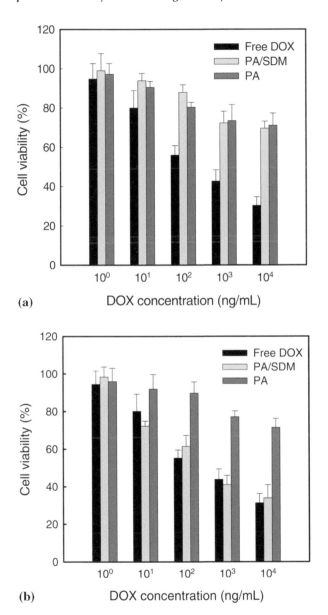

(a)

(b)

Figure 11 Cytotoxicity of doxorubicin (DOX) delivered as a free drug, in PA or PA/SDM nanoparticles against MCF-7 cells, as a function of drug concentration at (a) pH 7.4 and (b) 6.8. Data represent a mean ± S.D., $n = 7$. (Adapted from Ref. 92.)

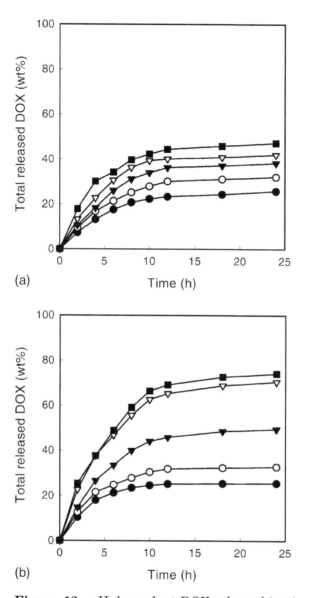

Figure 12 pH-dependent DOX release kinetics from (a) pullulan acetate (PA) and (b) PA/sulfadimethoxine (SDM) nanoparticles. pH 8.0 (●), 7.4 (○), 7.0 (▼), 6.8 (▽), and 6.4 (■). The doxorubicin (DOX)-loading efficiency of PA and PA/SDM nanoparticles were 23 ± 5 and 20 ± 4 wt%, respectively. (Adapted from Ref. 92.)

in Figure 13. PA nanoparticles minimally adhered to the MCF-7 cells and this interaction was not influenced by pH. However, in the case of PA/sulfadimethoxine, although the fluorescent intensity on the cell's surface was comparable to that of PA at a pH of 7.4, it sharply increased at a lower pH; in particular, at a pH of 6.4, the cell exhibited a three-fold increase in fluorescent intensity over that at a pH of 7.4. Thus, the degree of interaction between the PA/sulfadimethoxine and MCF-7 cells was augmented by decreasing pH, possibly by the pH-induced surface hydrophobicity (sulfadimethoxine de-ionization) of the nanoparticles. Furthermore, the nanoparticles along with the loaded drug were intracellularly localized, probably via an endocytic pathway. Therefore, the enhanced cytotoxicity of the PA/sulfadimethoxine nanoparticles at the tumor pH_e was

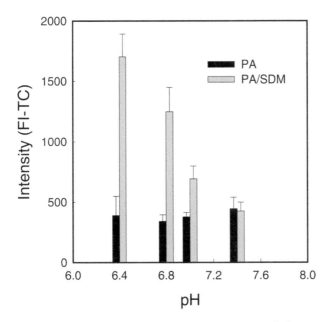

Figure 13 pH-Dependent interactions of fluorescene isothiocyanate-labeled pullulan acetate (PA) and PA/sulfadimethoxine (SDM) nanoparticles with MCF-7 cells. MCF-7 cells were incubated at 37°C for 1 h. The FITC-labeled polymer concentration was 10 μg/mL. Data represent a mean ± SD, $n = 7$. (Adapted from Ref. 92.)

induced by the combination of an accelerated drug release at pH_e and internalization into MCF-7 cells (92). These properties suggest a new mode for anticancer drug delivery systems.

5.1.2. Triggered Release by Polymeric Micelle Destabilization

A triggered release of an anticancer drug at tumor pH_e (<pH 7.0) by the physical destabilization of a carrier may constitute an advantage over the previous system. This would enhance carrier accumulation at target tumor sites. In general, stable nanocarriers stay in the vicinity of a leaky tumor vasculature after extravasation because of their size (93). Their location can form an obstacle to other nanoparticles, and further accumulation may be prevented. It is assumed that if the nanoparticles disintegrate completely by the time of the subsequent administration, a multiple dose scheme may result in effective tumor targeting.

According to a recent discovery by Dr. Bae's group (94,95), it was reported that the synthesis of the block copolymers of poly(L-histidine) (polyHis) and PEG, and the construction of polymeric micelles from the block copolymers responded to the local pH changes in the body. Because the polyHis block is a polybase, the effect of the pH during micelle fabrication has on the critical micelle concentration (CMC) was examined and the results are presented in Figure 14. At a diafiltration pH between 8.0 and 7.4, the CMC of the polyHis5K-b-PEG2K micelle increased slightly in relation to a decreasing pH. However, the CMC was significantly elevated below a pH of 7.2. It is evident that the protonation of the imidazole group in the copolymer at a lower pH level causes a reduction in hydrophobicity, and this outcome leads to an increase in CMC. In addition, below a pH of 5.0 (typically a pH of 4.8), the CMC of the polyHis5K-b-PEG2K micelle could not be detected. Together, these findings indicate that at a pH of 8.0, the less protonated polyHis constitutes the hydrophobic core in the micellar structure, but in a pH range of 5.0–7.4, the polymer produces less stable micelles. Figure 15 shows that the polyHis5K-b-PEG2K micelles prepared at

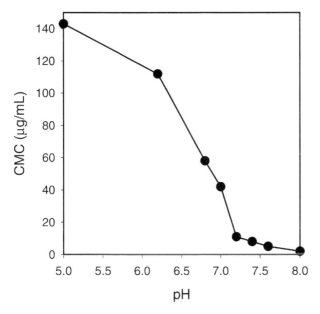

Figure 14 The pH effect on the critical micelle concentration (CMC) of polyHis5K-*b*-PEG2K polymeric micelles that were prepared at various pH values (pH 8.0–5.0). (Adapted from Ref. 94.)

a pH of 8.0 exhibited a sudden drop in fluorescence intensity at a pH of between 7.0 and 7.4, followed by a gradual fall to a pH of 6.0. It is apparent that this reduction of the fluorescence intensity was due to the instability of the hydrophobic core and that the cause of the dissociation of the micelles hinged on the protonation of the imidazole ring. It was concluded that the ionization of the polyHis block forming the micelle core determined the pH-dependent CMC and the stability (94). For further optimization of the pH sensitivity, a mixed micelle (75/25 weight ratio) of polyHis-*b*-PEG with PLLA-*b*-PEG was fabricated. The micelles were stable at a pH of 7.4, and their critical pH for destabilization was approximately 7.0; below the pH, DOX release was greatly accelerated (95).

Based on pH-sensitive biodegradable poly(β-amino ester), another approach to the topic of carrier dissociation

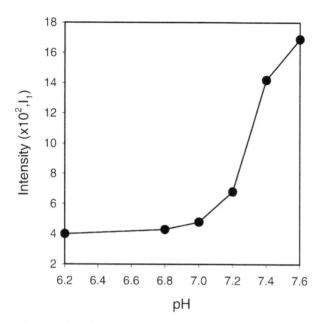

Figure 15   The change of pyrene fluorescence intensity (I_1) with the pH at constant micelle concentration (0.1 g/L). The poly-His5K-*b*-PEG2K polymeric micelles were prepared in NaOH(or HCl)–$Na_2B_4O_7$ buffer solution (pH 8.0) and exposed to each pH for 24 h. (Adapted from Ref. 94.)

at tumor pHe based on pH-sensitive biodegradable poly (β-amino ester) was reported by Lynn et al. (96,97). A poly(β-amino ester) (poly-1) (Fig. 16) that was used as a hydrophobic core was synthesized by a conjugate addition of 4,4′-trimethyldipiperidine with 1,4-butanediol diacrylate. The solubility profile of this hydrolytically degradable poly-

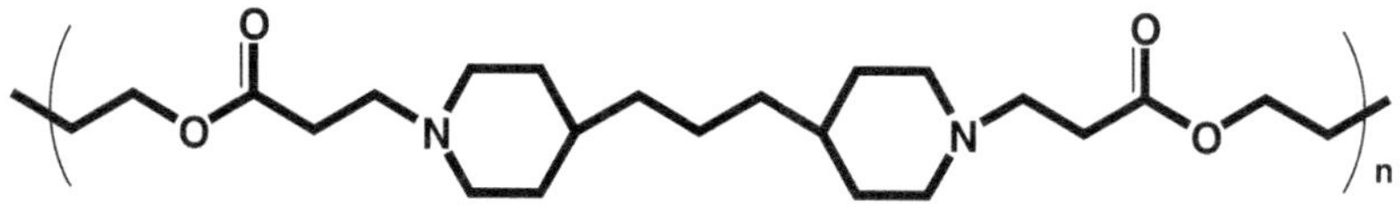

Figure 16   Chemical structure of poly(amino ester)(poly-1). (Adapted from Ref. 97.)

mer is directly influenced by a solution pH. Specifically, unprotonated polymer is insoluble in aqueous media in the pH range from 7.0 to 7.4, and the transition solubility occurs at a pH of about 6.5. The poly-1 microspheres that were suspended at a pH of 7.4 remained intact and stable when observed visually, but the spheres dissolved instantly when the pH of the suspending medium was between 5.1 and 6.5. The release profile of the labeled dextran from the poly-1 microspheres at a pH of 7.4 was characterized by a small initial burst in fluorescence (7–8%) which then reached a limiting value of about 15% after 48 h. This experiment demonstrated that the degradation of poly-1 was relatively slow under these conditions and that 85–90% of the encapsulated material could be retained in the polymer matrix for a long period of time at physiological pH values. Thus, the polymer system could provide increased therapeutic benefits by protecting its content during circulation in the body. It was also designed as in a nanoparticle form for tumor-selective paclitaxel delivery (98). However, no pH-dependent release pattern was reported.

5.2. Cytosolic Delivery

Another important use of pH-sensitive nanocarriers is the cytosolic delivery of a bioactive agent. The intracellular trafficking of drugs can seriously impact the efficacy of bioactive agents that are susceptible to lysosomal acidity and enzymes. It is therefore an important goal to design and synthesize polymers that can enhance the transport of endocytosed drugs from endosomal-lysosomal compartments to the cytoplasm.

To disrupt the endosomal-lysosomal membranes, exogenous agents such as chloroquine, glycerol, fusogenic peptides, and inactivated adenoviruses have often been employed. These agents enhance the efficiency of gene transfer but have some inherent problems. For example, whereas chloroquine has received FDA approval for the treatment of malaria, it is toxic to many cell types in vitro. In high doses, chloroquine may lead to a variety of undesirable side effects in the gastrointestinal and nervous systems (99). Its cellular toxicity and its side effects make chloroquine impractical for in vivo applications. Inactivated adenovirus particles conjugated to the polymer appear to be nontoxic in vitro but may be immunogenic and would introduce additional steps during synthesis and purification of the production process of such a gene delivery vehicle (100).

As described before, the pH of an endosome is acidified by proton-translocating ATPases (18) that depend on the stage of endosomal development. This pH gradient is a key trigger in the design of membrane-disruptive polymers, which could enhance the endosomal escape of bioactive agents.

5.2.1. The Triggering Polymeric Agent in the Liposomal Carrier

Acid-triggered conformational changes in polymers and the association of these polymers with liposomal bilayers often destabilize the lipid bilayer membranes.

Poly(2-ethylacrylic acid) (PEAA) (Fig. 17a) (101) at a concentration of 1 mg/mL can bind to a variety of multilamellar vesicles (MLVs) including negatively charged liposomes (102). The complexation was dependent on the external pH, the polymer's pK_a, and tacticity. When the complexation was attempted at a high pH (above 8.0), there was no complexation. However, a low pH resulted in greater binding. It was suggested that the hydrogen bonding between the polymer's un-ionized carboxyl groups and the lipid surface's phosphodiesters was a primary driving force for the complexation. The pH-responsible release profiles from various liposomal formulations were demonstrated in a number of studies (103–105). The transition of pH was modulated by adjusting the polymer's molecular weight (106) or by replacing PEAA with a more hydrophilic or hydrophobic AA derivative (105).

PEAA in solution, in general, experiences a coil-to-globule transition at a pH of 6.2 (107). The conformational transition of PEAA in the presence of dipalmitoylphosphatidylcholine (DPPC) vesicles shifted to 6.5 (107). Chen and colleagues

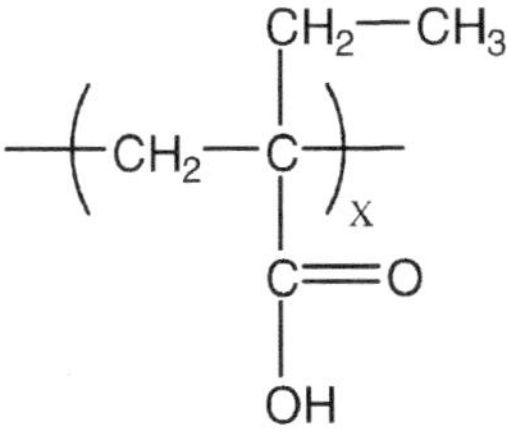

Figure 17 Chemical structures of (a) poly(2-ethylacrylic acid) (PEAA), (b) hydrophobically-modified succinylated poly(glycidol), and (c) poly(NIPPAm-*co*-MAA-*co*-ODA). (Adapted from Ref. 101.)

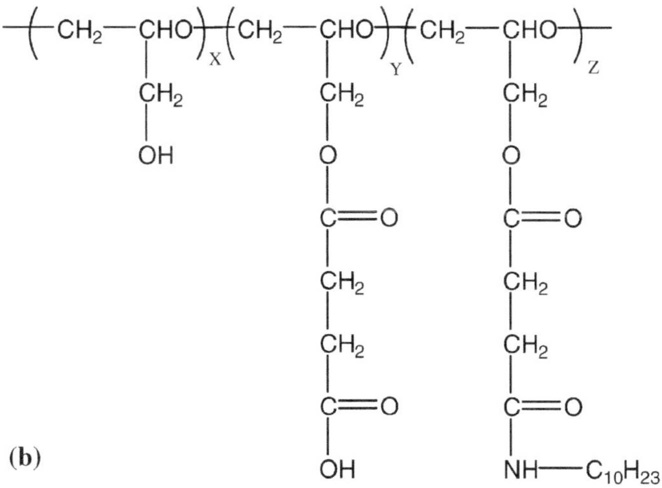

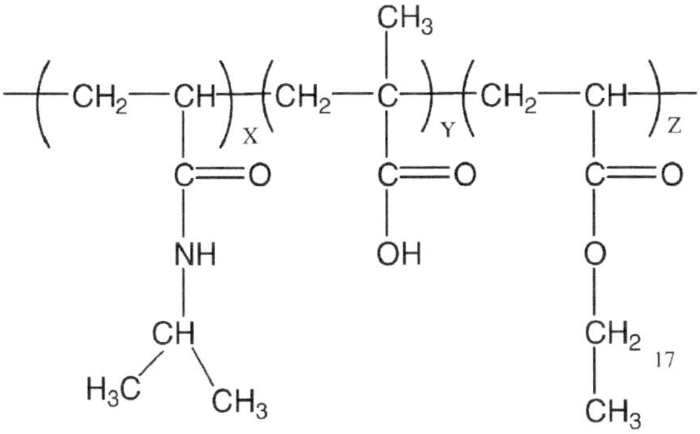

demonstrated that the PEAA covalently conjugated with small unilamellar eggPC liposomes (3.6 mol%) and that this conjugation caused a rapid release of calcein at a pH of 6.5. This finding showed that the surface-linked PEAA could destabilize the eggPC-cholesterol liposomes and encourage their fusion at a pH of 5.0 (104). Chung and associates (108) hypothesized that, at this pH, the penetration of hydrophobic segments of the PEAA into the membranes of neighboring liposomes leads to close vesicle-vesicle contacts, which, in this case, by facilitating local dehydration at the contact site, caused defects in the packing of the membrane lipids, which in turn eventually caused fusion. Drug release resulted in part from pore formation as the hydrophobic protonated polymer penetrated into the bilayer.

Succinylated poly(glycidol)s are structurally related to the PEG backbone (Fig. 17b). The polymer has hydroxyl and carboxylic groups (succinic acid) on side chains. Kono and colleagues (109–111) have synthesized various succinylated poly(glycidol)s bearing long alkyl chains ($M_w = 7600$) and incorporated the polymers in the structure of phosphatidylcholine derived from the small unilamellar vesicles of egg yolk (eggPC SUVs). The study involved efforts to observe leakage of the liposomal below a pH of 5.5 and more than 1.1 mol% polymer concentrations. Content leakage was accompanied by a concomitant intermembrane mixture. Fusion was enhanced by increasing amounts of polymers in the bilayer. Although the mechanisms by which the succinylated poly(glycidol) destabilized the lipid bilayers were not investigated, it was proposed that the protonated carboxylic acid groups interacted with the lipid membrane causing local dehydration and membrane fusion. The polymer was employed for the cytoplasmic delivery of calcein in CV-1 kidney cells via eggPC SUVs, and the efficiency of the delivery was correlated to the ability of polymer-coated liposomes that were fused with the endosomal-lysosomal membrane.

Several studies have shown that liposomes coated with the copolymers of NIPPAm and MAA bearing long alkyl chains (hydrophobic moiety) appeared to have a temperature-pH-responsive property. The alkyl substituent interacts with the liposome membrane and serves as an anchor that joins the polymer to the liposomes (112). The interaction of hydrophobically modified NIPPAm copolymers with the bilayers of dimyristoylphosphatidylcholine (DMPC) and 12-distearoyl-3-*sn*-phosphatidylcholine (DSPC) has been widely investigated (113–116).

Leroux's group (117–119) has proposed a new pH-sensitive liposomal system, and their study details the preliminary evidence that a pH-triggered release can benefit from both the pH phase transitions of hydrophobically modified poly(NIP-PAm-co-MAA) and the ability of this polymer to destabilize lipid bilayers. The studies focused on the pH-responsive release, after the incorporation of NIPPAm copolymers in liposomal membranes, of a highly water-soluble fluorescent aqueous content marker and the HPTS (pyranine) from egg phosphatidylcholine liposomes (EPC). The pH sensitivity of this system correlates with the precipitation of the copolymers at an acidic pH. The pH-triggered release, according to this study, was induced by the coil-globule transition (pH 5.1–5.7) of the NIPPAm copolymer because it depended on the polymer's molecular weight and the portion of hydrophobic moiety in the liposome. A low MW of the pH-sensitive NIP-PAm copolymer (M_w 9756) that bore 1 mol% of the hydrophobic anchor [octadecylacrylate (ODA)] was able to trigger the release of approximately 25 and 15% of the liposomal contents at a pH of 4.9 and 5.5, respectively. When the molecular mass (M_w 44,633) and the ODA content (1.6 mol%) were increased, a substantial improvement in the maximal release of the HPTS from the EPC liposomes was obtained; the released HPTS was 42% at a pH of 4.9 and 16% at a pH of 5.5 after the removal of the free polymers (see Fig. 18) (117). Also, the Mw and the ODA concentration of the copolymer did not influence its phase-transition pH.

A recent report characterized a randomly alkylated copolymer of NIPAAm, methacrylic acid, and *N*-vinyl-2-pyrrolidone in terms of its pH- and temperature-triggered conformational change (118). The polymer was complexed to liposomes so that pH-responsive vesicles would be produced. The polymer underwent a coil-to-globule phase transition

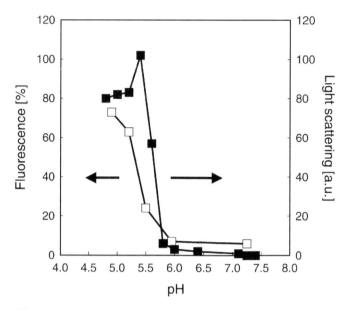

Figure 18 Comparison of the rate of HPTS [8-hydroxypyrene-1,3,6-trisulfonic acid (pyranine)] release from liposomes (eggPC) modified with NIPPAm-containing copolymers and the phase transition of the copolymer (NIPPAm/MAA/ODA 93:5:2 mol%). The copolymer phase transition was measured as an increase in light scattering at 480 nm. HPTS release was measured as an increase in HPTS fluorescence following release from liposomes ($\lambda_{ex} = 413$ nm, $\lambda_{em} = 512$ nm) and is expressed as a fraction of the fluorescence upon liposome solubilization with Triton X-100. (Adapted from Ref. 117.)

over a wide range of temperatures. At 37°C and at a pH of 7.4, it was partly dehydrated, although the polymer was water soluble. The liposome-copolymer complexes were stable at neutral pHs but rapidly released their contents under acidic conditions. The addition of PEG to the formulation had a detrimental effect on pH sensitivity but substantially enhanced the circulation time. It was recently reported that, at a copolymer/lipid mass ratio of 0.3, the large unilamellar niosome vesicles from a hydrophobically modified, pH-responsive copolymer consisting of NIPAAm, *N*-glycidylacrylamide, and

N-octadecylacrylamide exhibited good stability and pH sensitivity in buffer and human serum (119).

5.2.2. Endosomolytic Polymers for Gene Delivery

Macromolecules such as genes, oligonucleotides (ODN), and peptides-proteins are entrapped by endosomal sequestration after they experience a fluid-phase uptake or receptor-mediated endocytosis. When the macromolecules complete their vesicular uptake, the internalized carriers face degradative pathways that are a collection of hydrolases. The types of hydrolases include proteases (e.g. cathepsins), phosphatases, lipases, and glycosidases (120). Unless endosomes-lysosomes are target organelles, the success of the macromolecular therapeutics depends on the ability of such drugs to escape from the endosomal compartment into the cytosol (121).

The first polymers that were identified as having pH-dependent fusogenic properties (endosome membrane disruption) were the synthetic polypeptides polyHis (122). At a high pH, the polymer was neutral but developed a positive charge when the pH decreased. Midoux and coworkers (123,124) reported that the thing responsible for the presence of histidine in polymers was most likely a fusogenic material. The histidylated polylysine containing 84 histidyl residues (His$_{84}$-pLK) (Fig. 19) was complexed at a pH of 7.4 with a pUT650 plasmid encoding luciferase gene. The polyplexes were used to transfect HepG2 cells. The luciferase activity turned out to be 4.5 orders of magnitude higher than it was with the pLK/pDNA complexes. Additionally, it was 3.5 and 3 orders of magnitude higher than it was with the pLK/pDNA complexes in the presence of chloroquine and a fusogenic peptide, respectively (Fig. 20). They had also designed histidylated oligolysines, which increased the uptake for cytosolic delivery, and the nuclear accumulation of antisense ODN. A flow cytometric analysis showed a ten-fold enhancement of the ODN uptake in the presence of histidylated oligolysines. These results may be induced by a "proton sponge" mechanism of the imidazole group. The imidazole exhibits a pK_a around 6.0. Thus, it possesses a buffering capacity in the

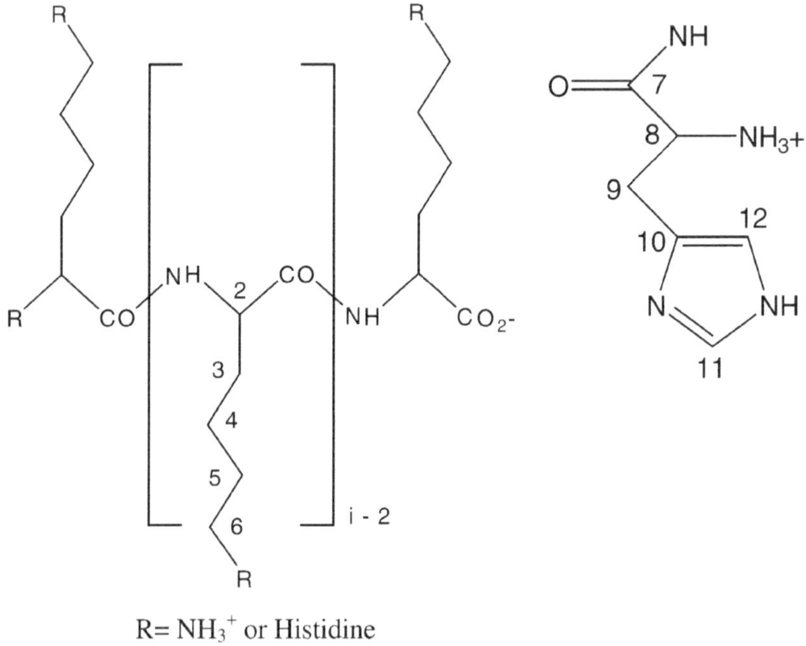

R= NH$_3^+$ or Histidine

Figure 19   Schematic structure of histidylated polylysine. i = DP; R = NH$_3^+$ or His. (Adapted from Ref. 123.)

endolysosomal pH range and possibly mediates a vesicular escape. When PEG-polyHis was used for gene delivery, the transfection efficiency, however, was approximately equivalent to the DNA:polylysine complex. At the same time, it showed a lower level of cytotoxicity relative to the macrophages that were cultured in vitro (100,125). The complex of the DNA:PEG-polyHis was formed in a 0.01 M acetate buffer (pH 5). This brought the histidine imidazole to an almost ionized state. The ionized imidazole cannot produce "proton sponge" material in the endosome. The incorporation of the histidine imidazole moieties represents a promising option for the improvement of both the endosomal escape and the efficiency of polymers. Moreover, these improvements do not increase the moieties' toxicity.

The polyethylenimines (PEI) containing primary, secondary, and tertiary amines have most often been used for

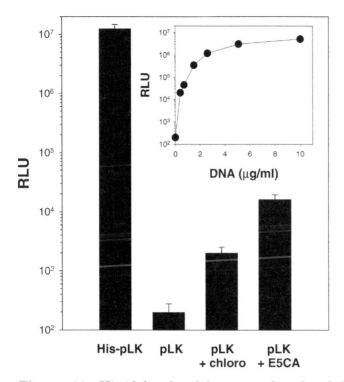

Figure 20 Histidylated polylysine mediated polyfection. Polyplexes were prepared by mixing either His_{84}-pLK ($40\,\mu g$) or pLK ($5\,\mu g$) in $0.3\,mL$ of serum-free DMEM with $10\,\mu g$ ($3\,pmol$) of pUT650 plasmid in $0.7\,mL$ of serum-free DMEM. The solution was kept for 30 min at 20°C. When relevant, the solution was made $100\,\mu M$ in chloroquine or $10\,\mu M$ in E5CA. HepG2 cells (3×10^5 cells plated in a $4\,cm^2$ well) were incubated for $4\,h$ at 37°C with polyplexes in the presence of 5% FBS. Then cells were washed and incubated in a complete culture medium containing 10% FBS. The gene expression was determined $48\,h$ later by assaying the luciferase activity in cell lysates. RLU, the number of relative light units, represents the luciferase activity in 10^6 cells. (Inset: Concentration dependence of polyfection with His_{84}-pLK/pDNA complexes.) (Adapted from Ref. 123.)

gene delivery that does not involve any exogenous agent. This is the case because of their superior efficacy in in vitro transfection. The amino groups' various pK_b values induce the high transfection property in question. Primary amines of PEI lead

to the construction of effective complexes, even with large
DNA molecules (126–128), and all of this leads to homogeneous spherical particles that have a size of approximately
100 nm or less and are capable of transfecting cells in vitro.
Other unprotonated amino moieties create the "proton
sponge" effect because the amino groups can protonate at
the endosomal pH. This effect induces the osmotic swelling
and the rupture of the vesicle membrane by the influx of coun-
terions. These are brought into the vesicle so that electroneutrality can be maintained (129). Thus, among the various
polymeric vectors, PEI has shown particularly promising efficacy in in vitro transfections and fusogenecity. The high density of positive charges, however, results in a rather high
toxicity, which is one of the major limiting factors especially
for its in vivo application (130). Additionally, extensive literature has reported various methods for the reduction of PEI
cytotoxicity (131).

Poly(amidoamine)s (MBI) are a class of strong cationic
polymers for endosomal membrane destabilization. The polymers are characterized by the presence of amido and tertiary
amino groups regularly arranged along the macromolecular
chain. Additional functional groups may be introduced as side
substituents (132). The polymers undergo a marked conformational change leading to an alteration from a relatively
coiled hydrophobic structure at a neutral pH (7.4) to a relaxed
hydrophilic structure at an acidic pH (5.5) (133). The conformational change is due to the protonation of the tertiary
amino groups and confers on these groups the ability to selectively damage biological membranes at a low pH. The polymers in a coil form appear to be rod-shaped as it displays a
much larger hydrodynamic volume than other hydrophilic
polymers (e.g., polyvinylpyrrolidone) of a similar molecular
weight (134). Viscosity measurements show that MBIs have
a compact structure at a pH of 7.4. This opens under the
acidic conditions, normally encountered following pinocytic
internalization into the endosomes and lysosomes (pH 6.5–
5.0). To investigate whether these characteristics should be
used to protect a drug payload in the circulation and to facilitate subsequent intracellular delivery, an MBI-Triton X-100

conjugate was synthesized as a model compound (133). Whereas Triton X-100 alone was highly hemolytic at all of the pHs (5.5–8.0) studied, the parent polymer MBI was not hemolytic at any pH and the MBI-Triton X-100 conjugate was hemolytic at pH 5.5, although not when a pH of either 7.4 or 8.0 was studied. This finding means that MBI has the potential to hide its Triton X-100 within its intramolecular core. These early observations initiated the search for new MBI structures that would be less toxic and would display inherent membrane activity at low pH values; in other words, they would be inherently endosomolytic.

New MBIs have subsequently been reported to be bio-compatible, cause pH-dependent hemolysis, and non-hepato-tropic when administered in vivo (135). Duncan and Ferruti and their coworkers demonstrated that the new MBIs (Fig. 21) have the property of pH-dependent membrane destabili-zation, which makes them potential endosomolytic polymers (135–137). The first evidence that MBIs can cause endosomal membrane destabilization came from the observation that they mediate the transfection of HepG2 cells using pSV-β-galactosidase as a marker (137). They characterized and synthesized four different MBIs (ISA 1, ISA 4, ISA 23, and ISA 22). To study the MBI-mediated transfection, polymers and various control vectors were used in combination with pSV-β-galactosidase and HepG2 cells. They used a Lipofec-tACE as a reference control from which to study the effects of the vector:DNA ratio (Fig. 22a). The greatest level of β-galactosidase expression was also mediated by the ISA 23 at a 1:9 ratio and with an excess polymer. The ISA 23 showed a higher level of β-galactosidase expression than the ISA 22. To allow the correlation with known nonviral vectors, the transfection efficiency was compared at a 5:1 vector excess (Fig. 22b). The ISA 23 resulted in a β-galactosidase expression that was equivalent to that seen using the PEI and LipofectIN. Both the ISA 22 and the ISA 23 showed significantly higher expressions than those corresponding to the LipofectACE experiment. The cytotoxicity of the ricin A-chain (RTA) was also tested so that direct evidence of high transfection efficiency could be obtained (138). RTA showed

ISA 1 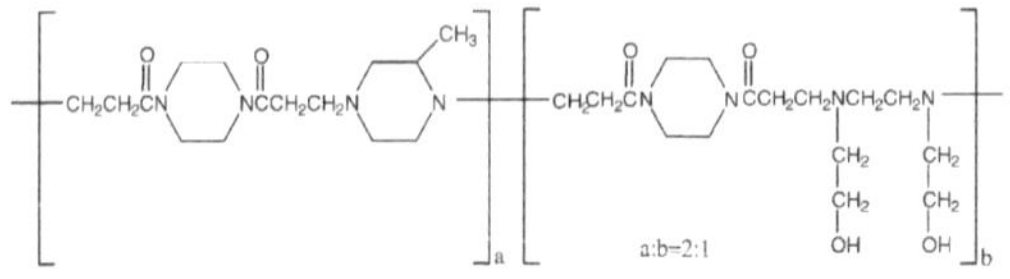   Mw 12,300

ISA 4 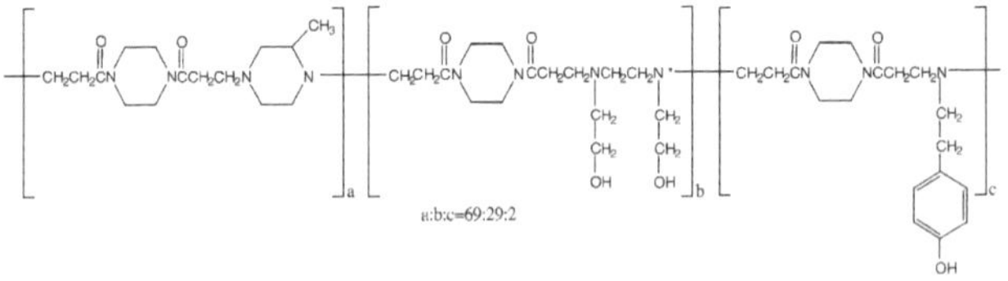   Mw 15,000

ISA 23 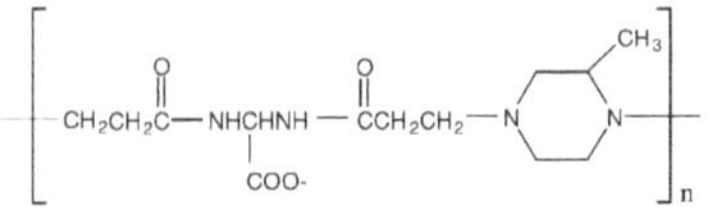   Mw 16,500

ISA 22 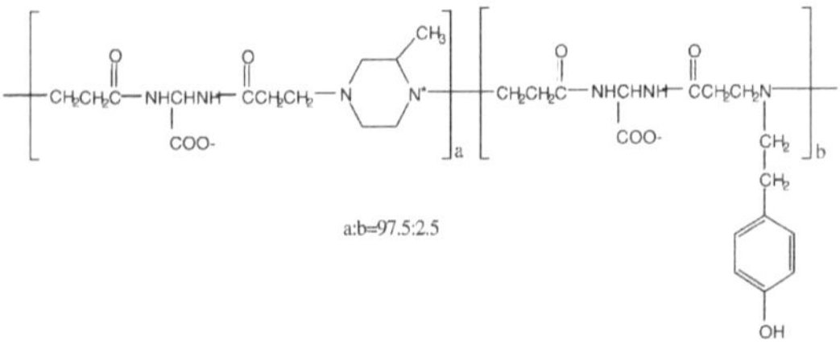   Mw 14,900

Figure 21 Chemical characteristics of the poly(amidoamine)s. M_n (number-average molecular weight) and M_w (weight-average molecular weight) were calculated by gel permeation chromatography against polyamidoamine standards. (Adapted from Ref. 137.)

a dose-dependent cytotoxicity that impacted the B16F10 cells in vitro ($IC_{50} = 1.4\,\mu g/mL$). When incubated with the B16F10 cells, ISA 1 and ISA 4 were relatively non-toxic, showing IC_{50} values of >2 and 1.8 mg/mL, respectively. In the case of ISA 1 and the RTA combination (250 ng/mL), the IC_{50} observed was 0.65 ± 0.05 mg/mL (Fig. 23).

Demonstrations have shown that acidic PEAA and poly-(propacrylic acid) (PPAA) have pH-dependent membrane-disruptive properties in liposomes depending on their molecular weight. A copolymer composed of AA and propacrylic

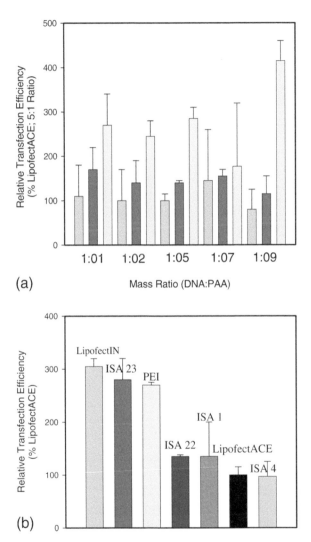

(a)

(b)

Figure 22 Transfection of HepG2 cells using pSV-β-galactosidase in the presence of various vectors. Panel (a) shows the β-galactosidase expression after incubation with square with left-vertical lines ISA 23, ISA 22 (☐), and LipofectACE (■) complexes at mass ratios between 1:1 and 1:10 using excess vector. In panel (b), the β-galactosidase expressions are compared after exposure of HepG2 cells to the plasmid alone, LipofectIN, LipofectACE, ISA 1, ISA 4, ISA 22, ISA 23, and PEI, and without plasmid. A mass ratio of 5:1 (excess vector) was used throughout. (Adapted from Ref. 138.)

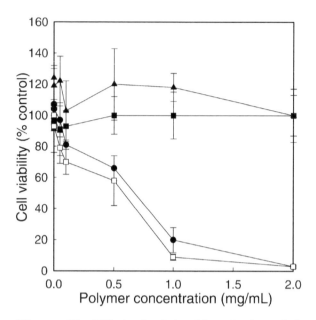

Figure 23 Effect of poly(amidoamine)s and dextran on the cytotoxicity of RTA. Assays were performed using B16F10 cells with a non-toxic concentration of RTA (250 ng/mL) and the data shown represent dextran + RTA (▲-▲), ISA 22 + RTA (■-■), ISA 1 + RTA (●-●), and ISA 4 + RTA (□-□). Data represent mean ± SD, $N = 18$. (Adapted from Ref. 137.)

acid has displayed a significant pH-dependent membrane-disruptive effect on red blood cells (139). Hoffman and his coworkers have investigated these cells' ability to disrupt eukaryotic cell membranes using a standard hemolytic assay (139–142). PEAA hemolyzes red blood cells with an activity of 10^7 molecules per red blood cell, which is as efficient on a molar basis as the peptide melittin. The membrane-disruption methods corresponding to these acidic polyelectrolytes differ from those of the polycations. The pH-dependent membrane-disruptive activities of the polymers are closely associated with the conformational change of the polymers. The transition from an expanded conformation at a high pH to a relatively hydrophobic, globular coil in acidic solution is triggered by the protonation of the free carboxylic groups of the polymer.

Consequently, the collapsed polymer chain provides an increased number of hydrophobic sites for enhanced polymer adsorption to the phospholipid membrane (143). The researchers also reported that the mechanism of RBC hemolysis by PEAA is consistent with the colloidal osmotic mechanism. PEAA's hemolytic activity rises rapidly as the pH decreases from 6.3 to 5.0; there is no hemolytic activity at pH 7.4. A related polymer, PPAA was synthesized, and the goal was to test whether making the pendant alkyl group more hydrophobic would, by adding one more methylene group, increase its hemolytic activity. PPAA was found to disrupt red blood cells 15 times more efficiently than PEAA at a pH of 6.1. According to this experiment, the PPAA was not active at a pH of 7.4 and displayed a pH-dependent hemolysis that shifted toward higher pHs. Table 4 shows the results of the transfection of NIH3T3 fibroblasts with ternary physical mixtures of the cationic lipid carrier DOTAP, pCMV β plasmid DNA, with or without PPAA at varying charge ratios of added DOTAP (+) to DNA (−), and PPAA (−). Table 4 shows that the addition of PPAA to DOTAP/DNA lipoplexes significantly enhances transfection efficiencies in the NIH3T3 mouse fibroblast cells at all of the charge ratios tested. The level of β-galactosidase gene expression with ternary DOTAP/DNA/PPAA carriers was increased approximately 20-fold over the control DOTAP/DNA lipoplexes prepared with a charge ratio of 1.3(±)(143). Recently, it was reported that, among the various

Table 4 Effect of Poly(propacrylic acid) on In Vivo Transfection[a]

	Charge ratio (+)			
Formulation	1.0	1.3	1.6	1.9
DOTAP/DNA	50 ± 10	103 ± 11	175 ± 3	346 ± 17
DOTAP[1]/DNA/PPAA	1514 ± 66	2197 ± 56	1276 ± 101	1076 ± 112
DOTAP[2]/DNA/PPAA	568 ± 55	1037 ± 76	876 ± 42	696 ± 30

[a]Transfection levels are reported as milliunits of β-galactosidase activity/mg of protein. Particles contain $10\,\mu g$ DOTAP[1] or DOTAP[2].
Source: Ref. 144.

PAA polymers studied, medium and low molecular PEAAc and PPAAc were identified as displaying significant pH-dependent disruptive activity. Relative to the disruptive cationic polymer poly(butyl methacrylate-co-[2-dimethyl aminoethyl]methacrylate-co-methyl methacrylate) (PDMAEM) the rank was PEAA < PPAA < PDMAEM (144).

5.2.3. Fusogenic Polymers for Anticancer Drug Delivery

Although various pH-sensitive polymers displaying fusogenic activity have been used for the cytosolic delivery of bioactive macromolecules, such polymers have seldom been used for anticancer drug delivery. Polymeric nanocarriers bearing tumor-specific ligands (Fig. 24a) have been utilized as cytosolic drug delivery carriers for active endocytosis and have improved cellular drug bioavailability. If the carrier simultaneously exhibits both a triggered release in early endosomes (approximately pH 6) and endosomal disruption after internalized, it will provide higher concentrations of the drug in the cytosol and the nucleus (potential drug acting sites) (Fig. 25). Such a system would be an effective delivery mode, particularly for MDR cancer cells. This approach will be useful for weakly basic drugs (most anticancer drugs are weak bases) whose partitioning in cytosol and subcellular organelles is greatly influenced by intracellular pH profiles (more sequestration of the drugs occurs in acidic subcellular compartments). Therefore, the development of polymeric nanoparticles that have targeting ligands and exhibit fusogenic activities is required so that more effective anticancer drug delivery systems—particularly for drug-resistant cancer cells—can be made available.

The fusogenic effect of polymers containing imidazoles was discussed earlier. If the polymeric micelles with the polyHis core were employed, it would become an effective delivery method against cancer cells. Commercial polyHis (M_w approximately 10,000) is available but is not soluble in most organic solvents owing to its high molecular weight and has difficulty in forming micelles or a self-assembled

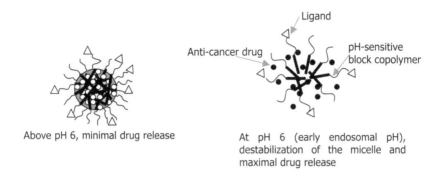

Above pH 6, minimal drug release

At pH 6 (early endosomal pH), destabilization of the micelle and maximal drug release

(a)

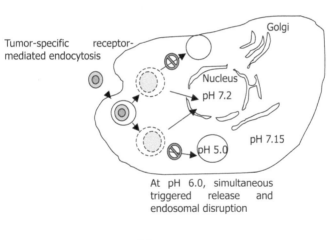

At pH 6.0, simultaneous triggered release and endosomal disruption

(b)

Figure 24 (a) A proposed new carrier for switching the release of an anticancer drug by early endosomal pH from polymeric micelles and (b) the polymeric micelles for effective anticancer drug delivery to MDR cells with receptor-mediated endocytosis, switching release rate, and fusogenic activity of polymeric micelle components.

nanostructure. Most recently, Bae's group reported a novel pH-sensitive polymer that is mixed with micelles and that is based on polyHis (polyHis; M_w 5000)/PEG (M_n 2000) (95). The polyHis/PEG micelles showed an accelerated DOX release as the pH decreased from 8.0. When the cumulative

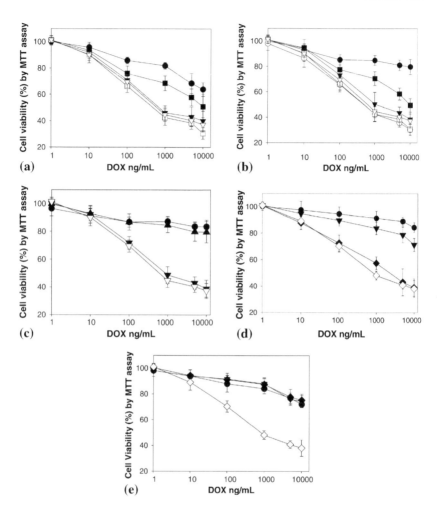

Figure 25 The cytotoxicity of DOX loaded mixed micelles with PLLA/PEG content (a) 0 wt.%, (b) 10 wt.%, (c) 25 wt.%, (d) 40 wt.%, and (e) 100 wt.% after 48 h incubation at varying pH [pH 7.4 (●); 7.2 (■); 7.0 (▲); 6.8 (▽); 6.6 (◇)]. Free ADR cytotoxicity is presented as a reference at pH 7.4 (○), 7.2 (□), 7.0 (△), 6.8 (▽), and 6.6 (◇). (Adapted from Ref. 95.)

release for 24 h was plotted as a function of pH, a gradual transition in the release rate appeared in a pH ranging from 8.0 to 6.8. In order to tailor the triggering pH to a more acidic tumor's extracellular pH, and to improve the micelle stability at a pH of 7.4, the PLLA/PEG block copolymer was blended with a polyHis/PEG: what formed was mixed micelles. The blending shifted the triggering pH to a lower value. Depending on the amount of PLLA/PEG in the blending, the mixed micelles were destabilized with a pH range of 7.2–6.6 (the triggering pH for the DOX release). The polyHis/PEG micelles that were loaded with DOX that exhibited a pH of 7.4 showed some cell-killing effect because the micelles had a certain degree of both instability and DOX release. Cell viability at a pH of 6.8 was greatly reduced under the influence of enhanced DOX release (Fig. 25a). The mixed micelles with 10 wt.% PLLA/PEG (Fig. 25b) caused less cytotoxicity at a pH of 7.4. Additionally, there was a high cytotoxicity below a pH of 7.2. The above results are distinguished from the cytotoxicity of free DOX (almost independent of pH). The 25 wt.% PLLA/PEG micelles (Fig. 25c) appeared to be more sensitive to the pH levels, especially when a distinction is made between a pH of 7.0 and a pH of 6.8. The 40 wt.% PLLA/PEG micelles (Fig. 25d) were distinguished by a pH of 6.8 and of 6.6 in cytotoxicity. Because both the 100 wt.% PLLA/PEG micelles (Fig. 25e) and the blank micelles had no pH-dependent cytotoxicity, the pH-dependent cytotoxicity solely relied on the release of DOX at a different pH. As a result, the change in pH from 8.0 to 6.6 may influence the cell's surface charges, cellular physiology, and cells' viability. The low pH is a favorable environment for tumor cells but has an opposite effect on normal cells. For the active internalization of the micelles, folic acid was introduced into the pH-sensitive mixed micelles. After the internalization, the micellar carrier actions could be combined with both the pH-triggered DOX release at an early endosomal pH and the fusogenic activity of polyHis, which helped the DOX to be released from the endosomal compartment to the cytosol. To visualize the effect of both the folate-mediated endocytosis and the fusogenic activity of the mixed micelles, the distribution of the DOX on MCF-7 cells was observed under a confocal microscopy. After 1 h, the cells' uptake of the folate-conjugated micelles proceeded rapidly. The folate-conjugated mixed micelles and the PLLA/PEG-folate micelles showed a high intracellular DOX concentration and a high distribution in both the cytosol and the nucleus; this was visualized by a red-intensity signifying the DOX. On the other hand, in the case of the PLLA/PEG-folate micelles, DOX was not broadly distributed in the cytosolic compartment but localized probably in the endosomes. This observation was clearly distinguished from the observation of the folate-conjugated mixed micelle that exhibited fusogenic activity. The DOX that was carried by the mixed micelles promoted—in a striking way—a cytosolic distribution of DOX. In contrast, for free DOX, only the low red intensity appeared in the peripheral region of the cells owing to the DOX's slow diffusion process into the cells, which amounted to a 1 h incubation period.

6. CONCLUSION

Traditionally, anionic pH-sensitive polymers have been extensively applied as an enteric-coating material for oral dosage forms. This was to protect labile drugs from dissolving too quickly in the stomach, prevent patients from experiencing irritation in their stomachs, and make the drug taste better. Recent works have demonstrated that the pH-dependent solubility of the polymers facilitated colon targeting and the oral delivery of peptide-protein drugs.

With the discovery of both new pH-sensitive functional groups and new polymerization technology, various anionic and cationic polymers, which are truly sensitive to small changes in physiological pH or responsive to a target pH, are now expanding in applications. These include enhanced drug release at tumor sites that is triggered by the tumor's extracellular pH, the cytosolic delivery of labile macromolecular pharmaceuticals that are combined with endosomal-lysosomal membrane disruption properties, and the delivery of anticancer drugs to MDR malignant cells. As a result, pH-sensitive polymeric drug carriers are becoming more feasible in the treatment by drug delivery systems of specific diseases.

REFERENCES

1. Chen G, Hoffman AS. Graft copolymers that exhibit temperature-induced phase transitions over a wide range of pH. Nature 1995; 373:49–52.

2. Brahim S, Narinesingh D, Guiseppi-Elie A. Release characteristics of novel pH-sensitive p(HEMA-DMAEMA) hydrogels containing 3-(trimethoxy-silyl) propyl methacrylate. Biomacromolecules 2003; 4:1224–1231.

3. Yu Y, Nakano M, Ikeda T. Photomechanics: Directed bending of a polymer film by light. Nature 2003; 425:145.

4. Malmstadt N, Yager P, Hoffman AS, Stayton PS. A smart microfluidic affinity chromatography matrix composed of poly(N-isopropylacrylamide)-coated beads. Anal Chem 2003; 75:2943–2949.

5. Bae YH, Okano T, Kin SW. Insulin permeation through thermosensitive hydrogels. J Control Release 1989; 9:271–279.

6. Soppimath KS, Aminabhavi TM, Dave AM, Kumbar SG, Rudzinski WE. Stimulus-responsive "smart" hydrogels as novel drug delivery systems. Drug Dev Ind Pharm 2002; 28:957–974.

7. Ramkissoon-Ganorkar C, Liu F, Baudys M, Kim SW. Modulating insulin-release profile from pH/thermosensitive polymeric beads through polymer molecular weight. J Control Release 1999; 59:287–298.

8. Patel VR, Amiji MM. Preparation and characterization of freeze-dried chitosan-poly(ethylene oxide) hydrogels for site-specific antibiotic delivery in the stomach. Pharm Res 1996; 13:588–593.

9. Horter D, Dressman JB. Influence of physicochemical properties on dissolution of drugs in the gastrointestinal tract. Adv Drug Deliv Rev 1997; 25:3–14.

10. Sasaki Y, Hada R, Nakajima H, Fukuda S, Munakata A. Improved localizing method of radiopill in measurement of entire gastrointestinal pH profiles: colonic luminal pH in normal subjects and patients with Crohn's disease. Am J Gastroenterol 1997; 92:114–118.

11. Bhagavan NV. Medical Biochemistry. Boston: Jones and Bartlett Publishers, 1992:15–16.

12. Tannock IF, Rotin D. Acid pH in tumors and its potential for therapeutic exploitation. Cancer Res 1989; 49:4373–4384.

13. Hobbs SK, Monsky WL, Yuan F, Roberts WG, Griffith L, Torchilin VP, Jain RK. Regulation of transport pathways in tumor vessels: Role of tumor type and microenvironment. Proc Natl Acad Sci USA 1998; 95:4607–4612.

14. van Sluis R, Bhujwalla ZM, Raghunand N, Ballesteros P, Alvarez JCS, Galons JP, Gillies RJ. In vivo imaging of extracellular pH using ^{1}H MRSI. Magn Reson Med 1999; 41: 743–750.

15. Ojugo ASE, McSheehy PMJ, McIntyre DJO, McCoy C, Stubbs M, Leach MO, Judson IR, Griffiths JR. Measurement of the extracellular pH of solid tumours in mice by magnetic resonance spectroscopy: a comparison of exogenous ^{19}F and ^{31}P probes. NMR Biomed 1999; 12:495–504.

16. Yamagata M, Hasuda K, Stamato T, Tannock IF. The contribution of lactic acid to acidification of tumours: studies of variant cells lacking lactate dehydrogenase. Br J Cancer 1998; 77:1726–1731.

17. Stubbs M, McSheehy PMJ, Griffiths JR, Bashford L. Causes and consequences of tumour acidity and implications for treatment. Opinion 2000; 6:15–19.

18. Maxfield FR, Yamashiro DJ. Intracellular Trafficking of Proteins. Cambridge, UK: Cambridge University Press, 1991: 157–182.

19. Florencia BS, Grinstein S. Determinants of the pH of the Golgi complex. J Biol Chem 2000; 275:21025–21032.

20. Demaurex N, Furuya W, D'Souza S, Bonifacino JS, Grinstein S. Mechanism of acidification of the trans-Golgi network (TGN). J Biol Chem 1998; 273:2044–2051.

21. Mellman I, Fuchs R, Helenius A. Acidification of the endocytic and exocytic pathways. Annu Rev Biochem 1986; 55:663–700.

22. Forgac M. Structure and properties of the vacuolar (H^{+})-ATPases. J Biol Chem 1999; 274:12951–12954.

23. Stevens TH, Forgac M. Structure, function and regulation of the vacuolar (H^+)-ATPase. Annu Rev Cell Dev Biol 1997; 13:779–808.

24. Gu F, Gruenberg J. ARF1 regulates pH-dependent COP functions in the early endocytic pathway. J Biol Chem 2000; 275:8154–8160.

25. Han X, Bushweller JH, Cafiso DS, Tamm LK. Membrane structure and fusion-triggering conformational change of the fusion domain from influenza hemagglutinin. Nat Struct Biol 2001; 8:715–720.

26. Gunther W, Luchow A, Cluzeaud F, Vandewalle A, Jentsch TJ. ClC-5, the chloride channel mutated in Dent's disease, colocalizes with the proton pump in endocytotically active kidney cells. Proc Natl Acad Sci USA 1998; 95:8075–8080.

27. Kornak U. Loss of the CLC-7 chloride channel leads to osteopetrosis in mice and man. Cell 2001; 104:205–215.

28. Schapiro FB, Grinstein S. Determinants of the pH of the Golgi complex. J Biol Chem 2000; 275:21025–21032.

29. Simon S, Roy D, Schindler M. Intracellular pH and the control of multidrug resistance. Proc Natl Acad Sci USA 1994; 91:1128–1132.

30. Simon S. Role of organelle pH in tumor cell biology and drug resistance. Drug Discov Today 1999; 4:32–38.

31. Qiu Y, Park K. Environment-sensitive hydrogels for drug delivery. Adv Drug Deliv Rev 2001; 53:321–339.

32. Higuchi S, Ozawa T, Maeda M, Inoue S. pH-induced regulation of the permeability of a polymer membrane with a transmembrane pathway prepared from a synthetic polypeptide. Macromolecules 1986; 19:2263–2267.

33. Kokufuta E, Shimizu N, Nakamura I. Preparation of polyelectrolyte-coated pH-sensitive poly(styrene) microcapsules and their application to initiation-cessation control of an enzyme reaction. Biotech Bioeng 1988; 32:289–294.

34. Chourasia MK, Jain SK. Pharmaceutical approaches to colon targeted drug delivery systems. Pharm Pharmaceut Sci 2003; 6:33–66.

35. Khan MZ, Prebeg Z, Kurjakovic N. A pH-dependent colon targeted oral drug delivery system using methacrylic acid copolymers. I. Manipulation of drug release using Eudragit® L100–55 and Eudragit® S100 combinations. J Control Release 1999; 58:215–222.

36. Azarmi S, Farid J, Nokhodchi A, Bahari-Saravi SM, Valizadeh H. Thermal treating as a tool for sustained release of indomethacin from Eudragit® RS and RL matrices. Int J Pharm 2002; 246:171–177.

37. Pignatello R, Bucolo C, Ferrara P, Maltese A, Puleo A, Puglisi G. Eudragit RS100® nanosuspensions for the ophthalmic controlled delivery of ibuprofen. Eur J Pharm Sci 2002; 16:53–61.

38. Comoglu T, Gönül N, Baykara T. Preparation and in vitro evaluation of modified release ketoprofen microsponges. Il Farmaco 2003; 58:101–106.

39. Weiß G, Knoch A, Laicher A, Stanislaus F, Daniels R. Simple coacervation of hydroxypropyl methylcellulose phthalate (HPMCP) I. Temperature and pH dependency of coacervate formation. Int J Pharm 1995; 124:87–96.

40. Chang RK. A comparison of rheological and enteric properties among organic solutions, ammonium salt aqueous solutions, and latex systems of some enteric polymers. Pharm Technol 1990; 14:62–70.

41. Guo HX, Heinämäki J, Yliruusi J. Amylopectin as a subcoating material improves the acidic resistance of enteric-coated pellets containing a freely soluble drug. Int J Pharm 2002; 235:79–86.

42. Schoneker DR, DeMerlis CC, Borzelleca JF. Evaluation of the toxicity of polyvinylacetate phthalate in experimental animals. Food Chem Toxicol 2003; 41:405–413.

43. Weldon MJ, Maxwell SD. Lymphocytes and macrophage interleukin receptors in inflammatory bowel disease: a more selective target for therapy. Gut 1994; 35:871–876.

44. Wenzel S. Antileukotriene drugs in the management of asthma. JAMA 1998; 280:38–39.

45. William WB. Leukotrienes and inflammation. Respir Crit Care Med 1998; 157:S210–S213.

46. Thomas P, Richards D, Richards A. Absorption of delayed release prednisolone in ulcerative colitis and Crohn's disease. J Pharm Pharmacol 1985; 37:757–758.

47. Van Saene JJM, Van Saene HFK, Geitz JN, Tarko-Smit NJP, Lerk CF. Quinolones and colonization resistance in human volunteers. Pharm Weekbl Sci Ed 1986; 8:67–71.

48. Azad Khan KA, Piris J, Truelove SC. An experiment to determine the active therapeutic moiety of sulphasalazine. Lancet 1977; 2:892–895.

49. Kim CK, Shin HJ, Yang SG, Kim JH, Oh Y. Once-a-day oral dosing regimen of cyclosporin A: combined therapy of cyclosporin A premicroemulsion concentrates and enteric coated solid-state premicroemulsion concentrates. Pharm Res 2001; 18:454–459.

50. Levine DS, Raisys VA, Ainardi V. Coating of oral beclomethasone dipropionate capsules with cellulose acetate phthalate enhances delivery of topically active antiinflammatory drug to the terminal ileum. Gastroenterology 1987; 92:1037–1044.

51. Hardy JG, Evans DF, Zaki I, Clark AG, Tonnesen HH, Gamst ON. Evaluation of an enteric coated naproxen tablet using gamma scintigraphy and pH monitoring. Int J Pharm 1987; 37:245–250.

52. Madajewicz S, Petrelli N, Rustum YM, Campbell J, Herrera L, Mittelman A, Perry A, Creaven PJ. Phase I–II trial of high-dose calcium leucovorin and 5-fluorouracil in advanced colorectal cancer. Cancer Res 1984; 44:4667–4669.

53. Lamprecht A, Yamamoto H, Takeuchi H, Kawashima Y. Microsphere design for the colonic delivery of 5-fluorouracil. J Control Release 2003; 90:313–322.

54. Ashford M, Fell JT, Attwood D, Sharma H, Woodhead PJ. An in vivo investigation into the suitability of pH-dependent polymers for colonic targeting. Int J Pharm 1993; 95:193–199.

55. Markus WR, Klein S, Beckert TE, Petereit H, Dressman JB. A new 5-aminosalicylic acid multi-unit dosage form for the therapy of ulcerative colitis. Eur J Pharm Biopharm 2001; 51:183–190.

56. Rodríguez M, Vila-Jato JL, Torres D. Design of a new multi-particulate system for potential site-specific and controlled drug delivery to the colonic region. J Control Release 1998; 55:67–77.

57. Takemura S, Watanabe S, Katsuma M, Fukui M. Human gastrointestinal transit study of a novel colon delivery system (CODES^{TM}) using γ-scintigraphy. Proceedings of the International Symposium on Controlled Release of Bioactive Material, Paris, Vol. 27, 2000.

58. Lorenzo-Lamosa ML, Remuñán-López C, Vila-Jato JL, Alonso MJ. Design of microencapsulated chitosan microspheres for colonic drug delivery. J Control Release 1998; 52:109–118.

59. Shimono N, Takatori T, Ueda M, Mori M, Higashi Y, Nakamura Y. Chitosan dispersed system for colon-specific drug delivery. Int J Pharm 2002; 245:45–54.

60. Park JW, Choi KH, Park KK. Acid-base equilibria and related properties of chitosan. Bull Kor Chem Soc 1983; 4:68–72.

61. Hosny EA, Khan Ghilzai MK, Al-Dhawalie AH. Effective intestinal absorption of insulin in diabetic rats using enteric-coated capsules containing sodium salicylate. Drug Dev Ind Pharm 1995; 21:1583–1589.

62. Hosny EA, Khan Ghilzai MK, Elmazar MM. Promotion of oral insulin absorption in diabetic rabbits using pH-dependent coated capsules containing sodium cholate. Pharm Acta Helv 1997; 72:203–207.

63. Choudhari KB, Labhasetwar V, Dorle AK. Liposomes as a carrier for oral administration of insulin: effect of formulation factors. J Microencapsul 1994; 11:319–325.

64. Shao Z, Li Y, Chermak T, Mitra AK. Cyclodextrin as mucosal absorption promoters of insulin. II. Effects of beta-cyclodextrin derivatives on alpha-chymotryptic degradation and enteral absorption of insulin in rats. Pharm Res 1994; 11:1174–1179.

65. Bird J, Best R, Lewis DA. The encapsulation of insulin in erythrocytes. J Pharm Pharmacol 1983; 35:246–247.

66. Hosny EA, Khan Ghilzai NM, Al-Najar TA, Elmazar MM. Hypoglycemic effect of oral insulin in diabetic rabbits using

pH-dependent coated capsules containing sodium salicylate without and with sodium cholate. Drug Dev Ind Pharm 1998; 24:307–311.

67. Touitou E, Rubinstein A. Targeted enteral delivery of insulin to rats. Int J Pharm 1986; 30:95–99.

68. Morishita I, Morishita M, Takayama K, Machida Y, Nagai T. Enteral insulin delivery by microspheres in three different formulations using Eudragit® L100 and S100. Int J Pharm 1993; 91:29–37.

69. Eaimtrakarn S, Itoh Y, Kishimoto J, Yoshikawa Y, Shibata N, Takada K. Retention and transit of intestinal mucoadhesive films in rat small intestine. Int J Pharm 2001; 224:61–67.

70. Trenktrog T, Müller BW, Specht FM, Seifert J. Enteric coated insulin pellets: development, drug release and in vivo evaluation. Eur J Pharm Sci 1996; 4:323–329.

71. Hosny EA, Al-Shora HI, Elmazar MA. Oral delivery of insulin from enteric-coated capsules containing sodium salicylate: effect on relative hypoglycemia of diabetic beagle dogs. Int J Pharm 2002; 237:71–76.

72. Kim YH, Bae YH, Kim SW. pH/temperature-sensitive polymers for macromolecular drug loading release. J Control Release 1994; 28:143–152.

73. Serres A, Baudys M, Kim SW. Temperature and pH-sensitive polymers for human calcitonin delivery. Pharm Res 1996; 13:196–201.

74. Ramkissoon-Ganorkar C, Baudys M, Kim SW. Effect of ionic strength on the loading efficiency of model polypeptide/protein drugs in pH-/temperature-sensitive polymers. J Biomater Sci Polym Ed 2000; 11:45–54.

75. Ramkissoon-Ganorkar C, Liu F, Baudys M, Kim SW. Effect of molecular weight and polydispersity on kinetics of dissolution and release from ph/temperature-sensitive polymers. J Biomater Sci Polym Ed 1999; 10:1149–1161.

76. Ramkissoon-Ganorkar C, Liu F, Baudys M, Kim SW. Modulating insulin-release profile from pH/thermosensitive polymeric beads through polymer molecular weight. J Control Release 1999; 59:287–298.

77. Torres-Lugo M, Garcia M, Record R, Peppas NA. Physico-chemical behavior and cytotoxic effects of poly(methacrylic acid-g-ethylene glycol) nanospheres for oral delivery of proteins. J Control Release 2002; 80:197–205.

78. Kim B, Peppas NA. Synthesis and characterization of pH-sensitive glycopolymers for oral drug delivery systems. J Biomater Sci Polym Ed 2002; 13:1271–1281.

79. Lowman AM, Peppas NA. Solute transport analysis in pH-responsive, complexing hydrogels of poly(methacrylic acid-g-ethylene glycol). J Biomater Sci Polym Ed 1999; 10:999–1009.

80. Lowman AM, Morishita M, Kajita M, Nagai T, Peppas NA. Oral delivery of insulin using pH-responsive complexation gels. J Pharm Sci 1999; 88:933–937.

81. Luessen HL, Lehr CM, Rentel CO, Noach ABJ, De Boer AG, Verhoef JC, Junginger HE. Bioadhesive polymers for the peroral delivery of peptide drugs. J Control Release 1994; 29:329–338.

82. Tozaki H, Komoike J, Tada C, Maruyama T, Terabe A, Suzuki T, Yamamoto A, Muranishi S. Chitosan capsules for colon-specific drug delivery: improvement of insulin absorption from the rat colon. J Pharm Sci 1997; 86:1016–1021.

83. Park SY, Bae YH. Novel pH-sensitive polymers containing sulfonamide group. Macromol Rapid Commun 1999; 20:269–273.

84. Kang SI, Bae YH. pH-Induced solubility transition of sulfonamide based polymers. J Control Release 2002; 80:145–155.

85. Kang SI, Bae YH. pH-induced volume-phase transition by reversible crystal formation. Macromolecules 2001; 34:8173–8178.

86. Kang SI, Na K, Bae YH. Sulfonamide containing polymers: A new class of pH-sensitive polymers and gels. Macromol Symp 2001; 172:149–156.

87. Bell PH, Richard O, Robin JR. Studies in chemotherapy. VII. A theory of the relation of structure to activity of sulfanilamide type compounds. J Am Chem Soc 1942; 64:2905–2917.

88. Bartlett MS, Shaw MM, Smith JW, Meshnick SR. Efficacy of sulfamethoxypyridazine in a urine model of pneumocystis

carinii pneumonia. Antimicrob Agents Chemother 1998; 42: 934–935.

89. Boriack PA, Christianson DW, Kingery-Wood J, Whitesides JM. Secondary interactions significantly removed from the sulfonamide binding pocket of carbonic anhydrase II influence inhibitor binding constants. J Med Chem 1995; 38: 2286–2291.

90. Han SK, Na K, Bae YH. Sulfonamide based pH-sensitive polymeric micelles: Physicochemical characteristics and pH-dependent aggregation. Colloids Surfaces A: Physicochem Eng Aspects 2003; 214:49–59.

91. Na K, Bae YH. Self-assembled hydrogel nanoparticles responsive to tumor extracellular pH from pullulan derivative/ sulfonamide conjugate: Characterization, aggregation and adriamycin release in vitro. Pharm Res 2002; 19:681–688.

92. Na K, Lee ES, Bae YH. Adriamycin loaded pullulan acetate/sulfonamide conjugate nanoparticles responding to tumor pH: pH-dependent cell interaction, internalization and cytotoxicity in vitro. J Control Release 2003; 87:3–13.

93. Yuan F, Leunig M, Huang SK, Berk DA, Papahadjopoulos D, Jain RK. Microvascular permeability and interstitial penetration of sterically stabilized (stealth) liposomes in a human tumor xenograft. Cancer Res 1994; 54:4564–4568.

94. Lee ES, Shin HJ, Na K, Bae YH. Poly(l-histidine)-PEG block copolymer micelles and pH-induced destabilization. J Control Release 2003; 90:363–374.

95. Lee ES, Na K, Bae YH. Polymeric micelle for tumor pH and folate-mediated targeting. J Control Release 2003; 91: 103–113.

96. Lynn DM, Langer R. Degradable poly(β-amino esters): synthesis, characterization, and self-assembly with plasmid DNA. J Am Chem Soc 2000; 122:10761–10768.

97. Lynn DM, Amiji MM, Langer R. pH-responsive biodegradable polymer microspheres: rapid release of encapsulated material within the range of intracellular pH. Angew Chem Int Ed 2001; 40:1707–1710.

98. Potineni A, Lynn DM, Langer R, Amiji MM. Poly(ethylene oxide)-modified poly(β-amino ester) nanoparticles as a pH-sensitive biodegradable system for paclitaxel delivery. J Control Release 2003; 86:223–234.

99. American Society of Health-System Pharmacists. AHFS 97 Drug Information. Besthesda, MD: American Society of Health-System Pharmacists, Inc., 1997; 557–562.

100. Pack DW, Putnam D, Langer R. Design of imidazole-containing endosomolytic biopolymers for gene delivery. Biotech Bioeng 2000; 67:217–223.

101. Drummond DC, Zignani M, Leroux JC. Current status of pH-sensitive liposomes in drug delivery. Prog Lipid Res 2000; 39:409–460.

102. Seki K, Tirrell DA. pH-dependent complexation of poly (acrylic acid) derivatives with phospholipid vesicle membrane. Macromolecules 1984; 17:1692–1698.

103. Kitano H, Akatsuka Y, Ise N. pH-responsive liposomes which contain amphiphiles prepared by using lipophilic radical initiator. Macromolecules 1991; 24:42–46.

104. Chen T, Choi LS, Einstein S, Klippenstein MA, Scherrer P, Cullis PR. Proton-induced permeability and fusion of large unilamellar vesicles by covalently conjugated poly (2-ethylacrylic acid). J Liposome Res 1999; 9:387–405.

105. Thomas JL, You H, Tirrell DA. Tuning the response of a pH-sensitive membrane switch. J Am Chem Soc 1995; 117: 2949–2950.

106. Schroeder UKO, Tirrell DA. Structural reorganization of phosphatidylcholine vesicle membranes by poly(2-ethylacrylic acid). Influence of the molecular weight of the polymer. Macromolecules 1989; 22:765–769.

107. Borden KA, Eum KM, Langley KH, Tirrell DA. Interactions of synthetic polymers with cell membranes and model membrane systems. On the mechanism of polyelectrolyte-induced structural reorganization in thin molecular films. Macromolecules 1987; 20:454–456.

108. Chung JC, Gross DJ, Thomas JL, Tirrell DA, Opsahl-Ong LR. pH-sensitive cation-selective channels formed by a simple

synthetic polyelectrolyte in artificial bilayer membranes. Macromolecules 1996; 29:4636–4641.

109. Kono K, Henmi A, Takagishi T. Temperature-controlled interaction of thermosensitive polymer-modified cationic liposomes with negatively charged phospholipid membranes. Biochim Biophys Acta 1999; 1421:183–197.

110. Kono K, Zenitani K, Takagishi T. Novel pH-sensitive liposomes: liposomes bearing a poly(ethylene glycol) derivative. Biochim Biophys Acta 1994; 1193:1–9.

111. Kono K, Igawa T, Takagishi T. Cytoplasmic delivery of calcein mediated by liposomes modified with a pH-sensitive poly(ethylene glycol) derivative. Biochim Biophys Acta 1997; 1325:143–154.

112. Ringsdorf H, Sackmann E, Simon J, Winnik FM. Interactions of liposomes and hydrophobically-modified poly-(N-isopropylacrylamides): an attempt to model the cytoskeleton. Biochim Biophys Acta 1993; 1153:335–344.

113. Polozova A, Winnik FM. Contribution of hydrogen bonding to the association of liposomes and an anionic hydrophobically modified poly(N-isopropylacrylamide). Langmuir 1999; 15: 4222–4229.

114. Bhattacharya S, Moss RA, Ringsdorf H, Simon J. A polymeric "flippase" for surface-differentiated dipalmitoylphosphatidylcholine liposomes. J Am Chem Soc 1993; 115:3812–3813.

115. Franzin CM, MacDonald PM, Polozova A, Winnik FM. Destabilization of cationic lipid vesicles by an anionic hydrophobically modified poly(N-isopropylacrylamide) copolymer: a solid-state ^{31}P NMR and ^{2}H NMR study. Biochim Biophys Acta 1998; 1415:219–234.

116. Winnik FM, Adronov A, Kitano H. Pyrene-labeled amphiphilic poly-(N-isopropylacrylamides) prepared by using a lipophilic radical initiator: synthesis, solution properties in water, and interactions with liposomes. Can J Chem 1995; 73:2030–2040.

117. Zignani M, Drummond DC, Meyer O, Hong K, Leroux JC. In vitro characterization of a novel polymeric-based pH-sensitive liposome system. Biochim Biophys Acta 2000; 1463: 383–394.

118. Roux E, Lafleur M, Lataste E, Moreau P, Leroux JC. On the characterization of pH-sensitive liposome/polymer complexes. Biomacromolecules 2003; 4:240–248.

119. Francis MF, Dhara G, Winnik FM, Leroux JC. In vitro evaluation of pH-sensitive polymer/niosome complexes. Biomacromolecules 2001; 2:741–749.

120. Juliano RL, Alahari S, Yoo H, Kole R, Cho MJ. Antisense pharmacodynamics: critical issues in the transport and delivery of antisense oligonucleotides. Pharm Res 1999; 16:494–502.

121. Braasch DA, Corey DR. Novel antisense and peptide nucleic acid strategies for controlling gene expression. Biochemistry 2002; 41:4503–4510.

122. Uster PS, Deamer DW. pH-dependent fusion of liposomes using titratable polycations. Biochemistry 1985; 24:1–8.

123. Midoux P, Monsigny M. Efficient gene transfer by histidylated polylysine/pDNA complexes. Bioconjugate Chem 1999; 10:406–411.

124. Pichon C, Roufaï MB, Monsigny M, Midoux P. Histidylated oligolysines increase the transmembrane passage and the biological activity of antisense oligonucleotides. Nucleic Acids Res 2000; 28:504–512.

125. Putnam D, Zelikin AN, Izumrudov VA, Langer R. Polyhistidine-PEG: DNA nanocomposites for gene delivery. Biomaterials 2003; 24:4425–4433.

126. Tang MX, Szoka FC. The influence of polymer structure on the interactions of cationic polymers with DNA and morphology of the resulting complex. Gene Ther 1997; 4:823–832.

127. Ogris M, Steinlein P, Kursa M, Mechtler K, Kircheis R, Wagner E. The size of DNA/transferrin-PEI complexes is an important factor for gene expression in cultured cells. Gene Ther 1998; 5:1425–1433.

128. Dunlap DD, Maggi A, Soria MR, Monaco L. Nanoscopic structure of DNA condensed for gene delivery. Nucleic Acids Res 1997; 25:3095–3101.

129. Boussif O, Lezoulach F, Zanta MA, Mergny MD, Scherman D, Demeneix B, Behr JP. A versatile vector for gene and oligonu-

cleotide transfer into cells in culture and in vivo polyethylenimine. Proc Natl Acad Sci USA 1995; 92:7297–7301.

130. Kircheis R, Schüller S, Brunner S, Ogris M, Heider KH, Zauner W, Wagner E. Polycation-based DNA complexes for tumor-targeted gene delivery in vivo. J Gene Med 1999; 1:111–120.

131. Kircheis R, Wightman L, Schreiber A, Robitza B, Röessler V, Kursa M, Wagner E. Polyethylenimine/DNA complexes shielded by transferrin target gene expression to tumors after systemic application. Gene Ther 2001; 8:28–40.

132. Ferruti P, Marchisio MA, Barbucci R. Synthesis, physicochemical properties and biomedical applications of poly (amido-amine)s. Polymer 1985; 26:1336–1348.

133. Duncan R, Ferruti P, Sgouras D, Tuboku-Metzger A, Ranucci E, Bignotti F. A polymer-Triton X-100 conjugate capable of pH-dependent red blood cell lysis: a model system illustrating the possibility of drug delivery within acidic intracellular compartments. J Drug Target 1994; 2:341–347.

134. Ferruti P, Manzoni S, Richardson SCW, Duncan R, Pattrick NG, Mendichi R, Casolaro M. Amphoteric linear poly(amido-amine)s as endosomolytic polymers: correlation between physico-chemical and biological properties. Macromolecules 2000; 21:7793–7800.

135. Richardson SCW, Ferruti P, Duncan R. Poly(amidoamine)s as potential endosomolytic polymers: evaluation in vitro and body distribution in normal and tumour bearing animals. J Drug Target 1999; 6:391–404.

136. Ranucci E, Spagnoli G, Ferruti P, Sgouras D, Duncan R. Poly (amidoamine)s with potential as drug carriers: degradation and cellular toxicity. J Biomater Sci Polym Ed 1991; 2:303–315.

137. Richardson SCW, Pattrick NG, Man YKS, Ferruti P, Duncan R. Poly(amidoamine)s as potential nonviral vectors: Ability to form interpolyelectrolyte complexes and to mediate transfection in vitro. Biomacromolecules 2001; 2:1023–1028.

138. Pattrick NG, Richardson SCW, Casolaro M, Ferruti P, Duncan R. Poly(amidoamine)-mediated intracytoplasmic

delivery of ricin A-chain and gelonin. J Control Res 2001; 13:225–232.

139. Murthy N, Robichaud JR, Tirrell DA, Stayton PS, Hoffman AS. The design and synthesis of polymers for eukaryotic membrane disruption. J Control Release 1999; 61: 137–143.

140. Murthy N, Campbell J, Fausto N, Hoffman AS, Stayton PS. Bioinspired polymeric carriers that enhance intracellular delivery of biomolecular therapeutics. Bioconj Chem 2003; 14:412–419.

141. Lackey CA, Murthy N, Press OW, Tirrell DA, Hoffman AS, Stayton PS. Hemolytic activity of pH-responsive polymer-streptavidin bioconjugates. Bioconj Chem 1999; 10:401–405.

142. Murthy N, Chang I, Stayton PS, Hoffman AS. pH-Sensitive hemolysis by random copolymers of alkyl acrylates acrylic acid. Macromol Symp 2001; 172:49–55.

143. Kyriakides TR, Cheung CY, Murthy N, Bornstein P, Stayton PS, Hoffman AS. pH-Sensitive polymers that enhance intracellular drug delivery in vivo. J Control Release 2002; 17:295–303.

144. Kusonwiriyawong C, van de Wetering P, Hubbell JA, Merkle HP, Walter E. Evaluation of pH-dependent membrane-disruptive properties of poly(acrylic acid) derived polymers. Eur J Pharm Biopharma 2003; 56:237–246.

4

Hydrogels for Oral Administration

SEONG HOON JEONG, YOURONG FU,
and KINAM PARK
Departments of Pharmaceutics and
Biomedical Engineering,
Purdue University, West Lafayette,
Indiana, U.S.A.

1. INTRODUCTION

1.1. Hydrogels

A hydrogel is a three-dimensional network of hydrophilic polymer chains held together by chemical bonds (i.e., covalent bonds) or physical bonding (e.g., hydrogen bonding, ionic interaction, and hydrophobic association) (1). Due to the hydrophilic nature of the polymer chains, the network is able to absorb water within its structure and swell without dissolving while maintaining the overall structure. Table 1 lists some examples of commonly used polymers that can be cross-linked to make hydrogels.

195

Table 1   Examples of Commonly Used Polymers for Making Hydrogels

Polymer name	Polymer structure
Poly(acrylic acid)	
Polyacrylamide	
Poly(N-isopropyl-acrylamide)	
Poly(ethylene oxide)	
Poly(hydroxyethyl methacrylate)	
Polyvinylpyrrolidone	

(Continued)

Table 1 (*Continued*)

Polymer name	Polymer structure

Poly(vinyl alcohol)

Carboxymethyl-
cellulose

Hydrogels can be classified into different groups, just for convenience, based on the source of origin, chemical structure, preparation method, electric charge, physical structure, crosslinking, or function. There are two types of crosslinking: chemical and physical crosslinking. In the chemical crosslinking, all polymer chains are crosslinked to each other by covalent bonds, and thus, strictly speaking, each hydrogel is one molecule. For this reason, a hydrogel is sometimes called a "supermacromolecule." While chemical crosslinking is well defined by each chemical bond, physical crosslinking is made through multiple, simultaneous interactions of weaker bonding, such as hydrogen bonding and hydrophobic interactions. The area of such physical bonding among laterally associated polymer chains is known as a junction zone (2).

Hydrogels have been used widely in the development of drug delivery systems and biomedical devices. The water content in a hydrogel is at least 10% of the total weight. If the water content exceeds 95% of the total weight, the hydrogel is called superabsorbent. Due to the presence of water, the interfacial tension between the hydrogel surface and aqueous solution is very low, and this is one of the reasons why protein adsorption and cell adhesion to the hydrogel surface is significantly reduced. The presence of water also minimizes the irritation to the surrounding tissue when a hydrogel is implanted into the body. The solid content in a swollen hydrogel ranges from about 90% to less than 1%, and thus the space, also known as an effective pore size, between polymer chains also varies. Hydrogels with an effective pore size in the ranges of 10–100 nm and 100 nm–10 μm are called microporous and macroporous hydrogels, respectively.

In the presence of abundant water, hydrogels absorb water and swell. This swelling process is the same as the dissolution of non-crosslinked hydrophilic polymers. Experimentally it is much easier to measure the weight of a swollen hydrogel than the volume, and thus the swelling ratio of hydrogels is usually expressed based on weights. The swelling ratio is defined as:

$$\text{Swelling ratio} = \frac{(\text{Weight of swollen gel})}{(\text{Weight of dried gel})}$$

One of the unique properties of hydrogels is that due to isotropic swelling, the original shape can be maintained during and after swelling.

1.2. Superporous Hydrogels

Hydrogels with the pore sizes larger than 10 μm are known as superporous hydrogels. The uniqueness of the superporous hydrogels is that the pores are interconnected to form an open cell structure, allowing extremely fast absorption of water into the center of the dried gel by capillary force. Figure 1 shows fast swelling property of a superporous hydrogel. The dried superporous hydrogel in Figure 1 swelled to 100 times of its dried weight (or volumes) within 30 s. This is something not possible with conventional dried hydrogels that do not have interconnected pores. Mechanical strength of a swollen gel can be improved by making superporous hydrogel composites (3,4) or superporous hydrogels with interpenetrating networks with the second polymer (5). The elastic property is useful in making mechanically strong superporous hydrogels more resilient to compression and elongation (6). The swollen

Figure 1 Swelling property of a dried hydrogel (left) to a larger size after swelling while maintaining the original shape (right) in aqueous solution.

hydrogel can be stretched to almost twice the original length without breaking.

2. APPLICATION OF HYDROGELS IN ORAL DRUG DELIVERY

Among all the routes of drug administration, oral drug delivery has been regarded as the most convenient method of drug administration. Drug delivery technologies are quite advanced enough to design dosage forms that can deliver drugs at a relatively constant rate for long periods of time ranging from days to months and even years. The time period of oral drug delivery, however, has been limited to a day. The maximum period of time available for drug absorption from an orally administered formulation is determined by the total transit time from mouth to colon, which is usually less than 24 h. Table 2 lists the residence time of both liquid and solid foods in each segment of the gastrointestinal (GI) tract (7). One of the properties of the GI tract is that the food contents remain in each part of the GI tract for different time periods. The values in Table 2 should be regarded as relative, not absolute, and are intended to emphasize general differences among different parts of the GI tract. Gastric emptying time ranges from 10 min to about 3 h depending on many factors,

Table 2 Transit Time in Each Segment of the GI Tract

	Type of food	
Segment	Liquid	Solid
Stomach	10–30 min	1–3 h
Duodenum	< 60 s	< 60 s
Jejunum and ileum	3 h ± 1.5 h	4 h ± 1.5 h
Colon		20–50 h

and when foods are present in the stomach, they have a tendency to delay the gastric emptying.

Development of once-a-day oral dosage forms is still a big challenge. Most drugs cannot be delivered for 24 h by a single administration, since oral dosage forms pass through the small intestine where most drug absorption occurs within a few hours. Once-a-day formulation is possible for some drugs, such as phenylpropanolamine and nifedifine, since they are absorbed well throughout the GI tract. Extending the residence time long enough for a drug to be absorbed from the upper small intestine remains an extensive research topic.

Recent advances in smart hydrogels have made it possible to exploit the changes in physiological uniqueness in different regions of the GI tract for the improved drug absorption as well as patient's convenience and compliance. Different hydrogels that are ideal in delivery of drugs to certain regions in the GI tract from oral cavity to colon are shown in Figure 2.

2.1. Oral Cavity

2.1.1. Fast-Melting Tablets

For more than a decade, fast-melting (also called fast-dissolving or fast-disintegrating) tablet technologies have been steadily advancing in the development of patient-friendly dosage forms. The fast-melting dosage forms, which can be administered easily without any water, are suitable for all age groups, but in particular for children, the elderly, and those who have difficulty in swallowing conventional tablets and capsules. The initial success of the first fast-melting

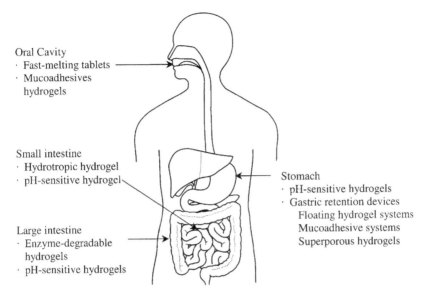

Oral Cavity
· Fast-melting tablets
· Mucoadhesives
  hydrogels

Small intestine
· Hydrotropic hydrogel
· pH-sensitive hydrogel

Large intestine
· Enzyme-degradable
  hydrogels
· pH-sensitive hydrogels

Stomach
· pH-sensitive hydrogels
· Gastric retention devices
  Floating hydrogel systems
  Mucoadhesive systems
  Superporous hydrogels

Figure 2 Different hydrogels and hydrogel formulations that can be used in different sites in the GI tract.

tablet formulation led to the development of different technologies. There are mainly three different technologies: freeze-drying, sublimation or heat molding, and direct compression. Freeze-drying technology produces tablets that can dissolve in a few seconds, while tablets produced by the sublimation and molding technology take a little bit longer to dissolve. The two technologies, however, are quite expensive and the prepared tablets are not mechanically strong. For this reason, direct compression technologies, which allow low cost of production and good mechanical roperties, are preferred, even though the melting time takes usually longer than 10 s, and sometimes more than 20 s.

To ensure the tablet's fast-melting property, water must be quickly absorbed into the tablet matrix. Current fast-melting tablet technologies are based on maximizing the porous structure of the tablet matrix and incorporating appropriate disintegrating agents and/or highly water-soluble excipients in the tablet formulation (8,9). Recently, microparticles of superporous hydrogels with fast swelling and super absorbent properties were applied to develop fast-melting tablets by direct compression method (10). The size and shape of superporous hydrogel particles can be varied. Superporous hydrogels can be ground in dry state to make porous super disintegrant microparticles. The hydrogel struts in the superporous hydrogels have numerous pores smaller than 1 mm, which are connected to each other to maintain open pore structures. This unique porous structure allows for transport of water through capillary forces, resulting in an extremely fast wicking effect into the tablet core. Tablets prepared by direct compression in the presence of superporous hydrogel microparticles disintegrate in about 10–20 s due to the fast uptake of water into the core of the tablet (11). Since a lubricant has to be added for mass production, the effect of different lubricants on the disintegration time was examined. It was found that the presence of lubricants, especially hydrophobic lubricants, significantly delays the disintegration time. As shown in Figure 3,

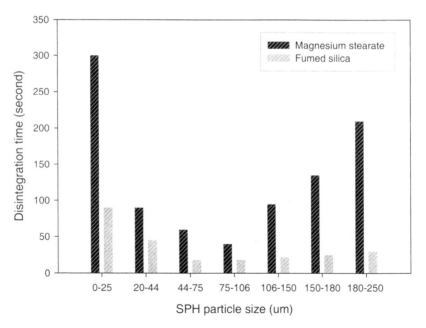

Figure 3   Effect of SPH particle size on the disintegration time of ketoprofen fast-disintegrating tablets.

the disintegration time is delayed to more than 100 s if magnesium stearate was used. On the other hand, fumed silica did not cause any such delays. In both cases, however, there were optimum particle size ranges for the fastest disintegration.

2.1.2. Buccal Drug Delivery Systems

In general the buccal tissue is regarded as more permeable to drugs than the skin. Buccal drug delivery may have faster onset of drug action and bypass the hepatic first-pass metabolism and degradation in the GI tract (12). One of the useful systems for increased patient compliance is bioadhesive dosage forms. Polymers that have been tested and used as mucoadhesives to the buccal tissue are hydroxyethylcellulose (HEC), hydroxypropylcellulose (HPC), polyvinylpyrrolidone, and poly(vinyl alcohol) (13), and mixtures of different polymers, such as HPC, hydroxypropylmethylcellulose (HPMC), karaya gum, and PEG 400 (14), or HPC and Carbopol 934 (15).

Mixing different polymers presents an advantage, since it can result in mucoadhesive hydrogel formulations even when each polymer alone does not form a hydrogel. For example, combining xanthan gum and locust bean gum resulted in formation of hydrogels, even though they do not form hydrogels individually (16). The formation of hydrogels was based on the double helix structure of xanthan gum and the straight chain of locus bean gum (14).

2.2. Stomach

The pH in the stomach, which is less than 3, is quite different from the neutral pH in the small or large intestine. This difference is large enough to be exploited in drug delivery using polyelectrolyte hydrogels. All the pH-sensitive hydrogels have pendant acidic (carboxylic or sulfonic acids) or basic (ammonium salts) groups that accept or release protons in response to changes in environmental pH. Poly(acrylic acid) becomes ionized as pH increases, while poly (N,N'-diethylaminoethyl methacrylate) (PDEAEM) becomes ionized as pH decreases (17).

Polycationic hydrogels swell less at neutral pH than at acidic pH; thus they minimize drug release at higher pH. This property was applied to mask poor tastes of some drugs in the neutral pH environment of the mouth. When caffeine was loaded into hydrogels made of copolymers of methyl methacrylate and N,N'-dimethylaminoethyl methacrylate (DMAEM), it was not released at neutral pH, but released with zero-order at pH 3–5 where DMAEM became ionized (18).

Polycationic hydrogels in the form of semi-interpenetrating polymer networks (semi-IPN) have also been used for drug delivery to the stomach. Semi-IPN of crosslinked chitosan and PEO showed more swelling under acidic conditions. This type of hydrogels could be applied for local delivery of antibiotics, such as amoxicillin and metronidazole, in the stomach for the treatment of *Helicobacter pylori* (19).

2.3. Gastric Retention Devices

Controlled-release drug delivery systems having prolonged gastric residence time have been studied extensively in both academia and industry (7). For many drugs that are mainly absorbed from the upper small intestine, such as drugs with absorption windows, controlled release in the stomach would result in improved bioavailability. Gastric emptying of oral dosage forms is known to be influenced most significantly by two main parameters: the physical properties (e.g., size and density) of the oral dosage form and the presence of food in the stomach (e.g., fasted or fed state). These parameters were exploited by various methods in the development of gastric retention devices.

2.3.1. Intragastric Floating Systems (Low-Density Systems)

In this system, the device is designed to float on top of the gastric juice due to its density being lower than that of water (7). One example of this type of devices is a single-unit hydrodynamically balanced system (HBS) which is composed of a drug, a hydrogel and other excipients (20). Commonly used excipients are gel-forming or highly swellable cellulose-type hydrocolloids and polysaccharides.

2.3.2. Mucoadhesive Systems

A mucoadhesive system is an oral dosage form that is designed to stick to the mucosal surface of the stomach. If the dosage form can stick to the mucosal surface, then in the ideal case its gastric residence time would be increased until it is removed by turnover of mucins. The best mucoadhesive known so far is crosslinked poly(acrylic acid) (21), which is commercially available under polycarbophil (22) and Carbopol. They are highly mucoadhesive at pH 1–3 of the stomach. This is because poly(acrylic acid) interacts with mucins and other biomolecules through numerous simultaneous hydrogen bondings provided by carboxyl acid groups of poly(acrylic acid) at acidic pH (7). This non-specific mucoadhesive property turns out to be a deterrent in developing a useful mucoadhesive formulation, since the gel sticks to almost all surfaces it is in contact with. It is necessary to find polymers with a specific selectivity that they can adhere only to the mucus layer in the stomach.

2.3.3. Superporous Hydrogel Systems

Superporous hydrogels swell to a very large size with the swelling ratio being as high as a few hundreds or more. They also possess the fast swelling property that is required to avoid premature emptying by the housekeeper waves (3,4). This fast swelling property was applied to develop a gastric retention device that can remain in the stomach due to its large size after swelling.

Superporous hydrogels can be made to have high mechanical strength even after swelling. The mechanical strength of swollen superporous hydrogels can be improved by adding composite materials (4). One of the first useful composite materials was Ac-Di-Sol, which is crosslinked carboxymethylcellulose sodium with a hollow microparticulate shape. The hollow microparticles provide physical entanglements of polymer chains around the microparticles. This increased the effective crosslinking density without making the super-porous hydrogels too brittle. When tested in dogs, superporous hydrogel composites showed long-term gastric retention, ranging from several hours to a day (3,4). Biodegradable superporous hydrogels can be prepared by adding biodegradable crosslinkers, such as functionalized albumin (23).

2.4. Small Intestine

The small intestine, which is about 7 m long, is divided into the duodenum, the jejunum, and the ileum. The pH of the GI tract increases to an average of 6.9 in the duodenum, rising to 7.5 in the distal small intestine. The significant pH change from that in the stomach to around 7 has been exploited to develop the pH-dependent squeezing hydrogel system (24). pH-responsive hydrogel particles based on poly-(acrylic acid) were used to develop pH-dependent silicone matrix (25). The release patterns of several model drugs having different aqueous solubilities and partitioning properties were correlated with the pH-dependent swelling pattern of the hydrogel particles. At pH 1.2, the network swelling was low and the release was limited to an initial burst. At pH 6.8, the network became ionized and higher swelling resulted in increased release. Sugar-based hydrogels, which are biodegradable copolymeric hydrogels based on sucrose acrylate, N-vinyl-2-pyrrolidinone, and acrylic acid, were investigated for oral drug delivery (26). A drug entrapped in the hydrogel showed a faster release profile in intestinal fluid than in the gastric fluid as a result of higher swelling in the intestinal fluid.

Hydrogels, especially superporous hydrogels, can also be used to develop peptide delivery systems through oral routes. Peptide drugs have been administered mostly by the parenteral route. Recently, peroral peptide delivery systems using superporous hydrogels were developed (27). Superporous hydrogels used in the systems increased their volume by about 200 times. Due to the volume increase, the gels were able to stick to the intestinal gut wall mechanically and deliver the incorporated drug directly to the gut wall. The acid groups of poly(acrylic acid) of the superporous hydrogels is known to extract calcium ions from the gut wall to result in an opening of the tight junctions and also to deactivate the deleterious gut enzymes. After the peptide drugs have been delivered across the gut wall, the superporous hydrogels become over-hydrated and their structure become broken down by the peristaltic forces of the gut, and the remnants of the delivery systems are easily excreted together with the feces.

2.5. Large Intestine and Colon

The large intestine is divided into several parts: cecum, appendix, colon, and rectum. The major function of the colon is to absorb water and electrolytes. Primarily, absorption occurs in the proximal half of the colon (28). The surface area of colon is small, but due to the large residence time in the colon, drug absorption can be significant for some drugs. One of the important applications of drug delivery to the colon is the delivery of protein and peptide drugs including vaccines. A number of different methods, which utilize specific enzymes, different pH, and transit time of GI tract in different parts of the body individually or in combination, have been used for colon-specific drug delivery. Microbial enzymes present predominantly in the colon can be used for site-specific drug delivery. pH-sensitive polymers can protect drugs until they reach the colon. Based on the average transit time of 5 h from mouth to colon, a controlled release dosage form that can deliver a drug in the colon can be developed (29). Poly(acrylamide-co-maleic acid) and poly(N-vinyl-2-pyrrolidone-co-acrylamide-co-maleic acid) hydrogels showed a strong pH-dependent drug release behavior; i.e., minimum release at pH 2.0 but maximum release at pH 7.4 (30).

Colon-specific biodegradable hydrogels were prepared using azoaromatic crosslinkers (30). As shown in Figure 4, the hydrogels do not swell very well at low pH, so the release of drug will be minimized in the stomach. As the hydrogels move down the GI tract, they swell at higher pH due to ionization of the carboxylic acid groups. Since azoreductase

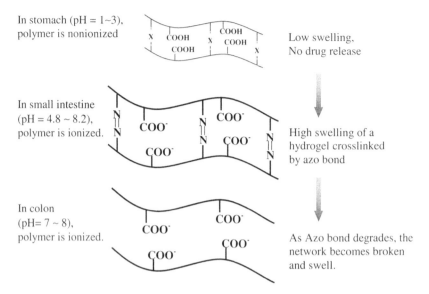

In stomach (pH = 1~3), polymer is nonionized

Low swelling, No drug release

In small intestine (pH = 4.8 ~ 8.2), polymer is ionized.

High swelling of a hydrogel crosslinked by azo bond

In colon (pH= 7 ~ 8), polymer is ionized.

As Azo bond degrades, the network becomes broken and swell.

Figure 4 Schematic illustration of colon-specific drug delivery from biodegradable and pH-sensitive hydrogels. The azoaromatic moieties in the crosslinks are designated by $-N=N-$. (From Ref. 30.)

produced by microbial flora of the colon degrades azoaromatic crosslinkers of hydrogels, the hydrogels can be degraded only in the colon (31). The swelling kinetics of hydrogels can be controlled by the polymer composition and by the crosslinking density. Knowing the transit time, pH in the GI tract, diffusion rate of drugs through hydrogels, the rate of enzymatic degradation, and degree of swelling, one may be able to design drug delivery devices with desired release profiles at the desired time (32).

3. HYDROGELS FOR SPECIFIC APPLICATIONS

3.1. Hydrotropic Hydrogels for Delivery of Poorly Soluble Drugs

Drug molecules have to be dissolved in the GI tract to be absorbed. Poor aqueous solubility of hydrophobic drugs, which usually results in low bioavailability, has been one of the limitations in designing clinically useful oral drug delivery formulations. Hydrotropes have been used to increase the aqueous solubility of poorly soluble drugs by 2–4 orders of magnitude (33,34). Application of low molecular weight hydrotropes in drug delivery, however, has not been practical, because it may result in absorption of a significant amount of hydrotropes themselves into the body along with the drug. One approach to prevent co-absorption of hydrotropes from the GI tract after oral administration is to make polymeric hydrotropic agents (hydrotropic polymers). Various polymeric solubilizing systems, such as polymeric micelles and micelle-like aggregates from polymeric amphiphilic agents, have been used to increase the solubility of poorly water-soluble drugs (35,36). Hydrotropic polymers and hydrogels are expected to provide an alternative method to increase the aqueous solubility of poorly soluble drugs for oral administration.

Hydrotropic hydrogels were examined to improve the aqueous solubility of paclitaxel (37). The loading of paclitaxel into the hydrogels was carried out by solubilizing paclitaxel in aqueous solutions of 2-(4-vinylbenzyloxy)-N-picolylnicotinamide) (2-VBOPNA) and 6-(4-vinylbenzyloxy)-N-picolylnicotinamide (6-VBOPNA), followed by the in situ crosslinking reaction to form hydrogels. As shown in Figure 5, paclitaxel solubility in hydrogels increased as the concentration of 2-VBOPNA or 6-VBOPNA used in hydrogel synthesis increased. The paclitaxel solubility was increased up to 1.62 mg/mL in the experiment, which is more than 5000 times that of paclitaxel solubility in water (37).

3.2. Alginate Hydrogels for Oral Vaccine Delivery

Peyer's patches, which are collections of lymphoid tissue containing B and T lymphocytes and macrophages, can be found in duodenum, jejunum, and ileum. These areas may be targets for protein and peptide drug delivery including oral vaccines (38). Lymphoid tissues analogous to Peyer's patches can be found in the large intestine as well. Specialized epithelial cells in the dome region of the patch, called follicle associated

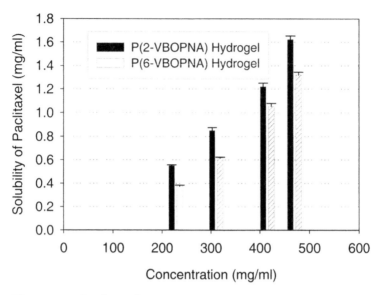

Figure 5   Paclitaxel solubility in *N*-picolylnicotinamide hydrogels as a function of the concentrations of 2-VBOPNA and 6-VBOPNA. (From Ref. 37.)

epithelial or microfold cells, pick up antigens and transport them to the underlying lymphoid tissue (38).

Immunization has been relied on in the induction of humoral immunity by parenteral administration of vaccines. Antibodies from this do not necessarily reach the mucosal surfaces of the GI tract where most infectious agents encounter the host first. Secretory IgA (sIgA) is the dominant antibody isotype present at mucosal sites. Since sIgA inhibits the attachment of bacteria and viruses to mucosa and neutralizes toxins, mucosal immunity can provide the first line of immunological defense. However, induction of immunity at mucosal surfaces needs administration of vaccines directly to the mucosal site.

Since vaccines are quite susceptible to low pH in the stomach and enzymatic degradation in the GI tract, various methods have been tried to improve the vaccine delivery. Alginate microspheres have been examined to deliver vaccines by oral administration (38). Vaccine-containing alginate hydrogel microspheres were shown to be effective for oral vaccination in several animal species. This method has several advantages as compared with other methods. The alginate vaccine delivery system removes the use of organic solvents or high temperatures, which are often needed for the preparation of microparticles using other methods. Since an aqueous environment is used throughout the preparation process, this method can also be used with live bacteria and viruses. The alginate microspheres are able to protect vaccines from degradation in the GI tract. Since microparticles with a diameter 1–10 μm are known to be absorbed well by the Peyer's patches, special attention has been paid to making small-sized particles (39).

REFERENCES

1. Kamath KR, Park K. Biodegradable hydrogels in drug delivery. Adv Drug Delivery Rev 1993; 11:59–84.

2. Burchard W. Networks in nature. Br Polym J 1985; 17: 154–163.

3. Chen J, Blevins WE, Park H, Park K. Gastric retention properties of superporous hydrogel composites. J Contr Rel 2000; 64:39–51.

4. Chen J, Park K. Synthesis and characterization of superporous hydrogel composites. J Contr Rel 2000; 65:73–82.

5. Qiu Y, Park K. Superporous IPN hydrogels having enhanced mechanical properties. Pharm Sci Tech 2003; 4:406–412.

6. Omidian H, Park K. Experimental design for the synthesis of polyacrylamide superporous hydrogels. J Bioact Compat Polym 2002; 17:433–450.

7. Hwang S-J, Park H, Park K. Gastric retentive drug delivery systems. Crit Rev Ther Drug Carrier Syst 1998; 15:243–284.

8. Habib W, Khankari R, Hontz J. Fast-dissolve drug delivery systems. Crit Rev Ther Drug Carrier Syst 2000; 17:61–72.

9. Sastry SV, Nyshadham JR, Fix JA. Recent technological advances in oral drug delivery. A review. Pharm Sci Technol Today 2000; 3:138–145.

10. Yang SY, Fu Y, Jeong SH, Park K. Application of poly(acrylic acid) superporous hydrogel microparticles as a super-disintegrant in fast-disintegrating tablets. J Pharm Pharmacol 2004; 56:429–436.

11. Park K. Superporous hydrogels for pharmaceutical and other applications. Drug Delivery Technol 2002; 2:38–44.

12. Veillard MM, Longer MA, Martens TW, Robinson JR. Preliminary studies of oral mucosal delivery of peptide drugs. J Contr Rel 1987; 6:123–131.

13. Anders R, Merkle HP. Evaluation of laminated muco-adhesive patches for buccal drug delivery. Int J Pharm 1989; 49:231–240.

14. Nagai T, Machida Y. Buccal delivery systems using hydrogels. Adv Drug Delivery Rev 1993; 11:179–191.

15. Ishida M, Machida Y, Nambu N, Nagai T. New mucosal dosage form of insulin. Chem Pharm Bull 1981; 29:810–816.

16. Watanabe K, Yakou S, Takayama K, Machida Y, Nagai T. Factors affecting prednisolone release from hydrogels prepared with water-soluble dietary fibers, xanthan and locust bean gums. Chem Pharm Bull 1992; 40:459–462.

17. Qiu Y, Park K. Environment-sensitive hydrogels for drug delivery. Adv Drug Delivery Rev 2001; 53:321–339.

18. Siegel RA, Falamarzian M, Firestone BA, Moxley BC. pH-controlled release from hydrophobic/polyelectrolyte copolymer hydrogels. J Contr Rel 1988; 8:179–182.

19. Patel VR, Amiji MM. Preparation and characterization of freeze-dried chitosan-poly(ethylene oxide) hydrogels for site-specific antibiotic delivery in the stomach. Pharm Res 1996; 13:588–593.

20. Singh BN, Kim KH. Floating drug delivery systems: an approach to oral controlled drug delivery via gastric retention. J Contr Rel 2000; 63:235–259.

21. Park K, Robinson JR. Bioadhesive polymers as platforms for oral-controlled drug delivery: method to study bioadhesion. Int J Pharm 1984; 19:107–127.

22. Khosla R, Davis SS. The effect of polycarbophil on the gastric emptying of pellets. J Pharm Pharmacol 1987; 39:47–49.

23. Park K. Enzyme-digestible swelling hydrogels as platforms for long-term oral drug delivery: synthesis and characterization. Biomaterials 1988; 9:435–441.

24. Gutowska A, Bark YS, Kwon IC, Bae YH, Cha Y, Kim SW. Squeezing hydrogels for controlled oral drug delivery. J Contr Rel 1997; 48:141–148.

25. Carelli V, Coltelli S, Di Colo G, Nannipieri E, Serafini MF. Silicone microspheres for pH-controlled gastrointestinal drug delivery. Int J Pharm 1999; 179:73–83.

26. Shantha KL, Harding DRK. Synthesis and evaluation of sucrose-containing polymeric hydrogels for oral drug delivery. J Appl Polym Sci 2002; 84:2597–2604.

27. Dorkoosh FA, Verhoef JC, Borchard G, Rafiee-Tehrani M, Verheijden JHM, Junginger HE. Intestinal absorption of human insulin in pigs using delivery systems based on super-porous hydrogel polymers. Int J Pharm 2002; 247:47–55.

28. Kunth K, Amiji M, Robinson JR. Hydrogel delivery systems for vaginal and oral applications. Formulations and biological considerations. Adv Drug Delivery Rev 1993; 11:137–167.

29. Friend DR. Colon-specific drug delivery. Adv Drug Delivery Rev 1991; 7:149–199.

30. Bajpai SK, Sonkusley J. Hydrogels for colon-specific oral drug delivery: synthesis and characterization. J Macromol Sci Pure Appl Chem A 2001; 38:365–381.

31. Akala EO, Kopeckova P, Kopecek J. Novel pH-sensitive hydrogels with adjustable swelling kinetics. Biomaterials 1998; 19:1037–1047.

32. Brondsted H, Kopecek J. Hydrogels for site-specific oral drug delivery: synthesis and characterization. Biomaterials 1991; 12:584–592.

33. Coffman RE, Kildsig DO. Effect of nicotinamide and urea on the solubility of riboflavin in various solvents. J Pharm Sci 1996; 85:951–954.

34. Yalkowsky SH. Solubilization by Complexation. Solubility and Solubilization in Aqueous Media. New York: Oxford University Press, 1999:321–396.

35. Kataoka K, Kwon GS, Yokoyama M, Okano T, Sakurai Y. Block copolymer micelles as vehicles for drug delivery. J Contr Rel 1993; 24:119–132.

36. Allen C, Maysinger D, Eisenberg A. Nano-engineering block copolymer aggregates for drug delivery. Colloid Surf B 1999; 16:3–27.

37. Lee SC, Acharya G, Lee J, Park K. Hydrotropic polymers: synthesis and characterization of polymers containing picolyl-nicotinamide moieties. Macromolecules 2003; 36:2248–2255.

38. Bowersock TL, Hogenesch H, Suckow M, Porter RE, Jackson R, Park H, Park K. Oral vaccination with alginate microsphere systems. J Contr Rel 1996; 39:209–220.

39. Mcghee JR, Mestecky J, Dertzbaugh MT, Eldridge JH, Hirasawa M, Kiyono H. The mucosal immune system: from fundamental concepts to vaccine development. Vaccine 1992; 10:75–88.

5

Hydrogels for the Controlled Release of Proteins

WIM E. HENNINK, CORNELUS F. VAN
NOSTRUM, and DAAN J. A.
CROMMELIN
Department of Pharmaceutics,
Utrecht Institute for Pharmaceutical
Sciences (UIPS), Utrecht University,
Utrecht, The Netherlands

JEROEN M. BEZEMER
OctoPlus Technologies B.V.,
Leiden, The Netherlands

1. INTRODUCTION

Modern biotechnology has resulted in the production of a great variety of pharmaceutically active proteins (1). Recent statistics show that the Food and Drug Administration (FDA) has approved 130 biotechnology-derived protein medicines and vaccines (2). The unfavorable biopharmaceutical properties of these protein drugs, however, have severely hampered their therapeutic and clinical application. First of

215

all, oral administration is hardly possible due to chemical and enzymatic degradation in the gastrointestinal tract. Second, proteins have a short half-life after parenteral administration (e.g., intravenous injection), which makes repeated injections or continuous infusion of the protein necessary to obtain a therapeutic effect (3). To overcome these problems, a large number of delivery systems have been designed and evaluated for the release of proteins (4–6). A frequently investigated polymer involved in the design of controlled-release systems for proteins is poly(lactic-co-glycolic acid), or PLGA. This polymer, however, has some intrinsic drawbacks as a protein-releasing matrix. Organic solvents have to be used to prepare pharmaceutical dosage forms (e.g., microspheres), and a low pH might be generated inside the matrix during degradation (7,8). Both factors adversely affect the structural integrity of the protein to be delivered. Moreover, it is difficult to manipulate the release of a protein from PLGA matrices. As an alternative for these biodegradable polyesters, hydrogels (crosslinked, hydrophilic polymeric networks) have been proposed as protein-releasing matrices. This chapter summarizes the work carried out on biodegradable hydrogels based on dextran (dex) and amphiphilic poly(ether ester) multiblock copolymers for protein delivery.

2. HYDROGELS: GENERAL FEATURES

Hydrogels are polymeric networks that absorb and retain large amounts of water. In the polymeric network, hydrophilic groups or domains are present that are hydrated in an aqueous environment, thereby creating the hydrogel structure. As the term "network" implies, crosslinks have to be present to avoid dissolution of the hydrophilic polymer chains/ segments into the aqueous phase. Because of their water-absorbing capacity, hydrogels have been found to have widespread applications in different technological areas; hydrogels have been used as materials for contact lenses and protein separation, as matrices for cell encapsulation, and as devices for the controlled release of drugs and proteins (9–14). As mentioned, crosslinks have to be present in a hydrogel in order to prevent dissolution of the hydrophilic polymer chains in an aqueous environment. A great variety of establishment of crosslinking has indeed been used to prepare hydrogels. Crosslinking can be established either by chemical means (e.g., by introducing chemical linkages between different polymer chains) or by physical interactions (15). When the protein-loaded gels are administered by injection, the advantages that correspond to hydrogels' biodegradability become obvious. Therefore, labile bonds are frequently introduced in the gels. The labile bonds can be broken, in most of the cases by chemical or enzymatic hydrolysis, under physiological conditions. But degradability as such is not the ultimate solution. Once the hydrogels are implanted, it is of the utmost importance that the gels have good biocompatibility and that the formed degradation products have a low toxicity. In general, hydrogels possess good biocompatibility because their hydrophilic surface has a low interfacial free energy in contact with body fluids, a characteristic that results in a low tendency for proteins and cells to adhere to these surfaces. Moreover, the soft and rubbery nature of hydrogels minimizes irritation to surrounding tissue (16,17). The nature of the formed degradation products can be tailored by a rational and proper selection of the hydrogel building blocks.

3. DEXTRAN HYDROGELS AS PROTEIN-RELEASING MATRICES

3.1. Introduction

During the past decades, many studies have been devoted to the release of proteins from degradable hydrogels based on dextran or dex. An important reason to use dex as building blocks for hydrogels concerns the low or absent toxicity; not surprisingly, then, dex has been used as a plasma expander (18). Initially, research was focused on the use of enzymatically degrading hydrogels that were crosslinked by gamma irradiation or by the free radical polymerization of (meth)acrylate functionalized dex and degraded by the action of the enzyme dextranase (19–23). An enzymatically degrading bovine serum albumin (BSA)-loaded hydrogel that was cross-linked by a condensation reaction between activated dex and 1,10-diaminodecane has also been reported (24). The required enzyme dextranase is of bacterial or fungal origin and present in the human colon (25). Therefore, dex-based gels are under investigation as a colon delivery system (24,26). For delivery at other sites, dextranase has to be co-encapsulated in the hydrogel system in order to control the release of the entrapped protein. However, dextranase is only active in hydrogels with a relatively low crosslink density (21,27,28). Moreover, the use of dextranase might evoke unwanted immune responses to this protein. In order to circumvent the use of dextranase in parenteral systems, current research is now focused on the non-enzymatic degradation of dex hydrogels, i.e., on hydrogels that degrade and release their contents upon hydrolytic cleavage of the crosslinks. In the following sections, we will differentiate between chemically and physically crosslinked hydrogels that are hydrolytically sensitive and discuss their use as protein-releasing matrices.

3.2. Chemically Crosslinked Dextran Hydrogels

Methacrylate groups coupled to water-soluble polymers such as dex (structure given in Fig. 1) are sensitive toward hydrolysis under physiological conditions (29). However, after polymerization, the methacrylate esters are very resistant to hydrolysis (30). This means that gels derived from these polymers can only degrade under physiological conditions once the polymer main chains are hydrolyzed by a matching enzyme (dextranase). In order to increase sensitivity toward hydrolysis, we introduced degradable units between the methacrylate groups and the dex backbone, i.e., a carbonate ester in dex derivatized with hydroxyethylmethacrylate (HEMA) (dex-HEMA) or a combination of a carbonate ester and lactic acid groups in dex derivatized with HEMA-lactate (dex-lactate-HEMA) (Fig. 1) (31). After polymerization of these derivatives, the gels degraded under physiological

(A)   (B)

(C)

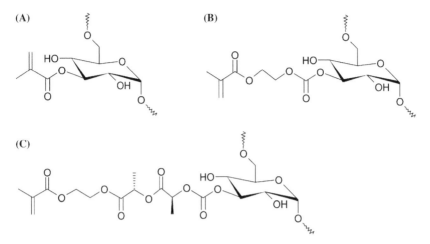

Figure 1   Structure of (A) dexMA, (B) dexHEMA and (C) dex(lactate)₂HEMA.

conditions, owing to the presence of (carbonate) ester groups in the crosslinks, and yielded dex, lactic acid, and short fragments of polyhydroxyethylmethacrylate (pHEMA) as degradation products (schematically shown in Fig. 2).

The degradation time varied from 1 day to more than 3 months and could be controlled by the type of ester group in the crosslinks, the crosslink density of the gel, and the length of the lactic acid spacer (30,32). Interestingly, the gels had good biocompatibility both in the form of implants and in the form of injectable microspheres, and the degradation time in vivo (after subcutaneous implantation in rats) was about the same as the degradation time in vitro (33,34). This means that, also in vivo, the degradation is most likely caused by chemical hydrolysis.

The release of a model protein [immunoglobulin G (IgG)] from degrading, cylinder-shaped (radius 0.23 cm, length 1 cm) dex-lactate-HEMA hydrogels with varying initial water content and degrees of substitution (DS, average number of HEMA-lactate groups per 100 glycopyranose residues) was investigated (30,35). Representative release profiles are shown in Figure 3. From this figure, it appears that, for gels with a high initial water content, a first-order release of the

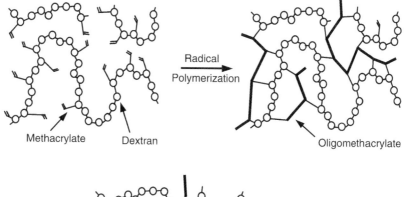

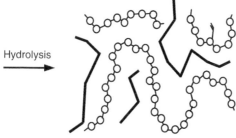

Figure 2 Schematic representation of the formation and degradation for dex(lactate)HEMA hydrogels. Gels are formed by radical polymerization of the methacrylate groups using KPS and TEMED as initiator system. Degradation occurs by hydrolysis of the carbonate and lactate ester.

protein is to be observed (diffusion-controlled release), whereas for gels with a lower initial water content, an almost zero-order release is to be observed for 35 days (degradation-controlled release).

Protein-loaded injectable microspheres can be prepared in an all-aqueous system as schematically shown in Figure 4 (36). Because the protein is present during the hydrogel formation, it can be entrapped in pores of the gel that are smaller than the protein.

Figure 5 shows the release of IgG from different degrading dex-HEMA microspheres at pH 7.0 and 37°C. Interestingly, dex-HEMA microspheres showed a delayed release of the entrapped protein. The delay time increased with an increasing degree of methacrylate substitution, a correspon-

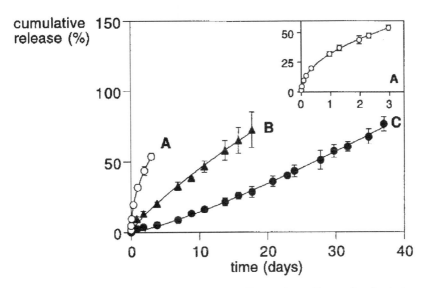

Figure 3 Cumulative release of IgG in time from dex-lactate-HEMA hydrogels (DS 10) with initial water content 90% (A), 80% (B), and 70 % (C). (From Ref.30.).

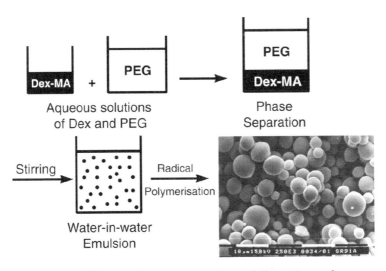

Figure 4 Schematic representation of the microsphere preparation process.

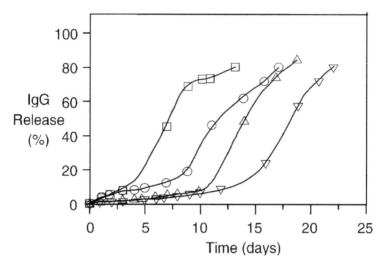

Figure 5 Cumulative release of IgG in time from degrading dex-HEMA microspheres [water content 50% (w/w)] in time DS 3 (□), DS 6 (○), DS 8 (△), and DS 11 (▽). (From Ref. 35.)

dence that demonstrates that the release of the protein was fully controlled by the degradation rate of the microspheres (35).

In vivo experiments using DBA/2 mice with Sutton lymphoma 2 (SL2), lymphosarcoma, revealed that the cytokine interleukin-2 (IL-2), which was slowly released from dex-HEMA microspheres over a period of 3–5 days, had comparable therapeutic effects as repeated injections of the free protein were administered for 5 consecutive days (37). This finding demonstrates that the biological activity of the released protein was preserved.

Another route to dex-based hydrogels was published by Zhang and colleagues. They modified dex with a polymerizable group, either acryloyl chloride or allyl isocyanate (38,39). The dex derivatives were dissolved in dimethyl formamide (DMF) together with a poly(D,L-lactic acid)diacrylate macromer (PDLLAM). A network was obtained by ultraviolet (UV)-induced polymerization. The swelling of the gels depended on, among other things, the ratio of dex/PDLLA in the network, the degree of the substitution of dex with the polymerizable group, and the UV irradiation time (38). These gels were investigated as a matrix for the release of albumin and insulin (40,41). The release was dependent on the gel composition and was governed by a combination of diffusion and the degradation of the matrix. Although Zhang and colleagues claim that these hydrogels are biodegradable, complete biodegradability was not observed and is unlikely owing to the choice of dex derivatives.

The chemically crosslinked dex hydrogels discussed here were prepared by the free radical polymerization of the methacrylate units under mild conditions (room temperature, pH 7), using either UV irradiation in the presence of a photoinitiator or an initiator system composed of a peroxydisulfate and N,N,N',N'-tetramethylenediamine (TEMED). However, unreacted peroxydisulfate and TEMED as well as their degradation products, have to be extracted from the gel before in vivo application. Moreover, these initiator systems can also damage proteins once they are present during preparation of the gels, as will be further discussed in Sec. 5. Therefore, we recently developed physically crosslinked dex hydrogels that avoid the use of harmful crosslinking agents, as will be described in the next section.

3.3. Physically Crosslinked Dextran Hydrogels

In physically crosslinked gels, dissolution of the polymer network is prevented by physical interactions that exist between the polymer chains. Physical crosslinking can be established by, for instance, ionic, hydrophobic, or coiled-coil interactions (15). A novel physical method by which to create hydrogel is the use of stereocomplex formation. This method has been recently investigated by us and others (42–46). In this section, the results obtained with these gels based on dex are summarized.

The novel hydrogel system, under investigation within our department, is schematically shown in Fig. 6.

Importantly, the hydrogel is formed in an all-aqueous environment in which the use of organic solvents is avoided. Crosslinking is established by stereocomplex formation

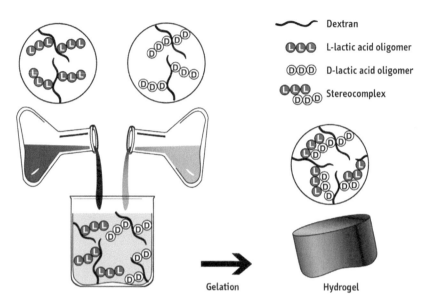

Figure 6   The concept of the stereocomplex hydrogel system. The hydrogel is simply obtained after mixing aqueous solutions of dex-(L)-lactate (L-lactic acid oligomer grafted to dextran) and dex-(D)-lactate.

between lactic acid oligomers of opposite chirality grafted to dex (42). Figure 7 shows the chemical structure of the synthesized dex-lactate products characterized by their degree of the polymerization (DP) of the lactate graft and by their DS. The hydrogel system is expected to be fully biodegradable, since the lactic acid oligomers (that will be degraded to lactic acid) are coupled to dex via a hydrolytically sensitive carbonate ester bond. These features make this hydrogel system very suitable for the controlled release of pharmaceutically active proteins.

Protein-loaded stereocomplex hydrogels were simply prepared by dissolving the protein in the dex-lactate solutions prior to mixing. It was shown that, under physiological conditions, the gels are fully degradable (46). The degradation time depended on both the pH and the composition of the hydrogel (i.e., on the number of lactate grafts, the length and polydispersity of the grafts, and the initial water content) and varied from 1 to 7 days (Fig. 8). Under non-degrading conditions

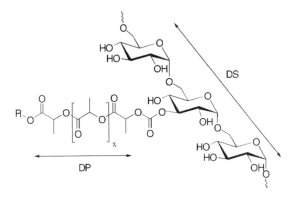

Figure 7 Dex-lactate product with $DP = x + 2$ and DS (degree of substitution; the number of lactate grafts per 100 glucopyranose units).

(pH 4), the hydrogels, having a water content of almost 90% in their swollen states, were stable for more than 1 month.

As shown in Figure 9, the gels showed a release of the entrapped model proteins (IgG and lysozyme) over 6 days,

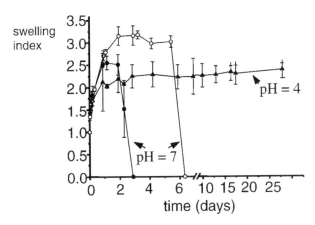

Figure 8 Swelling behaviour of dex-lactate stereocomplex hydrogels (DS = 6, 70% water 37°C): high polydispersity lactic acid grafts ($DP_{average} = 12$, $M_w/M_n \approx 1.25$, filled symbols) and low polydispersity grafts (DP = 11 to 14, $M_w/M_n \approx 1.01$, open circle). The filled triangles represent swelling under non-degrading conditions (pH 4). (From Ref. 46.)

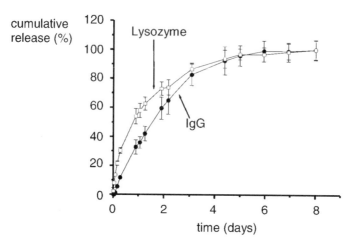

Figure 9   Release profiles of lysozyme (open squares) and IgG (filled circles) from dex-lactate stereocmplex hydrogel with low polydispersity grafts (DS = 6, DP = 11 to 14, 70% water pH 7, 37°C). (From Ref. 46.)

and the kinetics depended on the gel characteristics, such as the polydispersity of the lactate grafts and the initial water content. The release of lysozyme was by diffusion, whereas for the bigger IgG, whose hydrodynamic radius approaches the estimated mesh size of the hydrogels, swelling/degrada-degradation as well, played a role in the release. Importantly, the proteins were quantitatively released from the gels and with full preservation of the enzymatic activity of lysozyme, a fact that emphasizes the protein-friendly preparation method of the protein-loaded stereocomplex hydrogel.

In a recent study, we investigated the therapeutic effi-cacy of recombinant human IL-2 (rhIL-2)-loaded, in situ gel-ling, physically crosslinked dex hydrogels, locally applied to SL2 in mice (47). As a control, free rhIL-2 was administered locally in either a single injection or on 5 consecutive days. All mice received the same total dose of rhIL2. The rhIL2-loaded hydrogels released most IL-2 over a period of about 5 days. The biocompatibility and the biodegradability of the gels were excellent, as there was no acute or chronic inflam-matory reaction and as the gels were completely replaced by fibroblasts after 15 days. The therapeutic efficacy of rhIL2-loaded in situ gelled hydrogels is very good, as was demonstrated in DBA/2 mice bearing SL2. The therapeutic effect of a single application of gels loaded with 1×10^6 IU rhIL2 is at least comparable to the therapeutic effect of the injection of an equal dose of free rhIL2 (47).

4. AMPHIPHILIC POLY(ETHER ESTER) MULTIBLOCK COPOLYMERS AS PROTEIN-RELEASING MATRICES

4.1. Introduction

Physically crosslinked hydrogels can be obtained from multi-block copolymers or graft copolymers. Among them, multiblock copolymers consisting of repeating blocks of hydrophilic poly(-ethylene glycol) (PEG) and hydrophobic poly(butylene ter-ephthalate) (PBT) have been studied extensively for controlled-release applications (48–58). In vitro and in vivo studies have shown that PEG/PBT copolymers are well tolerated and do not cause adverse tissue or systemic effects (59–62). The poly(-ether ester)s are biodegradable and degrade by hydrolysis (ester bonds) and oxidation (ether bonds) (63). The copolymers, also known under the registered trademark PolyActive^{TM}, have been used in a wide range of biomedical applications, including bone (64), cartilage (65), and skin repair (62). Medical devices made of PEG/PBT copolymers have been approved by both the FDA in the United States and Notified Bodies in Europe.

The poly(ether ester) multiblock copolymers are synthe-sized by the polycondensation of dimethyl terephthalate (DMT) with PEG and 1,4-butanediol (BD). An important advantage of multiblock copolymers is the possibility of vary-ing the amount and the length of each of the building blocks, enabling the manufacturing of a wide range of customized polymers. In the case of PEG/PBT copolymers, this can be obtained by varying the feed ratio of PEG to BD or by adjust-ing the molecular weight of the hydrophilic PEG blocks. The copolymers are abbreviated as a**PEGTb**PBT**c**, in which **a** is the PEG molecular weight, **b** the weight % poly(ethylene gly-col)-terephthalate (PEGT), and **c** the weight % PBT (Fig. 10).

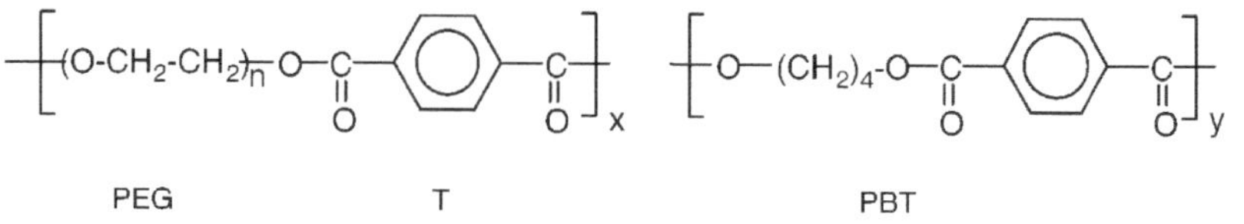

PEG   T   PBT

Figure 10   Chemical structure of the block that from PEG/PBT multiblock copolymers n determines the molecular weight of the PEG segment, x and y determine the weight ratio between PEGT and PBT.

Characteristics such as rate of degradation, swelling, and mechanical strength can be precisely controlled by a proper combination of the two copolymer segments. Thus, a tight control over the release rates of diverse molecules, including vitamins, peptides, and proteins, can be obtained (50,54,56,58).

4.2.   Protein Release from Poly(Ether Ester) Hydrogel Films

Protein-containing PEG/PBT matrices (films or microspheres) were prepared according to an emulsion method (50). Polymer solutions in chloroform or dichloromethane were mixed with aqueous protein solutions to create a water-in-oil (w/o) emulsion. The emulsions were either cast onto a glass plate to form a film or dispersed in an aqueous solution to produce microspheres according to a water-in-oil-in-water (w/o/w) emulsion method. As demonstrated for a model protein (lysozyme) that was entrapped in films (approximately 100 μm in thickness), the release rate increased with both the increasing PEGT content of the copolymer and an increase in the molecular weight of the PEG segments (Fig. 11). This was explained by the difference in the degree of the copolymers' swelling in water (50). Swelling, and consequently the permeation of solutes through the hydrogel, increased as the molecular weight of the PEG segments or the PEG content increased (51).

For highly swollen copolymers, the release of lysozyme was complete within 1 day and followed first-order kinetics. Polymers that were swollen to a lesser extent released

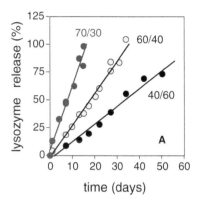 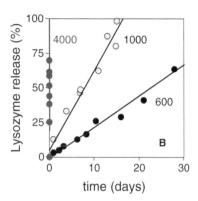

time (days)   time (days)

Figure 11 Cumulative release of lysozyme from PEG/PBT films in PBS (pH 7.4) at 37°C (A) M_w of the PEG segment is 1000 g/mole and PEGT/PBT ratio is 70/30; 60/40 or 40/60; (B) PEGT wt% is fixed between 70 and 80% and M_w of the PEG segment is 600, 1000 or 4000 g/mole. (Adapted from Ref. 50.)

lysozyme with an almost constant rate (near zero-order release). This release behavior was attributed to a combination of protein diffusion and polymer degradation (50). However, the mass loss observed during the in vitro release periods was very limited (51). Therefore, the effect of polymer chain scission on protein diffusion was studied. Polymers of different molecular weights were obtained by incubating 1000PEGT70PBT30 in phosphate buffered saline (PBS) solutions at 37°C for various time periods. Subsequently, the predegraded polymers were used to prepare protein-containing films. The lysozyme release rate of films increased with the decreasing molecular weight of the polymers. Figure 12 shows the diffusion coefficient of lysozyme (D), calculated from the release curves, as a function of the polymer molecular weight (M_n).

A linear relationship was observed between D and M_n^2. Taking into account the polymer degradation during release, M_n can be written as a function of time, resulting in a time-dependent diffusion coefficient. The new equation for the time-dependent diffusion coefficient was incorporated into solutions of the diffusion equation that could adequately describe lysozyme release from poly(ether ester) copolymers (50).

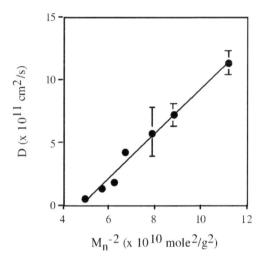

Figure 12   Effect of polymer molecular weght on lysozyme diffusion coefficient (1000PEGT70PBT30). (From Ref. 50.)

The release mechanism was explained by assuming that chain scission decreased the amount of effective crosslinks or entanglements present in the swollen copolymer matrix. Therefore, the resistance to the diffusion of proteins through the copolymer hydrogels was reduced. In the case of the observed zero-order release, the increasing permeability of the matrix in time may have compensated for the decline in the release rate caused by the reduced protein concentration in the matrix, to yield a constant release rate (50).

4.3.   Protein Release from Poly(Ether Ester) Microspheres

Protein-containing PEG/PBT microspheres were prepared according to the well-known double-emulsion method (52,53). In contrast to what is often observed for hydrophobic polymers like PLGA, dense poly(ether ester) microspheres were obtained. This outcome was explained by the amphiphilic character of the polymers, which created very stable emulsions. The latter is considered a prerequisite for the formation of non-porous microspheres. A drawback of applying the w/o/w emulsion technique for the production of protein-loaded microspheres is related to the entrapment efficiency. In the case of polymers with a relatively low degree of swelling (e.g., less than 100% swelling), entrapment efficiencies close to 100% were observed for lysozyme. However, for highly swollen microspheres, only 10% of the lysozyme was effectively entrapped within the microspheres. This finding was ascribed to a premature release of lysozyme during the 2 h, during which the solvent in PBS evaporated. The release during preparation increased with the increasing equilibrium water content of the matrix owing to a greater diffusivity. In order to encapsulate small hydrophilic drugs in highly swollen PEG/PBT microspheres, other methods should be applied— methods that avoid in-water hardening of the spheres.

In general, microspheres released proteins and peptides in a similar fashion as films (50,52–55). A continuous release could be obtained, without an initial burst, for periods ranging from days to months. Besides, by changing the copolymer composition or the molecular weight, the release profile could be altered by a variation of the w/o emulsion composition (56). An increase in the water content of the emulsion (the water/ polymer ratio, w/p) increased the water uptake of the microspheres. At low w/p, this had no effect on the release rate. However, when the w/p was above a critical value of 1–1.5 mL/g, the release kinetics was altered. Figure 13 shows that the rate of the BSA release from microspheres increased with increasing w/p. Interestingly, a delayed release was observed for the formulation with the lowest w/p. Such a delayed release may be used to develop vaccine formulations with intrinsic booster effects.

4.4. In Vivo Protein Release

In order to obtain an in vivo/in vitro correlation for PEG/PBT multiblock copolymers, radiolabeled lysozyme was incorporated into microspheres (57). Three different compositions were selected to produce microspheres with distinct release rates. The release of lysozyme in vitro (PBS, pH 7.4, 37°C) was essentially complete within 1 week for a 1000PEGT70PBT30 composition, whereas release continued for 3–4 weeks for both the

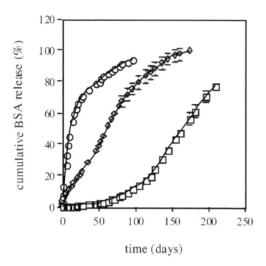

Figure 13 Effect of emulsion w/p on the release of BSA from 1000PEGT70PBT30 microspheres in PBS at 37°C. Matrices were prepared from emulsions with a w/p of 1.75 ml/g (O), 1.5 ml/g (△) and 1.25 ml/g (□). (From Ref. 56.)

1000PEGT60PBT40 and the 600PEGT77PBT23 microspheres. After rats were given subcutaneous injections,[14]C methylated lysozyme could be detected beyond 14 days after the dose for the 1000PEGT70PBT30 microspheres but were measurable up to the last sample point (28 days) for the 1000PEGT60PBT40 and the 600PEGT77PBT23 microspheres (Fig. 14).

On the basis of plasma radioactivity concentrations after injection of lysozyme-containing microspheres and the observed clearance rate of free ^{14}C lysozyme, the cumulative in vivo release was determined. A comparison of the release profiles showed an excellent congruence between release in vitro in PBS and in vivo in rats (Fig. 15). Very often for PLGA microspheres, in vivo release differs from in vitro release insofar as shape and/or time-course are concerned. In contrast to PLGA-based microspheres, protein diffusion through the swollen PEG/PBT matrix is the main rate-controlling factor, and this conclusion may explain the good correlation. However, the results for the release of lysozyme from PEG/PBT microspheres cannot be directly extrapolated to other

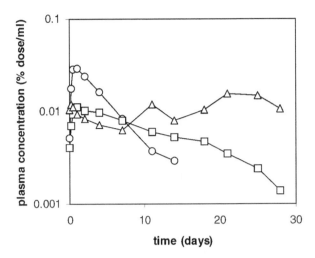

Figure 14 Mean plasma radioactivity concentration in male rats after a single subcutaneous dose of ^{14}C-methylated lysozyme, entrapped in 1000PEGT70PBT30 (○) 1000PEGT60PBT40 (△) 600PEGT77PBT23 (□) microspheres. (From Ref. 57.)

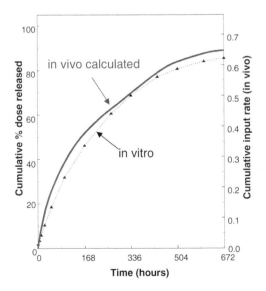

Figure 15 Comparision of in vitro cumulative release and in vitro cumulative input of radioactivity for 1000PEGT60PBT40 microspheres. (From Ref. 56.)

proteins. In vivo, the protein-dependent metabolism and immune response will affect the protein plasma levels. Furthermore, the model protein (lysozyme) that was studied was fairly small. Ultimately, the size of the protein may determine the route of transport through the tissue into the blood stream.

4.5. Tissue Engineering

At present, various groups study extensively the application of PEG/PBT copolymers in tissue engineering scaffolds (65,66). The controlled release of growth factors from such scaffolds could be important to enhance cell migration, proliferation, and differentiation. In order to prepare porous protein-loaded polymer scaffolds, prefabricated PEG/PBT scaffolds were used, of which the inner scaffold pores were coated with a protein-loaded polymeric film (67). To create the coating, a w/o emulsion was conducted through the pores of the scaffold by applying a vacuum. After freeze-drying, the resulting scaffolds contained a 10–40 μm coating, distributed over the scaffold pores. The release of a model protein (lysozyme) could be tailored by the emulsion composition. Figure 16 plots the release of a model protein, lysozyme, against time for three different w/p ratios. By decreasing

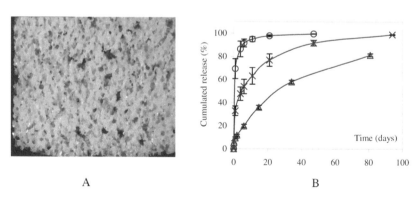

A   B

Figure 16 Light micrograph of a coated PEG/PBT scaffold (A) and lyozyme release from coated scaffolds (B). Coating were prepared from emulsions varying in ratio of aueous protein solution to polymer (✕: w/p = 2, ○: w/p = 1, Δ: w/p = 0.5). (From Ref. 65.)

the w/p ratios, the release rates could be varied from a complete release within a couple of days up to over 2 months. No extensive decrease of the activity of the protein was observed (66). However, in future studies, the effectiveness of this protein release system should be assessed using relevant growth factors like rhBMP-2 or TGF-β.

5. THE STABILITY OF HYDROGEL-ASSOCIATED PHARMACEUTICAL PROTEINS

Protein degradation reactions can be divided into two broad classes. Chemical degradation reactions occur because the amino acids forming the primary structure of the protein are liable to oxidation, deamidation, isomerization, racemization, and disulfide exchange reactions. Peptide bond hydrolysis can occur as well (68,69). Some amino acids are known to be sensitive to particular degradation reactions. For example, cysteine, methionine, tryptophan, and tyrosine are readily oxidized. Deamidation reactions occur with asparagine and glutamine. Peptide bond hydrolysis often occurs around asparagine or aspartic acid residues.

Preservation of the secondary, tertiary, and—where relevant—quaternary structures of proteins is crucial for the pharmacological activity of the molecule. Both the therapeutic and the toxic effects exerted by the protein in vivo depend on these structural characteristics. These higher-order structures are mainly stabilized by physical forces such as electrostatic interactions, hydrogen bonds, hydrophobic interactions, and van der Waals interactions. In general, these forces are much weaker than covalent forces and, therefore, easily disturbed by changes in environment like temperature, pH, shear, and ionic strength. The molecular structure itself may change, and aggregation of colliding molecules may occuras well.

It is difficult to ensure full stability of a pharmaceutical protein both on the shelf and after administration and also before its release from the carrier system, after which release the protein in question becomes pharmacologically active. An illustration of this point will be helpful: an Ig molecule from the IgG class has a molecular weight of 150 kDa and consists of over 1000 amino acids and of two sugar chains. Establishing the amino acid sequence and the chemical structure of all building blocks requires sophisticated modern analytical techniques like mass spectrometry. But the real and critical challenge is the assessment of the physical structure of the molecule. Our "toolbox" of physicochemical approaches with which to assess and monitor the higher-order structures is filled with chromatographic approaches, light scattering analyses, nuclear magnetic resonance (NMR) spectroscopy, x-ray crystallography, and different types of other spectroscopic approaches, ultracentrifugation, and calorimetric analyses (70). Immunoassays can also be used to assess structural aspects of proteins. And, finally, bioassays in vitro (cell systems) and in vivo (animal models) provide, of course, highly relevant information on a therapeutic or toxic effect of the molecule. But, in general, bioassays lack the sensitivity that is needed to pick up smaller changes in effect.

When monitoring the protein performance with all these techniques, the question arises which changes are relevant and which are irrelevant. Is $x\%$ deamidation or $y\%$ oxidation acceptable? Endogenous proteins secreted by the patients themselves, for example, are also present in different isoforms (in case of glycoproteins) or in partly degraded forms. Another issue is the sensitivity of proteins for degradation. Some proteins (e.g., lysozyme and albumin) tend to be rather robust. Interestingly, these compounds are rather often found as model proteins in stability studies. Other proteins are much more sensitive to external influences such as rhIL-2, growth hormones, and some interferons. In the discussion concerning acceptable and non-acceptable protein degradation, the potential immunogenicity of the therapeutic protein is of importance, as serious side effects have been ascribed to immune responses following the administration of pharmaceutical proteins (71). Aggregation is often seen as a major driver of immune responses, and serious efforts have been made to exclude aggregate formation in pharmaceutical formulations.

In light of the above findings, it is important, when considering protein-containing hydrogels as parenteral controlled-release systems, to pay attention to protein stability. Surprisingly, in the existing literature, generally only one or a few analytical approaches were used from the "toolbox" for monitoring protein integrity. A study of the stability of IL-2 in hydrogels based on modified dex was published by Cadée and colleagues (72). These protein-loaded hydrogels were prepared without exposure to organic solvents ("all-water-hydrogel-formation"; see also Fig. 4 and Sec. 3.2). Crosslinking was established by radical polymerization with potassium peroxodisulfate (KPS) and TEMED as catalyst and initiator, respectively. Sodium dodecyl sulfate polyacrylamide gel electrophoresis (SDS-PAGE), high-pressure liquid chromatography (HPLC), and bioassays were used to monitor protein integrity. Also found were oxidized protein and some degree of aggregation, both under reduced and non-reduced conditions in the SDS-PAGE test. The released product still retained 60–70% of the bioactivity of the original product. In a second study, Cadée and colleagues (73) investigated in great detail the oxidation reactions of human IL-2 in the presence of KPS and TEMED. Here SDS-PAGE, reversed-phase HPLC (RP-HPLC), and mass spectrometry were the major analytical techniques for the monitoring of protein. If KPS alone was used, oxidation occurred in the methionines in IL-2 but not in the (buried) tryptophan. The presence of TEMED or other reducing agents clearly reduced oxidative damage, but some dimer formation occurred. In conclusion, oxidative damage by KPS can be modulated by adding other excipients.

Woo and colleagues (74) worked with composite microsphere systems based on both hydrophobic PLGA and the hydrophilic acrylic ester of hydroxyethyl starch (acHES). Albumin or horseradish peroxidase (HRP) was encapsulated in the acHES core particles, which were pre-crosslinked with KPS and TEMED. These protein-loaded core particles were later taken up in the PLGA microspheres. This idea of encapsulating the hydrogel acHES particulates in the hydrophobic PLGA microspheres arose from the notion that the current, standard PLGA microsphere technology, based on w/o/w double emulsion formation, does not properly protect protein integrity because the protein is in intimate contact with the hydrophobic PLGA matrix. Sodium dodecyl sulfate polyacrylamide gel electrophoresis, size-exclusion chromatography/high-pressure liquid chromatography (SEC-HPLC), and enzyme activity (HRP) were used to monitor the fate of the protein. Protein material extracted from the microspheres showed no degradation with the chromatographic and electrophoresis techniques used. In the composite spheres, HRP enzyme activity was preserved for about 80%; for the conventional PLGA spheres, the activity dropped to about 60%. For both the PLGA/acHES and the PLGA microspheres, release patterns were very similar, showing large burst effects. Earlier, Wang and colleagues (75) explored a different route for the preservation of protein integrity in PLGA microspheres. They pre-encapsulated the model protein BSA in polyvinyl alcohol (PVA) microspheres and then dispersed these PVA spheres into PLGA microspheres. The authors conclude that the BSA was neither degraded nor aggregated. That may be true, but BSA integrity was monitored by SEC-HPLC only, and the validation of the analytical method was not reported. Recombinant interferon-α-2a (IFN-α) is one of the therapeutic proteins that has to be administered for long periods of time; it is a prime candidate for encapsulation in sustained-release systems. But, IFN-α is a tough candidate as it is a molecule that readily loses its structure. Zhou and colleagues (76) coated calcium alginate microcores containing IFN-α that had a poly-DL-lactide-poly(ethylene glycol) (PELA) film. The integrity of the protein was followed only by a monitoring of the biological activity, not by other means. Biological activity for IFN-α in the microspheres ranged from 6.5% in standard PLGA microspheres prepared via the o/w/o emulsion technique to 48% for the microcore-coated PELA microspheres. The authors indicate that further studies are under way to improve the performance of their composite systems.

Another approach for the minimization of the degradation of the protein during PLGA microsphere preparation and in the release phase is as follows: just mix PLA with PEG instead of water and make suspended BSA-containing microspheres. The hypothesis is that the hydrophilic PEG acts as an early pore former, avoiding unwanted pH drops and protein degradation during disintegration of the microspheres (77). Structural information on the released BSA was obtained through SDS-PAGE and isoelectric focusing (IEF) electrophoretic, circular dichroism (CD) spectrometric, and fluorescence spectrometric analyses. The PEG content of the microspheres was varied between 0 and 30%. At the 10% PEG level, aggregates were observed, both non-covalently bonded and disulfide bonded. With 20% PEG, no aggregation could be observed anymore. The authors conclude that this reduction in BSA aggregation behavior is the result of an increased water content of the PEG-containing PLA spheres during the release process.

A different approach to make PLGA more "protein-friendly" was proposed by Kissel and collaborators (78,79). They synthesized block copolymer of poly(L-lactic-co-glycolic acid) (LPLG) and polyethyleneoxide (PEO) (LPLG-PEO-LPLG) linear triblock copolymers for the encapsulation of erythropoietin (EPO) in microspheres. Interestingly, the occurrence of covalently bound aggregates (20% as measured by SDS-PAGE) tended to be higher for the LPLG-PEO-LPLG than for the PLGA microspheres prepared according to the same protocol. Co-encapsulation of protective additives reduced the aggregation level to less than 5%. AB star-branched block copolymers consisting of A blocks of PLG or PLGA and star-branched PEG were synthesized as well. Sodium dodecyl sulfate polyacrylamide gel electrophoresis showed aggregation levels between 3 and 10%, levels that are much higher than those found in marketed regular EPO formulations (80). So far, this approach has not been met with much success.

An interesting case study was published by Van Dijkhuizen-Radersma and colleagues (58). Poly(ether-ester)s composed of hydrophilic PEGT blocks and hydrophobic PBT blocks were used to develop a sustained release system for calcitonin. Calcitonin is a relatively small protein that readily forms aggregates, often fibrillar in shape, in aqueous media. To avoid this aggregation, a number of additives were screened with little success: calcitonin tends to form aggregates in the polymer and in solution, sooner or later. The calcitonin structure in the polymer in the dry solid state and upon hydration was monitored by Fourier transform infrared (FTIR). Interestingly, intermolecular beta sheets, suggesting non-covalent aggregation, could be observed both upon hydration of the hydrogel-forming polymer and before release of the protein from the hydrogel. This is one of the few examples where molecular properties of proteins inside the hydrogel polymer in the dry and hydrated states were established. Another example can be found in a short article by Van de Weert and colleagues (81) in which a description is given of the model protein lysozyme as it was monitored in PEGT/PBT films by FTIR. Similar non-covalently bonded aggregates were found in the films; but, when the protein was released, it was in its monomeric form, with full enzymatic activity.

6. CONCLUDING REMARKS

There is a need for reliable and versatile controlled-release systems for pharmaceutically active proteins. Hydrogel technology might meet these demands. This chapter demonstrates that the release profiles of proteins can be tailored by the hydrogel network structure. Importantly, both chemically and physically crosslinked hydrogels can be rendered biodegradable through the introduction of hydrolytically sensitive groups into the network. A major concern in protein carrier development is the integrity of the biopharmaceutical upon release. So far, protein-structure monitoring has only rarely been done, and if it is done, only a few groups make a real effort to collect basic information on protein degradation processes and ways to maintain integrity. However, from the information presently available, it is clear that hydrogels are, in terms of protecting the structure of the encapsulated protein, superior to alternative systems, like those based on hydrophobic polyesters. Although a number of attractive hydrogel systems are presently available, there is certainly room for novel systems with improved characteristics methods. It is expected that principles from the expanding research area of supramolecular chemistry will be applied to the design of novel types of hydrogels with tailored properties. Also, protein engineering might contribute to the development of hydrogel systems with a very precise control over their microstructure.

Abbreviations

acHES:	acrylic ester of hydroxyethyl starch
BD:	1,4-butanediol
BSA:	bovine serum albumin
CD:	circular dichroism
dex:	dextran
dex-HEMA:	dextran derivatized with HEMA
dex-lactate-HEMA:	dextran derivatized with HEMA-lactate
dex-MA:	dextran derivatized with methacrylic acid
DMT:	dimethyl terephthalate
DP:	degree of polymerization
DS:	degree of substitution (number of methacrylate groups or oligo lactate grafts per 100 glucopyranose residues)
EPO:	erythropoietin
FTIR:	Fourier transform infrared
HEMA:	hydroxyethylmethacrylate
HRP:	horseradish peroxidase
IEF:	isoelectric focusing
IFN-α:	interferon-α-2a
IgG:	immunoglobulin G
KPS:	potassium peroxodisulfate
LPLG-PEO-LPLG:	block copolymer of poly(L-lactic-co-glycolic acid) and polyethyleneoxide
PBS:	phosphate buffered saline
PBT:	poly(butylene terephthalate)
PEG:	poly(ethylene glycol)
PEGT:	poly(ethylene glycol)-terephthalate
PELA:	poly-DL-lactide-poly(ethylene glycol)
PLGA:	poly(lactic-co-glycolic acid)
PVA:	polyvinyl alcohol
IL-2:	interleukin-2

rhIL2:   recombinant human interleukin-2
SEC-HPLC: size-exclusion chromatography/high-pressure
                liquid chromatography
SL2:   Sutton lymphoma 2
TEMED: N,N,N',N'-tetramethylethylenediamine
w/o/w: water-in-oil-in water
w/p: water/polymer ratio

REFERENCES

1. Crommelin DJA, Sindelar RD. Pharmaceutical Biotechnology. The Netherlands:Harwood Academic Publishers, Amsterdam, 2002.

2. Crommelin DJA, Storm G, Verrijk R, De Leede L, Jiskoot W, Hennink WE. Shifting paradigms: biopharmaceuticals versus low molecular weight drugs. Int J Pharm 2003; 266:3–16.

3. Lee VHL. Peptide and Protein Drug Delivery. New York: Marcel Dekker Inc., 1990.

4. Gombotz WR, Pettit DK. Biodegradable polymers for protein and peptide drug delivery. Bioconjug Chem 1995; 6:332–351.

5. Sershen S, West J. Implantable, polymeric systems for modulated drug delivery. Adv Drug Deliv Rev 2002; 54:1225–1235.

6. Sinha VR, Trehan A. Biodegradable microspheres for protein delivery. J Control Release 2003; 90:261–280.

7. Van de Weert M, Hennink WE, Jiskoot W. Protein instability in poly(lactic-co-glycolic acid) microparticles. Pharm Res 2000; 17:1159–1167.

8. Schwendeman SP. Recent advances in the stabilization of proteins encapsulated in injectable PLGA delivery systems. Crit Rev Ther Drug Carrier Syst 2002; 19:73–98.

9. Hoffman AS. Hydrogels for biomedical applications. Adv Drug Deliv Rev 2002; 43:3–12.

10. Peppas NA, Bures P, Leobandung W, Ichikawa H. Hydrogels in pharmaceutical formulations. Eur J Pharm Biopharm 2000; 50:27–46.

11. Nguyen KT, West JL. Photopolymerizable hydrogels for tissue engineering applications. Biomaterials 2002; 23:4307–4314.

12. Kissel T, Youxin Li, Unger F. ABA-triblock copolymers from biodegradable polyester A-blocks and hydrophilic poly(ethylene oxide) B-blocks as a candidate for in situ forming hydrogel delivery systems for proteins. Adv Drug Deliv Rev 2002; 54:99–134.

13. Anseth KS, Metters AT, Bryant SJ, Martens PJ, Elisseeff JH, Bowman CN. In situ forming degradable networks and their application in tissue engineering and drug delivery. J Control Release 2002; 78:199–209.

14. Lee KY, Mooney DJ. Hydrogels for tissue engineering. Chem Rev 2001; 101:1869–1879.

15. Hennink WE, Van Nostrum CF. Novel crosslinking methods to design hydrogels. Adv Drug Deliv Rev 2002; 54:13–36.

16. Park H, Park K. Issue of implantable drug delivery systems. Pharm Res 1996; 13:1770–1776.

17. Andersson JM. In vivo biocompatibility of implantable delivery systems and biomaterials. Eur J Pharm Biopharm 1994; 40:1–8.

18. Thoren L. The dextrans, clinical data. Dev Biol Stand 1981; 48:157–167.

19. Edman P, Ekman B, Sjoholm I. Immobilization of proteins in microspheres of polyacryldextran. J Pharm Sci 1980; 69: 838–842.

20. Hennink WE, Talsma H, Borchert JCH, De Smedt SC, Demeester J. Controlled release of proteins from dextran hydrogels. J Control Release 1996; 39:47–55.

21. Franssen O, Vos OP, Hennink WE. Delayed release of a model protein from enzymatically degrading dextran hydrogels. J Control Release 1997; 44:237–245.

22. Hennink WE, Franssen O, Van Dijk-Wolthuis WNE, Talsma H. Dextran hydrogels for the controlled release of proteins. J Control Release 1997; 48:107–114.

23. Franssen O, Stenekes RJH, Hennink WE. Controlled release of a model protein from enzymatically-degrading dextran microspheres. J Control Release 1999; 59:219–228.

24. Chiu HC, Hsiue GH, Lee YP, Huang LW. Synthesis and characterization of pH-sensitive dextran hydrogels as a potential colon-specific drug delivery system. J Biomater Sci Polym Ed 1999; 10:591–608.

25. Schomburg D, Salzman M (eds.). Enzyme Handbook. Berlin: Springer, 1999.

26. Brondsted H, Andersen C, Hovgaard L. Crosslinked dextran— a new capsule material for colon targeting of drugs. J Control Release 1998; 53:7–13.

27. Franssen O, Van Rooijen RD, De Boer D, Maes RAA, Hennink WE. Enzymatic degradation of crosslinked dextrans. Macromolecules 1999; 32:2896–2902.

28. Meyvis TKL, De Smedt SC, Demeester J, Hennink WE. Rheological monitoring of long-term degrading polymer hydrogels. J Rheol 1999; 43:933–950.

29. Van Dijk-Wolthuis WNE, Van Steenbergen MJ, Underberg WJM, Hennink WE. Degradation kinetics of methacrylated dextrans in aqueous solution. J Pharm Sci 1997; 86:413–417.

30. Van Dijk-Wolthuis WNE, Tsang SKY, Van Steenbergen MJ, Hoogeboom C, Hennink WE. Degradation and release behaviour of dextran based hydrogels. Macromolecules 1997; 30:4639–4645.

31. Van Dijk-Wolthuis WNE, Tsang SKY, Kettenes-van den Bosch JJ, Hennink WE. A new class of polymerizable dextrans with hydrolyzable groups: hydroxyethyl methacrylated dextran with and without oligolactate spacer. Polymer 1997; 38:6235–6242.

32. Cadée JA, De Kerf M, De Groot CJ, Den Otter W, Hennink WE. Synthesis and characterization of 2-(methacryloyloxy)ethyl-(di)-L-lactate and their application in dextran-based hydrogels. Polymer 1999; 40:6877–6881.

33. Cadée JA, Van Luyn MJA, Brouwer LA, Plantinga JA, Van Wachem PB, De Groot CJ, Den Otter W, Hennink WE. In vivo biocompatibility of dextran-based hydrogels. J Biomed Mater Res 2000; 50:397–404.

34. Cadée JA, Brouwer LA, Den Otter W, Hennink WE, Van Luyn MJA. A comparative biocompatibility study of microspheres based on crosslinked dextran or poly(lactic-co-glycolic)acid

after subcutaneous injection in rats. J Biomed Mater Res 2001; 56:600–609.

35. Franssen O, Vandervennet L, Roders P, Hennink WE. Degradable dextran hydrogels: controlled release of a model protein from cylinders and microspheres. J Control Release 1999; 60:211–221.

36. Stenekes RJH, Franssen O, van Bommel EMG, Crommelin DJA, Hennink WE. The preparation of dextran microspheres in an all aqueous system: effect of the formulation parameters on particle characteristics. Pharm Res 1998; 15:557–561.

37. De Groot CJ, Cadée JA, Koten JW, Hennink WE, Den Otter W. Therapeutic efficacy of IL-2-loaded hydrogels in a mouse tumor model. Int J Cancer 2002; 98:134–140.

38. Zhang Y, Won CY, Chu CC. Synthesis and characterization of biodegradable network hydrogels having both hydrophobic and hydrophilic components with controlled swelling behavior. J Polym Sci Part A. Polym Chem 1999; 37:4554–4569.

39. Zhang Y, Won CY, Chu CC. Synthesis and characterization of biodegradable hydrophobic-hydrophilic hydrogel networks with a controlled swelling property. J Polym Sci Part A. Polym Chem 2000; 38:2392–2404.

40. Zhang Y, Chu CC. Biodegradable dextran-polylactide hydrogel network and its controlled release of albumin. J Biomed Mater Res 2001; 54:1–11.

41. Zhang Y, Chu CC. In vitro release behavior of insulin from biodegradable hybrid hydrogel networks of polysaccharide and synthetic biodegradable polyester. J Biomater Appl 2002; 16:305–325.

42. De Jong SJ, De Smedt SC, Wahls MWC, Demeester J, Kettenes-van den Bosch JJ, Hennink WE. Novel self-assembled hydrogels by stereocomplex formation in aqueous solution of enantiomeric lactic acid oligomers grafted to dextran. Macromolecules 2000; 33:3680–3686.

43. Lim DW, Choi SH, Park TG. A new class of biodegradable hydrogels stereocomplexed by enantiomeric oligo(lactide) side chains of poly(HEMA-g-OLA)s. Macromol Rapid Comm 2000; 21:464–471.

44. Fujiwara T, Mukose T, Yamaoka T, Yamane H, Sakurai S, Kimura Y. Novel thermo-responsive formation of a hydrogel by stereo-complexation between PLLA-PEG-PLLA and PDLA-PEG-PDLA block copolymers. Macromol Biosci 2001; 1:204–208.

45. Li SM, Vert M. Synthesis, characterization, and stereocomplex-induced gelation of block copolymers prepared by ring-opening polymerization of L(d)-lactide in the presence of poly(-ethylene glycol). Macromolecules 2003; 36:8008–8014.

46. De Jong SJ, Van Eerdenbrugh B, Van Nostrum CF, Kettenes-van den Bosch JJ, Hennink WE. Physically crosslinked dextran hydrogels by stereocomplex formation of lactic acid oligomers: degradation and protein release behavior. J Control Release 2001; 71:261–275.

47. Bos GW, Jacobs JJL, Koten JW, Van Tomme S, Veldhuis Th, Van Nostrum CF, Den Otter W, Hennink WE. In situ cross-linked biodegradable hydrogels loaded with IL-2 are effective tools for local IL-2 delivery. Eur J Pharm Sciences 2004; 21:561–567.

48. Goedemoed JH, Hennink WE. Polyetherester copolymers as drug delivery matrices. US patent 5980948, 1999.

49. Van Dijkhuizen-Radersma R, Hesseling SC, Kaim PE, De Groot K, Bezemer JM. Biocompatibility and degradation of poly(ether–ester) microspheres: in vitro and in vivo evalua-tion. Biomaterials 2002; 23:4719–4729.

50. Bezemer JM, Radersma R, Grijpma DW, Dijkstra PJ, Feijen J, Van Blitterswijk CA. Zero-order release of lysozyme from poly(ethylene glycol)/poly(butylene terephthalate) matrices. J Control Release 2000; 64:179–192.

51. Bezemer JM, Grijpma DW, Dijkstra PJ, Van Blitterswijk CA, Feijen J. A controlled release system for proteins based on poly(ether–ester) block-copolymers: polymer network charac-terization. J Control Release 1999; 62:393–405.

52. Bezemer JM, Radersma R, Grijpma DW, Dijkstra PJ, Van Blitterswijk CA, Feijen J. Microspheres for protein delivery prepared from amphiphilic multiblock copolymers: 1. Influence of preparation techniques on particle characteristics and protein delivery. J Control Release 2000; 67:233–248.

53. Bezemer JM, Radersma R, Grijpma DW, Dijkstra PJ, Van Blitterswijk CA, Feijen J. Microspheres for protein delivery prepared from amphiphilic multiblock copolymers: 2. Modulation of release rate. J Control Release 2000; 67:249–260.

54. Van Dijkhuizen-Radersma R, Péters FLAMA, Stienstra NA, Grijpma DW, Feijen J, De Groot K, Bezemer J. Control of vitamin B12 release from poly(ethylene glycol)/poly(butylene terephthalate) multiblock copolymers. Biomaterials 2002; 23:1527–1536.

55. Sohier J, Van Dijkhuizen-Radersma R, De Groot K, Bezemer J. Release of small water-soluble drugs from multiblock copolymer microspheres: a feasibility study. Eur J Pharm Biopharm 2003; 55:221–228.

56. Bezemer JM, Grijpma DW, Dijkstra PJ, Van Blitterswijk CA, Feijen J. Control of protein delivery from amphiphilic poly (ether ester) multiblock copolymers by varying their water content using emulsification techniques. J Control Release 2000; 66:307–320.

57. Van Dijkhuizen-Radersma R, Taylor LM, Wright SJ, John BA, De Groot K, Bezemer JM. In vitro/in vivo correlation for release of lysozyme from poly(ether–ester) microspheres. 30th Annual Meeting of the Controlled Release Society, Glasgow, Scotland, July 2003.

58. Van Dijkhuizen-Radersma R, Nicolas HM, Van de Weert M, Blom M, De Groot K, Bezemer JM. Stability aspects of salmon calcitonin entrapped in poly(ether–ester) sustained release systems. Int J Pharm 2002; 248:229–237.

59. Grote JJ, Bakker D, Hesseling SC, Van Blitterswijk CA. New alloplastic tympanic membrane material. Am J Otol 1991; 12:329–335.

60. Beumer GJ, Van Blitterswijk CA, Ponec M. Biocompatibility of degradable matrix induced as a skin substitute: An in vivo evaluation. J Biomed Mater Res 1994; 28:545–552.

61. Beumer GJ, Van Blitterswijk CA, Bakker D, Ponec M. Cell-seeding and in vitro biocompatibility evaluation of polymeric matrices of PEO/PBT copolymers and PLLA. Biomaterials 1993; 14:598–604.

62. Van Dorp AG, Verhoeven MC, Koerten HK, Van Der Nat-Van Der Meij TH, Van Blitterswijk CA, Ponec M. Dermal regeneration in full-thickness wounds in Yucatan miniature pigs using a biodegradable copolymer. Wound Repair Regen 1998; 6:556–568.

63. Deschamps AA, Grijpma DW, Feijen J. Poly(ethylene oxide)/poly(butylene terephthalate) segmented block copolymers: the effect of copolymer composition on physical properties and degradation behavior. Polymer 2001; 42:9335–9345.

64. Du C, Meijer GJ, Van de Valk C, Haan RE, Bezemer JM, Hesseling SC, Cui FZ, De Groot K, Layrolle P. Bone growth in biomimetic apatite coated porous Polyactive 1000PEG-T70PBT30 implants. Biomaterials 2002; 23:4649–4656.

65. Woodfield TBF, Bezemer JM, Pieper JS, Van Blitterswijk CA, Riesle J. Scaffolds for tissue engineering of cartilage. Crit Rev Eukaryot Gene Expr 2002; 12:209–236.

66. Deschamps AA, Claase MB, Sleijster MWJ, De Bruijn JD, Grijpma DW, Feijen J. Design of segmented poly(ether ester) materials and structures for tissue engineering of bone. J Control Release 2002; 78:175–186.

67. Sohier J, Haan RE, De Groot K, Bezemer JM. A novel method to obtain protein release from porous polymer scaffolds: emulsion coating. J Control Release 2003; 87:57–68.

68. Arakawa T, Philo JS. Biophysical and biochemical analysis of recombinant proteins. In: Crommelin DJA, Sindelar RD, eds. Pharmaceutical Biotechnology. London: Taylor and Francis, 2002.

69. Goolcharran C, Khossravi M, Borchardt RT. Chemical pathways of peptide and protein degradation. In: Frokjaer S, Hovsgaard L, eds. Pharmaceutical Formulation Development of Peptides and Proteins. London: Taylor and Francis, 2000.

70. Herron JN, Jiskoot W, Crommelin DJA. Physical methods to characterize pharmaceutical proteins. New York: Plenum, 1995.

71. Schellekens H. Bioequivalence and the immunogenicity of biopharmaceuticals. Nat Rev Drug Discov 2002; 1:457–462.

72. Cadée JA, De Groot CJ, Jiskoot W, Den Otter W, Hennink WE. Release of recombinant human interleukin-2 from dextran-based hydrogels. J Control Release 2002; 78:1–13.

73. Cadée JA, Van Steenbergen MJ, Versluis C, Heck AJR, Underberg WJM, Den Otter W, Jiskoot W, Hennink WE. Oxidation of interleukin-2 by potassium peroxodisulfate. Pharm Res 2001; 18:1461–1467.

74. Woo BH, Jiang G, Jo YW, DeLuca PP. Preparation and characterization of a composite PLGA and poly(acryloyl hydroxyethyl starch) microsphere system for protein delivery. Pharm Res 2001; 18:1600–1606.

75. Wang N, Wu XS, Li JK. A heterogeneously structured composite based on poly(lactic-co-glycolic acid) microspheres and poly(vinyl alcohol) hydrogel nanoparticles for long-term protein drug delivery. Pharm Res 1999; 16:1430–1435.

76. Zhou SB, Deng XM, He SY, Li XH, Jia WX, Wei DP, Zhang ZR, Ma JH. Study on biodegradable microspheres containing recombinant interferon-alpha-2a. J Pharm Pharmacol 2002; 54:1287–1292.

77. Jiang WL, Schwendeman SP. Stabilization of a model formalinized protein antigen encapsulated in poly(lactide-co-glycolide)-based microspheres. J Pharm Sci 2001; 90: 1558–1569.

78. Bittner B, Morlock M, Koll H, Winter G, Kissel T. Recombinant human erythropoietin (rhEPO) loaded poly(lactide-co-glycolide) microspheres: influence of the encapsulation technique and polymer purity on microsphere characteristics. Eur J Pharm Biopharm 1998; 45:295–305.

79. Pistel KF, Bittner B, Koll H, Winter G, Kissel T. Biodegradable recombinant human erythropoietin loaded microspheres prepared from linear and star-branched block copolymers: Influence of encapsulation technique and polymer composition on particle characteristics. J Control Release 1999; 59:309–325.

80. Hermeling S, Schellekens H, Crommelin DJA, Jiskoot W. Micelle-associated protein in epoetin formulations: a risk factor for immunogenicity?. Pharm Res 2003; 20:1903–1908.

81. Van de Weert M, van Dijkhuizen-Radersma R, Bezemer JM, Hennink WE, Crommelin DJA. Reversible aggregation of lysozyme in a biodegradable amphiphilic multiblock copolymer. Eur J Pharm Biopharm 2002; 54:89–93.

6

Thermosensitive Biodegradable Hydrogels for the Delivery of Therapeutic Agents

YOUNG MIN KWON and SUNG WAN KIM

Center for Controlled Chemical Delivery (CCCD),
Department of Pharmaceutics and
Pharmaceutical Chemistry, University of Utah,
Salt Lake City, Utah, U.S.A.

1. INTRODUCTION

Polymeric drug delivery is the most widely studied area of drug delivery in recent years (1). Polymers can be manipulated to possess certain properties that can meet specific criteria for the designing of suitable delivery systems. Polymeric drug delivery systems may provide advantages such as (i) increased efficacy, (ii) reduced side effects and toxicity, and (iii) convenience (1). Like small drugs in general, therapeutic macromolecular drugs require the use of polymeric systems, since these agents have very short half-lives in blood plasma and are susceptible to physical or chemical degradation. Although many therapeutic proteins, peptides, and DNA-based drugs are available due to advances in biotechnology, the conventional routes by which these biotechnology-derived drugs are administered—routes such as intravenous injection, intravenous infusion, subcutaneous injection, and so on—require frequent injections to achieve a therapeutic concentration in blood and may result in poor patient compliance stemming from frequent, painful injections as well as undesirable side effects. Significant research effort has been made to develop implantable or injectable parenteral devices for the sustained and controlled release of protein drugs in order to reduce the frequency of injection (1).

Substantial efforts have been made to develop injectable polymeric drug delivery systems. Because many biomedical polymers with molecular weights of pharmaceutical interest are not water-soluble, certain solvents under a given toxicity limit may be used to dissolve the polymer. For example, Kost and coworkers studied polymer implants of poly(lactide-co-glycolide) (PLGA) dissolved in a water miscible solvent, glycofurol. When a subcutaneous injection of the drug-polymer-solvent mixture is made, this solvent diffuses into the bloodstream so that its concentration in blood on a per day basis is under the FDA limit (2). Another example is poly-(ortho-ester) semisolid formulation (3). These systems use mild conditions and are suitable for protein delivery.

Polymers that exhibit physicochemical responses to stimuli have been widely explored as potential drug delivery systems. Kinds of stimuli investigated to date include chemical substances and changes in temperature, pH, and electric fields, and the like. Among these, polymers that, in an aqueous solution, exhibit dramatic changes upon a temperature change below or above the body temperature are of particular interest in drug delivery. For instance, low molecular weight triblock copolymers of PLGA and poly(ethylene glycol) (PEG) have been designed that are hydrophilic/hydrophobic balanced polymers. Jeong and coworkers first demonstrated a temperature-induced sol-gel transition of a PEG-PLGA-PEG aqueous solution upon subcutaneous injection in rats (4,5). More recently, the intratumoral injection of Oncogel™ (Macromed, Inc., Sandy, UT), an injectable formulation of PLGA-PEG-PLGA (ReGel™) and paclitaxel, has demonstrated both the solubilization of a water-insoluble drug and the slow clearance of paclitaxel from the injection site with minimal distribution into other organs (6). This system is currently undergoing Phase II human clinical trials.

The use of thermosensitive polymers in drug delivery and biomedical application is widespread. The application of poly(*N*-isopropylacrylamide) [poly(NIPAAm)] and its copolymers is probably the best example. As mentioned above, thermosensitive and biodegradable polymers are becoming very important in the development of nontoxic, injectable systems to tackle challenging problems in the delivery of bioactive agents, as will be discussed in this chapter.

2. POLY(*N*-ISOPROPYLACRYLAMIDE) AND ITS COPOLYMERS

In the past two decades, studies have been carried out by a number of research groups to design polymers in aqueous environments that respond dramatically around certain transition temperatures (7–24). There are polymers that exhibit lower critical solution temperatures (LCST) in water, above which the polymer is insoluble. Poly(NIPAAm) has been of particular interest since the polymer shows an LCST of $\sim 32°C$, which falls in the temperature range between body and room temperatures. Below the LCST, the enthalpy term contributed by the hydrogen bond between water and polar groups of the polymer causes the polymer to stay dissolved in water. Above the LCST, hydrophobic interaction, which is an entropic term, becomes a major factor that triggers the collapse of the polymers (7). The LCST of poly(NIPAAm) can be controlled via copolymerization with monomers having different degrees of hydrophobicity. It is known that a higher LCST can be obtained by incorporating a hydrophilic monomer, whereas copolymerizing with a hydrophobic monomer results in a lower LCST (7).

Han and Bae have demonstrated that an aqueous solution of a copolymer of 2–5 mol% acrylic acid and NIPAAm exhibits reversible gelation without significant hysteresis (8). The critical gel concentration (CGC) in this case was 4% by weight. Gelation, rather than precipitation, was attributed to polymer chain entanglements and to the weak physical association of polymer precipitates while hydration was maintained by an ionized segment of AAc (7,8).

Bae et al. have demonstrated the on/off thermo-control of solute release from poly(NIPAAm) networks (9,10). The swelling in water as a function of temperature for two series of NIPAAm polymer networks was investigated. In the first series, n-butylmethacrylate (BMA) was copolymerized with NIPAAm, and in the second, polytetramethylene ether glycol (PTMEG) was incorporated into a NIPAAm network as a chemically independent interpenetrating network (IPN). When an increase in the BMA content in the poly(NIPAAm-co-BMA) network occurred, a lowering of the gel collapse point was observed, and the gel's deswelling occurred in a more gradual manner as the temperature increased. The temperature dependence of equilibrium swelling in water was a function of the gel composition in both networks. The networks formed a dense skin layer as the temperature increased past the gel collapse point. This dense layer decreased the water efflux and formed water pockets at the surface (9).

In the next study poly(NIPAAm)/PTMEG IPNs were synthesized, and their feasibility as thermosensitive hydrogels for controlled drug release was addressed (10). The release of indomethacin incorporated into these matrices showed pulsatile patterns in response to temperature changes with sensitivity to a few degrees of fluctuation. The lag time and the release profile of indomethacin in the low-temperature region of each temperature cycle were influenced by the applied temperature and the gel composition. The results of this study demonstrated that solute release can be regulated by a rapid deswelling of the gel surface in response to an applied temperature change (10).

The unique solution behavior of poly(NIPAAm) has opened up a new array of applications. Hoffman and coworkers developed methods for biorecognition and immunoassay using a series of poly(NIPAAm) polymers (11–14). In particular, poly(NIPAAm) was conjugated to a monoclonal antibody (Mab), and a novel separation method for an immunoassay was designed (11). The poly(NIPAAm) precipitated out of water above an LCST of 31°C, enabling a polymer-bound immune complex to be separated from the solution. The principal advantages of this method are that it utilizes a homogeneous incubation for the antigen-antibody reaction and that it has the ability to assay large-molecular-weight antigens with sensitivities equivalent to other nonisotopic heterogeneous immunoassays. In addition, because the polymer-immune complex may be reversibly redissolved by cooling, the method may be used both to concentrate the signal and to isolate the target analyte (11).

Okano and coworkers have also investigated extensively the phase transition behavior of poly(NIPAAm) hydrogel (15–17) and its applications in tissue engineering (16) and bioseparation (17). Takei and coworkers have demonstrated reversible bioseparation based on a similar concept. Immunoglobulin G (IgG) was modified by poly(NIPAAm) to create a novel bioconjugate that exhibits reversible phase transition behavior at 32°C in aqueous media. A terminal carboxyl group introduced into PIPAAm molecules by the polymerization of IPAAm with 3-mercaptopropionic acid was used for the conjugation to IgG via a coupling reaction of activated ester with a protein amino group. These conjugates exhibited a rapid response to changes in solution temperature and significant phase separation above the LCST corresponding to that for the original poly(NIPAAm). These conjugates bound to antigen quantitatively in the aqueous system, and the antigen-bound complex also demonstrated phase separation and precipitation above a critical temperature, indicating that PIPAAm conjugated to a biomolecule can operate as a switching molecule (17).

Feil et al. have demonstrated the molecular separation of solutes of three different sizes by using a poly(NIPAAm-co-butyl methacrylate)(BMA) copolymer hydrogel membrane at several temperatures (18,19). The diffusion of urinine (MW 300) and dextran (MW 4400) was found to follow the free-volume theory in this hydrogel membrane although the partitioning of the solutes did not follow gel hydration due to partial exclusion and interactions between the hydrogel and the solutes. The screening effect of the polymer network played no major role for the diffusion of these solutes, but probably contributed significantly to the very low diffusion rate of dextran (MW 150,000) in the gel. Upon incorporating a pH-sensitive component (diethylaminoethyl methacrylate) they also found a mutual influence of temperature and pH on the swelling of the gel (18,19).

The copolymerization of NIPAAm with BMA and AAc rendered the copolymer pH sensitive as well as thermosensitive. This resulted in the design of copolymer beads as carriers for oral delivery, encapsulating polypeptide drugs such as calcitonin (20) and insulin (21). The copolymer beads prevent gastric degradation in the stomach while providing for a controlled release of a peptide drug later in the intestines. Serres et al. used linear terpolymers poly(NIPAAm-co-BMA-co-AAc) to fabricate beads loaded with a peptide drug, human calcitonin, which was dissolved in an aqueous phase. The polymeric beads were formed by the dropwise addition of a cold, aqueous solution of a temperature-sensitive polymer with human calcitonin into an oil bath kept at a temperature above the LCST of a polymer, precipitating the polymer and entrapping the peptide. The loading and the release of human calcitonin were also studied as a function of acrylic acid content in the terpolymers. As the acrylic acid content increased from 0 to 10 mol%, the loading efficiency and stability of calcitonin improved significantly. The same trend was observed for the quantity of released calcitonin. In vivo biological activity of the released hormone was preserved (20).

Taking a step further, Kim et al. have studied the loading, release, and the preservation of bioactivity of a polypeptide, insulin, entrapped in a microbead formation. The morphology of the beads studied by scanning electron microscopy (SEM) revealed that they consisted of a dense skin layer and a porous inner structure. This suggested that the critical step of bead preparation is the formation of this skin layer or "surface curing." The release of insulin was triggered by a change in pH from acidic to neutral pH, whereas little release occurred at low pH. The released insulin had a preserved secondary structure as shown by CD spectroscopy. Also, it had preserved bioactivity as the injection of released insulin in rats showed no difference from standard insulin (21).

Ramkissoon-Ganorkar et al. have designed a novel polymeric delivery system that utilizes linear, pH/temperature-sensitive terpolymers of NIPAAm, BMA, and AAc. The unique properties of the pH/temperature-sensitive polymeric beads make it a potential system for the oral drug delivery of peptide and protein drugs to different regions of the intestinal tract (22). The goal of this study was to investigate the effect of polydispersity and of the molecular weight (MW) of terpolymers of poly(NIPAAm-co-BMA-co-AA) with a feed mol ratio of NIPAAm/BMA/AA 85/5/10 on the polymer dissolution rate and on the release kinetics of a model protein, insulin. Varying the average molecular weight (Mw) and polydispersity of the polymer modulated the dissolution rate of the polymer and the release rate of insulin from pH/ temperature-sensitive polymeric beads. An increase in the polydispersity of the polymer through the addition of high MW polymer chains caused a decrease in the release rate of insulin and in the polymer dissolution rate. High MW polymer chains impose a certain degree of interaction between polymer chains, and this is due to chain entanglement. There is a limiting value of MW above which chain entanglement has no effect on the drug release rate (22).

As mentioned earlier, poly(NIPAAm-co-AAc) where AAc content is small (2–5 mol%) forms loose hydrogels above the LCST. This makes the system an attractive candidate for tissue engineering application as an extracellular matrix (ECM). Above a critical concentration, aqueous polymer solutions of N-isopropylacrylamide copolymers with small amounts of acrylic acid, synthesized in benzene by free radical polymerization, exhibited four distinct phases as the temperature increased: a clear solution, an opaque solution, a gel, and a shrunken gel. The transition between the opaque solution phase and the gel phase was in the range of 30–34°C and was reversible without syneresis and noticeable hysteresis under the experimental conditions used. Islets of Langerhans, isolated from a Sprague-Dawley rat pancreas and entrapped in the gel matrix, remained viable, with no significant decrease in the insulin secretion function in vitro for 1 month. When islets were encapsulated with the gel matrix in hollow fibers with a molecular weight cut-off (MWCO) of approximately 400,000 and were exposed to dynamic changes in glucose and theophylline concentrations, their insulin secretion patterns exhibited a smaller lag time and higher amplitude in insulin release than did the islets entrapped in a conventional alginate matrix under the same experimental conditions. From these two observations, that is, the gel reversibility and the islet functionality in the matrix observed in the in vitro experiments, the N-isopropylacrylamide copolymers with acrylic acid synthesized in this study seem to be optimum candidates for the ECM for the recharging of the entrapped cells when the decrease in the cell functionality in the system is noted (23).

The purpose of a second application of poly(NIPAAm) in tissue engineering concerns the grafting of poly(NIPAAm) onto the cell culture surface so that cells can both attach under the growing condition (above the LCST) and easily and reversibly detach below the LCST. This is important in that the conventional method of cell detachment using proteolytic enzymes can damage cells. This concept brought about the so-called "cell sheet engineering" in which this type of surface is used. Such technology may be useful in the repairing of damages in various organs (24).

3. PEO-PPO-PEO TRIBLOCK COPOLYMERS

Poly(ethylene oxide)-poly(propylene oxide)-poly(ethylene oxide) (PEO-PPO-PEO) is another kind of thermosensitive polymer, known as poloxamer (25,26). Poloxamers are a series of surfactant polymers that are commercially available. Among these, poloxamer 407 (Pluronic F127) has been of particular interest, since this amphiphilic triblock copolymer can form micelles at low concentration in water and, with increasing concentration, can form a hydrogel at body temperature via packing of the micelles. This means that the aqueous solution of this polymer may be used for a parenteral depot system. However, the gel depot of the PEO-PPO-PEO block copolymers disintegrated quickly (within 1 day) upon dilution and therefore is not suitable as a sustained release system. It was speculated that this surface dissolution may be related to its gelling mechanism, which involved the mere packing of micelles at high concentrations (25).

4. THERMOSENSITIVE AND BIODEGRADABLE POLYMER HYDROGELS

Biodegradable polymers used for drug delivery to date have mostly been in the form of injectable microspheres or implant systems, which require complicated fabrication processes using organic solvents. A disadvantage of such systems is that the use of organic solvents can cause denaturation when protein drugs are to be encapsulated. Furthermore, the solid form requires surgical insertion, which often results in tissue irritation and damage.

Synthesis of a thermosensitive, biodegradable hydrogel consisting of blocks of poly(ethylene oxide) and poly(L-lactic acid) was carried out (4). Aqueous solutions of these copolymers exhibited temperature-dependent reversible gel-sol transitions. The hydrogel can be loaded with bioactive molecules in an aqueous phase at an elevated temperature (around 45°C), where they form a solution. In this form, the polymer is injectable. On subcutaneous injection and subsequent rapid cooling to body temperature, the loaded copolymer forms a gel that can act as a sustained release matrix for drugs (4).

In the next series of studies, Jeong et al. found compositions of HPL/HPB balanced triblock copolymers, PEG-PLGA-PEG, that exhibit sol-to-gel (lower transition) and gel-to-sol (upper transition) transitions as temperature monotonically increases (27). The lower transition is important for drug delivery application because the solution both flows freely at room temperature and becomes a gel at body

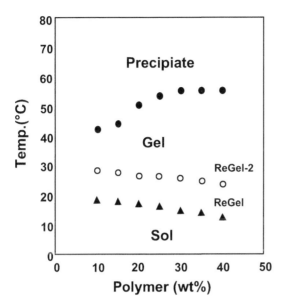

Figure 1 Phase diagram for PLGA-PEG-PLGA in aqueous solution (ReGel^{TM}). (From Refs. 6.)

temperature. The lower transition may be related to micellar growth and intra- and intermicelle phase mixing and packing, while the upper transition involves a breakage of the micellar structure. The CGC and critical gel temperature are controlled by polymer parameters, such as block length and composition of PEG-PLGA-PEG triblock copolymers, and by additives such as salts (27).

When aliquots of the 33 wt.% aqueous solutions of PEG-PLGA-PEG triblock copolymers were administered via subcutaneous injection into rats, transparent gels were observed. The gels showed good mechanical strength and the integrity of the gels persisted longer than 1 month. The gel underwent degradation by hydrolysis and became opaque, a development that was probably due to the preferential mass loss of PEG-rich segments from the in situ formed gel. The number-averaged molecular weight (Mn) determined by gel-

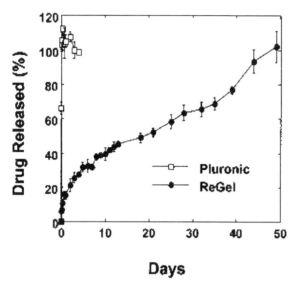

Figure 2 In vitro release of paclitaxel from OncoGel^{TM} (23%, w/w in water) and Pluronic F-127. (From Ref. 6.)

permeation chromatography (GPC) decreased from 3300 to 1900 and a 30% mass loss was observed over 1 month (5).

Two model drugs were released from the PEG-PLGA-PEG triblock copolymer hydrogel formed in situ at 37°C. While ketoprofen, a model hydrophilic drug, was released over 2 weeks with a first-order release profile, spironolactone, a model hydrophobic drug, was released over 2 months with an S-shaped release profile. The release profiles were simulated by models considering both degradation and diffusion, and were better described by a model assuming a domain structure of the gel (28).

The ABA-type triblock copolymer, PLGA-PEG-PLGA, is also a biodegradable, biocompatible polymer that demonstrates reverse thermal gelation properties (6,29). Its phase diagram is shown in Figure 1. The unique characteristics of the gel (ReGel^{TM}, Macromed, Inc.) lie in the following two key properties: (i) the triblock copolymer is a water soluble, biodegradable polymer at temperatures below the gel transi-

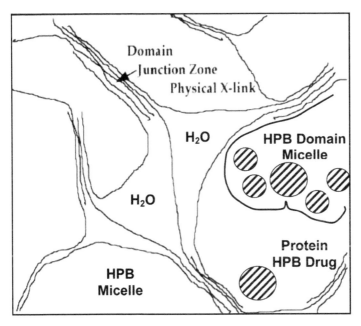

Figure 3 Physical crosslinks in the PLGA-PEG-PLGA in water (ReGel^{TM}) showing hydrophobic domain structures (Courtesy of Macromed, Inc., Sandy, UT.)

tion temperature; and (ii) the polymer also forms a water-insoluble gel once injected. This hydrogel formed above transition temperature possessed physical interactions between hydrophobic domains. A dynamic mechanical analysis revealed that an increase in viscosity of approximately 4 orders of magnitude accompanies the sol-gel transition. The gel forms a controlled release drug depot with delivery times ranging from 1 to 6 weeks. The inherent ability of the polymer to solubilize (400 to > 2000-fold) and stabilize poorly soluble and sensitive drugs, including proteins, is a substantial benefit. The gel provided excellent control of the release of paclitaxel for approximately 50 days, as shown in Figure 2. Paclitaxel loaded in the hydrophobic domain (Fig. 3) slowly diffuses out in the first phase of release followed by a more rapid release when the degradation of the polymer matrix

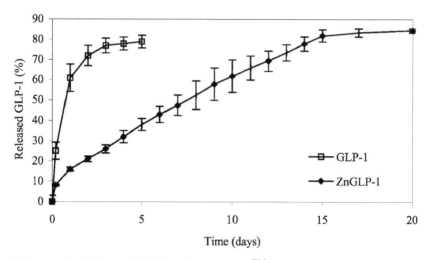

Figure 4 Released GLP-1 from ReGel™ formulation in vitro. The graph represents the average ± SE, and each group was composed of three samples. (From Ref. 32.)

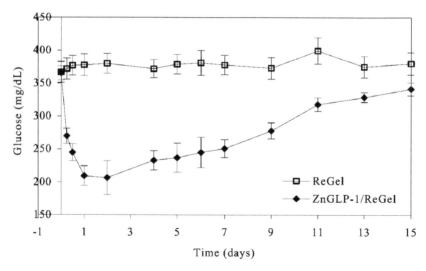

Figure 5 Blood glucose level in ZDF rats after injection. The graph represents the average ± SE, and each group was composed of five rats. (From Ref. 32.)

becomes more significant. Direct intratumoral injection of ReGel/paclitaxel (OncoGel^{TM}) was mentioned earlier. Efficacies equivalent to maximum tolerated systemic dosing were observed at OncoGel doses that were ten-fold lower (6). This ABA-type triblock copolymer was used as a drug release depot for continuous release of human insulin and of glucagons-like peptide-1 (GLP-1). The observation of both reduced initial burst and a constant release of human insulin from ReGel in vitro is due to the domain structure of the gel and to the modification of the association states of insulin by zinc. Animal studies using SD rats were performed to verify the release profile of insulin from this ABA block copolymer hydrogel. ReGel formulation maintained insulin release for up to 15 days, which could allow diabetic patients to reduce the number of insulin injections to two per month for basal insulin requirements (31).

The triblock copolymer hydrogel was used also as an injectable formulation for the controlled release of GLP-1 in vitro and in vivo (32). Because the aqueous solution of ReGel used in this study has a sol-gel transition around 32°C, the mixture of the GLP-1 and the aqueous free-flowing polymer solution spontaneously formed a gel at body temperature. GLP-1 was formulated into ReGel as an insoluble zinc complex to stabilize GLP-1 against aggregation and to sustain the release rate. The in vitro GLP-1 release from ReGel formulation at 37°C showed no initial burst, as shown in Figure 4. Further release was controlled by zero order kinetics. An animal study using Zucker Diabetic Fatty (ZDF) rats, as a type-2 diabetes animal model, showed that a blood glucose level was maintained at mild hyperglycemic level as shown in Figure 5. The glucose level dropped to a level significantly lower ($\sim$200 mg/dL) than the control ($\sim$400 mg/dL). A single injection of ZnGLP-1 loaded ReGel can be used for the controlled delivery of bioactive GLP-1 over a 2-week period (32).

5. BIODEGRADABLE MICROSPHERES BASED ON THE THERMOSENSITIVE PROPERTY OF PLGA-PEG-PLGA

Injectable controlled release systems based on biodegradable copolymers of lactic and glycolic acids (PLGAs) have become widely used for the delivery of both protein therapeutics and vaccine antigens (33–47). Although numerous protein therapeutics have been approved or are in clinical trials, the development of more sophisticated delivery systems for this rapidly expanding class of therapeutic agents has not kept pace. The short in vivo half-lives, the physical and chemical instability, and the low oral bioavailability of proteins currently necessitate their administration by frequent injections of protein solutions. This problem can be overcome through the use of injectable depot formulations in which the protein is encapsulated in, and released slowly from, microspheres made of biodegradable polymers. Although the first report of the sustained release of a microencapsulated protein was more than 20 years ago, the instability of proteins in these dosage forms has prevented their clinical use. Advances in protein stabilization, however, have aided the development of sustained release forms of several therapeutic proteins, and clinical testing of a monthly formulation human growth hormone was carried out. The obvious advantage of this method of delivery is that the protein is administered less frequently, sometimes at lower overall doses, than when formulated as a solution. More importantly, this method of delivery can justify commercial development of proteins that, for a variety of reasons, could not be marketed as solution formulations (36).

However, there are drawbacks observed in protein-loaded microspheres. The protein release kinetics exhibits an initial fast release followed by a slow release, resulting in an incomplete protein release despite significant degradation of microspheres (36). The very slow release kinetics was attributed to the protein aggregation and nonspecific adsorption within the microspheres. It was found that the protein

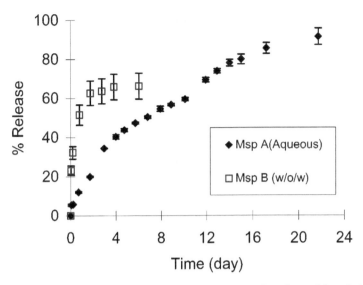

Figure 6   In vitro release of human insulin from Msp A (aqueous-based) and Msp B (dichloromethane-based) ($n = 3$, mean $\pm$ SD). (From Ref. 49.)

was significantly denatured and aggregated during the double emulsion formulation step (35). This problem is inherent in the fact that the polymer (e.g., PLGA) is not water soluble; thus a water-immiscible solvent (i.e., methylene chloride) has to be used. This also means that high mechanical energy input for the dispersal of the proteins in the organic phase and in this process w/o interface is created (37). Kwon et al. have observed that rapid aggregation of insulin is induced by the interface between water and methylene chloride (38).

   For example, Nutropin™ (Genentech, Inc., South San Francisco, CA), human growth hormone-loaded PLGA microspheres (40), was approved by FDA. However, the initial burst effect was not resolved. Burst release can result in acute overdosage and this could lead to fluid retention, headache, nausea, vomiting, or hyperglycemia.

   Over the last few years, improvements were suggested toward overcoming the difficulty of stabilizing PLGA-encapsulated proteins. In addition to the stabilization of proteins during encapsulation with anhydrous methods, protein complexation with zinc and the control of PLGA microclimate pH with basic excipients were also suggested (33).

One way to overcome the problems associated with both PLGA microspheres and the use of organic solvents is to utilize the thermosensitive property of PLGA-PEG-PLGA so that protein loading can be achieved without having to use organic solvents such as CH_2Cl_2, hence avoiding water/organic solvent interfaces (48,49).

Microspheres of the biodegradable, triblock copolymer (PLGA-PEG-PLGA, Mw = 4000, 1500–1000–1500 by NMR) were prepared in two methods: microsphere A (Msp A; aqueous-based) and microsphere B (Msp B; dichloromethane). For both microspheres, an equal amount of Zn-insulin was loaded (~4% of polymer mass). In vitro release studies were carried out with both. As shown in Fig. 6, Msp A exhibited a continuous and nearly complete release of insulin over 3 weeks. The first phase of insulin release (the first 10 days) from Msp A seems to be dependent more upon diffusion, indicated by the slight decrease in the release rate over time. Then, after day 10, the insulin release rate turned to an

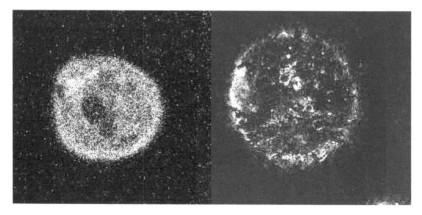

Figure 7 Confocal microscopy images of FITC-insulin loaded microspheres. (Left) Msp A; (right) Msp B. (From Ref. 49 with permission.)

increasing mode, and this is probably where the degradation of the matrix, at this time point, begins to play a more significant role in release than in the earlier phase. However, Msp B exhibited an initial burst release (~50% in 1 day), and release was discontinued at ~60% afterwards. In the preparation of Msp B, during the formation of primary emulsion where it involves high shear and heat generation in the creation of a large water/organic solvent interfacial area, proteins can undergo rapid aggregation, and thus the incomplete release of proteins from microspheres may be due to these trapped aggregates formed during microsphere fabrication. This accounts for the slow and incomplete release after the initial release phase with the burst effect (48).

In the case of Msp A, microspheres were prepared in a mild environment; that is, an organic solvent and a high shear were absent. A circular dichroism (CD) spectrum of insulin released from Msp A was virtually identical to that of a freshly prepared native insulin solution. This means that the released insulin preserved its secondary structure. In contrast, the CD spectrum for Msp B indicated a loss of secondary structure integrity due to both the use of dichloromethane and the harsh preparation conditions employed (48).

The observed release pattern from both types of microspheres lies in the distribution of the protein inside a microsphere, which is associated with the preparation method. In order to see this, FITC (fluorescein isothiocyanate)-insulin incorporated microspheres were observed under a confocal microscope, as shown in Fig. 7. For Msp A, a homogeneous distribution of fluorescence was observed while Msp B exhibited a rather heterogeneous distribution of FITC-insulin. In addition, Msp B shows significant surface fluorescence. These observations are, hence, consistent with the observed initial burst from Msp B and from the constant insulin release from Msp A over a prolonged period of time. It is reported that the constant release of insulin from triblock copolymer hydrogel may be attributed to the hydrophilic/hydrophobic domain structure of the gel. The incorporation of a significant fraction of insulin in the hydrophobic domain may have made possible the sustained release of insulin (49).

An animal study with streptozotocin-induced diabetic rats was carried out and involved the subcutaneous injection of both microspheres. While Msp B caused a burst effect (hypoglycemia) followed by a quick change in blood glucose and insulin levels, Msp A exhibited relatively sustained blood glucose levels and the release of insulin for 10 days. In vitro and in vivo insulin release profiles were found to be rather consistent (49).

6. CONCLUSIONS

The utility of thermosensitive hydrogels in drug delivery and biomedical application is immense. In particular, bio-degradable, thermosensitive triblock copolymers can be formulated into potentially useful drug delivery systems for therapeutic protein drugs or poorly water soluble drugs because such copolymers exhibit unique aqueous solution properties, biodegradability, and biocompatibility. Single injections of paclitaxel, insulin, and GLP-1 in the triblock copolymer hydrogel (ReGel) have demonstrated sustained and controlled release patterns of the agents both in vitro and in vivo. Based on the thermosensitive property of the triblock copolymer, PLGA-PEG-PLGA, biodegradable micro-spheres for protein delivery have been designed and contin-uous insulin release achieved, resulting in both a significant reduction in burst release and the preservation of bioactiv-ity. These systems can be potentially useful for the delivery of a wide range of therapeutically challenging agents such as DNA as well as protein and water-insoluble/toxic drugs.

REFERENCES

1. Heller J. Use of polymers in controlled release of active agents. In: Robinson JR, Lee VHL, eds. Controlled Drug Delivery. New York: Marcel Dekker, 1987:179–212.

2. Eliaz RE, Kost J. Characterization of polymeric PLGA-injectable implant delivery system for the controlled release of proteins. J Biomed Matter Res 2000; 50:388–396.

3. Van de Weert M, van Steenbergen MJ, Cleland JL, Heller J, Hennink WE, Crommelin DJ. Semisolid, self-catalyzed poly (ortho ester)s as controlled-release system: protein release and stability issues. J. Pharm Sci 2002; 91:1065–1074.

4. Jeong B, Bae YH, Lee DS, Kim SW. Biodegradable block copolymers as injectable drug-delivery systems. Nature 1997; 388:860–862.

5. Jeong B, Bae YH, Kim SW. In situ gelation of PEG-PLGA-PEG triblock copolymer aqueous solutions and degradation thereof. J Biomed Mater Res 2000; 50:171–177.

6. Zentner GM, Rathi R, Shih C, McRea JC, Seo M-H, Oh H, Rhee BG, Mestecky J, Moldoveanu Z, Morgan M, Weitman S. Biodegradable block copolymers for delivery of proteins and water-insoluble drugs. J Contr Rel 2001; 72:203–215.

7. Jeong B, Kim SW, Bae YH. Thermosensitive sol-gel reversible hydrogels. Adv Drug Delivery Rev 2002; 54:37–51.

8. Han CK, Bae YH. Inverse thermally reversible gelation of aqueous N-isopropylacrylamide copolymer solution. Polymer 1998; 39:2809–2814.

9. Bae YH, Okano T, Kim SW. "On-off" thermocontrol of solute transport. I. Temperature dependence of swelling of N-isopropylacrylamide networks modified with hydrophobic components in water. Pharm Res 1991; 8:531–537.

10. Bae YH, Okano T, Kim SW. "On-off" thermocontrol of solute transport. II. Solute release from thermosensitive hydrogels. Pharm Res 1991; 8:624–628.

11. Monji N, Hoffman AS. A novel immunoassay system and bioseparation process based on thermal phase separating polymers. Appl Biochem Biotechnol 1987; 14:107–120.

12. Miura M, Cole CA, Monji N, Hoffman AS. Temperature-dependent absorption/desorption behavior of lower critical solution temperature (LCST) polymers on various substrates. J Biomater Sci Polym Ed 1994; 5:555–568.

13. Chen G, Hoffman AS. Graft copolymers that exhibit temperature-induced phase transitions over a wide range of pH. Nature 1995; 373:49–52.

14. Hoffman AS. Bioconjugates of intelligent polymers and recognition proteins for use in diagnostics and affinity separations. Clin Chem 2000; 46:1478–1486.

15. Yoshida R, Sakai K, Okano T, Sakurai Y. Modulating the phase transition temperature and thermosensitivity in N-isopropylacrylamide copolymer gels. J Biomater Sci Polym Ed 1994; 6:585–598.

16. Okano T, Yamada N, Sakai H, Sakurai Y. A novel recovery system for cultured cells using plasma-treated polystyrene dishes grafted with poly(N-isopropylacrylamide). J Biomed Mater Res 1993; 27:1243–1251.

17. Takei YG, Matsukata M, Aoki T, Sanui K, Ogata N, Kikuchi A, Sakurai Y, Okano T. Temperature-responsive bioconjugates. 3. Antibody-poly (N-isopropylacrylamide) conjugates for temperature-modulated precipitations and affinity bioseparations. Bioconjug Chem 1994; 5:577–582.

18. Feil H, Bae YH, Feijen J, Kim SW. Mutual influence of pH and temperature on the swelling of ionizable and thermosensitive hydrogels. Macromolecules 1992; 25:5528–5530.

19. Feil H, Bae YH, Feijen J, Kim SW. Mechanism of solute diffusion in thermosensitive hydrogels. In: Proceedings of the 18th International Symposium on the Controlled Release of Bioactive Materials; Lee VHL, Ed; CRC Inc: Lincolnshire, IL, 1991:142.

20. Serres A, Baudys M, Kim SW. Temperature and pH-sensitive polymers for human calcitonin delivery. Pharm Res 1996; 13:196–201.

21. Kim Y-H, Bae YH, Kim SW. pH/temperature-sensitive polymers for macromolecular drug loading release. J Contr Rel 1994; 28:143–152.

22. Ramkissoon-Ganorkar C, Liu F, Baudys M, Kim SW. Effect of molecular weight and polydispersity on kinetics of dissolution and release from pH/temperature-sensitive polymers. J Biomater Sci Polym Ed 1999; 10:1149–1161.

23. Vernon B, Kim SW, Bae YH. Thermosensitive copolymer gels for extracellular matrix. J Biomed Mater Res 2000; 51:69–79.

24. von Recum H, Kikuchi A, Okuhara M, Sakurai Y, Okano T, Kim SW. Retinal pigmented epithelium cultures on thermally responsive polymer porous substrates. J Biomater Sci Polym Ed 1998; 9:1241–1253.

25. Katakam M, Ravis WR, Banga AK. Controlled release of human growth hormone in rats following parenteral administration of poloxamer gels. J Contr Rel 1997; 49:21–26.

26. Staratton LP, Dong A, Manning MC, Carpenter JF. Drug delivery matrix containing native protein precipitates suspended in a poloxamer gel. J Pharm Sci 1997; 86: 1006–1010.

27. Jeong B, Bae YH, Kim SW. Thermosensitive gelation of PEG-PLGA-PEG triblock copolymer aqueous solutions. Macromolecules 1999; 32:7064–7069.

28. Jeong B, Bae YH, Kim SW. Drug release from biodegradable injectable thermosensitive hydrogel of PEG-PLGA-PEG triblock copolymers. J Contr Rel 2000; 63:155–163.

29. Lee DS, Shim MS, Kim SW, Lee H, Park I, Chang T. Novel thermoreversible gelation of biodegradable PLGA-block-PEO-block-PLGA triblock copolymers in aqueous solution. Macromol Rapid Commun 2001; 22:587–592.

30. Jeong B, Kibbey MR, Birnbaum JC, Won Y-Y, Gutowska A. Thermogelling biodegradable polymers with hydrophilic backbones: PEG-g-PLGA. Macromolecules 2000; 33:8317–8322.

31. Kim YJ, Choi S, Koh JJ, Lee M, Ko KS, Kim SW. Controlled release of insulin from injectable biodegradable triblock copolymer. Pharm Res 2001; 18:548–550.

32. Choi S, Baudys M, Kim SW. Control of blood glucose by Novel GLP-1 delivery using biodegradable triblock copolymer of PLGA-PEG-PLGA in type 2 diabetic rats. Pharm Res 2004; 21: 827–831.

33. Schwendeman SP. Recent advances in the stabilization of proteins encapsulated in injectable PLGA delivery systems. Crit Rev Ther Drug Carrier Syst 2002; 19:73–98.

34. Putney SD, Burke PA. Improving protein therapeutics with sustained-release formulations. Nature Biotechnol 1998; 16: 153–157.

35. Lu W, Park TG. Protein release from poly(lactic-co-glycolic acid) microspheres: Protein stability problems. PDA J Pharm Sci Tech 1995; 49:13–19.

36. Hanes J, Cleland JL, Langer R. New advances in microsphere-based single-dose vaccines. Adv Drug Delivery Rev 1997; 28:97–119.

37. van de Weert M, Hennink WE, Jinkoot W. Protein instability in poly (lactic-co-glycolic acid) microparticles. Pharm Res 2000; 17:1159–1167.

38. Kwon YM, Baudys M, Knutson K, Kim SW. In situ insulin aggregation induced by water/organic solvent interface. Pharm Res 2001; 18:1754–1759.

39. Okada H. One-and three-month release injectable microspheres of the LH-RH superagonist leuprorelin acetate. Adv Drug Del Rev 1997; 28:43–70.

40. Cleland JL, Jones AJS. Stable formulations of recombinant human growth hormone and interferon-γ for mocroencapsulation in biodegradable microspheres. Pharm Res 1996; 13: 1464–1475.

41. Morlock M, Kissel T, Li YX, Koll H, Winter G. Erythropoietin loaded microspheres prepared from biodegradable LPLG-PEO-LPLG triblock copolymers: protein stabilization and in-vitro release properties. J Contr Rel 1998; 56:105–115.

42. Crotts G, Sah H, Park TG. Adsorption determines in-vitro protein release rate from biodegradable microspheres: quantitative analysis of surface area during degradation. J Contr Rel 1997; 47:101–111.

43. King TW, Patrick CW. Development and in vitro characterization of vascular endothelial growth factor-loaded poly(DL-lactic-co-glycolic acid)/poly(ethylene glycol) microspheres using a solid encapsulation/single emulsion/solvent extraction technique. J Biomed Mater Res 2000; 51:383–390.

44. Cho KY, Choi SH, Kim C-H, Nam YS, Park TG, Park J-K. Protein release microparticles based on the blend of poly

(D,L-lactic-co-glucolic acid) and oligo-ethylene glycol grafted poly(L-lactide). J Contr Rel 2001; 76:275–284.

45. Meinel L, Illi OE, Zapf J, Malfanti M, Merkle HP, Gander B. Stabilizing insulin-like growth factor-I in poly(D,L-lactide-co-glycolide) microspheres. J Contr Rel 2001; 70:193–202.

46. Carrasquillo KG, Stanley AM, Aponte-Carro JC, De Jesus P, Costantino HR, Bosques CJ, Griebenow K. Non-aqueous encapsulation of excipient-stabilized spray-freeze dried BSA into poly(lactide-co-glycolide) microspheres results in release of native protein. J Contr Rel 2001; 76:199–208.

47. Takenaga M, Yamaguchi Y, Kitagawa A, Ogawa Y, Mizushima Y, Igarashi R. A novel sustained-release formulation of insulin with dramatic reduction in initial rapid release. J Contr Rel 2002; 79:81–91.

48. Kwon YM, Kim SW. New biodegradable polymers for delivery of bioactive agents. Macromol Symp 2003; 201:179–186.

49. Kwon YM, Kim SW. Biodegradable triblock copolymer microspheres based on thermosensitive sol-gel transition. Pharm Res 2004; 21:339–343.

7

Hydrogels
Stimuli-Sensitive Hydrogels

AKIHIKO KIKUCHI and TERUO OKANO

Institute of Advanced Biomedical Engineering
and Science, Center of Excellence (COE) Program
for the 21st Century, Tokyo Women's Medical
University, and Core Research for Evolutional
Science and Technology (CREST), Japan Science
and Technology Agency, Tokyo, Japan

1. INTRODUCTION

Hydrogels are physically or chemically cross-linked polymer networks swollen with large amounts of water. Due to their crosslinked nature, these gels do not dissolve in aqueous media but contain an enormous amount of solvated water molecules within the entangled polymer chain matrix. Hydrogel properties are reviewed elsewhere in this book. This chapter is dedicated to a unique hydrogel family that responds to

275

externally applied stimuli that, in turn, alter these hydrogels' swelling properties. During the past two decades, much work has been dedicated to the development of stimuli-responsive hydrogel materials. These "intelligent materials" sense external stimuli, alter (depending on the degree or strength of stimulation) their physicochemical network properties, and release drug molecules or absorb water or both to reach an equilibrated state. Such auto-feedback systems are commonly observed in metabolic processes in the human body. Therefore, these intelligent materials are appropriately applied to the development of new drug delivery matrices responding to several physiological stimuli arising from disease states or metabolic events in the human body. One key strategy for drug delivery systems is the spatio-temporal control of drug release responding to any changes in body physiology at specific sites (1,2).

In the present chapter, several types of stimuli-responsive hydrogels are introduced, and their applications to drug delivery systems will be reviewed.

2. STIMULI-RESPONSIVE HYDROGELS AND THEIR APPLICATIONS IN DRUG DELIVERY SYSTEMS

2.1. Temperature Sensitive Polymeric Materials

Several types of hydrogels are known to undergo physical changes in response to changes in temperature. These include poly(vinylmethyl ether)s cross-linked by gamma ray irradiation (3), poly(N-isopropylacrylamide) (PIPAAm) and its derivatives (4–15), poly(ethylene oxide)-b-poly(propylene oxide)-b-poly(ethylene oxide) triblock copolymers (16–19), poly(alkyl vinyl ether)s and their block copolymers, hydroxymethylcellulose, gelatin, and other more exotic materials. All of these materials share in common a unique hydration chemical structure in aqueous milieu that is metastable and can be radically altered by increasing the thermal energy in the system. Changing temperature therefore often produces a dramatic and pseudo first-order phase change resulting from the dehydration and rehydration of the materials' chemistry, resulting in a collapse and expanding behavior in water. This phase transition is reversible, with some characteristic hysteresis, upon reversal of the temperature change. Many materials exhibit this property (20–23).

2.1.1. Poly(N-isopropylacrylamide) Hydrogels for Thermo-Responsive Drug Delivery

With several chemical features analogous to proteins, polyamides, and poly(amino acids), PIPAAm was synthesized as a model polypeptide analogue so that its solution behavior in water could be investigated (24). PIPAAm in an aqueous solution shows a lower critical solution temperature (LCST) at 32°C. This unique temperature response of PIPAAm in an aqueous solution has been extensively investigated for use in stimuli-responsive materials for biomedical applications in drug delivery systems, bioseparations, bioconjugates, and noninvasive cell manipulations (4–7,9,11,25–42). In cross-linked PIPAAm networks, the polymer's aqueous soluble-insoluble changes that occur in relation to temperature result in reversible swelling-deswelling in the hydrogels. Copolymerization of IPAAm with either hydrophilic or hydrophobic monomers produces an increase or decrease in the transition temperature of PIPAAm. Such a strategy can be exploited so that a hydrogel's transition temperature can be controlled and that the drug release behavior can, thus, be regulated. In the present section, recent advances in the development of thermo-responsive hydrogels and their applications in drug delivery systems are reviewed.

For PIPAAm hydrogels, a large temperature increase originating below the polymer transition temperature of 32°C induces an outside-in shrinking response in the gel: thermal transfer and polymer mass transfer kinetics compete for the determining of the polymer phase behavior. For this condition, the result is the formation of a dense shrunken collapsed polymer layer at the gel-water interface (a skin layer). These dehydrated polymer skin layers on the surface of shrunken PIPAAm hydrogels prevent even water molecules from readily diffusing through the gel. Thus, drug release from PIPAAm and their derivative hydrogels is governed by a drug's bulk water diffusion below the transition temperature, with the release of the drug completely impeded above the transition temperature (5,6,31). This temperature-dependent drug release mechanism produces a thermo-responsive on-off drug release with PIPAAm hydrogels that is useful for triggered thermal release control.

Rapid swelling-deswelling gel kinetics cannot be obtained with conventional hydrogels in which mass transfer kinetics are governed by the reciprocal of the squared dimension of the gels (23). Therefore, both the inclusion of macro- or microporous structures within gels and a reduction in gel size are common methods by which to accelerate gel shrinking behavior (3,13,43). To overcome this limited swelling-deswelling kinetics issue, we have introduced freely mobile grafted PIPAAm chains within multiple bonded three-dimensional cross-linked PIPAAm hydrogels (10,44,45). Using telomerization polymerization, we prepared a chain transfer agent in the form of PIPAAm chains that had one terminal amino end group and that were in the presence of aminoethanethiol. Terminal amino groups were then converted, through a reaction with acryloyl chloride, to polymerizable acrylamide moieties. PIPAAm macromonomers were then copolymerized with an IPAAm monomer in the presence of the cross-linker, N,N'-methylenebisacrylamide. The structural formula is shown in Fig. 1. These novel graft-type PIPAAm hydrogels exhibit the same transition temperature as conventional PIPAAm hydrogels at 32°C but also exhibit highly altered swelling-deswelling kinetics. The molecular weight of the grafted PIPAAm chains showed significant influences on gel deswelling. Comb-type grafted PIPAAm hydrogels with different graft chain lengths were examined for their swelling-deswelling characteristics. These kinetics were investigated with PIPAAm hydrogels containing grafted side chains of 2900, 4000, and 9000 g/mol. PIPAAm grafted gels with a graft molecular weight of 9000 (IGG9000) showed rapid shrinking kinetics upon a temperature increase above the transition temperature in buffer at 32°C. Such rapid gel shrinking is

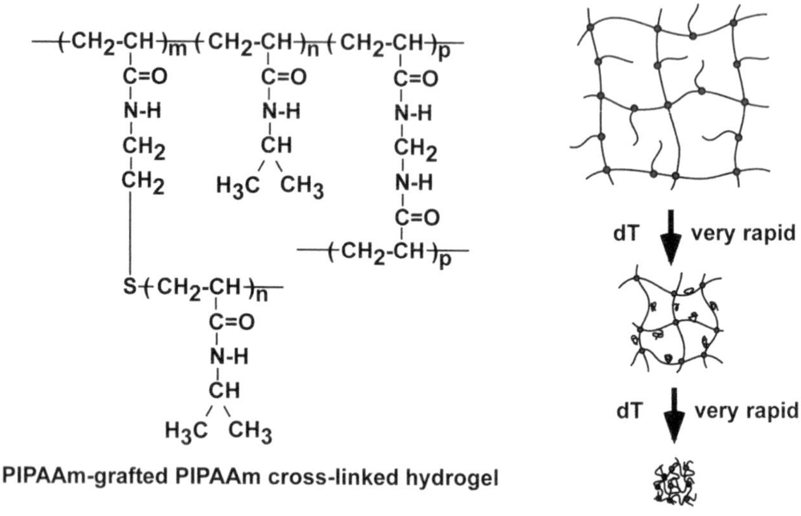

$$-(CH_2\text{-}CH)_m(CH_2\text{-}CH)_n(CH_2\text{-}CH)_p-$$

PIPAAm-grafted PIPAAm cross-linked hydrogel

Figure 1 Structural formula of linear PIPAAm-grafted PIPAAm hydrogels and schematic illustration of their shrinking behavior.

probably due to the existence of grafted PIPAAm chains that, owing to their mobile nature in the gel, dehydrate and then aggregate upon a temperature increase prior to bulk gel dehydration. Such hydrophobic aggregation of grafted chains induces and accelerates the entire gel-deswelling behavior. However, hydrogels with grafted chains of lower molecular weight showed relatively slow deswelling, with only partial skin layer formation, even though the overall grafted chain content was the same (approximately 30 wt.%) as for the PIPAAm gels having longer grafted chains. The cylinder-shaped, faster deswelling hydrogel, IGG9000, showed repetitive and large oscillations in length when gels were subjected to temperature cycles between 20 and 40°C with hysteresis accorded to collapse-rehydration dynamics and water mass transport in these bulk systems (44).

Such differences in bulk gel-deswelling characteristics have a significant influence on the release of incorporated drug molecules, especially for higher molecular weight drugs. Figure 2 shows the temperature-induced drug release behavior from conventional and PIPAAm-grafted hydrogels that

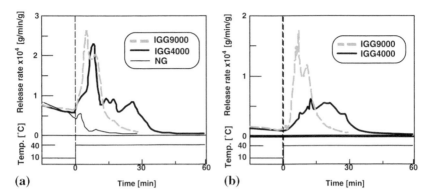

Figure 2 Thermo-responsive drug release behavior from PIPAAm-grafted PIPAAm hydrogels upon step temperature gradient for (a) sodium salicylate, and (b) dextran of MW 9300.

had different graft chain lengths. Figure 2a illustrates the release of small, water soluble sodium salicylate from these gels upon a temperature increase from 10 to 40°C. Drug release from conventional PIPAAm hydrogels was pulsed. Immediately after the temperature increased, a small amount of the drug was released, and then a substantial decrease in the amount of the drug being released took place so that the release was completely impeded after 15 min. This result is due to the impermeable, dense collapsed polymer skin layer formation at the gel's outer surface. Drug release from graft-type hydrogels exhibited rapid, pulsed release; 65% of drug molecules incorporated were released from IGG9000 within the first 15 min after temperature change. IGG4000, however, exhibited oscillating release profiles. For relatively lower molecular weight grafted chains, chain aggregation forces still operated within the gels above the transition temperature, and thus, surface skin layers were formed on these hydrogel surfaces. However, these lower molecular weight chains' collapse forces are weaker than those of longer chains: the increasing internal hydrostatic pressure resulted from the bulk gel shrinking of both the aggregated grafted chains and, subsequently, the main chain. These hydrostatic forces eventually overwhelm the weaker short-chain collapse forces on the device surface, an outcome that allows a pulse of outward water-drug flux to release the pressure gradient. This produces a cyclic, pulsatile drug release profile.

For higher molecular weight drugs (e.g., dextran MW 9300), a burst release was observed for IGG9000 hydrogels after a temperature increase to 40°C (Fig. 2b). By contrast, the dextran release from IGG4000 was relatively suppressed and distinctly different from the small molecule case. These differences in drug release behavior for the two grafted gels are probably due both to the chain aggregation forces operating within the hydrogels and to their influence on diffusing molecules of different size. That is, longer grafted chains form more hydrophobic cores within the hydrogels immediately after a temperature increase above the transition temperature, an outcome that facilitates faster hydrogel deswelling and hydrostatic pressure-driven outward transport. In contrast, the hydrophobic aggregation between the dehydrated, shrunken grafted chains of IGG4000 is comparatively relatively weak. Therefore, the thin collapsed polymer skin layers formed at the interface of IGG4000 gels limit and retard large drug diffusion from the gel's interior. These results strongly suggest that the drug release behavior of various drugs with a diverse range of molecular weights can be regulated with the grafted chain lengths of comb-type grafted PIPAAm hydrogels.

Similar fast deswelling characteristics were observed using poly(ethylene glycol) (PEG)-grafted PIPAAm hydrogels similar in structure to PIPAAm-grafted hydrogels above the characteristic transition temperature near 32°C (46). PEG-grafted chains did not alter the bulk hydrogels' thermoresponsive characteristics, although PEG itself is a highly hydrophilic polymer. Thus, hydrophilic PEG-based channels or pores remained after the bulk gel transition. This outcome is in sharp contrast to conventional PIPAAm hydrogels containing acrylamide (AAm) or acrylic acid (AAc) as the hydrophilic co-monomer, since the introduction of these co-monomers incorporates random compositional distribution throughout the bulk gel and is not polymer grafting per se. This hydrophilic monomer incorporation produces a significant increase in the gel transition temperature, and further co-monomer introduction eventually abolishes the thermo-responsive characteristics of PIPAAm hydrogels. Maintenance of the thermo-responsive characteristics of PEG-grafted hydrogels is probably due to the structural independence of PEG-grafted chains and PIPAAm cross-linked chains. The deswelling mechanism for PEG-grafted PIPAAm hydrogels is distinct from PIPAAm-grafted hydrogels, as PEG chains are highly hydrated at all of the temperature ranges examined. The PEG chains form water-releasing channels within the dehydrated and collapsed skin layers on the shrinking PIPAAm hydrogels owing to a microphase separation of PEG chains within the collapsing PIPAAm phase. This effect accelerates bulk gel-deswelling above the transition temperature. Therefore, rapid hydrogel dehydration is achieved. Using this unique deswelling mechanism that shrinks PEG-grafted PIPAAm hydrogels, researchers can achieve both rapid drug release and bulk water release. Furthermore, such hydrogels can be used to target temporal drug releases, activated only when stepwise temperature stimuli are applied at target sites.

2.1.2. Thermo-Responsive Sol-to-Gel Transitions and Exploitation for Injectable Drug Delivery

Poly(ethylene oxide)-b-poly(propylene oxide)-b-poly(ethylene oxide) (PEO-PPO-PEO) (Fig. 3) is a well-known thermo-responsive self-assembling polymer with the trade name of Pluronic®, (BASF, Florham, NJ., USA) (47). A variety of Pluronics have been developed with varying compositions of PEO and PPO blocks. Pluronics with relatively large PEO weight ratios are soluble in water through the self-association of relatively hydrophobic PPO segments, resulting in the formation of polymer micelles. These polymeric micelles exist in an equilibrated state between monomeric polymer molecules and micelles. Some types of Pluronics exhibit thermally reversible physical gelation at certain concentration ranges and temperatures (48). Pluronic F127, containing 70 wt.% PEO with

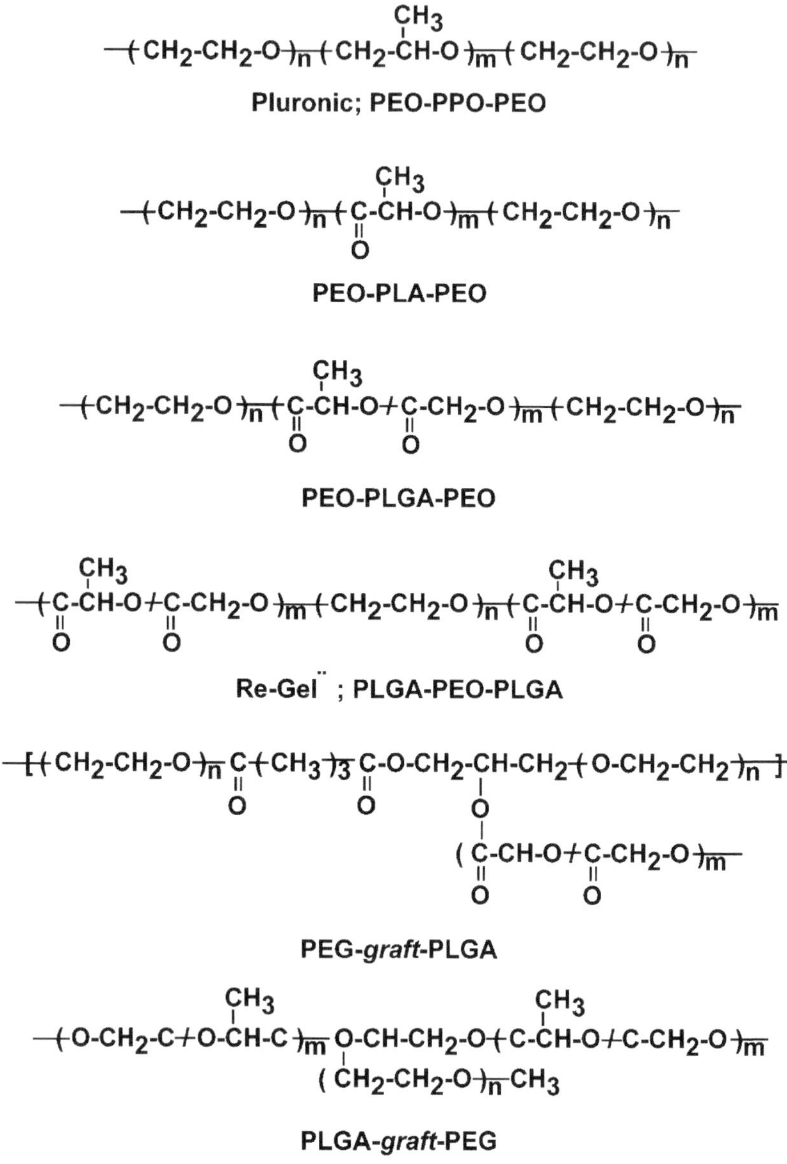

$$-\!(CH_2\text{-}CH_2\text{-}O\,)_{\overline{n}}(\,CH_2\text{-}\overset{\overset{\displaystyle CH_3}{|}}{CH}\text{-}O\,)_{\overline{m}}(\,CH_2\text{-}CH_2\text{-}O\,)_{\overline{n}}$$

Pluronic; PEO-PPO-PEO

$$-\!(CH_2\text{-}CH_2\text{-}O\,)_{\overline{n}}(\,\underset{\overset{\|}{O}}{C}\text{-}\overset{\overset{\displaystyle CH_3}{|}}{CH}\text{-}O\,)_{\overline{m}}(\,CH_2\text{-}CH_2\text{-}O\,)_{\overline{n}}$$

PEO-PLA-PEO

$$-\!(CH_2\text{-}CH_2\text{-}O\,)_{\overline{n}}(\,\underset{\overset{\|}{O}}{C}\text{-}\overset{\overset{\displaystyle CH_3}{|}}{CH}\text{-}O\!+\!\underset{\overset{\|}{O}}{C}\text{-}CH_2\text{-}O\,)_{\overline{m}}(\,CH_2\text{-}CH_2\text{-}O\,)_{\overline{n}}$$

PEO-PLGA-PEO

$$-\!(\underset{\overset{\|}{O}}{C}\text{-}\overset{\overset{\displaystyle CH_3}{|}}{CH}\text{-}O\!+\!\underset{\overset{\|}{O}}{C}\text{-}CH_2\text{-}O\,)_{\overline{m}}(\,CH_2\text{-}CH_2\text{-}O\,)_{\overline{n}}(\,\underset{\overset{\|}{O}}{C}\text{-}\overset{\overset{\displaystyle CH_3}{|}}{CH}\text{-}O\!+\!\underset{\overset{\|}{O}}{C}\text{-}CH_2\text{-}O\,)_{\overline{m}}$$

Re-Gel\u{·}\u{·} ; PLGA-PEO-PLGA

$$-\![(CH_2\text{-}CH_2\text{-}O\,)_{\overline{n}}\underset{\overset{\|}{O}}{C}(\,CH_3\,)_{\overline{3}}\underset{\overset{\|}{O}}{C}\text{-}O\text{-}CH_2\text{-}\underset{\overset{\displaystyle |}{O}}{CH}\text{-}CH_2\!+\!O\text{-}CH_2\text{-}CH_2\,)_{\overline{n}}\,]-$$

$$(\,\underset{\overset{\|}{O}}{C}\text{-}CH\text{-}O\!+\!\underset{\overset{\|}{O}}{C}\text{-}CH_2\text{-}O\,)_{\overline{m}}$$

PEG-*graft*-PLGA

$$-\!(O\text{-}CH_2\text{-}C\!+\!O\text{-}\overset{\overset{\displaystyle CH_3}{|}}{CH}\text{-}C\,)_{\overline{m}}O\text{-}CH\text{-}CH_2\text{-}O\!+\!C\text{-}\overset{\overset{\displaystyle CH_3}{|}}{CH}\text{-}O\!+\!C\text{-}CH_2\text{-}O\,)_{\overline{m}}$$

$$(\,CH_2\text{-}CH_2\text{-}O\,)_{\overline{n}}CH_3$$

PLGA-*graft*-PEG

Figure 3 Structural formulae for a variety of types of thermo-gelling materials.

12,000 MW triblock segments, is a typical example of a thermally reversible gelation material. Aqueous solutions of Pluronic F127 (>20 wt.%) gel spontaneously upon a temperature increase of ~20°C without syneresis of water molecules; that is, no volume change occurs during gelation. This outcome is due to the dense packing of polymeric micelles, which occupy the entire volume of the polymer solution and encapsulate all of the bulk water. Such gels achieve a thermally reversible solution state upon a decrease in temperature. This unique property has been utilized for injectable drug delivery systems, especially for labile drugs such as peptides and proteins. The Pluronic drug delivery formulation results from the mixing of polymers directly with a drug at low temperatures, followed by injection of the mixture into the body. At injection sites, the Pluronic solution—once placed in the body—gels owing to a temperature increase. Drug molecules then diffuse from the in situ-formed gels in a controlled release fashion (16,17,19,49,50). Pluronic gel monoliths slowly dissolve owing to their dilution with body fluids and are ultimately excreted from the body. Thus, drug release can be extended from several hours to a few days, with drug release for longer periods not achievable with this formulation.

Kim and his colleagues (51–55) designed ABA-type triblock polymers that, as do Pluronics, exhibit thermoresponsive gelation at body temperature. As this book has devoted one chapter for these thermogelling triblock copolymers and their applications in drug delivery systems, only limited information will be given here.

The researchers used methoxy-terminated PEG with a known molecular weight for macroinitiator, to initialize the ring-opening polymerization of lactide or glycolide and formed a PEG-polylactide diblock copolymer. Then, two diblock polymer molecules were coupled using hexamethylene diisocyanate to form PEG-PLLA-PEG (51,56) or PEG-PLGA-PEG (52,53,55) triblock copolymers that exhibited a relatively narrow molecular weight distribution ($Mw/Mn \sim 1.2$). As shown in Figure 3, these triblock copolymers contain biodegradable polyester segments. Researchers found the thermally induced gelation of triblock polymer aqueous solutions at 17–40 wt.%

near 30–36°C (52) (Fig. 4). A further increase in temperature affected gel appearance, from transparent to turbid (just above the physiological temperature), translucent, turbid, and then finally dissolving back to an opaque solution at a critical temperature ranging from 44 to 70°C.

Kim and his colleagues then investigated the in situ gelation of a PEG-PLGA-PEG triblock copolymer that, being in an aqueous solution (33 wt.%), was injected subcutaneously into rats (57). Transparent gels formed immediately after subcutaneous injection, and these formed gels became strong enough to handle with forceps 24 h after injection. PEG-PLGA-PEG triblock copolymer hydrogels remained at subcutaneous sites 1 month after incubation with turbid appearance. This outcome is in sharp contrast to Pluronic hydrogels, which dissolve and disappear from the injection site within a few days. The relatively hydrophobic core of PEG-PLGA-PEG triblock copolymers prevents permeation of water molecules and, thus, results in a relatively stable core-shell micelle structure. These gels lost their mass approximately 30% after 1 month of incubation in the body, probably due to hydrolysis.

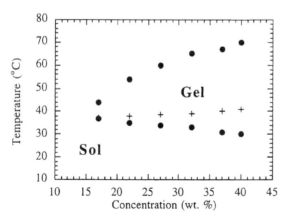

Figure 4 Phase diagram of PEG-PLGA-PEG (550-2810-550) triblock copolymer aqueous solution. Cross bar: temperature where the formed gels become turbid. (From Ref. 50.)

^{13}C NMR analyses revealed that degradation in the body caused PEG segments to be preferentially diffused out of the gels.

Kim and his colleagues (53) also investigated the drug release behavior from PEG-PLGA-PEG in situ formed gels. Two low molecular weight compounds, ketoprofen and spironolactone, were used as model drug molecules having different hydrophobicities. The relatively hydrophilic ketoprofen was released monotonously through diffusion mechanisms, with approximately 90% of the drug released within 5 days. In contrast, the more hydrophobic spironolactone was released and showed a sigmoid curve, with the release extending over 50 days. Given that the polymeric micellar structure was maintained within the triblock copolymer gels, the spironolactone molecules in PEG-rich shell layers were released mainly by a diffusion process, while drugs preferentially existing in the hydrophobic micelle core were released via diffusion and bulk micelle matrix degradation. Thus, the longer-term sustained release of drugs was achieved using PEG-PLGA-PEG triblock copolymer gels.

These triblock copolymer formulations are attractive as drug delivery depots because (i) their formulation requires no organic solvent; (ii) the triblock copolymer matrices can be stored as dry, solid forms before administration; (iii) drugs with a delivery vehicle can be injected directly by syringe so that no surgical operation is necessary; and (iv) the polymeric matrix is biodegradable, and the components come to have low molecular weight molecules so that those components are excreted by physiological mechanisms.

Zentner et al. (58) prepared PLGA-PEG-PLGA triblock copolymers via the bulk polymerization of PEG with lactide and glycolide in the presence of stannous 2-ethylhexanoate, for commercial delivery products, ReGel® (see structures in Fig. 3). These triblock copolymers have an analogous but inverted structure derived from the triblock systems reported by Kim and colleagues (51–55). Aqueous polymer solutions of more than 10 wt.% show three phases—solution, gel, and precipitate—depending on temperature, as shown in Figure 5. During the solution-to-gel transition, 23 wt.% ReGel showed

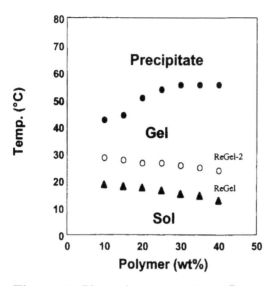

Figure 5 Phase diagram of ReGel®. (From Ref. 55.)

large viscosity changes (approximately four orders of magnitude) from 0.4 P in the sol state to 5700 P at the onset of gelation at 13.6°C. Such viscoelastic behavior is reproducible with repeated temperature changes. In vitro degradation of ReGel showed more rapid degradation at higher temperatures. Complete degradation of ReGel occurred after 6–8 weeks at 37°C, whereas, at low temperatures, polymers were stable for (for example) 20–30 weeks at 5°C and more than 2 years at −10°C. The in vivo degradation of ReGel after subcutaneous injection indicated that gel appearance dramatically changes between 2 and 4 weeks: initially gelled polymer matrices decrease in size during the first 2 weeks; then, they become a mixture of gel and a viscous liquid, and then a completely viscous liquid with no gel; finally, the matrices are completely absorbed into the body, and follow a simple hydrolysis mechanism.

Highly hydrophobic, and thus practically insoluble, drugs such as paclitaxel and cyclosporin A ($\sim$4 µg/mL), can be dissolved dramatically by mixing these drugs into a 23 wt.%

ReGel at 5°C, a solubility increase of 400 to >2000-fold. These drugs are stable in ReGel formulations, with more than 85% stability after 30 days of incubation at 37°C, and more than 99% after storage for one year at −10 and 5°C. Thus, drug stability is significantly improved by the use of ReGel formulations.

Zentner et al. (58) also evaluated in vitro paclitaxel release from the ReGel system and in comparison with the Pluronic F127 system. Drug molecules were completely released from the Pluronic F127 formulation within a 1-day period, indicating an unstable gel structure (Fig. 6). In sharp contrast, the paclitaxel release from ReGel systems showed a two-phase release pattern: a diffusion-governed mechanism for initial 14 days, and the combined mechanism of diffusion and polymer degradation for 50 days. Direct injection of paclitaxel containing ReGel solutions to solid tumor sites that had developed in mice resulted in the gelation of polymers only at injection sites, that is, within tumor tissues. Paclitaxel thus

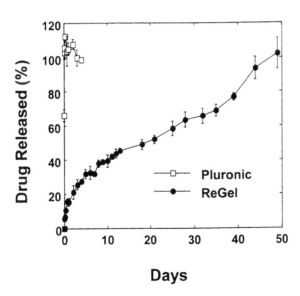

Figure 6 In vitro release of paclitaxel from ReGel® (23 w/w% in water) in comparison with that from Pluronic F-127. (From Ref. 55.)

affected not normal tissues but only tumor tissue. The drug remained at the tumor site 42 days after injection, by which time it had gradually decreased to 20%.

ReGel can also be utilized for the formulation of peptide drugs, like insulin (59), growth hormones (GH) and granulocyte colony-stimulating factor (G-CSF). In vitro protein drug release is monotonous and lasts for approximately 2 weeks after injection. Such differences may be due to the different molecular weights of drug molecules and the drug distribution within the gel matrix.

The degradation of the ReGel matrix results in the formation of lactic acid, glycolic acid, and PEG with a molecular weight of 1000 or 1450, which are all considered biocompatible. Thus, the ReGel formulation may be widely applicable to many types of drug molecules and peptides.

Jeong et al. (60) designed a relatively short-term drug delivery depot based on PEG grafted with biodegradable poly(lactide-co-glycolide) (PLGA) (see structure in Fig. 3). They initially reacted PEG (MW 1000) with glutaric anhydride to form carboxyl-terminated PEG, followed by chain extension with epoxy-PEG. During this reaction, pendant hydroxyl groups were also simultaneously introduced. These hydroxyl groups were then used to polymerize lactide and glycolide and, thus, to form hydrophobic, biodegradable grafted side chains. In water, PEG chains were hydrated and elongated, whereas PLGA chains were hydrophobic and condensed. These graft type polymers thus formed a polymer micelle structure 9 nm in diameter in water above their critical micelle concentration of 0.01–0.05 wt.%. Polymer solutions of more than 16 wt.% show a temperature-dependent sol-to-gel phase transition. These gels form two phases, water and precipitates, with an increase in temperature. Interestingly, polymer solutions at some concentrations remain in a gel phase at 37°C. Such properties allow for the use of the solutions as an injectable, in situ-gelling drug delivery depot. Aqueous polymer solutions of 22 wt.% show a lower viscosity of approximately 30 cP at 20°C, but induce gel formation when the temperature rises to 37°C. Gels remain unchanged for 7 days, after which point the gel matrix disintegrates, resulting in a clear solution. These data indicate that in situ-formed PEG-g-PLGA gels can be used for drug delivery systems that require only short-term duration.

Chung et al. (61) prepared graft copolymers comprising a PLGA backbone with PEG-grafted chains by a one-pot ring-opening polymerization of lactide and glycolide in the presence of methoxy-PEG and epoxy-PEG (see structure in Fig. 3). These PLGA-g-PEG polymer solutions showed a temperature-dependent sol-to-gel phase transition at concentrations of 20–30 wt.% and a transition temperature of around 30°C. Gel formation occurred immediately after a temperature increase, with subcutaneous injection into rats resulting in the formation of a round-shaped gel at injection sites. PLGA-g-PEG gels remained at injection sites for more than 2 months. Thus, the long-term release of drugs can be achieved with these gels.

Jeong et al. (62) also used these two types of copolymers as a protein delivery depot or tissue engineering matrices. Two different formulations were prepared: (i) PEG-g-PLGA/PLGA-g-PEG (50/50 by weight) and (ii) PLGA-g-PEG alone. These graft copolymers contained three graft chains and had a number average molecular weight of ~6000. The two formulations were dissolved in phosphate buffer at 25 wt.% with insulin, and injected subcutaneously (35.54 mg/kg-rat) into streptozotocin-induced diabetic rats. After injection, blood glucose levels were monitored periodically: normal glucose levels were observed for 5 days after the injection of the PEG-g-PLGA/PLGA-g-PEG mixture, and 2 weeks in the case of PLGA-g-PEG alone. The former formulations, consisting of two different graft copolymers, were mechanically weaker than the latter, and faster erosion occurred in the body for mixtures of PEG-g-PLGA/PLGA-g-PEG. Tissues around gel implantation sites were analyzed histologically and, after one month, were found to exhibit minimal chronic inflammation. Thus, these in situ-formed gels can be used as delivery devices for peptides and proteins.

In situ formation of gels can also be applied so that target cells are fixed at desired sites and that the gel structures are maintained as long as the repaired tissues replace the gel matrices. Such an approach, termed *cell delivery*, has recently received a great deal of attention from researchers. Results indicate that injected cells can, depending on the signal strength, respond to external signals and produce bioactive compounds (63–65).

Aoshima and colleagues (66–72) recently investigated the cationic living polymerization of a variety of vinyl ethers having oxyethylene side chains with ω-alkyl or hydroxyl groups in the presence of added bases. Owing to the characteristics of ionic polymerization, polymers synthesized in this fashion have a narrow molecular weight distribution. Thus, in these block copolymers, block lengths and polymer molecular weights are precisely regulated. The unique characteristics of these synthesized poly(vinyl ether)s are (i) phase separation temperatures controlled by changing side oxyethylene units or ω-alkyl groups, (ii) narrow molecular weight distribution resulting in a highly sensitive phase separation temperature, and (iii) block copolymers of two types of thermo-responsive poly(vinyl ether)s showing multiple phase separation temperatures, corresponding to that of each of the blocks. This is all in sharp contrast to the properties of random copoly(vinyl ether)s. While these polymers were being characterized, stimuli-responsive soluble-insoluble changes were observed. Further investigation revealed that the block copolymers formed micelles and physical gels similar to PEO-PPO-PEO triblock copolymers or PEO-*b*-poly(alkylene oxide)s. Figure 7 shows several types of poly(vinyl ether)s synthesized. They are classified as having (i) hydrophilic, (ii) thermo-responsive, and (iii) hydrophobic segments. Thermo-responsive physical gelation of polymer solutions can be obtained by the inclusion of hydrophilic segments and thermo-responsive segments, or of thermo-responsive segments and hydrophobic segments, in block copolymers.

Figure 8 (66) shows the typical viscosity changes of a poly[2-(2-ethoxy)ethoxyethyl vinyl ether]-b-poly(methoxyethyl vinyl ether) [(EOEOVE)$_{200}$-*b*-MOVE$_{400}$] solution upon temperature increase. Poly(EOEOVE) exhibits a phase separation temperature of 40°C, whereas polyMOVE shows a transition temperature of 70°C. In the block copolymer solution, the transparent solution forms a clear gel at 40°C,

a) HYDROPHILIC

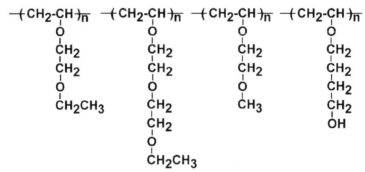

HOVE

b) THERMO-RESPONSIVE

EOVE; 20°C   EOEOVE; 41°C   MOVE; 72°C   HOBVE; 42°C

c) HYDROPHOBIC

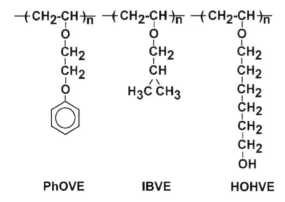

PhOVE   IBVE   HOHVE

Figure 7 Structural formulae of functional poly(vinyl ether)s.

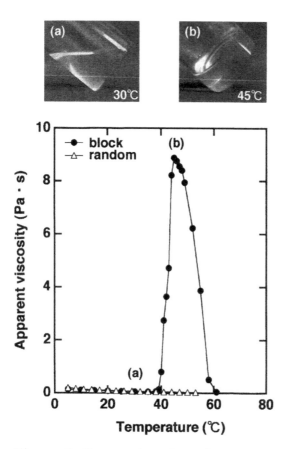

Figure 8 Temperature dependent apparent viscosity changes of 20 wt.% $EOEOVE_{200}$-b-$MOVE_{400}$. Solution appearances (a) at 30°C and (b) at 40°C are also indicated. (From Ref. 60.)

resulting in a sharp increase in solution viscosity. This gel phase is maintained between 40 and 55°C. A further temperature increase results in a hot clear liquid until 63°C is reached. Then, solution viscosity suddenly drops, and the solution becomes opaque to turbid above 63°C. In sharp contrast, random copolymers of EOEOVE and MOVE show not viscosity changes but precipitation at 53°C. Therefore, the structure of block copolymers is an important factor in temperature-responsive physical gelation.

To elucidate the physical gelation mechanisms of block poly(vinyl ether)s, Aoshima and colleagues utilized small-angle neutron scattering (SANS)(71) and differential scanning calorimetry (72) to study poly(ethoxyethyl vinyl ether)-*b*-poly(hydroxyethyl vinyl ether)(EOVE-HOVE). The block copolymers formed a micelle structure in aqueous solutions above the transition temperature of EOVE (20°C). At higher concentrations above 10 wt.%, micelles formed a crystal-like structure with a macrolattice formation, which should be the primary factor underlying physical gel formation at elevated temperatures. This unique property could be utilized to incorporate drug molecules from aqueous solutions and, thus, to formulate an injectable drug delivery system. Although such an application of these poly(vinyl ether)s and their block copolymers has not yet been reported to date, it could be beneficial to some researchers trying to develop new drug delivery devices because the appropriate selection of block components can result in the design of polymers that feature particular desired properties.

2.2. pH-Responsive Polymeric Hydrogel Systems

The gastrointestinal tract is known to possess a wide pH range, from a gastric pH of 1–2 to an intestinal tract pH of 7–8. Such significant changes could be utilized for the formation of pH-responsive drug delivery devices. Additionally, tumor sites and some sites of infection are known to have local acidic pH values amenable to pH-sensitive release methods.

Dong et al. (73) prepared pH- and thermally responsive gels for applications in protein delivery. They prepared macroporous hydrogels that consisted of *N*-isopropylacrylamide (IPAAm), acrylic acid, and divinyl-terminated polydimethylsiloxane in the presence of small amounts of the cross-linker methylenebisacrylamide. The resulting gels remained in an unswollen state at pH 1.4 and 37°C. However, at pH 7.4, gels swelled significantly owing to the dissociation of carboxyl groups of acrylic acids and formed carboxylate anions, which, in turn, formed large pores within the gel matrixes from the repulsion of the anionic groups. Protein drugs were then released through the large pores formed by swelling. Incorporation of amylase as a model protein drug into these gels resulted in a high retention of enzymatic activity. Therefore, Dong and colleagues concluded that the thermo- and pH-sensitive macroporous hydrogels could be utilized for the gastrointestinal delivery of protein drugs.

Kang and Bae (74) prepared various types of sulfonamide-containing copolymers with N,N-dimethylacrylamide (DMAAm) as the main chain components. Because sulfonamide units have pH sensitivity, pH-dependent polymer solubility changes as well as the pKa shifts of precursor compounds and monomeric sulfonamide units after polymerization were studied. The introduction of methacryloyl groups and further polymerization induced an increase in the pKa values of the analogous units of sulfonamides. The pH-dependent solubility changes revealed that, with increasing sulfonamide units in the copolymers, the pH-dependent solubility transition increased from approximately pH 5 to pH 7, and at higher pH ranges, the copolymers were completely soluble in water. Interestingly, such solubility changes occurred within a narrow range of pH changes (ca. 0.2–0.3 pH units). Furthermore, the solubility transition was completely reversible without hysteresis. Such sulfonamide-containing polymers were used to form cross-linked hydrogels (75). The introduction of a methacryloyl group to sulfadimethoxine resulted in a polymerizable sulfonamide moiety. This sulfonamide monomer was then polymerized with DMAAm in the presence of the cross-linking agent, N,N'-methylenebisacrylamide so that hydrogels with different sulfonamide contents would form (Fig. 9). Kang and Bae then investigated the pH-responsive gel-swelling behavior. All the prepared gels swelled at higher pH regions, and as the sulfonamide content in the copolymer gels increased, the swelling ratio increased. A sharp swelling transition was observed, within a narrow pH range, for hydrogels that had higher sulfonamide contents (see Fig. 10). At lower pH ranges, hydrogen bonding between hydrogen atoms in sulfonamide groups and oxygen atoms of sulfonyl groups may stabilize weaker dispersive interactions from surrounding phenyl groups in sulfonamide units. Such

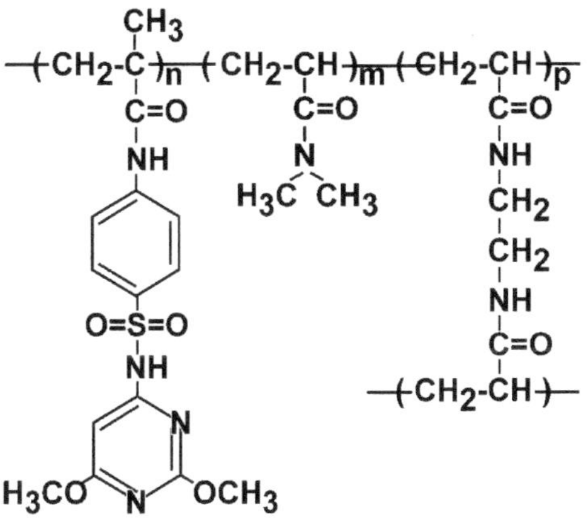

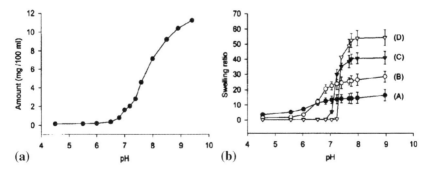

Figure 9 Structural formula of pH-responsive sulfonamide copolymer gels.

interactions induce the crystallization of sulfonamide units at low pH regions. This finding is confirmed by the x-ray diffraction measurements and apparent effective cross-linking density measurements of the sulfonamide gels at pH 6.8 and 7.4. Seeking possible applications in insulin delivery systems, the researchers further utilized pH-responsive sulfonamide-containing hydrogels for glucose-responsive

Figure 10 (a) pH-dependent solubility changes of sulfonamide monomer at 37°C. (b) pH-dependent equilibrium swelling ratio of the sulfonamide containing gels with (A) 10, (B) 20, (C) 30, (D) 40 mol%, respectively at 37°C. (From Ref. 69.)

swelling-deswelling control. They prepared sulfonamide-containing hydrogels in the presence of polymerizable glucose oxidase (GOx) and polymerizable catalase. GOx worked to form glucuronic acid from glucose and to maintain an acidic microenvironmental pH. Catalase stabilizes GOx by reducing the concentration of hydrogen peroxide, an outcome that has adverse effects on GOx activity. The prepared hydrogels showed pH-responsive gel-swelling changes, with linear swelling changes between pH 6.9 and 7.4. The researchers tested these gels for glucose-responsive swelling changes by immersing the gels in glucose solutions of different concentrations. With increasing glucose concentration, the microenvironment in the hydrogels became acidic owing to the formation of glucuronic acid within the gels, inducing gel shrinking through the neutralization of sulfonamide groups. By reducing the glucose concentration, re-swelling occurred. The glucose-responsive hydrogel swelling-deswelling changes were reproducible. The gels shrink with glucose concentration, a finding that warrants further investigation for applications in insulin delivery devices; moreover, the gel itself has the potential to improve the use of stimuli-responsive devices.

Traitel et al. (70) prepared poly(2-hydroxyethyl methacrylate-co-*N,N*-dimethylaminoethyl methacrylate) [poly(HEMA-co-DMAEMA)] hydrogels which entrapped GOx, catalase, and insulin. This hydrogel system proved to be the reverse of that presented by Kang and Bae (77). Glucose reacted with GOx to form glucuronic acid, which protonates hydrogel amino groups to induce gel-swelling. As the gel-swelling increased, the entrapped insulin was released. Furthermore, preliminary in vivo tests with rats revealed that released insulin shows the ability to decrease blood glucose levels. There was no observed tissue encapsulation around the hydrogels after implantation in the peritoneal cavity. Although confirmation remains necessary that no entrapped GOx and catalase leaked from the hydrogels, Gox-entrapped pH-responsive hydrogel systems show unique characteristics of insulin release.

Ma et al. (78) recently reported the pH-responsive micelle and physical gel formation of ABA-type triblock copolymers. They prepared ABA-type triblock copolymers

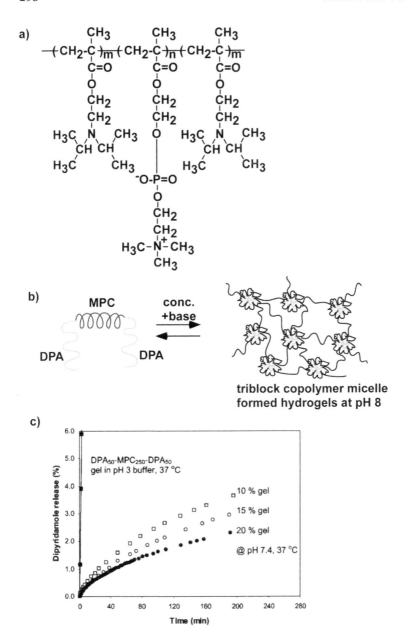

Figure 11 *(Caption on facing page)*

through an atom transfer radical polymerization (ATRP) technique to obtain polymers with narrower molecular weight distributions of 1.12–1.20. The A block consisted of either poly[2-(diisopropylamino) ethyl methacrylate] (DPA) or poly[2-(diethylamino)ethyl methacrylate] (DEA), and the B block contained poly([2-methacryloyloxyethyl phosphorylcholine]) (MPC). At low pH regions, amino groups in the A blocks were protonated and highly soluble in water, whereas these blocks deprotonated at neutral or higher pH ranges. The resulting triblock copolymers were soluble in water at acidic pH and became micelles of A blocks in ABA triblock copolymers to form hydrophobic aggregated cores, and the neutral hydrophilic B blocks formed the micelle corona (Figure 11a,b). At higher polymer concentrations in basic pH solution, physical gels are formed, as shown in Figure 11b. Thus, at physiological pH, drugs can be incorporated in the micelle cores, and a slow release of the drugs is achieved. Figure 11c shows the concentration-dependent release profiles of a model cardiovascular drug, dipyridamole, from DPA-MPC-DPA triblock copolymer gels. At pH 2, the polymer gels immediately dissolved and released drugs rapidly. Thus, pH-dependent controlled release can be achieved with the triblock copolymers indicated here.

2.3. Glucose-Responsive Hydrogel Systems for Possible Insulin-Release Devices

Insulin-dependent diabetes mellitus patients lack the pancreatic function that releases insulin in response to blood glucose levels. These patients require daily self-injections of

---

Figure 11 (*Facing page*) (a) Structural formula of pH-responsive triblock copolymer consisting of poly[(diisopropylamino)ethyl methacrylate]-*b*-poly(methacryloyloxyethyl phosphorylcholine)-*b*-poly[(diisopropylamino)ethyl methacrylate] prepared by ATRP technique. (b) Formation of macroscopic gels of concentrated solution of triblock copolymer. (c) Drug release behavior from triblock copolymer gels at 37°C and at pH 7.4. (From Ref. 72.)

an appropriate amount of insulin that help them to avoid hyperglycemia. Diabetic patients suffer from a gradual decline in the efficiency of various organs, leading to vision loss and long-term diseases. Severe conditions may even lead to patient death. Thus, injection of properly dosed insulin at proper times is required for insulin-dependent diabetes mellitus therapy. Self-injection of insulin, however, results in patient discomfort, varied bioavailability, and sometimes a hypoglycemic coma due to an overdose of insulin. Alternatively, insufficient insulin induces hyperglycemia and related complications. Therefore, the precise control of blood glucose levels with an effective, stimuli-responsive insulin release would be of great utility. Here, several formulations incorporating hydrogels for glucose concentration-dependent insulin release are reviewed.

Several insulin-release systems have already been introduced, and those systems utilize responsive gels and membranes with glucose (79,80). Ishihara et al. (81) prepared two types of gel membranes that respond to glucose and that regulate insulin permeability through the gel membranes. They prepared gel membranes which immobilized independently with GOx and nicotinamide. Glucose molecules were oxidized upon reaction with GOx, resulting in hydrogen peroxide oxidized gel-immobilized nicotinamide molecules to give positive charges in the gel membrane. These sequential reactions induced hydrophilic changes in the gel membranes and, thus, enhanced the membranes' permeability for the release of insulin. Kost et al. (82) and Albin et al. (83) prepared hydrogels consisting of hydroxyethyl methacrylate and N,N-dimethylaminoethyl methacrylate immobilized with GOx. Glucose molecules were converted, by means of the immobilized GOx, to glucuronic acid, and amino groups in the gels became protonated owing to the increased microenvironment acidity. Thus, the gels became swollen and induced insulin permeability that, at 400 mg/dl glucose, was 2.2–5.5 times higher than that at 0 mg/dl.

Obaidat and Park (84,85) prepared sol-gel transition polymers responsive to glucose. They prepared a water-soluble copolymer of acrylamide and allyl glucose. The resulting polymers cross-linked in the presence of the sugar-recognizing lectin, concanavalin A (Con A), at side chain glucose moieties. Because binding constants of native glucose molecules are higher than those of glucose moieties on the copolymer side chains, an exchange reaction occurred between added glucose and copolymer glucose moieties, inducing a gel-to-solution phase transition. Such changes can be utilized for the permeation control of insulin. Thus, the authors used two-chamber cells separated with copolymer-Con A hydrogel membranes in order to investigate the glucose-dependent permeation control of insulin. Copolymer-Con A membranes showed glucose-dependent changes, though the response was slow, indicating that further optimization is needed to achieve sensitive membrane property alterations that can effectively facilitate the release of insulin in appropriate amounts and at appropriate times.

The above-mentioned examples all used proteins such as GOx and Con A. Exposure of these proteins and peptides to the body may cause an undesirable immune response upon contact. Therefore, these naturally derived proteins and peptides, and their whole systems, should be separated from the body using semi-permeable membranes. Utilization of totally synthetic polymer systems would be a versatile choice for the construction of new glucose-responsive insulin-release devices.

Kitano et al. (86–88) and Shiino et al. (89–92) prepared totally synthetic polymers with glucose-responsive functions. They focused on the unique characteristics of phenylboronic acid as a glucose-responsive moiety. Boronate is known to form reversible bondings with polyols such as cis diol sugar compounds like glucose (Fig. 12). The researchers prepared water-soluble copolymers containing phenylboronic acid side chains using m-acrylamidophenylboronic acid (AAPBA) and various water soluble monomers, including N-vinylpyrrolidone, acrylamide, and DMAAm. The resulting copolymers formed reversible complexes with poly(vinyl alcohol) as polyol compounds (86–88). These complexes dissociate with the addition of glucose in a concentration-dependent manner (Fig. 13) (88). Such complex formation and dissociation occurred owing to the different dissociation constants of phenylboronate anions with PVA or glucose. Utilizing the

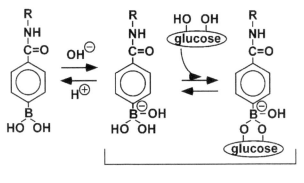

uncharged   **anionic**

Figure 12   Schematic illustration of equilibria of alkylphenylboronic acid in aqueous solution in the presence of sugar.

polyol-binding characteristics of boronate anions, the researchers prepared polymeric microgel beads containing phenylboronic acid moieties (with diameters ranging from 100 to 400 μm) (90). At the same time, the researchers modified insulin with glucuronic acid to form a complex with boronate anions on the microgel beads. Insulin-bound microgel beads were then packed into mini-columns, and a glucose solution in phosphate-buffered saline was passed through the mini-column. Insulin was released as effluent in a glucose concentration-dependent manner. Insulin release from the mini-column can be modulated by the stepwise modulation of a glucose concentration from 80 to 200 mg/dl (anything above this range was called *hyperglycemic*). A superior characteristic of the mini-column systems is the recharge capability of gluconate insulin onto the phenylboronate anions in the column matrix beads. The system, however, had the disadvantage that glucose-responsive insulin release can be achieved only at pH 8.5, where boronate anions complex with gluconated insulin or blood glucose. Thus, it is necessary to reduce the working pH to physiological conditions, that is, to pH 7.4. To overcome this disadvantage, amino groups were incorporated proximal to phenylboronic acid moieties (91). The lone electron pair of the amino nitrogen atom coordinated with the unoccupied orbital of the boron atom so as to allow the formation of stable complexes between polyol compounds and boronate anions, even at physiological pH 7.4. Shiino et al. (91) have prepared amine-containing phenylboronic acid microgel beads incorporating gluconated insulin. Amine-containing phenylboronic acid microgel beads were designed to release insulin in a glucose concentration-dependent manner at pH 7.4. If the daily insulin requirement in adult diabetes mellitus patients is 1 mg, only 1.36 cm^3 of insulin conjugated phenylboronic acid gel beads would be required to maintain normal blood glucose levels. The researchers compared their newly developed systems with the microcapsulated Con A-glucosylated insulin systems that Makino et al. (93) have created. In the case of microencapsulated Con A-glucosylated insulin, over 8000 cm^3 (8 liters!) of microcapsules were required to maintain similar therapeutic effects. This amount is too large for use in the human body. Thus, the amine-containing phenylboronic acid microgel beads systems not only display the realistic dosage capacity required for clinical applications, but also possess the feasible rechargeability of insulin once exhausted.

Kataoka et al. (94) developed dual stimuli-responsive hydrogels using 3-AAPBA as glucose-responsive moieties and IPAAm as thermo-responsive units (the prepared gels were termed *NB hydrogels*). The prepared gels with 10 mol % boronate units and in the absence of glucose in the medium showed a transition temperature at 22°C. The gels remained in a swollen state below the transition temperature of 22°C. The glucose addition to the medium induced an increase in the transition temperature to 36°C, that is, almost close to the physiological temperature, at 5 g/L glucose concentration, which is an extremely higher glucose concentration. Phenylboronic acid moieties were in an equilibrium state between the uncharged (undissociated) and charged anionic (dissociated) states. As the glucose concentration increased, the equilibrium shifted to produce more anionic charged boronate groups, inducing a hydrophilic and swollen gel (Fig. 12). Thus, the gel swelling-deswelling can be regulated with a glucose concentration at fixed temperatures. Insulin was then incorporated into dual

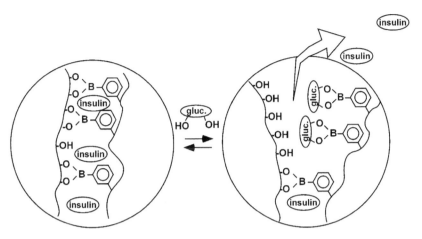

Figure 13 Schematic illustration of glucose-responsive insulin release from phenylboronic acid containing hydrogels cross-linked with poly(vinyl alcohol).

stimuli-responsive polymer gel beads and investigated for their glucose-responsive insulin-release behavior at pH 9.0 and at 28°C, where the gels showed significant and repetitive swelling-deswelling changes in glucose concentration. This finding is probably due to the bulk hydrophobicity of the polymer matrix, as well as the higher pKa value of phenylboronate anions. Insulin release was significantly suppressed below 10–20% at glucose concentrations ranging from 0 to 1.0 g/L during a 24-h incubation. More rapid insulin release was observed at glucose levels of 3.0 g/L, where approximately 80% of incorporated insulin was released during the initial 10 h. NB hydrogels were in a relatively hydrophobic and shrunken state below glucose concentrations of 1.0 g/L, and then swelled above this concentration owing to an increase in hydrophilicity arising from an increased number of boronate anions bound with glucose. Electrostatic charge was an additional driving force for hydrogel swelling. These results clearly indicate that a threshold in glucose concentration exists that alters hydrogel swelling-deswelling changes and thus the release of insulin from the swollen gels. The repetitive insulin-release characteristics were then investigated using thermo- and glucose-responsive NB hydrogels. Figure 14 shows glucose-responsive insulin-release behavior from NB hydrogels at 28°C. These data show that insulin release is, without delay, synchronized with glucose-concentration changes. Interestingly, the gels stopped releasing insulin because decreases in glucose concentrations resulted in the formation of dense shrunken layers on the gel surfaces. The obtained gels were sensitive to glucose concentration changes, which affect gel-swelling states, although temperature ranges and operating pH must be optimized before clinical application can be achieved.

Matsumoto et al. (95) recently synthesized a new glucose-sensitive monomer, 4-(1,6-dioxo-2,5-diaza-7-oxamyl)phenyl-boronic acid (DDOPBA), possessing a low pKa value of ~7.8. They chose poly(N-isopropylmethacrylamide) (PNIPMAAm) because this polymer shows a phase transition temperature near the physiological temperature of 37°C. Then, they copolymerized PNIPMAAm with the newly developed monomer, DDOPBA (structural formulae shown in Fig. 15a). The researchers investigated the pKa values of the obtained copolymers by changing the pH, as well as the temperature of the solution. The apparent pKa value of DDOPBA is 7.79, which is approximately 0.4 units lower than that of previously used

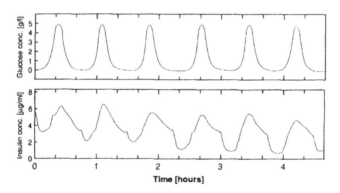

Figure 14 Glucose responsive insulin release from poly(N-isopropylacrylamide-co-acrylamidophenylboronic acid) copolymer gels. (From Ref. 88.)

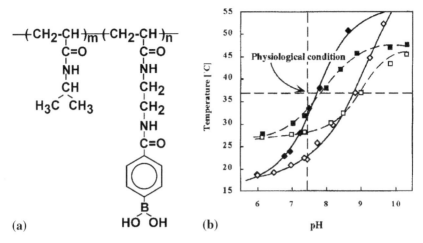

Figure 15 (a) Structural formula of poly[*N*-isopropylacrylamide-co-4-(1,6-dioxo-2,5-diaza-7-oxamyl]phenylboronic acid) [P(NIPAAm-DDOPBA)]. (b) Glucose and pH dependent changes in the LCST of 10% DDOPBA (square plot) and of 18.4% DDOPBA (diamond) in the presence (closed plots) and absence (open plots) of 5 g/L glucose. (From Ref. 89.)

(AAPBA) over a wide range of glucose concentrations. Thus, the researchers synthesized copolymers of NIPMAAm with 10 mol% of DDOPBA. Owing to the restricted main chain rotation by α-methyl groups on PNIPMAAm, the transition temperature increased relative to the PNIPAAm, although the hydrophobicity was higher in the PNIPMAAm than in the PNIPAAm. Figure 15b shows the pH- and glucose-responsive transition temperature changes of the newly developed copolymer of NIPMAAm and DDOPBA. The copolymer showed transition temperature changes that were similar to those observed in the control copolymer NIPAAm and the AAPBA, yet a distinctive difference was seen in the transition temperature ranges, shifting to higher regions to include the physiological temperatures range. Furthermore, it is noteworthy that the NIPMAAm-DDOPBA copolymer containing higher PBA moieties resulted in a decrease in the transition temperature; at the same time, however, more drastic changes in the transition temperature were observed as pH increased. Moreover, a significant transition temperature change occurred in the presence of glucose near physiological pH 7.4. These results strongly suggest that the newly developed copolymers could be a promising material for the preparation of glucose-responsive insulin delivery devices operating at physiological temperature and pH ranges.

2.4. Other Stimuli-Responsive Hydrogel Systems

2.4.1. Electrically Responsive Hydrogel Systems

Several technologies are currently under investigation for the development of new drug delivery devices that can achieve chronotherapy, specifically, iontophoresis, infusion pumps, and sonophoresis (96).

Using polyelectrolyte gels as drug containers, Kishi et al. (97) prepared drug delivery devices responding to electrical stimuli to alter gel-swelling behavior. The gels responded to the on-off stimulus of electrical currents, which induced the gel swelling-shrinking that contributes to the release of drug molecules. The researchers prepared poly(sodium acrylate) microparticle gels containing pilocarpine as a model drug. During application of a direct electric current, the pilocarpine release increased in a current-dependent manner. However, the pilocarpine release did not stop upon termination of the electrical stimulus, since the prepared gels themselves maintained a highly swollen state. Thus, a complete on-off release regulation of drugs cannot be achieved with this polymer gel system.

Kwon et al. (98–100) prepared cross-linked poly(2-acrylamide-2-methylpropanesulfonic acid-co-butyl methacrylate) [P(AMPS-co-BMA)] hydrogels and evaluated the applicability of these hydrogels for electrical stimuli-responsive drug delivery devices. They used a cationic drug molecule, edrophonium chloride, within the negatively charged hydrogel. Rapid drug release from the hydrogels resulted from an application of electric fields through the ion exchange between positively charged drug molecules and protons at the cathode. The squeezing effects arising from the electric field application induced rapid drug release from the gels, which increased as the voltages increased in a dose-dependent manner. Using the P(AMPS-co-BMA) hydrogels, an on-off drug release regulation was achieved under an on-off application of electric current.

Kwon et al. (101) further investigated the electric current-induced release of anionic heparin from a positively charged polyallylamine polyion complex. Rapid structural changes and an apparent dissociation of the polyion complex occurred upon application of an electric current. During the electric current application, the positively charged polyallylamine was neutralized at the cathode owing to the microenvironmental pH changes, and apparent dissociation of the polyion complex occurred. Although bioactive heparin was released by electric current application, polyallylamine was also released. Such positively charged molecules are incompatible in vivo, and these complications must be addressed before clinical applications can be considered.

2.4.2. Inflammation-Responsive Hydrogel Systems

Inflammatory reactions are commonly observed at injury sites. Inflammation-responsive cells like macrophages and polymorphonuclear cells play a key role in normal healing processes after injury. Hydroxyl radicals (•OH) are produced from the inflammation-responsive cells at the injured sites. Yui et al. (102,103) recently designed a hydroxyl radical-responsive drug delivery system that utilized the hydroxyl radicals produced at inflammation sites. The researchers utilized hyaluronic acid (HA), a linear mucopolysaccharide consisting of repeating units of N-acetyl-D-glucosamine and D-glucuronic acid. In the human body, HA is usually degraded by the specific enzyme hyaluronidase or by hydroxyl radicals. However, in a healthy state, the hyaluronidase-driven degradation of HA is unusual, and the majority of degradation occurs by hydroxyl radical exposure. For the preparation of inflammation-responsive hydrogels, HA was cross-linked with

ethylene glycol diglycidylether or polyglycerol polyglycidy-
lether. A Fenton reaction was used to produce hydroxyl radicals, and the hydroxyl radical-induced degradation behavior
of cross-linked HA hydrogels was observed. HA degradation
in response to hydroxyl radicals was observed only at the surface of the gel, indicating that these gels exhibit surface erosion
degradation. Further utilization of these hydrogels involved
the introduction of microspheres as model drug carriers in
the hydrogels, and the release behavior of microspheres from
the gels was monitored. It was noted that the release of microspheres from this system also followed the surface erosion
characteristics of the gels. In vivo degradation was also investigated. Surgery-induced inflammation and thus hydroxyl
radical production at the surgery sites also degraded the HA
hydrogels. Control HA gels did not degrade after long incubation times in the body. Thus, the prepared HA gels can be used
in vivo for inflammation-induced drug delivery systems, such
as incorporation with anti-inflammatory drugs and specifically
for chronic inflammatory problems including rheumatoid
arthritis.

2.4.3. Antigen-Responsive Hydrogel Systems

Recently Miyata et al. focused on naturally occurring bioactive proteins to develop new hydrogels having antigens
and entrapped antibodies within the gel matrices (104–106).
The obtained poly(acrylamide) (PAAm) hydrogels had chemically cross-linked points of MBAAm; moreover, the vinyl-con-
jugated antigen and antibody complex formation through
non-covalent multiple bonds served to further cross-link the
gel. When free rabbit IgG antigen molecules were added to
the solution with immersed hydrogels, an exchange reaction
occurred between the antigen molecules bound to the hydrogels' main chains and the added free antigen owing to the difference in the binding constants to the goat anti-rabbit IgG
antibodies. This antigen competitive exchange resulted in a
decreased number of cross-linking points in the hydrogels,
and thus promoted the swelling of hydrogels (Fig. 16). Such
changes are very antigen specific, so that the addition of other

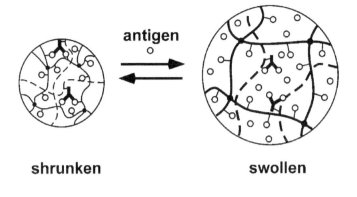

antigen

shrunken   **swollen**

⌒⌒ : antigen-bound cross-linked polymer

⌒⌒ : antibody-bound polymer

o   : free antigen

Figure 16 Swelling-deswelling changes of antigen-responsive hydrogels.

antigens including goat IgG did not alter hydrogel swelling, since there is no specific binding between goat IgG and goat anti-rabbit IgG and, thus, no decrease in gel cross-linking points. However, hydrogel shrinking did not occur because the newly formed antigen-antibody complexes were no longer mutually connected within the gel matrices. Thus, Miyata et al. then prepared semi-interpenetrating polymer network (semi-IPN) gels of antigen-conjugated cross-linked hydrogels synthesized in the presence of a polymerized antibody. The polymerized antibody does not easily diffuse out of these semi-IPNs, since the polymer chains are entangled with three-dimensionally cross-linked antigen polymer networks. Through the semi-IPN gels, antigen concentration-dependent repetitive gel swelling-deswelling kinetics was obtained. Antigen addition induced the dissociation of the polymerized antigen-antibody complex within the semi-IPN hydrogels, and the gels swelled. On the contrary, the swollen gels began to shrink gradually in the absence of free antigen molecules.

Antigen containing polymer chains were entangled within the cross-linked gels, and thus the micro-environmental antigen concentration increased when in the presence of polymerized antigens and in the absence of free antigens. Therefore, gel deswelling occurred repeatedly. Using two-chamber cells separated with antigen-antibody semi-IPN hydrogels, the researchers then investigated the antibody concentration-dependent permeation control of hemoglobin. Stepwise antigen exposure of the antigen-antibody complex-forming semi-IPN showed repeating swelling-deswelling changes, and on-off permeation control of hemoglobin was synchronized through the semi-IPN gel membranes. The prepared semi-IPN hydrogel systems can be used as possible drug delivery systems for the antigen-dependent release of drug molecules.

3. CONCLUSIONS

In the present chapter, stimuli-responsive hydrogels and their applications in stimuli-responsive drug delivery systems were reviewed. A variety of stimuli can be exploited for the design of new drug delivery devices that can be used to satisfy a variety of options for controlled release. In all cases, effective gel systems should include (i) sensor functions specific to the desired stimulus, (ii) signal processing with which to alter appropriate gel physicochemical properties and subsequent drug release profiles, (iii) the release of an adequate amount of a drug (dose) over an appropriate time frame, and (iv) a reliable halt to drug release after the stimulus signal returns to normal. Synthetic polymers possess attractive potential because experts in the field can easily alter their design in order to achieve the desired characteristics of hydrogels that respond to many stimuli of physiological and technical relevance. A more precise design of new materials and innovative gel architectures will facilitate novel fine-tuned drug delivery devices that, by responding to desired stimuli, offer the medical community more effective and reliable treatments of a variety of diseases.

ACKNOWLEDGMENTS

The authors are thankful to Professors Glen S. Kwon (University of Wisconsin) and David W. Grainger (Colorado State University) for their scientific and technical discussions throughout the preparation of this review.

REFERENCES

1. Okano T, Yoshida R. Intelligent polymeric materials for drug delivery. In: Tsuruta T, Hayashi T, Kataoka K, Ishihara K, Kirmura Y, eds. Biomedical Applications of Polymeric Materials. Boca Raton, FL: CRC Press, 1993:407–428.

2. Okano T, ed. Biorelated Polymers and Gels: Controlled Release and Applications in Biomedical Engineering. New York: Academic Press, Inc, 1998.

3. Hirasa O, Morishita Y, Onomura R, Ichijo H, Yamauchi A. Preparation and mechanical properties of thermo-responsive fibrous hydrogels made from poly(vinyl methyl ether)s. Kobunshi Ronbunshu 1989; 46(11):661–665.

4. Bae YH, Okano T, Kim SW. Temperature dependence of swelling of crosslinked poly(N,N-alkyl substituted acrylamides) in water. J Polym Sci B 1990; 28:923–936.

5. Bae YH, Okano T, Kim SW. "On-off" thermocontrol of solute transport. I. Temperature dependence of swelling of N-isopropylacrylamide networks modified with hydrophobic components in water. Pharm Res 1991; 8(4):531–537.

6. Bae YH, Okano T, Kim SW . "On-off" thermocontrol of solute transport. II. Solute release from thermosensitive hydrogels. Pharm Res 1991; 8(5):624–628.

7. Yoshida R, Sakai K, Okano T, Sakurai Y. Drug release profiles in the shrinking process of thermoresponsive poly(N-isopropylacrylamide-co-alkyl methacrylate) gels. Indust Eng Chem Res 1992; 31(10):2339–2345.

8. Yoshida R, Okuyama Y, Sakai K, Okano T, Sakurai Y. Sigmoidal swelling profiles for temperature-responsive

poly(N-isopropylacrylamide-co-butyl methacrylate) hydrogels. J Membr Sci 1994; 89:267–277.

9. Yoshida R, Sakai K, Okano T, Sakurai Y. Modulating the phase transition temperature and thermosensitivity in *N*-isopropylacrylamide copolymer gels. J Biomater Sci Polym Ed 1994; 6(6):585–598.

10. Yoshida R, Uchida K, Kaneko Y, Sakai K, Kikuchi A, Sakurai Y, Okano T. Comb-type grafted hydrogels with rapid temperature responses. Nature 1995; 374(6519):240–242.

11. Dong L-C, Hoffman AS. Synthesis and application of thermally reversible heterogels for drug delivery. J Contr Rel 1990; 13:21–31.

12. Dong L-C, Hoffman AS. Characterization of water in thermally reversible hydrogels using differential scanning calorimetry (DSC). Proc Int'l Symp Control Rel Bioact Mater 1990; 17:116–117.

13. Wu XS, Hoffman AS, Yager P. Synthesis and characterization of thermally reversible macroporous poly(*N*-isopropylacrylamide) hydrogels. J Polym Sci A 1992; 30:2121–2129.

14. Wu XS, Hoffman AS, Yager P. Synthesis and characterization of thermally reversible macroporous poly(*N*-isopropylacrylamide) hydrogels. J Polym Sci A 1992; 30:2121–2129.

15. Hoffman AS. "Intelligent" polymers in medicine and biotechnology. Macromol Symp 1995; 98:645–664.

16. Gilbert JC, Hadgraft J, Bye A, Brookes L. Drug release from Pluronic F-127 gels. Int J Pharm 1986; 32:223–228.

17. Fults KA, Johnston TP. Sustained-release of urease from a poloxamer gel matrix. J Parent Sci Technol 1990; 44(2): 58–65.

18. Gilbert JC, Richardson JL, Davies MC, Palin KJ, Hadgraft J. The effect of solutes and polymers on the gelation properties of Pluronic F-127 solutions for controlled drug delivery. J Contr Rel 1987; 5:113–118.

19. Johnston TP, Punjabi MA, Froelich CJ. Sustained delivery of interleukin-2 from a poloxamer 407 gel matrix following intraperitoneal injection in mice. Pharm Res 1992; 9(3):425–434.

20. Marchetti M, Prager S, Cussler EL. Thermodynamic predictions of volume changes in temperature-sensitive gels. 1. Theory. Macromolecules 1990; 23(6):1760–1765.

21. Marchetti M, Prager S, Cussler EL. Thermodynamic predictions of volume changes in temperature-sensitive gels. 2. Experiments. Macromolecules 1990; 23(14):3445–3450.

22. Tanaka T. Kinetics of phase transition in polymer gels. Physica 1986; 140A:261–268.

23. Sato-Matsuo E, Tanaka T. Kinetics of discontinuous volume-phase transition of gels. J Chem Phys 1988; 89(3):1695–1703.

24. Heskins M, Guillet JE, James E. Solution properties of poly(N-isopropylacrylamide). J Macromol Sci Chem A 1968; 2:1441–1445.

25. Cole C-A, Schreiner SM, Priest JH, Monji N, Hoffman AS. N-isopropylacrylamide and N-acryloxysuccinimide copolymer; a thermally reversible, water-soluble, activated polymer for protein conjugation. In: Russo P, ed. Reversible Polymeric Gels and Related Systems. Washington, DC: American Chemical Society, 1987:245–254.

26. Monji N, Cole C-A, Tam M, Goldstein L, Nowinski RC, Hoffman AS. Application of a thermally-reversible polymer-antibody conjugate in a novel membrane-based immunoassay. Biochem Biophys Res Commun 1990; 172(2):652–660.

27. Chen G, Hoffman AS. Preparation and properties of thermo-reversible, phase-separating enzyme-oligo(N-isopropylacrylamide) conjugates. Bioconj Chem 1993; 4:509–514.

28. Chilkoti A, Chen G, Stayton PS, Hoffman AS. Site-specific conjugation of a temperature-sensitive polymer to a genetically-engineered protein. Bioconj Chem 1994; 5:504–507.

29. Stayton PS, Shimoboji T, Long C, Chilkoti A, Chen G, Harris JM, Hoffman AS. Control of protein-ligand recognition using a stimuli-responsive polymer. Nature 1995; 378: 472–474.

30. Yamada N, Okano T, Sakai H, Karikusa F, Sawasaki Y, Sakurai Y. Thermo-responsive polymeric surfaces; control of attachment and detachment of cultured cells. Makromol Chem Rapid Commun 1990; 11:571–576.

31. Okano T, Bae YH, Jacobs H, Kim SW. Thermally on-off switching polymers for drug permeation and release. J Control Rel 1990; 11:255–265.

32. Kanazawa H, Yamamoto K, Matsushima Y, Takai N, Kikuchi A, Sakurai Y, Okano T. Temperature-responsive chromatography using poly(N-isopropylacrylamide)-modified silica. Anal Chem 1996; 68(1):100–105.

33. Kanazawa H, Kashiwase Y, Yamamoto K, Matsushima Y, Kikuchi A, Sakurai Y, Okano T. Temperature-responsive liquid chromatography. 2. Effect of hydrophobic groups in N-isopropylacrylamide copolymer-modified silica. Anal Chem 1997; 69(5):823–830.

34. Kanazawa H, Yamamoto K, Kashiwase Y, Matsushima Y, Takai N, Kikuchi A, Sakurai Y, Okano T. Analysis of peptides and proteins by temperature-responsive chromatographic system using N-isopropylacrylamide polymer-modified columns. J Pharmaceu Biomed Anal 1997; 15:1545–1550.

35. Kanazawa H, Sunamoto T, Matsushima Y, Kikuchi A, Okano T. Temperature-responsive chromatographic separation of amino acid phenylthiohydantoins using aqueous media as the mobile phase. Anal Chem 2000; 72(24):5961–5966.

36. Yakushiji T, Sakai K, Kikuchi A, Aoyagi T, Sakurai Y, Okano T. Effects of cross-linked structure on temperature-responsive hydrophobic interaction of PIPAAm hydrogel modified surfaces with steroids. Anal Chem 1999; 71(6):1125–1130.

37. Kobayashi J, Kikuchi A, Sakai K, Okano T. Aqueous chromatography utilizing pH-/temperature-responsive polymer stationary phases to separate ionic bioactive compounds. Anal Chem 2003; 73:2027–2033.

38. Kobayashi J, Sakai K, Kikuchi A, Okano T. Aqueous chromatography utilizing pH-/temperature-responsive polymer stationary phases to separate ionic bioactive compounds. Anal Chem 2001; 73(9):2027–2033.

39. Kobayashi J, Kikuchi A, Sakai K, Okano T. Aqueous chromatography utilizing hydrophobicity-modified anionic temperature-responsive hydrogel for stationary phase. J Chromatogr A 2002; 958:109–119.

40. Okano T, Yamada N, Sakai H, Sakurai Y. A novel recovery system for cultured cells using plasma-treated polystyrene dishes grafted with poly(N-isopropylacrylamide). J Biomed Mater Res 1993; 27:1243–1251.

41. Kikuchi A, Okuhara M, Karikusa F, Sakurai Y, Okano T. Two-dimensional manipulation of confluently cultured vascular endothelial cells using temperature-responsive poly(N-isopropylacrylamide)-grafted surfaces. J Biomater Sci Polym Ed 1998; 9(12):1331–1348.

42. Yamato M, Kushida A, Konno C, Kikuchi A, Sakurai Y, Okano T. A novel tool of temperature-responsive cell culture surfaces and its application to matrix biology. Connect Tiss 1999; 31:13–16.

43. Kabra BG, Gehrke SH. Synthesis of fast response, temperature-sensitive poly(N-isopropylacrylamide) gel. Polym Commun 1991; 32(11):322–323.

44. Kaneko Y, Sakai K, Kikuchi A, Sakurai Y, Okano T. Fast swelling/deswelling kinetics of comb-type grafted poly(N-isopropylacrylamide) hydrogels. Macromol Chem Macromol Symp 1996; 109:41–53.

45. Kaneko Y, Sakai K, Kikuchi A, Yoshida R, Sakurai Y, Okano T. Influence of freely mobile grafted chain length on dynamic properties of comb-type grafted poly(N-isopropylacrylamide) hydrogels. Macromolecules 1995; 28(23):7717–7723.

46. Kaneko Y, Nakamura S, Sakai K, Aoyagi T, Kikuchi A, Sakurai Y, Okano T. Rapid deswelling response of poly(N-isopropylacrylamide) hydrogels by the formation of water release channels using poly(ethylene oxide) graft chains. Macromolecules 1998; 31(18):6099–6105.

47. Schmolka IR. Poloxamers in the pharmaceutical industry. In: Tarcha PJ, ed. Polymers for Controlled Drug Delivery. Boca Raton, FL, CRC Press, 1991:189–214.

48. Vadnere M, Amidon G, Lindenbaum S, Haslam JL. Thermodynamic studies on the gel-sol transition of some Pluronic polyols. Int J Pharmaceu 1984; 22:207–218.

49. Illum L, Davis SS. The organ uptake of intravenously administered colloidal particles can be altered using a non-ionic surfactant (poloxamer 338). FEBS Lett 1984; 167(1):79–82.

50. Miyazaki S, Takeuchi S, Yokouchi C, Takada M. Pluronic F-127 gels as a vehicle for topical administration of anticancer agents. Chem Pharm Bull 1984; 32(10):4205–4208.

51. Jeong B, Bae YH, Lee DS, Kim SW. Biodegradable block copolymers as injectable drug-delivery systems. Nature 1997; 388(28 August):860–862..

52. Jeong B, Bae YH, Kim SW. Thermoreversible gelation of PEG-PLGA-PEG triblock copolymer aqueous solutions. Macromolecules 1999; 32:7064–7069.

53. Jeong B, Bae YH, Kim SW. Drug release from biodegradable injectable thermosensitive hydrogel of PEG-PLGA-PEG triblock copolymers. J Contr Rel 2000; 63:155–163.

54. Jeong B, Kim SW, Bae YH. Thermosensitive sol-gel reversible hydrogels. Adv Drug Delivery Rev 2002; 54(1):37–51.

55. Jeong B, Bae YH, Kim SW. Biodegradable thermosensitive micelles of PEG-PLGA-PEG triblock copolymers. Colloid Surf B 1999; 16(1–4):185–193.

56. Jeong B, Lee DS, Shon J-i, Bae YH, Kim SW. Thermoreversible gelation of poly(ethylene oxide) biodegradable polyester block copolymers. J Polym Sci A 1999; 37:751–760.

57. Jeong B, Bae YH, Kim SW. In situ gelation of PEG-PLGA-PEG triblock copolymer aqueous solutions and degradation thereof. J Biomed Mater Res 2000; 50:171–177.

58. Zentner GM, Rathi R, Shih C, McRea JC, Seo M-H, Oh H, Rhee BG, Mestecky J, Moldoveanu Z, Morgan M, Weitman S. Biodegradable block copolymers for drug delivery of proteins and water-insoluble drugs. J Contr Rel 2001; 72: 203–215.

59. Kim YJ, Choi C, Koh JJ, Lee M, Ko KS, Kim SW. Controlled release of insulin from injectable biodegradable triblock copolymer. Pharm Res 2001; 18(4):548–550.

60. Jeong B, Kibbey MR, Birnbaum JC, Won Y-Y, Gutowska A. Thermogelling biodegradable polymers with hydrophilic backbones: PEG-g-PLA. Macromolecules 2000; 33:8317–8322.

61. Chung Y-M, Simmons KL, Gutowska A, Jeong B. Sol-gel transition temperature of PLGA-g-PEG aqueous solutions. Biomacromolecules 2002; 3(3):511–516.

62. Jeong B, Lee KM, Gutowska A, An YH. Thermogelling biodegradable copolymer aqueous solutions for injectable protein delivery and tissue engineering. Biomacromolecules 2002; 3(4):865–868.

63. Sun AM, O'Shea GM. Microencapsulation of living cells – a long-term delivery system. J Contr Rel 1985; 2:137–141.

64. Iwata H, Kobayashi K, Takagi T, Oka T, Yang H, Amemiya H, Tsuji T, Ito F. Feasibility of agarose microbeads with xenogeneic islets as a bioartificial pancreas. J Biomed Mater Res 1994; 28:1003–1011.

65. Bae YH, Kim SW, Kikuchi A, Chong S-J, Song SC. Sulfonylurea-grafted polymers for langerhans islet stimulation. In: Ottenbrite RM, Huang SJ, Park K, eds. Hydrogels and Biodegradable Polymers for Bioapplications. Washington, DC: American Chemical Society, 1996:42–57.

66. Aoshima S, Sugihara S. Syntheses of stimuli-responsive block copolymers of vinyl ethers with side oxyethylene groups by living cationic polymerization and their thermosensitive physical gelation. J Polym Sci A 2000; 38:3962–3965.

67. Aoshima S, Hashimoto K. Stimuli-responsive block copolymers with polyalcohol segments: Syntheses via living cationic polymerization of vinyl ethers with a silyloxyl group and their thermoreversible physical gelation. J Polym Sci A 2001; 39:746–750.

68. Sugihara S, Matsuzono S-i, Sakai H, Abe M, Aoshima S. Syntheses of amphiphilic block copolymers by living cationic polymerization of vinyl ethers and their selective solvent-induced physical gelation. J Polym Sci A 2001; 39: 3190–3197.

69. Sugihara S, Aoshima S. Syntheses of various stimuli-responsive diblock copolymers by living cationic polymeriza-

tion – thermally-induced micellization and physical gelation. Kobunshi Ronbunshu 2001; 58(6):304–310.

70. Aoshima S, Ikeda M, Nakayama K, Kobayashi E, Ohgi H, Sato T. Synthesis of poly(vinyl alcohol) graft copolymers by living cationic polymerization in the presence of added bases i. Design and synthesis of poly(vinyl alcohol) graft copolymers with well-controlled poly(vinyl ether) grafts. Polym J 2001; 33(8):610–616.

71. Okabe S, Sugihara S, Aoshima S, Shibayama M. Heat-induced self-assembling of thermosensitive block copolymer. 1. Small-angle neutron scattering study. Macromolecules 2002; 35(21):8139–8146.

72. Okabe S, Sugihara S, Aoshima S, Shibayama M. Heat-induced self-assembling of thermosensitive block copolymer. Rheology and dynamic light scattering study. Macromolecules 2003; 36:4099–4106.

73. Dong L-c, Yan Q, Hoffman AS. Controlled release of amylase from a thermal and pH-sensitive, macroporous hydrogel. J Contr Rel 1992; 19:171–178.

74. Kang SI, Bae YH. pH-induced solubility transition of sulfonamide-based polymers. J Contr Rel 2002; 80:145–155.

75. Kang SI, Bae YH. pH-induced volume-phase transition of hydrogels containing sulfonamide side group by reversible crystal formation. Macromolecules 2001; 34(23):8173–8178.

76. Traitel T, Cohen Y, Kost J. Characterization of glucose-sensitive insulin release systems in simulated in vivo conditions. Biomaterials 2000; 21:1679–1687.

77. Kang SI, Bae YH. A sulfonamide based glucose-responsive hydrogel with covalently immobilized glucose oxidase and catalase. J Contr Rel 2003; 86:115–121.

78. Ma Y, Tang Y, Billingham MC, Armes SP. Synthesis of biocompatible, stimuli-responsive, physical gels based on ABA triblock copolymers. Biomacromolecules 2003; 4(4):864–868.

79. Chung D-J, Ito Y, Imanishi Y. An insulin-releasing membrane system on the basis of oxidation reaction of glucose. J Contr Rel 1992; 18:45–54.

80. Klumb LA, Horbett TA. Design of insulin delivery devices based on glucose sensitive membranes. J Contr Rel 1992; 18:59–80.

81. Ishihara K, Kobayashi M, Shinohara I. Control of insulin permeation through a polymer membrane with responsive function for glucose. Makromol Chem Rapid Commun 1983; 4:327–331.

82. Kost J, Horbett TA, Ratner BD, Singh M. Glucose-sensitive membranes containing glucose oxidase: activity, swelling, and permeability studies. J Biomed Mater Res 1985; 19:1117–1133.

83. Albin G, Horbett TA, Ratner BD. Glucose-sensitive membranes for controlled delivery of insulin: Insulin transport studies. J Contr Rel 1985; 2:153–164.

84. Obaidat AA, Park K. Characterization of glucose dependent gel-sol phase transition of the polymeric glucose-concanavalin A hydrogel system. Pharm Res 1996; 13(7):989–995.

85. Obaidat AA, Park K. Characterization of protein release through glucose-sensitive hydrogel membranes. Biomaterials 1997; 18(11):801–806.

86. Kitano S, Kataoka K, Koyama Y, Okano T, Sakurai Y. Glucose-responsive complex formation between poly(vinyl alcohol) and poly(N-vinyl-2-pyrrolidone) with pendant phenylboronic acid moieties. Makromol Chem Rapid Commun 1991; 12:227–233.

87. Kitano S, Hisamitsu I, Koyama Y, Kataoka K, Okano T, Sakurai Y. Effect of the incorporation of amino groups in a glucose-responsive polymer complex having phenylboronic acid moieties. Polym Adv Technol 1991; 2:261–264.

88. Kitano S, Koyama Y, Kataoka K, Okano T, Sakurai Y. A novel drug delivery system utilizing a glucose responsive polymer complex between poly(vinyl alcohol) and poly(N-vinyl-2-pyrrolidone) with a phenylboronic acid moiety. J Contr Rel 1992; 19:162–170.

89. Shiino D, Kataoka K, Koyama Y, Yokoyama M, Okano T, Sakurai Y. A self-regulated insulin delivery system using boronic acid gel. J Intelligent Mater Syst Struct 1994; 5:311–314.

90. Shiino D, Murata Y, Kataoka K, Koyama Y, Yokoyama M, Okano T, Sakurai Y. Preparation and characterization of a glucose-responsive insulin-releasing polymer device. Biomaterials 1994; 15(2):121–128.

91. Shiino D, Murata Y, Kubo A, Kim YJ, Kataoka K, Koyama Y, Kikuchi A, Yokoyama M, Okano T, Sakurai Y. Amine containing phenylboronic acid gel for glucose-responsive insulin release under physiological pH. J Contr Rel 1995; 37:269–276.

92. Shiino D, Kubo A, Murata Y, Koyama Y, Kataoka K, Kikuchi A, Sakurai Y, Okano T. Amine effect on phenylboronic acid complex with glucose under physiological pH in aqueous solution. J Biomater Sci Polym Ed 1996; 7(8):697–705.

93. Makino K, Mack EJ, Okano T, Kim SW. A microcapsule self-regulating delivery system for insulin. J Contr Rel 1990; 12:235–239.

94. Kataoka K, Miyazaki H, Bunya M, Okano T, Sakurai Y. Totally synthetic polymer gels responding to external glucose concentration: their preparation and application to on-off regulation of insulin release. J Am Chem Soc 1998; 120: 12694–12695.

95. Matsumoto A, Ikeda S, Harada A, Kataoka K. Glucose-responsive polymer bearing a novel phenylborate derivatives as a glucose-sensing moiety operating at physiological pH conditions. Biomacromolecules 2003; 4(4):1410–1416.

96. Berner B, Dinh SM. Electronically assisted drug delivery: an overview. In: Berner B, Dinh SM, eds. Electronically Controlled Drug Delivery. Boca Raton, FL: CRC Press, 1998:3–7.

97. Kishi R, Hara M, Sawahata K, Osada Y. Conversion of chemical into mechanical energy by synthetic polymer gels (chemomechanical system). In: DeRossi D, Kajiwara K, Osada Y, Yamauchi A, eds. Polymer Gels–Fundamentals and Biomedical Applications. New York: Plenum Press, 1991:205–220.

98. Kwon IC, Bae YH, Okano T, Berner B, Kim SW. Stimuli sensitive polymers for drug delivery systems. Makromol Chem Macromol Symp 1990; 33:265–277.

99. Kwon IC, Bae YH, Kim SW. Electrically erodible polymer gel for controlled release of drugs. Nature 1991; 354:291–293.

100. Kwon IC, Bae YH, Okano T, Kim SW. Drug release from electric current sensitive polymers. J Contr Rel 1991; 17: 149–156.

101. Kwon IC, Bae YH, Kim SW. Heparin release from polymer complex. J Contr Rel 1994; 30:155–159.

102. Yui N, Okano T, Sakurai Y. Inflammation responsive degradation of crosslinked hyaluronic acid gels. J Contr Rel 1992; 22:105–116.

103. Yui N, Nihira J, Okano T, Sakurai Y. Regulated release of drug microspheres from inflammation responsive degradable matrices of crosslinked hyaluronic acid. J Contr Rel 1993; 25:133–143.

104. Miyata T, Asami N, Uragami T. Preparation of an antigen-sensitive hydrogel using antigen-antibody bindings. Macromolecules 1999; 32:2082–2084.

105. Miyata T, Asami N, Uragami T. A reversibly antigen-responsive hydrogel. Nature 1999; 399:766–769.

106. Miyata T, Uragami T, Nakamae K. Biomolecule-sensitive hydrogels. Adv Drug Delivery Rev 2002; 54(1):79–98.

8

Treatment of Malignant Brain Tumors with Controlled-Release Local-Delivery Polymers

ALEX A. KHALESSI and PAUL P. WANG

Department of Neurological Surgery, The Johns Hopkins School of Medicine, Baltimore, Maryland, U.S.A.

ROBERT S. LANGER

Department of Chemical Engineering and Division of Bioengineering, Massachusetts Institute of Technology, Cambridge, Massachusetts, U.S.A.

HENRY BREM

Departments of Neurological Surgery and Oncology, The Johns Hopkins School of Medicine, Baltimore, Maryland, U.S.A.

1. INTRODUCTION

Central nervous system (CNS) tumors account for tremendous morbidity and mortality. In the United States, the annual incidence of primary brain tumors approaches 16,800 people and corresponds with 13,100 lesion-related deaths (1). Histopathological classification deconstructs primary brain tumors into several categories based on grade and cell of origin. Gliomas, for example, derive from glial cells and constitute half of all primary brain tumors. Astrocytomas, in turn, encompass three-quarters of all gliomas, and define a heterogeneous tumor class ranging from low-grade to high-grade anaplastic variants. Glioblastoma multiforme (GBM) represents the most common and aggressive form of astrocytoma.

Unfortunately, high-grade malignant gliomas carry a dismal prognosis, and conventional therapy minimally alters the natural history of disease. The historic standard of care involves three phases: (i) surgical biopsy for appropriate pathologic diagnosis (2); (ii) operative debulking of accessible tumor, and (iii) adjuvant radiation and chemotherapy (3–7). Surgical resection alone offers a 6-month median survival with only 7.5% of patients surviving 2 years. Additional radiation extends median survival to 9 months and post-operative chemotherapy offers minimal benefit (8,9). Recent advances in neurosurgical techniques and neuroradiology, while improving the quality of life, likewise fail to confer long-term significant survival benefit. The continued disease burden inflicted by malignant brain tumors underscores the urgent need for novel therapeutic approaches.

Surgery, radiation, and chemotherapy comprise the three dimensions of conventional brain tumor therapy, and therefore offer possible intensive and extensive margins for treatment enhancement. Each modality carries its own set of management challenges and technical limitations. For example, the highly invasive nature of malignant gliomas prevents gross total resection without microscopic tumor remnants. Broader surgical margins, however, endanger functional brain tissue and increase the attendant likelihood of immediate morbidity and neurological deficit. Likewise, dose or field augmentation of radiation protocols presents unacceptable risks to healthy brain tissue. Surgery and radiation, in the absence of heretofore unrealized techniques to distinguish tumor from healthy brain tissue, confront significant technical limitations to improving patient survival. Thus, current efforts largely target adjuvant chemotherapy as a means of enhancing brain tumor therapy. Ongoing pharmacological research focuses on the tandem development of more effective agents and the advancement of drug delivery methods.

This chapter explores the development of biodegradable polyanhydride polymers as a local chemotherapeutic delivery system in brain tumor patients. In so doing, the authors detail the unique pharmacokinetic considerations inherent in CNS drug delivery, and the resultant advantages of local administration. Subsequent sections chronicle the development of biocompatible technologies, preclinical and clinical experience with polymer-based tumor treatments, and potential future advancements in local antineoplastic drug delivery.

2. DRUG DELIVERY CONSIDERATIONS IN THE CNS

With rare exceptions, the blood-brain barrier (BBB) physiologically isolates the CNS, and greatly complicates brain tumor pharmacotherapy. Tight junctions join endothelial cells within CNS-supplying capillaries (10), thereby posing a selective pharmacological barrier between the CNS and the remainder of the circulation. The chemical properties of many antineoplastic agents compound the drug delivery difficulties presented by the BBB. Though small, electrically neutral, lipid-soluble molecules readily access the CNS, chemotherapeutic moieties are often large, ionically charged or hydrophilic (11). Figure 1 illustrates the resultant poor cerebrovascular permeability associated with a broad spectrum of antineoplastic drugs (12). Thus, the achievement of therapeutic CNS doses by peripheral administration often requires

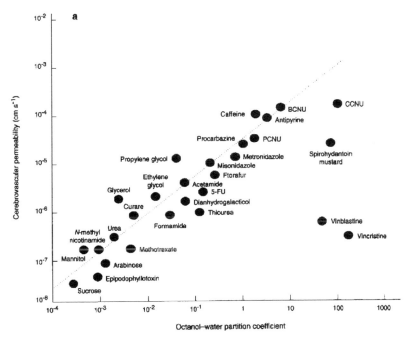

Figure 1 Cerebrovascular permeability vs. octanol-water partition coefficient of selected chemicals and drugs. (From Ref. 12.)

intolerable systemic concentrations and the introduction of dose-limiting toxicities.

Improved drug delivery in brain tumor patients therefore requires resolution of the practical obstacles posed by the BBB. Three overarching strategies characterize current efforts to enhance chemotherapeutic delivery: (i) chemical modification of existing antineoplastic agents to optimally accommodate natural BBB permeability characteristics; (ii) mechanical disruption of the BBB followed by sequential systemic drug administration; and (iii) BBB circumvention by local CNS drug delivery. The remainder of this section broadly surveys progress within each area, and proceeds to a more detailed examination of local drug delivery.

Given BBB permeability characteristics, pharmacological alteration of existing drugs to produce more lipophilic variants would theoretically increase CNS chemotherapy delivery. Early work with carmustine (BCNU), a known chemotherapeutic agent with modest efficacy against malignant brain tumors, led to the isolation of more lipophilic variants, lumustine (CCNU) and semustine (methyl-CCNU) (13). Later clinical trials, however, failed to realize the promise of improved CNS penetration; neither systemic agent demonstrated significant efficacy over BCNU in treating glial tumors (5). Carrier linkage represents an alternative means of enhancing BBB permeability to chemotherapeutic agents. Dihydropyridine, for example, serves as a lipophilic carrier for a variety of antibiotics, neurotransmitters and antineoplastic agents; recent work suggests increased intracranial drug concentrations with dihydropyridine coupling (14). Furthermore, the development of new transport vectors, typified by modified protein or receptorspecific monoclonal antibodies (15), defines ongoing efforts to increase CNS penetration with systemic chemotherapeutic administration.

The second general approach to CNS chemotherapeutic delivery divides into two phases: (i) initial disruption of the BBB; and (ii) systemic delivery of antineoplastic agents to exploit the compromised BBB. Intra-arterial hyperosmolar mannitol, for example, employs diuretic properties to acutely dehydrate and shrink endothelial cells, thereby widening the tight junctions associated with adjacent cell membranes. Williams et al. (16) studied mannitol premedication in 34 intracranial tumor patients receiving carboplatin and etoposide therapy. Though primitive neuroectodermal tumors (PNETs: four out of four patients) and CNS lymphomas (two out of four patients) demonstrated some response to mannitol administration, GBM, oligodendrogliomas, and metastatic carcinomas were refractory to the premedication regimen. Notably, improved BBB passage of hydrophilic substances with mannitol does not imply increased drug delivery to the actual tumor site, and represents a potential explanation for the observed results (17). The bradykinin agonist RMP-7 provides an alternative means of BBB disruption. Experimental brain tumor models demonstrate increased carboplatin uptake with prior RMP-7 administration (18). Unfortunately, RMP-7 likely shares mannitol's limited in vivo efficacy against more aggressive intracranial malignancies; antineoplastic agents lack established tumor site penetration after crossing the BBB by these means.

The third general approach to CNS drug delivery circumvents the BBB by local tumor site delivery. Three interventional strategies reflect the majority of local tumor site pharmacotherapy: (i) cerebrospinal fluid (CSF) exposure; (ii) catheter-based administration; and (iii) sustained release via implanted local delivery polymers. From a teleological standpoint, local delivery methods facilitate the sustained maintenance of significant tumor site drug concentrations without toxic systemic exposure. Moreover, 80–90% of malignant glioma recurrence occurs within 2 cm of the resection site, rendering such tumors particularly appropriate targets for local chemotherapy administration.

CSF chemotherapy infusion requires the placement of an intraventricular catheter or lumbar puncture. Given that the ependymal lining of the ventricular system lacks tight junctions, CSF drug delivery would physiologically bypass the BBB. Regrettably, CSF chemotherapy infusion has met with disappointing results primarily due to poor parenchymal drug penetration (19).

Catheter-based systems for drug administration rely on two central elements: (i) accurate and secure placement of the catheter tip within the tumor site; and (ii) a drug reservoir to provide an infusion source. The Ommaya reservoir, an early example of a catheter system, delivers intermittent bolus injections of chemotherapy into the tumor (20). Recent development of implantable pumps permits constant drug infusion over an extended time period, a more favorable alternative to bolus delivery. The Infusaid pump (Infusaid Corp., Norwood, MA) employs compressed vapor pressure to deliver a solution at a constant rate (21); the MiniMed PIMS system (MiniMed, Sylmar, CA) (22) and the Medtronic SynchroMed system (Medtronic, Minneapolis, MN) (23) rely on a solenoid pump and peristaltic mechanism, respectively, to maintain constant drug delivery. None of the existing catheter-based variants currently demonstrate superior treatment efficacy against malignant gliomas. Overall, catheter-based systems share common limitations of mechanical failure, obstruction by clot or tissue debris, and infection risk.

A variant of catheter-based systems is convection delivery, which relies on a simple pressure gradient rather than diffusion to disseminate drugs intraparenchymally. Unlike diffusion, convection is independent of drug molecular weight. Drug infusion into cerebral white matter generates a pressure gradient and thereby allows the convective introduction of high drug concentration into the brain without attendant structural or functional side effects (24–26). Primate trials, for example, examine the treatment of Parkinsonian symptoms using convection-enhanced drug delivery (CEDD) (27). A recent study employed CEDD in Taxol treatment of three brain tumor patients; diffusion-weighted MRI monitored drug delivery effects (28). Preclinical and clinical studies using CEDD to delivery immune toxins and chemotherapy agents are ongoing (29,30).

Biocompatible sustained-release polymers, the final method of local drug delivery, divide into biodegradable and non-biodegradable variants. Both involve surgical resection/debulking of malignant gliomas followed by polymer implantation in the residual tumor bed. Biodegradable polymers release drug by degradation of the polymer matrix. Non-biodegradable polymers, in contrast, remain intact after chemotherapeutic release (31). Both polymer designs slowly release drug over time.

3. BIOCOMPATIBLE POLYMER DEVELOPMENT: AN HISTORICAL PERSPECTIVE

Langer and Folkman (32) first reported in 1976 the sustained and predictable release of macromolecules from a non-biodegradable polymer. Ethylene vinyl acetate (EVAc) copolymer facilitates local diffusion of incorporated drug through matrix micropores. Chemical properties of the agent, namely molecular weight, charge and water solubility, determine the drug diffusion rate. Having demonstrated unprecedented local release properties, EVAc underwent safety assessments. Rabbit cornea studies established EVAc biocompatibility (33) and rat brain research supported polymer inertness (34–38). Clinically, EVAc polymers currently contribute to glaucoma treatment, dental care, insulin therapy, asthma treatment, contraception, and chemotherapy. Recent work in a rat brachial plexus injury model demonstrates successful EVAc polymer delivery of neurotrophic growth factors to treat peripheral nerve injuries (39). EVAc polymer inertness, however, remains a primary obstacle to approval for intracranial use (40). Aside from the permanent foreign body left by an intact EVAc matrix, the polymer drug release rate decreases over time, following first-order release kinetics (41). Later discussion emphasizes the advantage of constant drug diffusion rate or zero-order release kinetics for local delivery.

The development of biodegradable systems naturally evolved from concerns regarding polymer inertness. By coupling drug diffusion and polymer degradation, sustained polymer-based drug release was possible without the attendant risks of a permanent foreign body (Figs. 2 and 3). The lactide/glycolide polyesther family, characterized by lactic and

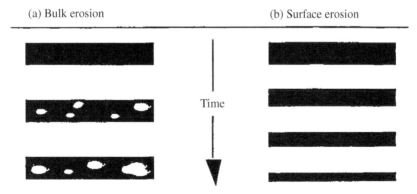

Figure 2 (a) Bulk erosion of biodegradable controlled-release polymer implants leads to unpredictable release profiles. (b) Polymers exhibiting surface erosion release drug at nearly constant rate (zero-order kinetics) as they dissolve in water. (From Ref. 61.)

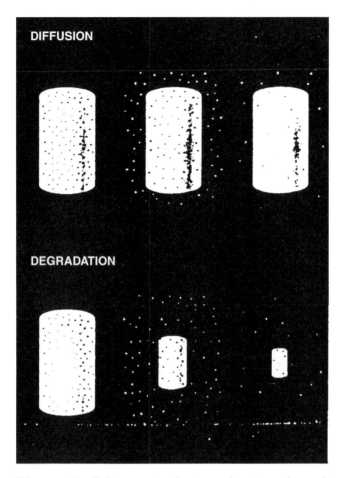

Figure 3 Polymer implants releasing drug by degradation provide more predictable release profiles compared to diffusion. (From Ref. 61.)

glycolic monomers polymerized with ester bonds, represented the first foray into biodegradable polymers. The polylactic-co-glycolic acid (PLGA) polymers enjoyed several advantages over their non-biodegradable EVAc predecessors. Aside from biodegradability, PLGA polymer drug release rates vary with the ratio of lactic and glycolic acid monomers, thereby allowing tailored delivery over time (42). Rat brain studies

(43–45) and PLGA fashioned into sutures (now in wide clinical use) (46,47) demonstrated polymer biocompatibility. PLGA polymers, often in an injectable microsphere form (48), have successfully delivered a variety of drugs, ranging from steroids, anti-inflammatory agents, antibiotics, anesthetics, and narcotics to antineoplastic agents (35,49–58). Notably, a PLGA microsphere design exists specifically for stereotactic implantation into the brain (59). The development of polyethylene glycol coating further advances the potential use of PLGA polymers. By covalent linkage to the polymer matrix, the coating significantly reduces opsonization and elimination by the immune system (60). Despite the great promise of PLGA polymers, associated bulk erosion (like a sugar cube) (Fig. 2a) connotes an unfavorable release characteristic for chemotherapy for the brain; sporadic drug dumping, suboptimal tissue exposure, and unexpected toxicity represent possible sequelae of bulk erosion and erratic drug release (61).

Leong et al. reported on the formulation of the polyan hydride poly[bis(p-carboxyphenoxyl)] propane-sebacic acid (pCPP:SA) matrix in 1985. pCPP:SA spontaneously breaks down to dicarboxylic acids in aqueous environments, and therefore equals PLGA polymer biodegradability. Moreover, the pCPP:SA matrix possesses several important qualities critical to effective local drug delivery. The extreme hydrophobicity of the pCPP:SA matrix, for example, shields incorporated drug from the aqueous media, thereby sparing compounds with particularly short biological half-lives from premature degradation. Additionally, pCPP:SA undergoes macroscopic breakdown by surface erosion (like peeling an onion) (Fig. 2b), and therefore avoids the pitfalls encountered with PLGA bulk erosion (like a sugar cube). Surface erosion further confers the theoretical advantage of stable zero-order release kinetics, implying stable drug delivery rates unrealized by earlier methods. Furthermore, the pCPP:SA polymer matches the manufacturing flexibility of earlier polymer systems, and is compatible with a variety of shapes, including microspheres, sheets, rods, and wafers (62–68). Analogous to the PLGA polymer system, the pCPP:SA matrix

breakdown rate varies with the ratio of its constituent monomers. The variance in pCPP:SA drug delivery time can be
dramatic: a 1 mm thick pure CPP polymer degrades in 3
years while a 20% CPP and 80% SA [pCPP:SA (20:80)]
requires 3 weeks (58).

Multiple studies confirm pCPP:SA biocompatibility. Standard assays attribute no mutagenic, carcinogenic, or
teratogenic effects to the dicarboxylic acid breakdown products
of pCPP:SA polymers (69). In vitro proliferation assays with
endothelial and smooth muscle cells showed no growth inhibition by the pCPP:SA matrix (62). Rabbit cornea assays with
pCPP:SA polymers (26,60) identified no reactive inflammation.
Rat (70), rabbit (71), and monkey (72) brain studies established
neural tissue biocompatibility. Given its enumerated drug
release characteristics and biodegradability, pCPP:SA enjoys
widespread clinical use; the following section explores current
neuro-oncologic applications of pCPP:SA polymers.

Other local drug delivery systems developed for sustained intracranial release merit discussion. The fatty acid
dimer-sebaceic acid (FAD:SA) polymers (73–75), for example,
specifically facilitate delivery of hydrophobic chemotherapeutic agents (76–78). Polyethylene glycol coated liposomes successfully deliver anthracyclines (79) and gelatin-chondroitin
sulfate coated microspheres serve as depots for cytokine
release (80). Various surgical materials, namely fibrin glue
(81), gelatin sponges (82), Surgicel (oxidized regenerated
cellulose) (75), polymethylmethacralate (83), and silastic tubing (84), have served as vectors for local intracranial drug
delivery.

4. CLINICAL APPLICATIONS OF POLYANHYDRIDE POLYMERS FOR INTRACRANIAL DRUG DELIVERY

4.1. BCNU (Gliadel®) Development and Clinical Use

Carmustine (BCNU), a member of the nitrosurea family, was
widely used as a systemic agent for treating malignant brain
tumors. Given its established mechanism of action against

glioma cells, BCNU was a natural candidate for the initial development of intracranial polymer-based chemotherapy. BCNU specifically chloroethylates guanine at the O6 position, alkylating the nitrogen bases of DNA. Low molecular weight and lipid solubility allows BCNU passage across the BBB at potentially tumorcidal concentrations (13), and facilitates its use as a systemic agent. However, BCNU efficacy as systemic chemotherapy confronts dual limitations: (i) a relatively short half-life (<15 min); and (ii) a systemic toxicity profile that includes bone marrow suppression and pulmonary fibrosis. Clinical trials devoted to systemic BCNU administration indeed demonstrated only modest improvements in survival (4,6). BCNU incorporation into local delivery polymers potentially minimizes half-life and systemic toxicity limitations and maximizes clinical efficacy

4.1.1. Preclinical Trials

In Vivo Release Kinetics and Biodistribution

Preclinical studies of BCNU-polymer preparations proceeded in four systematic stages. The first series of experiments examined in vivo release kinetics of BCNU loaded polymers. The initial study involved EVAc copolymer implantation in the rat brain (31). Subsequent to polymer placement, a Bratton-Marshall assay measured BCNU concentrations in both cerebral hemispheres, and serum samples were collected at prescribed time points. The hemisphere ipsilateral to polymer placement corresponded with peak BCNU levels at 4 h; clinically significant concentrations persisted at day 7. In contrast, both contralateral hemisphere and serum BCNU levels were at least an order of magnitude lower throughout the experiment. Thus, the study served as proof of principle of the ability of polymer technology to simultaneously achieve sustained release and local delivery of chemotherapy within the CNS.

A second experiment employed the same rat intracranial model to compare BCNU release kinetics and biodistribution using pCPP:SA (20:80) copolymer delivery vs. direct stereotactic injection (85). Following implantation or injection, tritiated BCNU allowed drug distribution assessment by quantitative radiography in brain sections collected at various time points. High performance liquid chromatography confirmed the correspondence of radioactivity with active BCNU. Polymer delivery afforded BCNU exposure to 50% of the ipsilateral hemisphere at day 3 and 10% exposure at day 14. Considered in absolute terms, a polymer containing 600 μg of BCNU, on days 3 and 7 post-implantation, maintained drug tissue concentrations of 6 mM within a 10 mm radius. Direct stereotactic injection failed to match the sustained delivery of the pCPP:SA polymer system; a 1–3 h broad, post-injection spike of BCNU distribution rapidly dissipated over time.

A third experiment examined 20% (w/w) BCNU loaded pCPP:SA polymer release kinetics and biodistribution in a primate intracranial model. Tumorcidal concentrations of BCNU spread up to 4 cm from the polymer implantation site 24 h post-operatively (86). Taken together, these animal studies established the drug delivery capabilities of polymer technology, and confirmed the viability of local, sustained, and clinically significant in vivo delivery of chemotherapeutic agents within the CNS.

Preliminary Efficacy of BCNU-Loaded
Polymers in Glioma Models

Using rat flank and intracranial 9L gliosarcoma models, Tamargo et al. (70) initially compared the efficacy of polymer and systemically based BCNU. EVAc polymer delivery in the flank model produced significant tumor growth delay relative to systemic administration (15.3 vs. 11.2 days, $p < 0.05$). In the intracranial model, a 10 mg polymer with 20% (w/w) BCNU polymers dramatically improved survival in animals with established 9L gliosarcoma. EVAc and pCPP:SA polymers conferred respective survival advantages of 7.3-fold and 5.4-fold over controls. Systemic BCNU, in contrast, increased survival only 2.4-fold compared to controls.

A second study in the same established rat intracranial 9L gliosarcoma model evaluated the efficacy of 20% (w/w) BCNU-loaded pCPP:SA polymers vs. direct stereotactic injection of an equivalent BCNU dose (87). Compared to controls, the polymer group enjoyed a 271% median survival improvement; stereotactic injection extended median survival only 36%. Moreover, the polymer group contained twice as many long-term survivors.

Polymer Design: Monomer Ratio and BCNU
Percentage Optimization

Following the original promise of polymer in vitro release kinetics and efficacy data, the third series of studies established the optimal pCPP:SA monomer ratio and BCNU dosing percentage (88). Initially, Sipos et al. assessed the in vitro release kinetics for 50:50 and 20:80 pCPP:SA formulations of both 4% and 32% BCNU-loaded polymers (81). Theoretically, a lower CPP proportion should slow polymer degradation, thereby enhancing sustained drug release. Though release kinetic studies demonstrated minimal differences between the 50:50 and 20:80 variants loaded at 4% BCNU, the 32% BCNU-loaded polymers revealed a 150% increased release duration (18 vs. 7 days) with the 50:50 formulation. Thus, empirical data confirmed the proposed relationship between drug release duration and the pCPP:SA monomer ratio.

Subsequently, an efficacy study examined multiple permutations of BCNU dose percentage and monomer ratio in the established rat intracranial 9L gliosarcoma model, testing polymer loads of 0%, 4%, 8%, 12%, 20% and 32% (w/w) in both 50:50 and 20:80 pCPP:SA formulations. The 20% (w/w) BCNU-loaded pCPP:SA (20:80) combination corresponded with maximal longevity, conferring 63% survival at 200 days relative to a control group median survival less than 20 days. Toxicity studies in the cynomoglus monkey intracranial model found no systemic or local morbidities associated with 20% BCNU-loaded pCPP:SA (20:80) polymer implantation; MRI images obtained up to 150 days post-implantation documented no evidence of edema or mass effect (81).

Preliminary Efficacy of BCNU-Loaded
Polymers in Brain Metastasis Models

Brain metastases constitute the majority of newly diagnosed brain tumors (89), and represent a major cause of mortality in patients with metastatic cancer (90). Metastatic
disease therefore merits separate consideration for local polymer-based chemotherapy. A final set of preclinical studies
investigated local polymer delivery of various neoplastic
agents, including BCNU, to combat brain metastases (91).
The experimental series followed a logical progression and
employed a mouse intracranial model. Initially, toxicity studies with each agent loaded in pCPP:SA polymers established
maximum non-toxic dosages. Subsequently, experiments
with several metastatic tumor lines evaluated treatment
efficacy in the presence and absence of concurrent radiation
therapy. Cell lines included B16 melanoma, RENCA renal cell
carcinoma, CT26 colon cancer, and Lewis lung carcinoma.
Though both BCNU-loaded polymers and radiation therapy
demonstrated independent efficacy, combination treatment
conferred a synergistic survival benefit against B16 melanoma (median survival 35 days vs. 21.5 days for controls;
$p = 0.0005$), RENCA cell carcinoma (38.5 vs. 12 days; $p <$
0.007), and Lewis lung carcinoma (23 vs. 21 days; $p =$
0.001). Later work with the EMT-6 breast cancer line in mice
showed intracranial BCNU polymers efficacy in the presence
(41 vs. 17 day median survival; $p = 0.02$) and absence (> 200
vs. 17 days; $p < 0.0001$) of radiation therapy (92). Encouraged
by these findings, clinical trials of BCNU-loaded pCPP:SA
treatment for metastatic brain tumors are in the planning
stages.

4.1.2. Clinical Experience for Recurrent Gliomas

Having confirmed pCPP:SA polymers are (i) biocompatible
and non-toxic with BCNU-loading and possible radiation
therapy, (ii) capable of broad BCNU distribution from the
implantation site, and (iii) more effective delivery systems
than systemic BCNU in animal survival studies, research
development progressed to a multicenter phase I–II clinical

trial in humans (93). Enrollment criteria required standard therapy failure; only patients with recurrent malignant gliomas after a previous craniotomy for surgical debulking qualified. Other eligibility requirements included an indication for reoperation, a unilateral single tumor focus with ≥ 1 cm^3 of

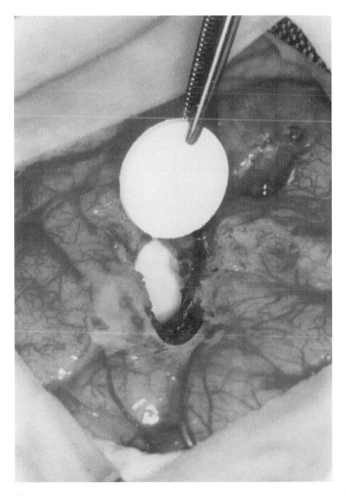

Figure 4 After surgical debulking, the tumor resection cavity is lined with up to eight 200 mg 3.8% (w/w) BCNU pCPP:SA (20:80) (Gliadel®) polymer, where the loaded drug is gradually released as they dissolve.

enhancing volume on MRI or CT, completion of external beam radiotherapy, a Karnofsky Performance Scale (KPS) score ≥60, and no nitrosurea exposure during the 6 weeks prior to polymer implantation. Twenty-one patients received treatment; the study tested polymer BCNU loads of 1.93, 3.85, and 6.35% (w/w). Each polymer weighed 200 mg, and most patients received a maximum of eight wafers in the tumor cavity following debulking (Fig. 4). Initial tumor volumes were similar in all groups.

The first human trial of BCNU-loaded polymers provided a safety and radiological data. Frequent blood chemistry and urinalysis panels uncovered no evidence of systemic toxicity. KPS scores fell slightly in the immediate postoperative period, but returned to baseline and remained stable for at least 7 weeks, suggesting preserved quality of life during the chemotherapeutic course. Implanted polymers were detectable on postoperative CT and MRI, appearing as areas of decreased signal intensity on T1-weighted MRI, and some remaining visible by CT for as long as 49 days after surgery. In 13 of the 21 patients, routine protocol scans revealed some areas of marked enhancement around the implant sites. Significantly, the observed enhancement at no time corresponded with clinical findings, namely neurological impairment or other toxic sequelae, and generally resolved spontaneously.

Over the course of the study, 10 patients experienced declining neurological status with increased MRI or CT enhancement, and therefore underwent reoperation. Intraoperatively, a rim of necrotic tissue up to 1 cm thick, similar to that described in interstitial brachytherapy, was found with general KPS score improvement subsequent to tissue removal.

From an efficacy standpoint, overall median survival times were 46 weeks from implantation and 87 weeks after initial diagnosis; 86% of patients lived more than 1 year after diagnosis. Based on this work, the 3.85% BCNU-loaded polymer was chosen for further clinical study.

The resultant clinical trial was a rigorous, multicentered, prospective, randomized, double-blinded, and placebo-controlled phase III study (94). The investigators examined the efficacy of 3.85% BCNU (w/w) pCPP:SA polymers in treating recurrent malignant gliomas; enrollment included 222 patients from 27 North American medical centers. Randomized patients received either a BCNU treatment polymer or a blank placebo. Selection criteria matched the phase I–II with one exception: the phase III trial prohibited any chemotherapy 4 weeks preoperatively and permitted systemic administration 2 weeks postoperatively. Individual patients and their surgeons decided upon additional operations for tumor recurrence during the study period.

In terms of study population characteristics, mean age was 47.8 years, and randomization rendered the treatment

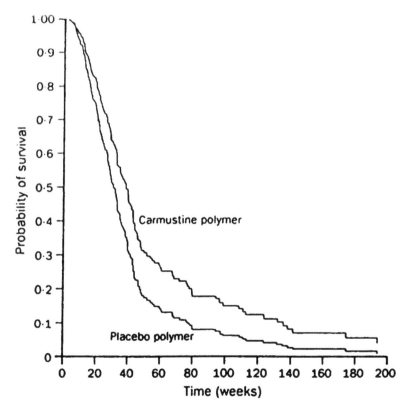

Figure 5 Overall survival for patients receiving BCNU-loaded polymers vs. controls at the time of the operation for recurrent malignant gliomas. (From Ref. 94.)

and placebo groups well matched for age, tumor type and preoperative KPS scores. All patients underwent antecedent external beam radiotherapy; 48.2% of control and 52.7% of treatment group patients underwent chemotherapy prior to trial enrollment.

Median postoperative survival of patients receiving BCNU-loaded polymers was 34 weeks compared to 23 weeks with placebo (hazard ratio 0.67, $p = 0.006$) (Fig. 5). The 6-month survival rate for treatment and control groups was 60% and 47% respectively. Remarkably, the difference in 6-month survival rate was quite dramatic in the patients with the most malignant tumors; i.e., GBM patients ($n = 145$); BCNU patients were 50% more likely to survive 6 months postoperatively than controls. Study results supported BCNU-loaded polymers as safe and well tolerated. Significantly, polymer delivery showed no evidence of bone marrow suppression or other toxicities associated with systemic BCNU administration. Within 6 months of polymer placement, 11.8% of BCNU patients and 11.6% of control patients underwent reoperation. The BCNU-treated group experienced a higher incidence of intracranial infection than the placebo group (4/110 or 3.6% vs. 1/112 or 0.9%), but the difference did not reach clinical significance. Postmortem examination of 11 brains from the treatment group patients revealed mild inflammatory reactions with no marked necrosis.

Taken together, the enumerated series of clinical studies established BCNU-loaded polymers as a safe and effective treatment of recurrent malignant gliomas. In 1996, the Food and Drug Administration (FDA) approved 3.85% BCNU (w/w) pCPP:SA polymer (Gliadel) for the treatment of recurrent GBM. Gliadel was the first treatment for malignant gliomas to receive FDA approval in 23 years.

4.1.3. Clinical Experience as Initial Therapy

Further research into possible Gliadel applications coincided with a general oncologic principle: treatments effective at recurrence invariably prove more efficacious as initial therapy. Thus, given Gliadel's established role in treating recurrent malignant gliomas, attention naturally turned to its therapeutic potential during initial disease presentation. A phase I–II trial, enrolling 22 patients, addressed the safety of Gliadel polymers at initial surgery (95). Concordant with earlier studies, polymer wafers weighed 200 mg and most patients received a maximum of eight wafers. Inclusion criteria included a single enhancing tumor focus $\geq 1\,cm^3$, age > 18, and a KPS score ≥ 60. The study population's mean age was 60. All patients received external beam radiation therapy averaging 5000 rads, and no patient received additional chemotherapy during the 6 months following surgery.

Phase I–II trial results confirmed Gliadel as a safe treatment option for patients newly diagnosed with malignant gliomas who received post-operative radiation therapy. The study was free of perioperative mortality, and furnished no evidence of polymer-induced local or systemic toxicity. Of 22 patients, 21 received a pathological diagnosis of GBM; median survival was 44 weeks from implantation with four patients surviving beyond 18 months.

With the safety of locally implanted Gliadel documented, a phase III, multicenter, randomized, double-blinded, placebo-controlled study was undertaken to assess polymer efficacy against newly diagnosed malignant gliomas (96). Similar to previous clinical studies, patient inclusion required a single tumor focus with $\geq 1\,cm^3$ of enhancement, age between 18 and 65 years, a KPS score ≥ 60, and a histopathological diagnosis of either anaplastic astrocytoma or GBM on intraoperative frozen section. Unfortunately, temporary unavailability of the drug required premature study termination at 32 subjects, rather than the intended 100 patients. Each randomized cohort included 16 patients; median age (55.5 yrs vs. 53 yrs) and KPS scores (75 vs. 90) were comparable between the BCNU and placebo groups. However, a discrepancy in tumor pathology persisted despite randomization. All placebo patients harbored GBMs, but only 11/16 BCNU patients received a GBM frozen section diagnosis.

Considering all 32 patients in the analysis, the treatment group's median survival of 58.1 weeks significantly outpaced the 39.9 weeks associated with placebo ($p = 0.012$) (Fig. 6).

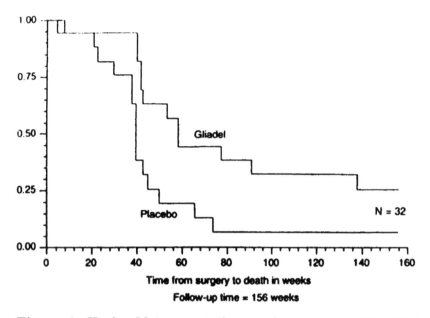

Figure 6 Kaplan-Meier survival curve for patients with initial therapy for grade III and grade IV gliomas treated with BCNU-loaded polymer implants vs. placebo. (From Ref. 96.)

Further subset evaluation of GBM patients revealed a median survival advantage of 53.3 weeks vs. 39.9 weeks with placebo ($p = 0.008$). Interestingly, three of the study's four long-term survivors (still alive 3 years after study termination) received BCNU-loaded polymers. Age and KPS score significantly impacted survival while the survival effect of tumor type failed to achieve statistical significance. The phase III trial reinforced earlier research documenting no local or systemic morbidities with polymer implantation.

Recently, a larger phase III, multicenter, randomized, double-blinded, placebo-controlled study of 240 patients definitively assessed the efficacy of Gliadel as initial malignant glioma therapy (97). Gliadel increased median survival (13.9 vs. 11.6 months, $p = 0.03$) relative to controls receiving blank polymers. Again, the effect for some patients was sustained with patients surviving 3 years in the Gliadel group compared to patients surviving 3 years in the control group.

In 2003, the FDA therefore approved Gliadel for the treatment of initially diagnosed malignant gliomas. The FDA approval decision marks another watershed in brain tumor therapy, and the continued evolution of polymer systems for local chemotherapeutic delivery.

BCNU-loaded polymers remain the basis for several ongoing clinical investigations. A phase I–II dose-escalation study recently established the maximal non-toxic BCNU loading dose in 20:80 pCPP:SA polymers. Analogous to findings from earlier animal models (81), the maximal tolerated dose in humans was 20% BCNU (w/w) (98). Given Gliadel loading at 3.85%, future clinical trials will include phase III assessment of an improved 20% BCNU wafer. Moreover, ongoing research efforts expand Gliadel application to metastatic tumors, and consider Gliadel inclusion in combination chemotherapy and with drug resistance modifiers.

4.2. Other Neuro-Oncologic Applications of Polymer Delivery

The successful development of BCNU-loaded sustained-release polymers from the laboratory to standard clinical use provides a model for future investigation. Local delivery opens neuro-oncology to chemotherapeutics previously prevented by systemic toxicity. The following section reviews other agents successfully incorporated into polymers or employed as adjuvant therapy to polymer-based treatment. O6-Benzylguanine and paclitaxel are in clinical trials, and camptothecin, for example, represents alternative polymer-based chemotherapeutic agents approaching the clinical research phase. Additionally, the section chronicles local polymer delivery of immunotherapy and angiogenesis inhibitors. Finally, the section provides a broad survey of various other agents receiving preclinical research attention as candidates for local polymer delivery.

4.2.1. O6-Benzylguanine

Over the course of Gliadel therapy, many brain tumors acquire resistance through the expression of AGT, a DNA repair protein found in the majority of brain tumors treated with BCNU (99). Mechanistically, BCNU exerts its antineoplastic effects by chloroethylation of DNA at the O6 position of guanine. AGT removes adducts at this position, thereby protecting tumor cells from BCNU activity. O6-Benzylguanine (O6-BG) irreversibly inactivates AGT. Acting as a substrate analog, O6-BG transfers a benzyl group to a cysteine residue at the AGT active site (100). Naturally, interruption of a mechanism responsible for tumor resistance represents a potentially useful adjuvant to existing Gliadel treatment.

O6-BG mediated AGT inhibition indeed increases tumor sensitivity to BCNU. However, concomitant O6-BG and systemic BCNU administration in animal models reduced the maximum tolerated dose 6-fold due to bone marrow toxicity (101). Presumably, O6-BG imparts the same increased BCNU sensitivity to tumor cells and those vulnerable to systemic toxicity. Rhines et al. postulated systemic O6-BG administration with intracranial BCNU polymer delivery, not systemic BCNU, would significantly reduce the side effects of combination therapy (102). In the established rat intracranial F98 glioma model (a tumor line with high AGT activity), systemic O6-BG with BCNU-loaded pCPP:SA polymers improved median survival over either O6-BG alone (34 vs. 22 days, $p = 0.0002$) or BCNU polymer alone (34 vs. 25 days, $p = 0.0001$). Moreover, Rhines et al. showed the reduction in BCNU polymer load was not necessary with the systemic introduction of O6-BG. The animals further demonstrated no gastrointestinal or bone marrow toxicity from combination therapy (95).

These results suggest that concurrent use of O6-BG and BCNU polymers may serve as an important addition to the treatment of malignant brain tumors. To this end, recently completed NIH-funded phase I clinical trials investigated O6-BG as an adjuvant to BCNU treatment. Ongoing efforts center on the method of O6-BG delivery. Incorporation of O6-BG within polymers for local delivery and dose-escalation studies for systemic and local-delivered O6-BG represent active areas of development.

4.2.2. Paclitaxel

Paclitaxel (Taxol), a microtubule-binding agent, exerts established tumorcidal activity against non-small cell lung cancer, breast cancer, and ovarian cancer. In vitro studies demonstrate potent paclitaxel activity against rat and human glioma lines (103). Blood-brain barrier impermeability to paclitaxel (104) and its anti-tumor properties render it an excellent candidate for local polymeric delivery.

In vitro kinetics studies with 20–40% (w/w) paclitaxel-loaded pCPP:SA (20:80) polymers demonstrated sustained drug release approaching 1000 h. Biodistribution studies concordantly revealed tumorcidal paclitaxel concentrations in the rat brain for > 30 days post-implantation. Median survival in the established rat 9L gliosarcoma model improved from 19.5 days in rats treated with blank polymers to 61.5 days with 20% paclitaxel-loaded polymers ($p < 0.001$) (105).

Clinical trials employing a paclitaxel-loaded PACLIMER (106) microsphere delivery system for treatment of ovarian cancer are currently underway. In terms of intracranial applications, initial toxicity trials in dogs demonstrate no early mortality or morbidity attributable to the paclitaxel polymer (107). Pending completion of the canine toxicity studies and demonstration of safety, phase I–II clinical trials for paclitaxel-loaded polymer treatment of intracranial tumors are planned. Paclitaxol in loaded polymer has already been utilized clinically for recurrent ovarian cancer.

4.2.3. Camptothecin

The camptothecins inhibit topoisomerase I, a DNA replicating enzyme, and thereby exhibit antineoplastic properties (108). Though in vitro and in vivo studies offered great promise, clinical trials exposed unexpected toxicities with systemic administration (109). Though toxicity prevented its use as a systemic agent for gliomas, camptothecin remained suitable for local polymer delivery (30). Among the family of camptothecins, sodium camptothecin possessed chemical properties suitable for facile polymer loading.

Initially incorporated into EVAc polymers, sodium camptothecin demonstrated sustained release by in vitro kinetics experiments. Efficacy studies in the established rat intracranial 9L gliosarcoma model showed dramatically increased median survival from 19 days in control animals to > 120 days with 50% (w/w) polymer administration ($p < 0.001$). Additionally, 59% of treatment animals survived > 120 days; no control rats lived beyond 32 days. Systemic camptothecin, in contrast, conferred no survival benefit. Animals undergoing polymer implantation suffered no associated local or systemic toxicity.

Studies examining sodium camptothecin-loaded pCPP:SA polymers include the previously described mouse metastatic tumor study (84) and recent work in the established rat intracranial 9L gliosarcoma model (110). In the metastatic study, camptothecin-loaded pCPP:SA polymers conferred a survival benefit only against the B16 melanoma and only when administered in conjunction with radiation therapy (median survival 27.5 vs. 19 days; $p = 0.043$). Against the 9L gliosarcoma, 50% (w/w) camptothecin-loaded polymers extended median survival from 17 days in controls to 69 days ($p < 0.001$). Polymer implantation corresponded with no notable local or systemic toxicity. The study confirmed active camptothecin release for 1000 h. Ongoing research involves preclinical efficacy studies of various camptothecin analogs (111).

4.2.4. Immunotherapy

Immunotherapy to combat malignant gliomas requires the generation and maintenance of an anti-tumor immune response. Genetic mutations within tumor cells produce protein expression patterns foreign to the host, thereby offering appropriate targets for immune system recognition. Cytokines, namely interleukins (IL), interferons (IFN), and colony-stimulating factors, mediate immune system activation, and therefore represent the focus of current research. Cytokines carry systemic toxicity and are largely unable to permeate the BBB, necessitating the development of local cytokine delivery systems for brain tumor immunotherapy.

Two general strategies exist for local cytokine delivery. Irradiated tumor cells, transduced to secrete the cytokine in a paracrine fashion, provide the first means of sustained cytokine release at the tumor site. Using IL-2 transduced tumor cells in the C57BL/6 mouse intracranial melanoma model, Thompson et al. generated an immune response to wild-type tumor by direct tumor site injection; flank injection of IL-2 transduced cells failed to yield a similar immune response (112). Granulocyte-macrophage colony-stimulating factor (GM-CSF) transduced cells, conversely, elicited an immune response with flank, not intracranial, injection. Notably, simultaneous intracranial injection of IL-2 transduced cells and subcutaneous flank injection of GM-CSF transduced cells produced a synergistic immune response.

Additional study of IL-2 transduced cells in the same animal model offered several interesting findings (113). In mice challenged with intracranial or more distal tumor burden, intracranial implantation of IL-2 transduced cells enhanced survival relative to controls. Moreover, after successful rejection of an initial challenge, treated animals exhibited immunological memory by mounting an immune response to a second tumor challenge, and enjoyed the attendant survival advantage. To underscore the importance of local IL-2 delivery, identical or 10-fold increased doses of subcutaneously injected IL-2 transduced tumor cells did not produce a comparable memory response. Mechanistically, gene-knockout mice identified natural killer (NK) cells, not CD4+ T-cells, as most responsible for the anti-tumor immune response. The failure of flank injected IL-2 paracrine cells to generate a similar immune response prompted the following postulate: immune cells within the CNS possess different cytokine requirements than their peripheral counterparts.

A final study of local paracrine delivery identified IL-12, a cytokine with antiangiogenesis properties in addition to its immunoregulatory effects, as a potential substrate candidate (114). Using the established rat intracranial 9L gliosarcoma model, the study examined intracranial implantation of IL-12 transduced tumor cells. Reverse transcriptase-poly merase chain reaction confirmed IL-12 expression. Local paracrine delivery prolonged survival and induced immunological memory in treated animals. Thus, a second injection of wild-type 9L gliosarcoma tumor cells successfully produced an immune response.

Several experiments recently explored the interaction between local paracrine immunotherapy and polymer-delivered chemotherapy. A study in the mouse intracranial F16-B10 melanoma model, for example, combined local paracrine delivery of IL-2 via transduced tumor cells with either 10% (w/w) BCNU-loaded or 1% (w/w) carboplatin-loaded pCPP:SA polymers (115). Of the animals receiving BCNU-loaded polymers and paracrine immunotherapy, 70% survived > 72 days; control group median survival was 15.8 days with none living past 72 days ($p = 0.0023$). Carboplatin-loaded polymers with immunotherapy produced comparable synergies: 80% of treatment animals vs. zero controls survived > 72 days (control group median survival of 20.6 days; $p = 0.0001$). Histological examination of combination therapy specimens revealed rare degenerating tumor cells with a marked mixed inflammatory reaction on postimplantation day 14, and no tumor cells with resolution of the inflammatory reaction on day 72. Overall, the study suggested combined paracrine immunotherapy and polymer-based chemotherapy merits further attention as a treatment for malignant gliomas.

The second general strategy for local cytokine delivery involves direct loading onto polymers. In 1998, Wiranowska et al. established, as proof of principle, that polymers were capable of the sustained release of biologically active cytokines (116). Both in vitro and in vivo experiments documented the release of active murine IFN-α/β by loaded EVAc polymers. In vitro assays determined the majority of activity release occurred in the first 4 days; in vivo trials suggested most of activity release spanned the first 24 h with gradual tapering over the next 3 days.

Gulambek et al. in 1993, introduced the gelatin-chondroitin sulfate (GCS) microsphere system for local delivery of drugs in an injectable mixture (73). Hanes et al. applied the GCS system to cytokine delivery, encapsulating IL-2 and confirming sustained release of activity over 2 weeks in vitro and up to 3 weeks in vivo (117). Subsequently, studies in the rat intracranial 9L gliosarcoma model, the mouse intracranial B16-F10 melanoma model, and the mouse liver CT26 carcinoma model compared immune response generation from direct intratumoral GCS-IL-2 mixture injections, local paracrine delivery, and negative controls. In the B16-F10 model, 42% of animals receiving the GCS-IL-2 mixture exhibited protection on a second tumor challenge. Recently, and perhaps most promisingly, Rhines et al. assessed the combination of IL-2 loaded microspheres and BCNU-loaded pCPP:SA polymers; combined therapy increased survival over either modality alone, achieving statistical significance (118).

4.2.5. Angiogenesis Inhibitors

Angiogenesis, the formation of new blood vessels, constitutes a critical element of tumor growth (119). In the absence of angiogenesis, nutrients diffuse only a few millimeters from the tumor periphery, and tumor size arrests at an equilibrium between peripheral cell proliferation and central cell death. Blood vessel formation facilitates nutrient delivery to central cells, resulting in exponential tumor growth with potential metastatic spread (120). GBM, among the most angiogenic of all neoplasms, represents a natural target for angiogenesis inhibitors.

Historically, the development of local polymer-delivered antiangiogenesis agents began with heparin and cortisone (121), a combination shown to inhibit angiogenesis (122). In the rabbit cornea VX2 carcinoma model, EVAc copolymer administration of heparin and cortisone reduced angiogenesis 60% at 21 days postimplantation ($p < 0.05$). The same study assessed heparin- and cortisone-loaded pCPP:SA polymer implantation in the rat flank 9L gliosarcoma model; the combination inhibited tumor growth by 78% ($p < 0.05$).

Squalamine, an aminosterol isolated from the dogfish shark, inhibits tumor mitogen-induced endothelial cell proliferation, and thereby exhibits antiangiogenesis properties (123). EVAc copolymers loaded with 20% (w/w) squalamine inhibit vascular growth in the rabbit cornea VX2 carcinoma model. Currently, phase I clinical trials are underway to evaluate squalamine efficacy against a variety of advanced cancers (124). Ongoing work explores the potential advantages of local polymer-based squalamine delivery.

Minocycline, a broad-spectrum antibiotic with established anticollagenase properties (125), offers great promise as an antiangiogenesis agent. Minocycline-loaded EVAc polymers inhibited neovascularization in the rabbit cornea VX2 carcinoma model by a factor of 4.5, 4.4, and 2.9 on days 7, 14, and 21 respectively ($p < 0.05$ at all time points) (126). Follow-up study confirmed sustained minocycline delivery; polymers released 55% of the drug at 90 days (29). From an efficacy standpoint, 50% (w/w) minocycline-loaded EVAc polymers, when implanted simultaneously with tumor injection, extended survival from 13 to 69 days ($p < 0.001$) in the rat intracranial 9L gliosarcoma model. Though minocycline-loaded polymers alone failed to impact survival against the established rat intracranial 9L gliosarcoma, combined surgical resection and polymer implantation extended median survival by 43% ($p < 0.002$). Systemic BCNU administration further improved median survival by 90% over controls. Moreover, recent work with 40–50% (w/w) minocycline-loaded pCPP:SA polymers suggests even greater efficacy gains relative to the studied EVAc copolymers (127).

4.2.6. Preclinical Survey of Other Chemotherapeutic Agents

This section surveys a broad range of compounds in various stages of preclinical development, and concludes with two agents, quinacrine and mitaxantrone, that represent areas of active research. Though not exhaustive, the review provides further insight into the rational development of local polymer-based chemotherapy.

Cyclophosphamide

Cyclophosphamide (Cytoxan), an alkylating agent widely used for the treatment of various malignancies, offers an instructive example. Hydroxycyclophosphamide, its active metabolite, poorly crosses the BBB, and therefore possesses limited potential as a systemic treatment for malignant gliomas. Furthermore, cyclophosphamide necessitates activation by the hepatic p450 cytochrome oxidase system, limiting its potential for local intracranial delivery (128).

4-Hydroxyperoxycyclophosphamide (4-HC), a derivative of cyclophosphamide, spontaneously converts into the active metabolite (129), rendering it a better option for local polymeric delivery. Given 4-HC hydrophobicity, the FAD:SA system was chosen for delivery. Pharmacokinetic and biodistribution studies confirmed favorable release kinetics and intracranial distribution (130). Toxicity studies established 20% (w/w) as the maximum tolerable polymer dose. In the established rat intracranial 9L gliosarcoma and F98 glioma models, 4-HC-loaded polymers offered a median survival of 77 days with 40% surviving beyond 80 days; blank polymer controls corresponded with a 14 day median survival and no long-term survivors ($p = 0.004$).

L-Buthionine sulfoximine (BSO), an inhibitor of the glutathione S-transferase (GST) enzyme pathway, potentiates 4-HC antitumor effects in the 9L gliosarcoma model (131). Recent work suggests the GST pathway plays a role in the inactivation of alkylating agents (132). Analogous to O6-BG development as an adjuvant to BCNU-loaded polymers, a series of studies examined BSO augmentation of 4-HC efficacy. Systemic BSO delivery in conjunction with 4-HC polymer implantation failed to enhance survival. Separate experiments assessed the safety and efficacy of local BSO delivery. Notably, rat brain implantation of BSO-loaded EVAc polymers depleted intracranial glutathione levels without impacting hepatic levels; systemic BSO administration achieved the converse. Moreover, co-release of 4-HC and BSO from the FAD:SA polymer boosted median survival 4.6-fold in the established rat intracranial 9L gliosarcoma model (61.5 days vs. 13 days, $p < 0.001$). Relative to blank polymer controls, 4-HC alone improved survival only 2.3-fold ($p < 0.001$).

Adriamycin

Adriamycin, an anthracylcine antibiotic, intercalates with DNA. By causing strand scission and double-stranded crossbreaks, adriamycin exhibits tumorcidal activity on breast cancer, acute leukemia, lymphoma, and other malignancies (133). Following successful Ommaya reservoir implantation in 1983 (134), needle-shaped EVAc polymers were developed for local adriamycin delivery (135). Pharmacokinetic studies confirmed adriamycin release followed zero-order kinetics. In nude mice, the needle polymers significantly inhibited the growth of brain tumor xenografts. More recent experiments assessed adriamycin loading in pCPP:SA polymers (136). In vitro assays documented sustained release, and in vivo studies in the established rat intracranial 9L glioma model showed improved median survival (33 days vs. 13 days in controls, $p < 0.0006$) with polymer implantation.

5-Fluorouracil

5-Fluorouracil (5-FU), a thymidine analog, blocks the conversion of deoxyuridylic acid to thymidilic acid, and thereby eliminates an essential precursor for DNA synthesis. Though effective in treating various non-intracranial malignancies, systemic toxicities including myelosuppression and gastrointestinal mucosal injury limit 5-FU efficacy against brain tumors (126). Early efficacy trials evaluating 5-FU as a systemic brain tumor therapy were indeed disappointing (137–139).

Local 5-FU delivery benefits from a long history of scientific investigation. Gelatin sponges (75), Surgicel (75), and silastic tubing loaded with a "chemotherapy cocktail" including 5-FU (140,141) represented early modes of delivery. The first polymer-based delivery approach involved a matrix consisting of "glassified monomers" with 10% polymetacrylic methyl acid. In 1986, 55 patients received the prototype polymers loaded with 5-FU, adriamycin, and mitomycin C; 47% of treated malignant glioma patients survived 1 year (142). Successful 5-FU loading onto PLGA microspheres occurred in 1995 (143). 5-Fluorouracil-loaded microspheres decreased mortality in the established rat intracranial 9L gliosarcoma model ($p = 0.017$); placebo and bolus 5-FU injections conferred no benefit (144). Notably, microsphere implantation corresponded with no observed toxicities, and histological examination showed only mild tissue reaction.

Menei et al. treated eight newly diagnosed GBM patients with surgical debulking and 5-FU-loaded PLGA microsphere implantation (145); patients also underwent postoperative adjuvant external beam radiation therapy. 5-Fluorouracil concentrations in the CSF were clinically significant up to 1 month after surgery, but serum levels remained low and transitory, confirming local delivery without systemic 5-FU exposure. Median survival time was 98 weeks with two patients exhibiting disease remission at 139 and 153 weeks.

Methotrexate

Methotrexate (MTX), a folate antagonist, enjoys limited efficacy against brain tumors due to BBB impermeability and systemic toxicities including myelosuppression and gastrointestinal necrosis (126). Though cisternal or intraventricular MTX infusion offered poor drug dissemination (19,146), broad parenchymal distribution with direct MTX injection (147) suggested promise for local MTX delivery.

Methotrexate modification improved stability and biodistribution. Covalent linkage of MTX to dextran enhanced penetration of a three-dimensional collagen lattice designed to simulate extracellular matrix, and did not adversely effect MTX tumorcidal properties against the human H80 glioma line (148). In the established rat intracranial 9L gliosarcoma model, FAD:SA polymers loaded with the MTX-dextran conjugate offered modest, though significant, survival enhancement over controls.

Encouraged by early clinical trials (149,150) documenting minimal toxicity with intratumoral MTX, other local delivery efforts include polymethylmethacrolate pellets (76), PLGA copolymer matrices (151), and fibrin glue systems (74). However, some case reports describe neurological side effects. One patient developed an abulic-hypokinetic syndrome and left hemiparesis after intraventricular MTX treatment of meningeal carcinoma (152). Large catheter tip cysts in two patients complicated Ommaya reservoir MTX administration (153).

Platinum Drugs

Carboplatin, a second generation platinum analog, offers a less neurotoxic alternative to cisplatin, its parent compound (154). Though systemic administration causes myelosuppression, local FAD:SA polymer delivery optimally addresses toxicity and solubility concerns. Olivi et al. determined a maximum non-toxic dose of 5% (w/w) in the rat intracranial F98 glioma model (69). Efficacy studies of carboplatin-loaded polymers demonstrated 52 day median survival relative to 16 days (with all dead by day 19) in controls. A separate experiment assessed carboplatin polymers against various mouse metastatic brain tumors (84). Combined with radiation therapy, carboplatin-loaded polymers prolonged survival against the CT26 colon carcinoma (median survival 33 days vs. 20.5 days for controls, $p = 0.013$) and RENCA cell carcinoma (15 vs. 12 days, $p < 0.01$). Against the B16 melanoma, carboplatin polymers alone enhanced survival (27 vs. 16.5 days, $p = 0.043$); combination therapy was ineffective.

Carboplatin coupling to α-cyclodextrin delays decomposition and increases bioavailability (155). Over 110 days, ethylcellulose microcapsules loaded with both agents at 2.2% (w/w) afforded 56% carboplatin release. Against the established rat intracranial F98 glioma, microcapsule delivery of α-cyclodextrin alone conferred a median survival of 20 days. Comparatively, carboplatin polymers offered 34 day median survival ($p < 0.001$) and dual agent polymer therapy corresponded with a 51 day median survival ($p < 0.01$ vs. carboplatin alone). Thus, α-cyclodextrin enhanced carboplatin efficacy.

Its neurotoxicity notwithstanding, several studies explore the potential efficacy of local cisplatin delivery. In the established rat intracranial 9L gliosarcoma model, biodegradable polylactic acid polymers loaded with cisplatin extended median survival. Compared to 24 days in the control group, 32 days with systemic cisplatin treatment, and 39 days with local bolus infusions of cisplatin, polymer delivery offered a median survival > 60 days ($p < 0.00006$ vs. systemic group; $p < 0.001$ vs. bolus group). Moreover, cisplatin-loaded polymers provided histopathological "cure" in 8 of 12 animals vs. 3 of 13 local bolus infusion animals ($p < 0.01$) and no cures in other study groups. Follow-up study confirmed cisplatin-loaded polymer biocompatibility (36).

Dexamethasone

Vasogenic edema, a major source of brain tumor-associated morbidity, results from malignant glioma degradation of the BBB (126). Though high dose corticosteroid therapy significantly alleviates the edema (156), sustained systemic exposure carries a significant toxicity profile. Diabetes mellitus, skin atrophy, "Cushingoid" features, weight gain, hemorrhagic gastrointestinal ulcers, myopathies, osteoporosis, and pathologic fractures represent a high price for systemic treatment (157). Local delivery, however, remains a viable option. When loaded 35% (w/w) onto the EVAc polymer, dexamethasone delivery in the rat brain was clinically significant for 21 days (158). Notably, concurrent plasma levels were low.

Efficacy studies in the rat intracranial 9L gliosarcoma model compared systemic and EVAc polymer dexamethasone administration (28). Water weight percentage served as the edema endpoint; both intracranial dexamethasone-loaded polymers (79.15%; $p < 0.05$) and intraperitoneal dexamethasone injections (79.16%; $p 0.05$) were superior to controls (79.45%) and intraperitoneal polymer implantation (79.39%). Thus, intracranial polymer implantation achieved equivalent edema reduction without the theoretical potential for systemic toxicity.

Quinacrine

Quinacrine, an early antimalarial drug rarely used in the clinical setting, exhibits a wide variety of intracellular actions. Though the mechanism remains elusive, quinacrine reduces mutagenicity in leukemia and glial cell lines (159,160). Given that the development of tumor resistance represents the most common cause of BCNU treatment failure, quinacrine administration could potentially maintain tumor sensitivity to BCNU, and retard mutation into resistant tumor variants. In the rat flank C6 glioma model, oral quinacrine and systemic BCNU therapy indeed reduced tumor growth relative to systemic BCNU alone. Preliminary

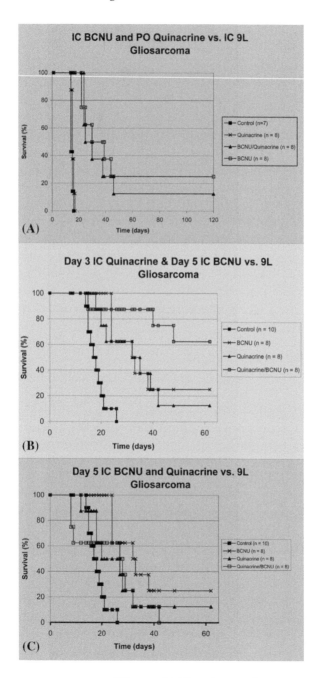

Figure 7 (*Caption on facing page*)

work by the authors evaluated oral quinacrine and concurrent intracranial implantation of BCNU-loaded 3.85% (w/w) pCPP:SA polymers in the established rat intracranial 9L gliosarcoma model. Unfortunately, combination therapy conferred no appreciable survival advantage (Fig. 7a). Poor CNS delivery of quinacrine, only 10% by oral administration, likely explains the disappointing results.

Local quinacrine delivery therefore offers a natural alternative to systemic administration. Figure 7b illustrates that the implantation of 15% (w/w) quinacrine-loaded pCPP:SA polymers and BCNU-loaded polymers significantly extend survival. Interestingly, the synergistic benefit of combined quinacrine and BCNU local delivery demonstrates time sensitivity; quinacrine polymer implantation 2 days prior to BCNU polymer placement enhances survival whereas concurrent tumor bed insertion offers no benefit (Fig. 7c). The importance of ordered administration of adjuvant quinacrine therapy provides some mechanistic insight, and remains the subject of ongoing research.

Mitaxantrone

Mitaxantrone, a dihydroxyanthracenedione derivative approved for the treatment of hepatic and ovarian cancer, exhibits poor CNS penetration and dose-limiting myelosuppression with systemic delivery. Recently, Dimeco et al. docu-

---

Figure 7 (*Facing page*) (A) Kaplan-Meier survival curve for animals treated with intracranial 10 mg 3.8% (w/w) BCNU pCPP:SA (20:80) polymers placed on day 5 after 9L gliosarcoma implantation with or without daily oral quinacrine gavages (20 mg/kg) starting day 5 and lasting 14 days. (B) Kaplan-Meier survival curve for animals treated with intracranial BCNU polymers placed on day 5 after 9L gliosarcoma implantation with or without intracranial 10 mg 15% (w/w) quinacrine pCPP:SA (20:80) polymers placed on day 3 after tumor implantation. (C) Kaplan-Meier survival curve for animals treated with intracranial BCNU polymers placed on day 5 after 9L gliosarcoma implantation with or without intracranial 10 mg 15% (w/w) quinacrine pCPP:SA (20:80) polymers concurrently on day 3 after tumor implantation.

mented significant survival improvement over controls (median 50 vs. 19 days; $p < 0.0001$) with intracranial 10% (w/w) mitaxantrone-loaded pCPP:SA wafer placement (161). Further animal studies, including the assessment of combination therapies, represent areas of active research.

Other Neuro-Oncology Related Agents

Polymer applications to brain tumor treatment extend beyond chemotherapy. Radiosensitizers, as the designation implies, serve as adjuvants to radiation therapy. 5-Iodo-2'-deoxyuridine (IudR), for example, replaces thymidine in replicating DNA. Animal studies examine IudR efficacy and delivery by pCPP:SA polymer systems (162,163). Additionally, EVAc copolymer administration of phenytoin, an anticonvulsant, decreased cobalt-induced seizures in Sprague-Dawley rats (164). Though neither drug class has yet seen clinical use within the context of a local delivery system, polymer versatility offers a broad extensive margin of novel therapeutic possibilities.

5. FUTURE DIRECTIONS

Biodegradable polymer-based delivery systems significantly enhance treatment of various neurosurgical pathologies, malignancies included. The reviewed literature therefore serves as proof of principle, confirming the value and potential of sustained local chemotherapy. Though ongoing research using biodegradable polymers demonstrates continued promise, other exciting approaches to CNS drug administration merit review.

For example, newly developed microchips (Fig. 8a) (165) represent a novel drug delivery method with significant clinical potential. Microchip systems, based on solid-state silicon technology, provide multiple microreservoirs for controlled single or multiple agent delivery. Each microreservoir corresponds with a thin anode membrane cover; electrochemical dissolution facilitates drug release in the form of solids, liquids, or gels. Thus, reservoirs are subject to independent release programming, and collectively offer a seemingly end-

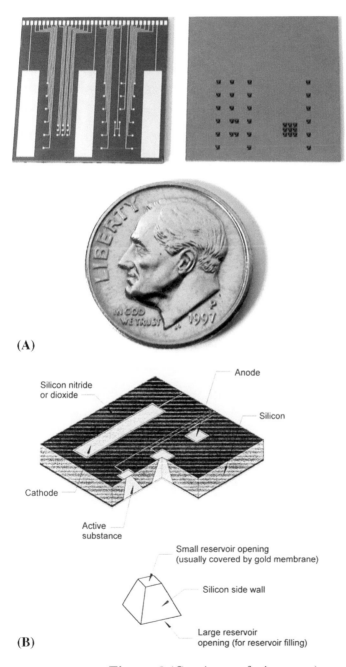

(A)

(B)

Figure 8 (*Caption on facing page*)

less permutation of release profiles and therapy combinations. As an integrated circuit, the delivery system possesses self-contained memory, processing, and microbattery capabilities. Furthermore, microchips offer an array of initial implantation options, ranging from surgery to oral ingestion. Current research studies examine microchip biocompatibility and efficacy.

Biodegradable "passive chips" (Fig. 8b) (166), based on biodegradable polymer technology, represent an alternative microchip development strategy. Biodegradable microchips rely on slow degradation of a thin polymeric membrane, not an integrated silicon-based circuit, for individual microreservoir release. Similar to its "active" counterpart, biocompatibility and efficacy studies of biodegradable microchips are currently underway. Both "pharmacy-on-a-chip" systems potentially offer a 1000 drug delivery capacity.

6. CONCLUSIONS

CNS tumors represent a significant pharmacological and clinical challenge. The BBB physiologically isolates the CNS, and limits antineoplastic drug delivery. Thus, systemic chemotherapy levels required for tumorcidal CNS concentrations often carry prohibitive systemic toxicity. Biodegradable poly-

---

Figure 8 (*Facing page*) (A) Microchip with dime caption. Front (left) and back views of a new microchip for controlled local release of chemicals. The dots between the three large bars (cathodes) on the front are the caps (anodes) covering the reservoirs holding the chemicals. Electrical voltage applied between the cap and cathode causes a reaction that dissolves the cap, thus releasing the reservoir's contents. The back view shows the larger openings through which the contents of the reservoirs are deposited. (These openings are sealed after filling.) Photograph by Paul Horwitz, Atlantic Photo Service, Inc. (B) Schematic of the passive microchip. Initial models will use PLGA and other existing polymer matrices for the substrate. The entire chip will be biodegradable.

mer technology, by circumventing the BBB, offers a new approach to treating brain tumors. Gliadel, the 3.85% (w/w) BCNU-loaded pCPP:SA (20:80) polymer, stands as the first FDA-approved treatment for malignant gliomas in 23 years, and provides an encouraging precedent for further polymer development. Multiple clinical trials and experiences demonstrate Gliadel safety and efficacy.

Local polymer-based drug delivery, by limiting systemic toxicity, expands the battery of potential chemotherapeutics against malignant gliomas. Many drugs therefore remain under active investigation as candidates for polymer administration. Moreover, advances in combined immunotherapy and local chemotherapy suggest further polymer-based applications. Microchip and CEDD technologies offer a broad investigative horizon, hopefully supplanting the established contributions of polymeric drug delivery systems.

The future holds exciting possibilities for both patient and neurosurgeon. New drug delivery technologies and treatment modalities extend therapy beyond surgical resection/debulking of a malignant tumor. Postoperatively, patients may soon benefit from programmed microchips sequentially delivering chemotherapeutics tailored to the intraoperative frozen pathology diagnosis. Microreservoirs loaded with adjuvant dexamethasone and dilantin could treat cerebral edema and prevent seizure complications. Immunotherapy, namely cytokine-loaded microspheres and irradiated tumor cells from the resected specimen, further expands potential tumor site interventions. Moreover, in vitro testing of resected tumor against a wide battery of chemotherapeutic agents will offer a unique sensitivity profile; promising agents will be loaded onto polymer microspheres and injected during a separate stereotactic procedure. Overall, the development of biodegradable polymer systems opens the CNS to pharmacological intervention, and heralds great progress in the treatment of malignant brain tumors.

DISCLOSURE

Under a licensing agreement between Guilford Pharmaceuticals and the Johns Hopkins University, Dr. Brem is entitled to a share of royalty received by the University on sales of products described in this work. Dr. Brem and the University own Guilford Pharmaceuticals stock, which is subject to certain restrictions under University policy. Dr. Brem also is a paid consultant to Guilford Pharmaceuticals. The terms of this arrangement are being managed by the Johns Hopkins University in accordance with its conflict of interest polices.

REFERENCES

1. Legler LM, Gloeckler Ries LA, Smith MA, Warren LL, Heineman EF, Kaplan RS, Linet MS. Brain and other central nervous system cancers: recent trends in incidence and mortality. J Natl Cancer Inst 1999; 91:1382–1390.

2. Rostomily R, Keles GE, Berger MS. Radical surgery in the management of low-grade and high-grade gliomas. Baillieres Clin Neurol 1996; 5:345–369.

3. Chang CH, Horton J, Schoenfeld D, Salazer O, Perez R -Tamayo, Kramer S, Weinstein A, Nelson JS, Tsukada Y. Comparison of postoperative radiotherapy and combined postoperative radiotherapy and chemotherapy in the multidisciplinary management of malignant gliomas. A joint Radiation Therapy Oncology Group and Eastern Cooperative Oncology Group study. Cancer Res 1983; 52:997–1007.

4. Green SB, Byar DP, Walker MD, Pistenmaa DA, Alexander EJ, Batzdorf U, Brooks WH, Hunt WE, Mealey JJ, Odom GL, Paoletti P, Ransohoff JI, Robertson JT, Selker RG, Shapiro WR, SKR Jr, Wilson CB, Strike TA. Comparisons of carmustine, procarbazine, and high-dose methylprednisolone as additions to surgery and radiotherapy for the treatment of malignant glioma. Cancer Treat Rep 1983; 67: 121–132.

5. Kornblith PL, Walker M. Chemotherapy for malignant gliomas. J Neurosurg 1988; 68:1–17.

6. Walker MD, Green SB, Byar DP, Alexander EJ, Batzdorf U, Brooks WH, Hunt WE, MacCarty CS, Mahaley MSJ, Mealey JJ, Owens G, Ransohoff JI, Robertson JT, Shapiro WR, Smith KRJ, Wilson CB, Strike TA. Randomized comparisons of radiotherapy and nitrosoureas for the treatment of malignant glioma after surgery. N Engl J Med 1980; 303: 1323–1329.

7. Walker MD, Alexander EJ, Hunt WE, MacCarty CS, Mahaley MSJ, Mealey JJ, Norrell HA, Owens G, Ransohoff J, Wilson CB, Gehan EA, Strike TA. Evaluation of BCNU and/or radiotherapy in the treatment of anaplastic gliomas. A cooperative clinical trial. J Neurosurg 1978; 49:333–343.

8. Barker FG, Chang SM, Gutin PH, Malek MK, McDermott MW, Prados MD, Wilson CB. Survival and functional status after resection of recurrent glioblastoma multiforme. Neurosurgery 1998; 42:709–720.

9. Mohan DS, Suh JH, Phan JL, Kupelian PA, Cohen BH, Barnett GH. Outcome in elderly patients undergoing definitive surgery and radiation therapy for supratentorial glioblastoma multiforme at a tertiary institution. Int J Rad Oncol Biol Phys 1998; 42:981–987.

10. Reese TS, Karnovsky MJ. Fine structural localization of a blood-brain barrier to exogenous peroxidase. J Cell Biol 1967; 34:207–217.

11. Grieg NH. Opitimizing drug delivery to brain tumors. Cancer Treat Rev 1987; 14:1–28.

12. Abbott NJ, Romero IA. Transporting therapeutics across the blood-brain barrier. Mol Med Today 1996; 2:106–113.

13. Loo TL, Dion RL, Dixon RL, Rall DP. The antitumor agent, 1,3-bis(2-chloroethyl)-1-nitrosourea. J Pharm Sci 1966; 55: 492–497.

14. Prokai L, Prokai-Tatrai K, Bodor N. Targeting drugs to the brain by redox chemical delivery systems. Med Res Rev 2000; 20:367–416.

15. Pardridge WM. Vector-mediated drug delivery to the brain. Adv Drug Deliv Rev 1999; 36:299–321.

16. Williams PC, Henner WD, Roman-Goldstein S, Dahlborg SA, Brummett RE, Tableman M, Dana BW, Neuwelt EA. Toxicity and efficacy of carboplatin and etoposide in conjunction with disruption of the blood-brain barrier in the treatment of intracranial neoplasms. Neurosurgery 1995; 37:17–28.

17. Warnke PC, Blasberg RG, Groothuis DR. The effect of hyperosmotic blood-brain barrier disruption on blood-to-tissue transport in ENU-induced gliomas. Ann Neurol 1987; 22: 300–305.

18. Elliott PJ, Hayward NJ, Dean RL, Blunt DG, Bartus RT. Intravenous RMP-7 selectively increases uptake of carboplatin in experimental brain tumors. Cancer Res 1996; 56:3998–4005.

19. Blasberg RG. Methotrexate, cytosine arabinoside, and BCNU concentration in the brain after ventriculocisternal perfusion. Cancer Treat Rep 1977; 61:625–631.

20. Ratcheson RA, Ommaya AK. Experience with the subcutaneous cerebrospinal-fluid reservoir. Preliminary report of 60 cases. New Engl J Med 1968; 279:1025–1031.

21. Chandler WF, Greenberg HS, Ensminger WD, Diaz RF, Junck LR, Hood TW, Gebarski SS, Page MA. Use of implantable pump systems for intraarterial, intraventricular, and intratumoral treatment of malignant brain tumors. Ann NY Acad Sci 1988; 531:206–212.

22. Lord P, Allami H, Davis M, Diaz R, Heck P, Fischell R. MiniMed Technologies programmable implantable infusion system. Ann NY Acad Sci 1988; 531:66–71.

23. Heruth KT. Medtronic SynchroMed drug administration system. Ann NY Acad Sci 1988; 531:72–75.

24. Bobo RH, Laske DW, Akbasak A, Morrison PF, Dedrick RL, Oldfield EH. Convection-enhanced delivery of macromolecules in the brain. Proc Natl Acad Sci USA 1994; 91:2076–2080.

25. Laske DW, Morrison PF, Lieberman DM, Corthesy ME, Reynolds JC, Stewart-Henney PA, Koong SS, Cummins A, Paik CH, Oldfield EH. Chronic interstitial infusion of protein to primate brain: determination of drug distribution and clear-

ance with single-photon emission computerized tomography imaging. J Neurosurg 1997; 87:586–594.

26. Lieberman DM, Corthesy ME, Cummins A, Oldfield EH. Reversal of experimental parkinsonism by using selective chemical ablation of the medial globus pallidus. J Neurosurg 1999; 90:928–934.

27. Lonser RR, Gogate N, Morrison PF, Wood JD, Oldfield EH. Direct convective delivery of macromolecules to the spinal cord. J Neurosurg 1998; 89:616–622.

28. Mardor Y, Roth Y, Lidar Z, Jonas T, Pfeffer R, Maier SE, Faibel M, Nass D, Hadani M, Orenstein A, Cohen JS, Ram Z. Monitoring response to convection-enhanced taxol delivery in brain tumor patients using diffusion-weighted magnetic resonance imaging. Cancer Res 2001; 61:4971–4973.

29. Lidar Z, Mardor Y, Jonas T, Pfeffer R, Faibel M, Nass D, Hadani M, Ram Z. Convection-enhanced delivery of paclitaxel for the treatment of recurrent malignant glioma: a Phase I/II clinical study. J Neurosurg 2004; 100:472–479.

30. Husain SR, Puri RK. Interleukin-13 receptor-directed cytotoxin for malignant glioma therapy: from bench to bedside. J Neurooncol 2003; 65:37–48.

31. Langer R. New methods of drug delivery. Science 1990; 249:1527–1533.

32. Langer R, Folkman J. Polymers for the sustained release of proteins and other macromolecules. Nature 1976; 263: 797–800.

33. Langer R, Brem H, Tapper D. Biocompatibility of polymeric delivery systems for macromolecules. J Biomed Mater Res 1981; 15:267–277.

34. Tamargo RJ, Myseros JS, Epstein JI, Yang MB, Chasin M, Brem H. Interstitial chemotherapy of the 9L gliosarcoma: controlled release polymers for drug delivery in the brain. Cancer Res 1993; 53:329–333.

35. Tamargo RJ, Sills AKJ, Reinhard CS, Pinn ML, Long DM, Brem H. Interstitial delivery of dexamethasone in the brain for the reduction of peritumoral edema. J Neurosurg 1991; 74:956–961.

36. Weingart JD, Sipos EP, Brem H. The role of minocycline in the treatment of intracranial 9L glioma. J Neurosurg 1995; 82:635–640.

37. Weingart JD, Thompson RC, Tyler B, Colvin OM, Brem H. Local delivery of the topoisomerase I inhibitor camptothecin sodium prolongs survival in the rat intracranial 9L gliosarcoma model. Int J Cancer 1995; 62:605–609.

38. Yang MB, Tamargo RJ, Brem H. Controlled delivery of 1,3-bis(2-chloroethyl)-1-nitrosourea from ethylene-vinyl acetate copolymer. Cancer Res 1989; 49:5103–5107.

39. Aszmann OC, Korak KJ, Kropf N, Fine E, Aebischer P, Frey M. Simultaneous GDNF and BDNF application leads to increased motoneuron survival and improved functional outcome in an experimental model for obstetric brachial plexus lesions. Plast Reconstr Surg 2002; 110:1066–1072.

40. Langer RS, Wise DL. . Medical Applications of Controlled Release. Boca Raton, Fl: CRC Press, 1984.

41. Brem H, Walter K, Langer R. Polymers as controlled drug delivery devices for the treatment of malignant brain tumors. Eur J Pharm Biopharm 1993; 27:2–7.

42. Lewis DH. Controlled release of bioactive agents from lactide/glycolide polymers. In: Chasin M, Langer R, eds. Biodegradable Polymers as Drug Delivery Systems. New York: Marcel Dekker, 1990:1–41.

43. Kong Q, Kleinschmidt-Demasters BK, Lillehei KO. Intralesionally implanted cisplatin cures primary brain tumor in rats. J Surg Oncol 1997; 64:268–273.

44. Lillehei KO, Kong Q, Withrow SJ, Kleinschmidt-Demasters BK. Efficacy of intralesionally administered cisplatin-impregnated biodegradable polymer for the treatment of 9L gliosarcoma in the rat. Neurosurgery 1996; 39:1191–1199.

45. Menei P, Daniel V, Montero-Menei C, Brouillardm M, Pouplard-Barthelaix A, Benoit JP. Biodegradation and brain tissue reaction to poly(D,L-lactide-co-glycolide) microspheres. Biomaterials 1993; 14:470–478.

46. Frazza EJ, Schmitt EE. A new absorbable suture. J Biomed Mater Res 1971; 5:43–58.

47. Brady JM, Cutright DE, Miller RA, Barristone GC. Resorption rate, route, route of elimination, and ultrastructure of the implant site of polylactic acid in the abdominal wall of the rat. J Biomed Mater Res 1973; 7:155–166.

48. Spenlehauer G, Vert M, Benoit JP, Boddaert A. In vitro and in vivo degradation of poly(D,L lactide/glycolide) type microspheres made by solvent evaporation method. Biomaterials 1989; 10:557–563.

49. Burt HM, Jackson JK, Bains SK, Liggins RT, Oktaba AM, Arsenault AL, Hunter WL. Controlled delivery of taxol from microspheres composed of a blend of ethylene-vinyl acetate copolymer and poly (d,l-lactic acid). Cancer Lett 1995; 88:73–79.

50. Sato H, Wang YM, Adachi I, Horikoshi I. Pharmacokinetic study of taxol-loaded poly(lactic-co-glycolic acid) microspheres containing isopropyl myristate after targeted delivery to the lung in mice. Biol Pharm Bull 1996; 19:1596–1601.

51. Peyman GA, Conway M, Khoobehi B, Soike K. Clearance of microsphere-entrapped 5-fluorouracil and cytosine arabinoside from the vitreous of primates. Int Ophthalmol 1992; 16:109–113.

52. Torres AI, Boisdron-Celle M, Benoit JP. Formulation of BCNU-loaded microspheres: influence of drug stability and solubility on the design of the microencapsulation procedure. J Microencapsul 1996; 13:41–51.

53. Williams RC, Paquette DW, Offenbacher S, Adams DF, Armitage GC, Bray K, Caton J, Cochran DL, Drisko CH, Fiorellini JP, Giannobile WV, Grossi S, Guerrero DM, Johnson GK, Lamster IB, Magnusson I, Oringer RJ, Persson GR, Van Dyke TE, Wolff LF, Santucci EA, Rodda BE, Lessem J. Treatment of periodontitis by local administration of minocycline microspheres: a controlled trial. J Periodontol 2001; 72:1535–1544.

54. Lamprecht A, Ubrich N, Yamamoto H, Schafer U, Takeuchi H, Maincent P, Kawashima Y, Lehr CM. Biodegradable nanoparticles for targeted drug delivery in treatment of inflammatory bowel disease. J Pharmacol Exp Ther 2001; 299:775–781.

55. Lee M, Browneller R, Wu Z, Jung A, Ratanawong C, Sharifi R. Therapeutic effects of leuprorelin microspheres in prostate cancer. Adv Drug Deliv Rev 1997; 28:121–138.

56. Benoit JP, Faisant N, Venier-Julienne MC, Menei P. Development of microspheres for neurological disorders: from basics to clinical applications. J Control Rel 2000; 65:285–296.

57. Sugiyama T, Kumagai S, Nishida T, Ushijima K, Matsuo T, Yakushiji M, Hyon SH, Ikada Y. Experimental and clinical evaluation of cisplatin-containing microspheres as intraperitoneal chemotherapy for ovarian cancer. Anticancer Res 1998; 18:2837–2842.

58. Singh M, Saxena BB, Singh R, Kaplan J, Ledger WJ. Contraceptive efficacy of norethindrone encapsulated in injectable biodegradable poly-dl-lactide-co-glycolide microspheres (NET-90): phase III clinical study. Adv Contracept 1997; 13:1–11.

59. Menei P, Benoit JP, Boisdron-Celle M, Fournier D, Mercier P, Guy G. Drug targeting into the central nervous system by stereotactic implantation of biodegradable microspheres. Neurosurgery 1994; 34:1058–1064.

60. Gref R, Minamitake Y, Peracchia MT, Trubetskoy V, Torchilin V, Langer R. Biodegradable long-circulating polymeric nanospheres. Science 1994; 263:1600–1603.

61. Brem H, Langer R. Polymer-based drug delivery to the brain. Sci Med 1996; July/August:52–61.

62. Mathiowitz E, Saltzman M, Domb A. Polyanhydride microspheres as drug carriers. II Microencapsulation by solvent removal. J Appl Polym Sci 1988; 35:755–774.

63. Mathiowitz E, Langer R. Polyanhydride microspheres as drug carriers. I Hot-melt microencapsulation. J Control Rel 1987; 5:13–22.

64. Bindschaedler C, Leong K, Mathiowitz E, Langer R. Polyanhydride microsphere formulation by solvent extraction. J Pharm Sci 1988; 77:696–698.

65. Chasin M, Domb A, Ron E. Polyanhydrides as drug delivery systems. In: Chasin M, Langer R, eds. Biodegradable

Polymers as Drug Delivery Systems. New York, NY: Marcel Dekker, 1990:43–70.

66. Howard III MA, Gross A, Grady MS, Langer RS, Mathiowitz E, Winn HR, Mayberg MR. Intracerebral drug delivery in rats with lesion-induced memory deficits. J Neurosurg 1989; 71:105–112.

67. Leong KW, Kost J, Mathiowitz E, Langer R. Polyanhydrides for controlled release of bioactive agents. Biomaterials 1986; 7:364–371.

68. Mathiowitz E, Kline D, Langer R. Morphology of polyanhydride microsphere delivery systems. Scanning Microsc 1990; 4:329–340.

69. Leong KW, D'Amore PD, Marletta M, Langer R. Bioerodible polyanhydrides as drug-carrier matrices. II Biocompatibility and chemical reactivity. J Biomed Mater Res 1986; 20: 51–64.

70. Tamargo RJ, Epstein JI, Reinhard CS, Chasin M, Brem H. Brain biocompatibility of a biodegradable, controlled-release polymer in rats. J Biomed Mater Res 1989; 23:253–266.

71. Brem H, Kader A, Epstein JI, Tamargo RJ, Domb A, Langer R, Leong KW. Biocompatibility of a biodegradable, controlled-release polymer in the rabbit brain. Sel Cancer Ther 1989; 5:55–65.

72. HBrem, Tamargo RJ, Olivi A, Pinn M, Weingart JD, Wharam M, Epstein JI. Biodegradable polymers for controlled delivery of chemotherapy with and without radiation therapy in the monkey brain. J Neurosurg 1994; 80:283–290.

73. Dang W, Saltzman WM. Dextran retention in the rat brain following release from a polymer implant. Biotechnol Prog 1992; 8:527–532.

74. Domb A, Bogdansky S, Olivi A, Judy K, Dureza C, Lenartz D, Pinn MI, Colvin M, Brem H. Controlled delivery of water soluble and hydrolytically unstable anti-cancer drugs for polymeric implants. Polymer Preprints 1991; 32: 219–220.

75. Shieh L, Tamada J, Chen I, Pang J, Domb A, Langer R. Erosion of a new family of biodegradable polyanhydrides. J Biomed Mater Res 1994; 28:1465–1475.

76. Olivi A, Ewend MG, Utsuki T, Tyler B, Domb AJ, Brat DJ, Brem H. Interstitial delivery of carboplatin via biodegradable polymers is effective against experimental glioma in the rat. Cancer Chemother Pharmacol 1996; 39:90–96.

77. Judy KD, Olivi A, Buahin KG, Domb A, Epstein JI, Colvin OM, Brem H. Effectiveness of controlled release of a cyclophosphamide derivative with polymers against rat gliomas. J Neurosurg 1995; 82:481–486.

78. Shikani AH, Eisele DW, Domb AJ. Polymer delivery of chemotherapy for squamous cell carcinoma of the head and neck. Arch Otolaryngol Head Neck Surg 1994; 120:1242–1247.

79. Gabizon AA. Liposomal anthracyclines. Hematol Oncol Clin North Am 1994; 8:431–450.

80. Golumbek PT, Azhari R, Jaffee EM, Levitsky HI, Lazenby A, Leong K, Pardoll DM. Controlled release, biodegradable cytokine depots: a new approach in cancer vaccine design. Cancer Res 1993; 53:5841–5844.

81. Hirakawa W, Kadota K, Asakura T, Niiro M, Yokoyama S, Hirano H, Yatsushiro K, Kubota Y, Shimodozono Y. [Local-chemotherapy for malignant brain tumors using methotrexate-containing fibrin glue]. Gan To Kagaku Ryoho 1995; 22: 805–809.

82. Ringkjob R. Treatment of intracranial gliomas and metastatic carcinomas by local application of cytostatic agents. Acta Neurol Scand 1968; 44:318–322.

83. Rama B, Mandel T, Jansen J, Dingeldein E, Mennel HD. The intraneoplastic chemotherapy in a rat brain tumour model utilizing methotrexate-polymethylmethacrylate-pellets. Acta Neurochir 1987; 87:70–75.

84. Oda Y, Tokuriki Y, Tsuda E, Handa H, Kieler J. Trial of anticancer pellet in malignant brain tumours; 5 FU and urokinase embedded in silastic. Acta Neurochir Suppl 1979; 28:489–490.

85. Grossman SA, Reinhard C, Colvin OM, Chasin M, Brundrett R, Tamargo RJ, Brem H. The intracerebral distribution of BCNU delivered by surgically implanted biodegradable polymers. J Neurosurg 1992; 76:640–647.

86. Fung LK, Ewend MG, Sills A, Sipos EP, Thompson R, Watts M, Colvin OM, Brem H, Saltzman WM. Pharmacokinetics of interstitial delivery of carmustine, 4-hydroperoxycyclophosphamide, and paclitaxel from a biodegradable polymer implant in the monkey brain. Cancer Res 1998; 58:672–684.

87. Buahin KG, Brem H. Interstitial chemotherapy of experimental brain tumors: comparison of intratumoral injection versus polymeric controlled release. J Neurooncol 1995; 26:103–110.

88. Sipos EP, Tyler B, Piantadosi S, Burger PC, Brem H. Optimizing interstitial delivery of BCNU from controlled release polymers for the treatment of brain tumors. Cancer Chemother Pharmacol 1997; 39:383–389.

89. Galicich JH, Sundaresan N, Thaler HT. Surgical treatment of single brain metastasis. Evaluation of results by computerized tomography scanning. J Neurosurg 1980; 53:63–67.

90. Sawaya R, Ligon BL, Bindal AK, Bindal RK, Hess KR. Surgical treatment of metastatic brain tumors. J Neurooncol 1996; 27:269–277.

91. Ewend MG, Williams JA, Tabassi K, Tyler BM, Babel KM, Anderson RC, Pinn ML, Brat DJ, Brem H. Local delivery of chemotherapy and concurrent external beam radiotherapy prolongs survival in metastatic brain tumor models. Cancer Res 1996; 56:5217–5223.

92. Ewend MG, Sampath P, Williams JA, Tyler BM, Brem H. Local delivery of chemotherapy prolongs survival in experimental brain metastases from breast carcinoma. Neurosurgery 1998; 43:1185–1193.

93. Brem H, Mahaley Jr MS, Vick NA, Black KL, Schold Jr SC, Burger PC, Friedman AH, Ciric IS, Eller TW, Cozzens JW, Kenealy JN. Interstitial chemotherapy with drug polymer implants for the treatment of recurrent gliomas. J Neurosurg 1991; 74:441–446.

94. Brem H, Piantadosi S, Burger PC, Walker M, Selker R, Vick NA, Black K, Sisti M, Brem S, Mohr G, Muller P, Morawetz R, Schold SC. Placebo-controlled trial of safety and efficacy of intraoperative controlled delivery by biodegradable polymers of chemotherapy for recurrent gliomas. The Polymerbrain Tumor Treatment Group. Lancet 1995; 345:1008–1012.

95. Brem H, Ewend MG, Piantadosi S, Greenhoot J, Burger PC, Sisti M. The safety of interstitial chemotherapy with BCNU-loaded polymer followed by radiation therapy in the treatment of newly diagnosed malignant gliomas: phase I trial. J Neurooncol 1995; 26:111–123.

96. Valtonen S, Timonen U, Toivanen P, Kalimo H, Kivipelto L, Heiskanen O, Unsgaard G, Kuurne T. Interstitial chemotherapy with carmustine-loaded polymers for high-grade gliomas: a randomized double-blind study. Neurosurgery 1997; 41:44–48.

97. Westphal M, Hilt DC, Bortey E, Delavault P, Olivares R, Warnke PC, Whittle IR, Jääskeläinen J, Ram Z. A phase 3 trial of local chemotherapoy with biodegradable carmustine (BCNU) wafers (Gliadel wafers) in patients with primary malignant glioma. Neuro-oncol 2003; 5.

98. Olivi A, Grossman SA, Tatter S, Barker II FG, Judy K, Olson J, Hilt D, Fisher JD, Piantadosi S. Results of a phase I dose escalation study using BCNU impregnated polymers in patients with recurrent malignant gliomas. J Clin Oncol 2003; 29(9):1845–1849.

99. Silber JR, Bobola MS, Ghatan S, Blank A, Kolstoe DD, Berger MS. O^6-methylguanine-DNA methyltransferase activity in adult gliomas: relation topatient and tumor characteristics. Cancer Res 1998; 58:1068–1073.

100. Pegg AE, Boosalis M, Samson L, Moschel RC, Byers TL, Swenn K, Dolan ME. Mechanism of inactivation of human O^6-alkylguanine-DNA alkyltransferase by O^6-benzylguanine. Biochemistry 1993; 32:11998–12006.

101. Page JG, Giles HD, Phillips W, Gerson SL, Smith AC, Tomaszewski JE. Preclinical toxicology study of O^6-benzylguanine (NSC-637037) and BCNU (carmustine, NSC-409962) in male and female beagle dogs. Proc Am Assoc Cancer Res 1993; 35:328.

102. Rhines LD, Sampath P, Dolan ME, Tyler BM, Brem H, Weingart J. O^6-benzylguanine potentiates the antitumor effect of locally delivered carmustine against an intracranial rat glioma. Cancer Res 2000; 60:6307–6310.

103. Cahan MA, Walter KA, Colvin OM, Brem H. Cytotoxicity of taxol in vitro against human and rat malignant brain tumors. Cancer Chemother Pharmacol 1994; 33:441–444.

104. Klecker R, Jamis-Dow C, Egorin M. Distribution and metabolism of 3-H Taxol in the rat (Abstract). Proc Am Assoc Cancer Res 1992; 34:381.

105. Walter KA, Cahan MA, Gur A, Tyler B, Hilton J, Colvin OM, Burger PC, Domb A, Brem H. Interstitial taxol delivered from a biodegradable polymer implant against experimental malignant glioma. Cancer Res 1994; 54:2207–2212.

106. Harper E, Dang W, Lapidus RG, Garver Jr RI. Enhanced efficacy of a novel controlled release paclitaxel formulation (PACLIMER delivery system) for local-regional therapy of lung cancer tumor nodules in mice. Clin Cancer Res 1999; 5:4242–4248.

107. Gabikian P, Li KW, Magee C, Morrell C, Tyler BM, Foster F, Dang W, Brem H, Walter KA. Safety of Intracranial Paclitaxel: Polilactofate Microspheres in Dogs (Poster). Proceedings of the American Association of Neurological Surgeons Annual Meeting, Chicago, IL, 2002.

108. Hsiang YH, Liu LF. Identification of mammalian DNA topoisomerase I as an intracellular target of the anticancer drug camptothecin. Cancer Res 1988; 48:1722–1726.

109. Slichenmyer WJ, Rowinsky EK, Donehower RC, Kaufmann SH. The current status of camptothecin analogues as antitumor agents. J Natl Cancer Inst 1993; 85:271–291.

110. Storm PB, Moriarity JL, Tyler B, Burger PC, Brem H, Weingart J. Polymer delivery of camptothecin against 9L gliosarcoma: Release, distribution, and efficacy. Journal of Neuro-Oncology 2002; 56:209–217.

111. Sampath P, Amundson E, Wall ME, Tyler BM, Wani MC, Alderson LM, Colvin M, Brem H, Weingart JD. Camptothe-

cin analogs in malignant gliomas: Comparative analysis and characterization. J Neurosurg 2003; 98:570–577.

112. Thompson RC, Pardoll DM, Jaffee EM, Ewend MG, Thomas MC, Tyler BM, Brem H. Systemic and local paracrine cytokine therapies using transduced tumor cells are synergistic in treating intracranial tumors. J Immunother Emphasis Tumor Immunol 1996; 19:405–413.

113. Ewend MG, Thompson RC, Anderson R, Sills AK, Staveley-O'Carroll K, Tyler BM, Hanes J, Brat D, Thomas M, Jaffee EM, Pardoll DM, Brem H. Intracranial paracrine interleukin-2 therapy stimulates prolonged antitumor immunity that extends outside the central nervous system. J Immunother 2000; 23:438–448.

114. DiMeco F, Rhines LD, Hanes J, Tyler BM, Brat D, Torchiana E, Guarnieri M, Colombo MP, Pardoll DM, Finocchiaro G, Brem H, Olivi A. Paracrine delivery of IL-12 against intracranial 9L gliosarcoma in rats. J Neurosurg 2000; 92:419–427.

115. Sampath P, Hanes J, DiMeco F, Tyler BM, Brat D, Pardoll DM, Brem H. Paracrine immunotherapy with interleukin-2 and local chemotherapy is synergistic in the treatment of experimental brain tumors. Cancer Res 1999; 59:2107–2114.

116. Wiranowska M, Ransohoff J, Weingart JD, Phelps C, Phuphanich S, Brem H. Interferon-containing controlled-release polymers for localized cerebral immunotherapy. J Interferon Cytokine Res 1998; 18:377–385.

117. Hanes J, Sills A, Zhao Z, Suh KW, Tyler B, DiMeco F, Brat DJ, Choti MA, Leong KW, Pardoll DM, Brem H. Controlled local delivery of interleukin-2 by biodegradable polymers protects animals from experimental brain tumors and liver tumors. Pharm Res 2001; 18:899–906.

118. Rhines LD, DiMeco F, Lawson HC, Tyler BM, Hanes J, Olivi A, Brem H. Local immunotherapy with interleukin-2 delivered from biodegradable polymer microspheres combined with interstitial chemotherapy: A novel treatment for experimental malignant glioma. Neurosurgery 2003; 52:1–8.

119. Folkman J. Tumor angiogenesis: therapeutic implications. N Engl J Med 1971; 285:1182–1186.

120. Gimbrone Jr MA, Leapman SB, Cotran RS, Folkman J. Tumor dormancy in vivo by prevention of neovascularization. J Exp Med 1972; 136:261–276.

121. Tamargo RJ, Leong KW, Brem H. Growth inhibition of the 9L glioma using polymers to release heparin and cortisone acetate. J Neurooncol 1990; 9:131–138.

122. Folkman J, Langer R, Linhardt RJ, Haudenschild C, Taylor S. Angiogenesis inhibition and tumor regression caused by heparin or a heparin fragment in the presence of cortisone. Science 1983; 221:719–725.

123. Sills Jr AK, Williams JI, Tyler BM, Epstein DS, Sipos EP, Davis JD, McLane MP, Pitchford S, Cheshire K, Gannon FH, Kinney WA, Chao TL, Donowitz M, Laterra J, Zasloff M, Brem H. Squalamine inhibits angiogenesis and solid tumor growth in vivo and perturbs embryonic vasculature. Cancer Res 1998; 58:2784–2792.

124. Bhargava P, Marshall JL, Dahut W, Rizvi N, Trocky N, Williams JI, Hait H, Song S, Holroyd KJ, Hawkins MJ. A phase I and pharmacokinetic study of squalamine, a novel antiangiogenic agent, in patients with advanced cancers. Clin Cancer Res 2001; 7:3912–3919.

125. Golub LM, Wolff M, Lee HM, McNamara TF, Ramamurthy NS, Zambon J, Ciancio S. Further evidence that tetracyclines inhibit collagenase activity in human cervicular fluid and from other mammalian sources. J Periodontal Res 1985; 20:12–23.

126. Tamargo RJ, Bok RA, Brem H. Angiogenesis inhibition by minocycline. Cancer Res 1991; 51:672–675.

127. Frazier JL, Wang PP, Case D, Tyler BM, Pradilla G, Weingart JD, Brem H. Local delivery of minocycline and systemic BCNU have synergistic activity in the treatment of intracranial glioma. J Neurooncol 2003; 64:203–209.

128. Chabner BA, Collins JM, eds . Cancer Chemotherapy: Principles and Practice. Philadelphia, PA: JB Lippencott, 1990.

129. Colvin M, Hilton J. Pharmacology of cyclophosphamide and metabolites. Cancer Treat Rep 1981; 65(suppl 3):89–95.

130. Buahin KG, Judy KD, Hartke C, Domb A, Maniar M, Colvin O, Brem H. Controlled release of 4-hydroperoxycyclopho-

sphamide from the fatty acid dimer-sebacic acid copolymer. Polym Adv Technol 1992; 3:311.

131. Sipos EP, Witham TF, Ratan R, Burger PC, Baraban J, Li KW, Piantadosi S, Brem H. L-buthionine sulfoximine potentiates the antitumor effect of 4-hydroperoxycyclophosphamide when administered locally in a rat glioma model. Neurosurgery 2001; 48:392–400.

132. Colvin OM, Friedman HS, Gamcsik MP, Fenselau C, Hilton J. Role of glutathione in cellular resistance to alkylating agents. Adv Enzyme Regul 1993; 33:19–26.

133. Goodman LS, Hardman JG, Limbird LE, Gilman AG. Goodman and Gilman's the pharmacological basis of therapeutics New York: McGraw-Hill, 2001.

134. Nakazawa S, Itoh Y, Shimura T, Matsumoto M, Yajima K. [New management of brain neoplasms. 2. Local injection of adriamycin]. No Shinkei Geka 1983; 11:821–827.

135. Lin SY, LF C, Lui WY, Chen CF, Han SH. Tumoricidal effect of controlled-release polymeric needle devices containing adriamycin HCl in tumor-bearing mice. Biomater Artif Cells Artif Organs 1989; 17:189–203.

136. Watts MC, Lesniak MS, Burke M, Samdani AF, Tyler BM, Brem H. Controlled release of adriamycin in the treatment of malignant glioma (Poster). American Association of Neurological Surgeons Annual Meeting. Denver, CO, April 1997, p. 305.

137. Levin VA, Edwards MS, Wara WM, Allen J, Ortega J, Vestnys P. 5-Fluorouracil and 1-(2-chloroethyl)-3-cyclohexyl-1-nitrosourea (CCNU) followed by hydroxyurea, misonidazole, and irradiation for brain stem gliomas: a pilot study of the Brain Tumor Research Center and the Childrens Cancer Group. Neurosurgery 1984; 14:679–681.

138. Shapiro WR. Studies on the chemotherapy of experimental brain tumors: evaluation of 1-(2-chloroethyl)-3-cyclohexyl-1-nitrosourea, vincristine, and 5-fluorouracil. J Natl Cancer Inst 1971; 46:359–368.

139. Shapiro WR, Green SB, Burger PC, Selker RG, VanGilder JC, Robertson JT, Mealey Jr J, Ransohff J, Mahaley Jr MS.

A randomized comparison of intra-arterial versus intravenous BCNU, with or without intravenous 5-fluorouracil, for
newly diagnosed patients with malignant glioma. J Neurosurg 1992; 76:772–781.

140. Oda Y, Uchida Y, Murata T, Mori K, Tokuriki Y, Handa H,
Kobayashi A, Hashi K, Kieler J. [Treatment of brain tumors
with anticancer pellet—experimental and clinical study
(author's trans.)]. No Shinkei Geka 1982; 10:375–381.

141. Oda Y, Kamijyo Y, Okumura T, Tokuriki Y, Yamashita J,
Handa H, Aoyama I, Hashi K, Mori K. [Clinical application
of a sustained release anticancer pellet]. No Shinkei Geka
1985; 13:1305–1311.

142. Kubo O, Himuro H, Inoue N, Tajika Y, Tajika T, Tohyama T,
Sakairi M, Yoshida M, Kaetsu I, Kitamura K. [Treatment of
malignant brain tumors with slowly releasing anticancer drug-
polymer composites]. No Shinkei Geka 1986; 14:1189–1195.

143. Boisdron-Celle M, Menei P, Benoit JP. Preparation and characterization of 5-fluorouracil-loaded microparticles as biodegradable anticancer drug carriers. J Pharm Pharmacol
1995; 47:108–114.

144. Menei P, Boisdron-Celle M, Croue A, Guy G, Benoit JP.
Effect of stereotactic implantation of biodegradable 5-
fluorouracil-loaded microspheres in healthy and C6 glioma-
bearing rats. Neurosurgery 1996; 39:117–124.

145. Menei P, Venier MC, Gamelin E, Saint-Andre JP, Hayek G,
Jadaud E, Fournier D, Mercier P, Guy G, Benoit JP. Local
and sustained delivery of 5-fluorouracil from biodegradable
microspheres for the radiosensitization of glioblastoma: a
pilot study. Cancer 1999; 86:325–330.

146. Blasberg RG, Patlak CS, Shapiro WR. Distribution of methotrexate in the cerebrospinal fluid and brain after intraventricular administration. Cancer Treat Rep 1977; 61:633–641.

147. Sendelbeck SL, Urquhart J. Spatial distribution of dopamine,
methotrexate and antipyrine during continuous intracerebral
microperfusion. Brain Res 1985; 328:251–258.

148. Dang W, Colvin OM, Brem H, Saltzman WM. Covalent coupling of methotrexate to dextran enhances the penetration

of cytotoxicity into a tissue-like matrix. Cancer Res 1994; 54:1729–1735.

149. Diemath HE. [Local use of cytostatic drugs following removal of glioblastomas]. Wien Klin Wochenschr. 1987; 99:674–676.

150. Weiss SR, Raskind R. Treatment of malignant brain tumors by local methotrexate. A preliminary report. Int Surg 1969; 51:149–155.

151. Zeller WJ, Bauer S, Remmele T, Wowra B, Sturm V, Stricker H. Interstitial chemotherapy of experimental gliomas. Cancer Treat Rep 1990; 7:183–189.

152. Uldry PA, Teta D, Regli L. [Focal cerebral necrosis caused by intraventricular chemotherapy with methotrexate]. Neurochirurgie 1991; 37:72–74..

153. Shimura T, Nakazawa S, Ikeda Y, Node Y. [Cyst formation following local chemotherapy of malignant brain tumor: a clinicopathological study of two cases]. No Shinkei Geka 1992; 20:1179–1183.

154. Olivi A, Gilbert M, Duncan KL, Corden B, Lenartz D, Brem H. Direct delivery of platinum-based antineoplastics to the central nervous system: a toxicity and ultrastructural study. Cancer Chemother Pharmacol 1993; 31:449–454.

155. Utsuki T, Brem H, Pitha J, Loftsson T, Kristmundsdottir T, Tyler BT, Olivi A. Potentiation of anti-cancer effects of microencapsulated carboplatin by hydroxypropyl α-cyclodextrin. J Contr Rel 1996; 40:251–260.

156. Maxell RE, Long DM, French LA. The clinical effects of a synthetic glucocorticoid used for brain edema in the practice of neurosurgery. In: Reulen HJ,Schurmann K, eds. Steroids and Brain Edema. Berlin: Springer-Verlag, 1972:219–232.

157. Melby JC. Drug spotlight program: systemic corticosteroid therapy: pharmacology and endocrinologic considerations. Ann Intern Med 1974; 81:505–512.

158. Reinhard CS, Radomsky ML, Saltzman WM, Brem H. Polymeric controlled release of dexamethasone in normal rat brain. J Contr Rel 1991; 16:331–340.

159. Giampietri A, Fioretti MC, Goldin A, Bonmassar E. Drug-mediated antigenic changes in murine leukemia cells: antagonistic effects of quinacrine, an antimutagenic compound. J Natl Cancer Inst 1980; 64:279–301.

160. Reyes S, Herrera LA, Ostrosky P, Sotelo J. Quinacrine enhances carmustine therapy of experimental rat glioma. Neurosurgery 2001; 49(4):969–973.

161. DiMeco F, Li KW, Tyler BM, Wolf AS, Brem H, Olivi A. Local delivery of mitoxantrone for the treatment of malignant brain tumors in rats. J Neurosurg 2002; 97:1173–1178.

162. Williams JA, Dillehay LE, Tabassi K, Sipos E, Fahlman C, Brem H. Implantable biodegradable polymers for IUdR radiosensitization of experimental human malignant glioma. J Neurooncol 1997; 32:181–192.

163. Williams JA, Yuan X, Dillehay LE, Shastri VR, Brem H, Williams JR. Synthetic, implantable polymers for local delivery of IUdR to experimental human malignant glioma. Int J Radiat Oncol Biol Phys 1998; 42:631–639.

164. Tamargo RJ, Rossell LA, Tyler BM, Aryanpur JJ. Interstitial delivery of diphenylhydantoin in the brain for the treatment of seizures in the rat model (Poster). American Association of Neurological Surgeons Annual Meeting, San Diego, CA. April, 1994, p. 377.

165. Santini Jr JT, Richards AC, Scheidt RA, Cima MJ, Langer RS. Microchip technology in drug delivery. Ann Med 2000; 32:377–379.

166. Richards Grayson AC, Choi IS, Tyler BM, Wang PP, Brem H, Cima MJ, Langer R. Multi-pulse drug delivery from a resorbable polymeric microchip device. Nat Mater 2003; 2:767–772.

9

Improving the Stability of PLGA-Encapsulated Proteins

STEVEN P. SCHWENDEMAN

Department of Pharmaceutical Sciences,
The University of Michigan, Ann Arbor,
Michigan, U.S.A.

JICHAO KANG

Formulation Development, Protein
Design Labs, Fremont, California,
U.S.A.

1. OVERVIEW AND HISTORICAL PERSPECTIVE

In the 1980s, the first long-term controlled-release peptide products were born, establishing the feasibility of controlling the release of small biomolecules. The polymer that was chosen for these products was the safe and biodegradable copolymer derived from lactic and glycolic acids [poly(lactic-co-glycolic acids) or PLGAs], which had an established safety record in clinically used resorbable sutures as well as numerous other desirable characteristics, including the ability to control the time-scale of bioerosion. Shortly thereafter, to meet the rapid growth in the number of protein

381

drugs entering clinical trials (1,2), a strong scientific interest followed to understand how to control the delivery of this special class of molecules from PLGAs. Frequent injections are standard practice for protein delivery because of the poor bioavailability from non-invasive routes (e.g., oral and transdermal) and the rapid clearance of proteins from the bloodstream.

In the 1990s during more intense research on the controlled-release of proteins from PLGA delivery systems, if scientists bothered to check, by-and-large ubiquitous instability of the encapsulated protein was observed. It was clearly recognized that in order to overcome protein instability in PLGAs, new approaches alternative to simple empiricism, such as those based on determining the molecular mechanism of instability, were necessary (3–5). In addition, detailed physical-chemical analysis and simplification of the problem, e.g., by the use of model proteins, has proven to be extremely useful (3,5). Several key stresses responsible for the instability of the encapsulated protein have been identified, such as the use of organic solvent/water mixtures during protein microencapsulation, and the moist and acidic microclimate in the polymer during long-term controlled-release under physiological conditions. Similarly, novel ways to bypass these stresses have been uncovered, i.e., anhydrous encapsulation methods, immobilizing the protein with a divalent counterion, and neutralizing the polymer microclimate pH with antacid excipients (3–5). It is now possible to slowly release stable proteins surpassing the 1-month benchmark. Despite these important advancements, much more work needs to be done in order to extend these successes to the large number of therapeutic proteins completing clinical trials.

2. EVIDENCE OF PROTEIN INSTABILITY

Protein stability issues generally pose much more difficult formulation challenges than do small molecules. The level of difficulty increases further still when encapsulating proteins in a biodegradable polymeric delivery system, such as PLGA, to achieve a controlled-release depot effect. A preponderance of evidence demonstrates that protein instability is the single most difficult obstacle in the development of PLGA-based delivery systems (4–6) for either model proteins, such as bovine serum albumin (BSA) and α-chymotrypsin, or therapeutic proteins, such as tissue plasminogen activator (t-PA) and basic fibroblast growth factor. Table 1 lists numerous examples of reported instability of proteins encapsulated in PLGA delivery systems.

3. WHAT IS THE INSTABILITY PATHWAY?

From the numerous publications during the early evaluation of protein delivery from PLGAs including several of those demonstrated in Table 1, it became clear that many traditional empirical approaches were not getting at the heart of the problem. Additional developmental hurdles involved the extensive analytical challenges to protein characterization as well as the fact that the protein was encapsulated in a polymer (5). Moreover, many scientists feared removal from the polymer would damage the protein. Our approach has been to develop an understanding of the pathways of protein instability, consisting of both the environmental stresses (e.g., organic solvent/water interface and extremes of pH) and molecular mechanisms (e.g., protein unfolding and deamidation) leading to irreversible alteration of the protein molecule (16,33). The groundwork for this approach was laid by (i) examining in detail the physical-chemical events that occur during the lifetime of the encapsulated protein (3–5,34), which allowed numerous potential deleterious stresses on the protein to be revealed; and (ii) reviewing the known molecular mechanisms of protein instability established in neighboring scientific disciplines (33,35).

Many of the latest developments have been described recently (5), and will not be repeated here. Instead, we will focus on our research directed at tackling likely the most

Table 1 Examples of Protein Instability in PLGA Delivery Systems[a]

Protein	Report of instability	Reference
	Peptide-bond fragmentation during release	(7,8)
Bovine serum albumin	Non-covalent aggregation during release (with negligible to minor disulfide-bonded component in aggregates)	(8,9)
	Non-covalent aggregation during encapsulation by solvent evaporation	(10)
Hen egg-white lysozyme	Covalent dimerization and formation of unknown products during release	(7)
Ribonuclease A	Non-covalent aggregation during encapsulation by solvent evaporation	(11)
	Soluble aggregation in the absence of zinc acetate and zinc carbonate during release	(12)
Growth hormone	Deamidation, oxidation, and aggregation observed at rates similar to those in solution during release	(13)
	Aggregation during encapsulation and release	(14)
Tetanus toxoid	Incomplete release; loss in immunoreactive antigen during release	(15–17) (18–20)
Erythropoietin	Covalent aggregation during solvent evaporation; aggregation during release	(21)
Insulin-like growth factor-I	Incomplete protein release over 25 days in the absence of zinc carbonate	(22)
Vascular endothelial growth factor	Heparin affinity decreased by 13% after 8 days of release	(23)
Bone morphogenetic protein	Incomplete protein release; 30% immunoreactive protein recovered after 28 days	(8)
Basic fibroblast growth factor	Incomplete protein release; 38% immunoreactive protein recovered after 28 days with heparin stabilizer	(8)
Carbonic anhydrase	Adsorption, aggregation, and denaturation during microencapsulation; incomplete release; fragmentation, aggregation, and loss of activity during microencapsulation	(24,25)
Human	Aggregation during microencapsulation	(26)

(Continued)

Table 1 (*Continued*)

Protein	Report of instability	Reference
interferon-gamma		
Human nerve growth factor	Aggregation and loss of biological activity during release	(22,27)
Porcine insulin	Deamidation, fragmentation, and covalent dimerization during release	(28)
Tumor necrosis factor-alpha (TNF-α)	Loss of biological activity during microencapsulation	(29)
Horseradish peroxidase	Loss of activity during microencapsulation and release	(30)
α-Chymotrypsin	Aggregation, denaturation, and loss of activity during microencapsulation and release	(31,32)

[a]Some examples were amended from Ref. 5.

difficult phase of the encapsulated protein, that is, during long-term controlled-release. The initial studies were performed with BSA, and these are described below.

3.1. Evaluating the Kinetics of Instability and Denatured State of Encapsulated BSA

Because protein instability was ubiquitous to PLGA delivery systems (Table 1), a somewhat "stable" protein was selected in order to reduce the number of potential instability mechanisms that might occur and would require specific biochemical analysis. BSA was a logical choice because of its low cost and extremely well-defined structure and stability (36). As shown in Figure 1, the protein was loaded into simple injectable PLGA delivery systems, which appear in shape very similar to common mechanical pencil leads. The so-called millicylinders were originally developed and marketed for peptide delivery (i.e., for Zoledax by AstraZeneca). In contrast to smaller geometries such as common injectable microspheres, this geometry for the study of protein delivery had several advantages: (i) encapsulation was anhydrous using a mild organic solvent, acetone; (ii) little micronization needed to

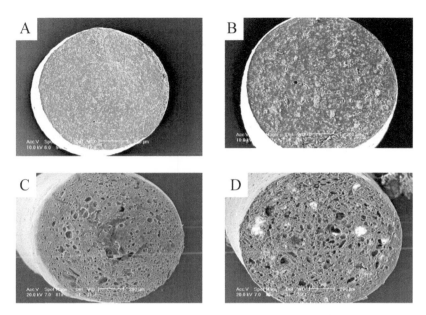

Figure 1 The SEM images of 15% BSA/PLGA 50/50 (i.v.=0.63 dl/g) millicylinders without salts (A,C) and with 3% Mg(OH)$_2$ (B,D). (A,B) before incubation; (C,D) after 7-day incubation in PBST at 37°C. (From Ref. 40.)

be performed; (iii) encapsulation was essentially 100% efficient, simple, and reproducible; and (iv) release studies were also simplified both in vitro and in vivo by releasing from a single macroscopic device.

After BSA loading into PLGA 50/50 millicylinders at 15% (w/w), the protein instability kinetics was monitored (8). In order to assess the stability of encapsulated protein, the protein was removed from the polymer by a mild extraction procedure, which minimized the undesirable situation of protein contact with the organic solvent when hydrated (16). As shown in Figure 2, the release of BSA was incomplete and a steady growth of insoluble aggregates in the polymer was observed. Standard biochemical analysis of proteins extracted from the polymer revealed two salient features of the denatured state, as shown in Table 2: (i) peptide-bond hydrolysis and (ii) noncovalent aggregation mediated by hydrophobic interactions.

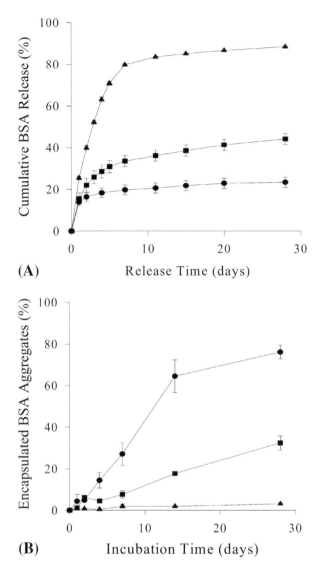

Figure 2 Effect of $Mg(OH)_2$ content on BSA release kinetics (A) and encapsulated BSA aggregation kinetics (B) during incubation of the PLGA implants at $37°C$ in PBST. Millicylinders were loaded with 15% BSA and 0% (●), 0.5% (■), and 3.0% (▲) $Mg(OH)_2$ (average ± S.E., $n = 3$). (From Ref. 8.)

Table 2 Comparison of BSA Instability under Simulated and
Encapsulated Conditions

	Encapsulated[a]	Simulated[b]
Tme to 50% aggregation	12 days	7 days
Aggregates soluble in denaturing solvent[c]	> 98%	> 94%
Peptide fragmentation[d]	25, 40, and 55 kDa	25, 40, and 55 KDa

[a]15% BSA in PLGA millicylinders incubated in PBST at 37°C.
[b]Lyophilized BSA at pH 2 incubated under 86% RH at 37°C.
[c]PBST containing 6 M urea and 1 mM EDTA.
[d]From SDS-PAGE of BSA samples treated with SDS and β-mercaptoethanol.
Source: From Ref. 8.

3.2. Simulating the Encapsulated Denatured State of BSA Outside the Polymer

In order to completely elucidate the instability pathways(s), the destabilizing stress(es) needed to be uncovered. This was handled by simulating the instability of the protein outside the polymer, and then determining which simulated stresses produced the same denatured state of the protein, as observed when encapsulated in the polymer (8). Numerous conditions consisting of varying pH, level of moisture, and the presence or absence of a polymer surface were examined. Among those simulations, BSA incubated in the presence of moisture and an acidic pH of 2, but not more than pH 3 (by controlling the pH of the protein solution before lyophilization), was required to match the denatured state of simulated BSA with that of encapsulated BSA. The pH required for BSA aggregation coincided with the lowest pH unfolding transition of the protein to the expanded form at pH 2.7 (36).

As shown in Table 2, the simulated instability of BSA at the low pH was matched with the instability of BSA encapsulated in PLGA, in terms of time to 50% aggregation ($\sim$10 days), aggregation type, and distribution of hydrolysis fragments. It was found that moisture affected the aggregation rate but not the mechanism. For water/protein weight ratios of 0–0.2 and 0.8–1.0, no aggregates and $\sim$70% non-covalent aggregates were produced, respectively (8). To simulate BSA adsorption, BSA solutions (pH 2–7) were incubated at 37°C for 1 week with blank PLGA microspheres or fine polymer powder to provide a polymer surface. Negligible adsorption (<2%) was recorded in any of these simulations (8). These data proved that the acidic and moist microclimate inside the PLGA pores caused BSA instability during the release, and completed the elucidated pathways of encapsulated BSA instability.

4. EXAMPLES OF STABILIZING PLGA-ENCAPSULATED PROTEINS

Some general rules to protein stabilization in PLGAs have been enumerated in a recent review (5). The general approach to protein stabilization is taken from (33) following elucidation of the instability pathway(s), as described for BSA above. Stabilization can be accomplished in two general ways: (i) directly inhibiting the mechanism (e.g., by adding competitors for the reactive site, see addition of free histidine below) or (ii) removing the deleterious stress (e.g., adjusting the pH; see addition of antacid excipients below). Common examples such as the addition of disaccharides, sucrose, and trehalose, which inhibit protein unfolding, would fall under the first category, whereas substituting anhydrous encapsulation in place of common double emulsion encapsulation methods would fall under the second. It is also worth mentioning the notable result of complexation of proteins as performed with Zn^{2+} for stabilization of growth hormone (12) or with heparin for stabilization of basic fibroblast growth factor (37). These two examples also fit under the first stabilization category.

Our group has focused on the conditions necessary to stabilize proteins by attenuating the deleterious and ubiquitous acidic microclimate in PLGAs; that is, to remove the acidic stress in the polymer. As described in Figure 3, the acidic microenvironment for encapsulated protein can in principle be ameliorated by at least three strategies: (i) increasing the permeability of the polymer to facilitate the escape of the

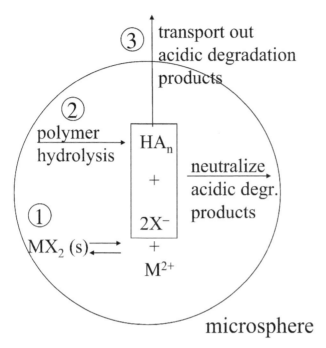

Figure 3 Description of formulation strategies for preventing development of highly acidic pores in PLGA microspheres: (1) addition of an antacid excipients (MX_2), which dissolves in response to the liberation of acidic degradation products (HA_n); (2) decrease polymer hydrolysis (e.g., by using higher lactide content); and (3) increase transport of acidic degradation products by increasing the permeability of the polymer [e.g., by blending poly(ethylene glycol) with the PLGA]. (From Ref. 5.)

acidic polymer degradation products; (ii) retarding the degradation rate of the polymer, so less acidic species are generated during the effective lifetime of the polymeric delivery system; and (iii) neutralizing the soluble acidic species in the polymer with basic excipients. Examples of improved protein stability by these approaches are described and discussed below.

4.1. Stabilization of BSA in PLGA Millicylinders

The deleterious effect of ubiquitous PLGA acidity on encapsulated peptides has been understood for many years (38). As described above, a careful analysis of PLGA-encapsulated BSA also implicated the important role of microclimate acidity on the damage to proteins. Recently, our group for the first time devised a rational approach to uniformly raise the microclimate pH in PLGA 50/50 to neutral levels by the addition of a poorly soluble basic additive, i.e., an antacid. This important effect was demonstrated by adding to BSA/PLGA 50/50 millicylinders $Mg(OH)_2$, which had been used for many years to inhibit acid-catalyzed biodegradation of poly(ortho esters) (8,39,40). The morphology of PLGA millicylinders encapsulating BSA before and after in vitro release at $37°C$ is shown in Fig. 1. BSA aggregation in PLGA millicylinders was essentially inhibited when 3% $Mg(OH)_2$ was added and continuous release was subsequently achieved. As seen in Figure 2, more than 80% of BSA was released continuously from PLGA millicylinders when 3% of $Mg(OH)_2$ was added in comparison to only 20% BSA release for specimens without base. Insoluble base was also found to reduce general acid-catalyzed hydrolysis of the polymer, and to enhance the water uptake of PLGA during release presumably by increasing the osmotic pressure in PLGA by the creation of magnesium carboxylate salts (8).

Despite this remarkable result, under certain formulations [such as when only 0.5% $Mg(OH)_2$ was added in Fig. 2], the incorporation of base was only moderately successful. To better understand the necessary criteria for successful neutralization by antacids, both the base and protein loading was varied (40). As shown in Table 3, at a constant $Mg(OH)_2$ loading of 3%, as BSA loading was decreased from 15% to 10% to 5%, the BSA aggregation in PLGA 50/50 millicylinders rose from 6.8% to 21% to 48%, respectively. Similar effects were observed as base loading was increased at constant BSA loading. A percolation hypothesis was developed based on (i) these data and (ii) confocal images of microspheres containing $Mg(OH)_2$ and the pH-sensitive fluorescent probe, 10-hydroxycamptothecin (41), which demonstrated heterogeneous pH neutralization by 3% $Mg(OH)_2$ without additional excipients. This hypothesis states that successful neutralization afforded by the basic additives requires a percolating network of pores connecting both the base and protein, and

Table 3 Effect of $Mg(OH)_2$ and BSA Content on BSA Aggregation in BSA/PLGA (0.63 dl/g) Millicylinders after 4-Week Release Study in PBST at 37°C

	$Mg(OH)_2$ content (15% BSA loading)					BSA content [3% $Mg(OH)_2$ loading]		
	0	0.5	1.5	3[c]	6	5	10	15
Soluble BSA[a] (%)	2.0 ± 0.3	9.4 ± 0.4	1.9 ± 0.1	4.6 ± 0.2	3.4 ± 0.1	10 ± 1	5.6 ± 0.3	4.6 ± 0.2
Non-covalent aggregate[b] (%)	74 ± 2	45 ± 3	20 ± 3	6.8 ± 0.3	1.0 ± 0.1	48 ± 6	21 ± 3	6.8 ± 0.3

Note: Average ± S.E., $n = 3$.
[a] Soluble in PBST.
[b] Soluble in PBST containing 6M urea and 1mM EDTA. Less than 2% were covalent aggregates soluble in PBST/6M urea/1mM EDTA/10mM DTT.
[c] Columns duplicated to show the trend.
Source: From Ref. 40.

homogeneous acid-neutralization by the base requires (i) selection of the appropriate base, and (ii) manipulation of the polymer microstructure (e.g., via appropriate combination of base and protein loading), to facilitate diffusion of the base to all the acidic pores in the polymer (40).

Because the co-encapsulation of basic additive often increases water content in the polymer matrix, which in turn can inhibit protein aggregation, the contributions of moisture and acidic microenvironment on BSA aggregation in PLGA millicylinders were more closely differentiated (42). Sucrose increases the water uptake of millicylinders, but does not neutralize the soluble acidic polymer degradation products. By comparing BSA stability in millicylinders encapsulating different levels of sucrose and millicylinders encapsulating different levels of $Mg(OH)_2$, it was possible to differentiate the contribution of moisture and acidic microenvironment to BSA aggregation. As seen in Fig. 4, BSA-PLGA 50/50 millicylinders with either 10% sucrose or 3% $Mg(OH)_2$ actually followed identical water uptake kinetics. However, the $Mg(OH)_2$-containing millicylinders showed significantly increased release rate and enhanced stability relative to those polymer containing 10% sucrose but without base. Considering the well-recognized stabilization effect of sucrose, this result provided strong evidence that $Mg(OH)_2$ stabilizes BSA mainly by neutralizing the acidic microclimate instead of simple water uptake. The small increase in release and stability of the sucrose-containing preparation relative to the no-excipient preparation is expected based on the influence of water uptake and presence of sucrose on BSA stability.

4.2. Stabilization of BSA in PLGA Microspheres by Adding Insoluble Base

The stabilization effect of insoluble base on protein in PLGA microspheres has also been demonstrated (8), as shown in Table 4. BSA also was shown to form non-covalent aggregates ($\sim$25–70%) when encapsulated in w/o/w PLGA microspheres, confirming that an acidic microclimate also commonly develops in PLGA 50/50 microspheres (41,43). Whereas the BSA

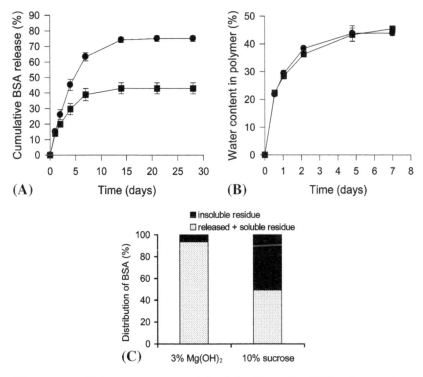

Figure 4 Polymer water uptake kinetics (A), BSA release from (B) and stability in (C) PLGA millicylinders containing 10% sucrose (■) or 3% Mg(OH)₂ (•). Released BSA, soluble residue, and insoluble residue indicate total BSA release, PBST-soluble, and -insolube BSA recovered from the polymer after 4-week incubation, respectively. BSA loading was 10% (average ± S.E., $n = 3$). (From Ref. 42.)

instability mechanism in microspheres was similar to that observed in millicylinders, co-encapsulation of Mg(OH)₂ afforded only moderate inhibition of BSA aggregation in microspheres. For example, the soluble fraction of BSA in PLGA microspheres (0.64 dl/g) decreased from 68% without base to 26% with Mg(OH)₂. This modest BSA stability increase afforded by Mg(OH)₂ was consistent with the aforementioned percolation hypothesis and suggested that Mg(OH)₂ could not diffuse to all acidic protein pores in the polymer.

Table 4 Stabilization Effect of Basic Salts on BSA Delivery from PLGA Microspheres

Formulations		Released[c] (%)	Soluble residue[d] (%)	Insoluble residue[e] (%)	Recovery (%)
BSA/PLGA	No base	4.4 ± 0.1	17 ± 2	68 ± 6	90
microspheres	3% $Mg(OH)_2$	6.9 ± 0.2	65 ± 2	26 ± 1	98
$(0.64 \, l/g)^a$	3% $MgCO_3$	17 ± 2	59 ± 1	13 ± 2	89
BSA/ PLGA	No base	16 ± 2	0.9 ± 0.1	24 ± 3	41^f
microspheres	3% $Mg(OH)_2$	37 ± 2	2.1 ± 0.1	30 ± 2	69^f
$(0.20 \, dl/g)^b$	3% $MgCO_3$	68 ± 2	24 ± 1	1.5 ± 0.2	94

[a]BSA loading was ~4% and the release study was carried out for 28 days.
[b]BSA loading was ~4% and the release study was carried out for 51 days.
[c]All the data represent mean $\pm$ S.E., $n = 3$.
[d]Soluble in PBST.
[e]Insoluble in PBST but soluble in 6 M urea.
[f]Less recovery was probably due to protein hydrolysis.
Source: From Ref. 8.

To overcome the poor BSA-$Mg(OH)_2$ co-percolation, another base, $MgCO_3$, which is about 10-fold more water soluble than $Mg(OH)_2$, was used to facilitate base diffusion. It was found that the more soluble base, $MgCO_3$, inhibited BSA aggregation at a level similar to the inhibition attained in millicylinders with $Mg(OH)_2$. The better percolation of basic additive in PLGA millicylinders was presumably attributed to the high amount of BSA (15% in millicylinders compared to 4–8% in microspheres) and the more porous polymer structure produced by solvent extrusion method. The tendency of PLGA microspheres to form spontaneously isolated pores, e.g., as recently documented by dextran-dye diffusion studies in microspheres produced by w/o/w emulsion-solvent evaporation method (44), could be another cause for the low percolation of basic additives in microspheres. As demonstrated in Table 4, for the medium-molecular-weight PLGA (0.64 dl/g) microspheres, aggregation was held to just 13% over 28 days with 89% recovery. Remarkably, co-encapsulation of $MgCO_3$ in low-molecular-weight PLGA (0.20 dl/g) resulted in a reduction of BSA aggregation to just 1.5% over

51 days with 94% recovery. This preparation controlled the release of BSA slowly and continuously over the entire experiment after a 32% burst (data not shown).

4.3. Stabilization of BSA in PLGA Microspheres by Combination of High Molecular Weight PLA and PEG

Another approach to ameliorate the harmful acidic microenvironment in PLGA is using high molecular weight PLA (poly-lactic acid), which degrades more slowly and will produce fewer acidic species per unit time. The permeability of polymer can also be elevated by using anhydrous microencapsulation relative to w/o/w, and by blending a pore-forming agent in the polymer to facilitate diffusion of acidic species out of the polymer. This approach was demonstrated by Jiang and Schwendeman (9) by encapsulation of the model protein, BSA, in PLA (i.v. = 1.07 dl/g) by an anhydrous microencapsulation.

In the anhydrous microencapsulation, protein and excipients were suspended/dissolved in PLA/acetonitrile solution and then added to cottonseed oil to form an o/o emulsion with Span 85 as an emulsifier. Petroleum ether was then added to extract the acetonitrile and the microspheres were hardened. The microspheres were then recovered by filtration and dried under vacuum. As shown in Table 5 and Fig. 5, without the pore-forming PEG, only 36% BSA was released from PLA microspheres in 1-month of incubation with a total recovery (released+all soluble and aggregated residue in polymer after release) of 76%. Blending in 30% of 35 kDa PEG with the PLA eliminated the BSA aggregation in polymer completely, with 82% of encapsulated BSA released in 1 month. The improved BSA stability in PLA/PEG microspheres could be attributed to a less acidic and more hydrophilic microenvironment in the polymer. As seen in Fig. 6, unlike PLGA 50/50, which caused a dramatic pH drop in the release medium after a 4-week incubation (41), a relatively neutral pH was retained in the release medium for both PLA and PLA/PEG microspheres. A slightly lower pH in the release medium incubated with PLA/PEG microspheres relative to that in PLA was also

Table 5 Effect of PEG on BSA Aggregation in O/O Microspheres after 29-day Release Study in PBST at 37°C

Form code	MW of PEG in blend[a] (kD)	PEG wt.%	Released BSA (%)	Soluble BSA[b] (%)	Non-covalent aggregates[c] (%)	Non-covalent and disulfide bonded aggregates[d] (%)	Recovery(%)
o	–	0	36 ± 1[e]	15 ± 1[e]	22 ± 1[e]	25 ± 2[e]	76 ± 2[e]
a	10	5	44.6 ± 0.3	30 ± 2	41 ± 2	41 ± 7	116 ± 7
b	10	10	40.4 ± 0.3	39 ± 2	26 ± 4	36 ± 4	115 ± 4
c	10	20	73.2 ± 0.3	37 ± 4	–	–	110 ± 4
d	35	20	75.5 ± 0.3	30.2 ± 0.2	–	–	106 ± 1
e	35	30	82.3 ± 0.2	30 ± 2	–	–	112 ± 2

[a]Theoretical protein loading was 5%; PLA (i.v. = 1.07 dl/g) was used in the formulation and the total polymer concentration (w/v) was 20%.
[b]Soluble in PBST.
[c]Soluble in denaturing agents (8 M urea).
[d]Soluble in combined denaturing and reducing agents (8 M urea and 10 mM DTT).
[e]The BSA release study from the PLA specimen was performed over 45 days and the residue was recovered after this time.
Source: From Ref. 9.

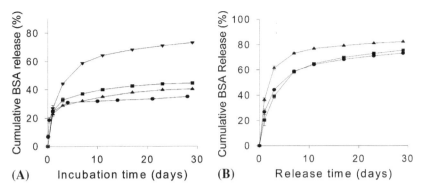

(A)   Incubation time (days)   (B)   Release time (days)

Figure 5   The effect of PEG content and molecular weight in the PLA/PEG blend on the release kinetics of BSA. (A) PEG 10,000 content was 0% (•), 5% (■), 10% (▲), and 20% (▼); (B) PEG molecular weight and content were 20% PEG 10,000 (■) and 20% PEG 35,000(•), 30% PEG 35,000 (▲), (average ± S.D., $n = 3$). (From Ref. 9.)

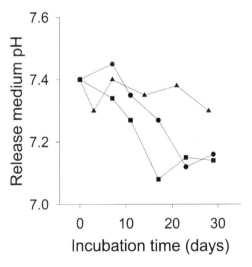

Incubation time (days)

Figure 6   pH change of release medium in PLA (▲), PLA/PEG 35,000 (80/20) (■), and PLA/PEG 10,000 (80/20) (•) microspheres containing 5% BSA when incubated at pH 7.4 PBST and 37°C. (From Ref. 9.)

observed ($\sim$0.1–0.2 pH unit difference), indicating that some acidic degradation products were able to diffuse out of the polymer device through the water channels formed by PEG in PLGA/PEG microspheres. In addition, the pa_H^* [see (41) for definition of this pH measurement in acetonitrile-H_2O mixture] inside PLA/PEG (20% of 35 kDa) microspheres before and after 30 days of incubation were determined to be 6.5 and 5.4, respectively, in contrast to $pa_H^* \sim 3$ in PLGA 50/50 microspheres after a similar incubation time (41). These results confirmed that the acid build-up was remarkably reduced in the PLA/PEG blend formulation.

The water content in PLA microspheres was also greatly enhanced by PEG. As seen in Figure 7, the higher the PEG content, the higher the increase of water uptake. Microspheres containing 20% PEG had almost twice the amount of water uptake relative to those with 10% PEG in the humid environment. The presence of 5% BSA did not increase water uptake rate significantly in the PLA/PEG blend microspheres, which was likely overwhelmed by the strong water sorption induced by the PEG.

Therefore, the less acidic microenvironment in PLA/PEG microspheres, which consequently stabilized BSA in polymer microspheres can be attributed to (i) a slower PLA degradation rate as compared to PLGA 50/50. The rate constant of PLA degradation at $37°C$ in water has been reported to be roughly 0.012 day^{-1}, in comparison to 0.103 day^{-1} for PLGA 50/50 (45); (ii) addition of PEG significantly increased the water content in the formulation, which is expected to inhibit protein aggregation; and (iii) blending PEG in PLGA likely increases the effective diffusion coefficient of the acidic degradation products in the polymer matrix, facilitating acid release from the polymer.

4.4. Stabilization of f-BSA in PLGA Microspheres

Although protein stability can be improved by neutralizing the detrimental acidic microenvironment, and without adjustment of this environment it may be difficult to stabilize an acid-labile protein, most proteins do require additional

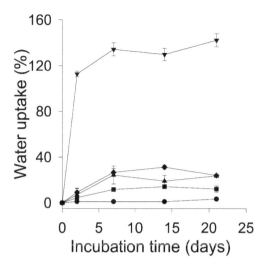

Figure 7 Water uptake kinetics of microspheres after incubation at 97% R.H. and 37°C: blank PLA microspheres (•), blank PLA/PEG 35,000 (90/10) microspheres (■), blank PLA/PEG 35,000 (80/20) microspheres (▲), 5%BSA/PLA/PEG (80/20) microspheres (◇), and blank PEG microspheres (▼). (From Ref. 9.)

formulation considerations to prevent degradation pathways specific to the protein of interest. For example, formalde-hyde-treated protein antigens undergo a unique form of for-maldehyde-mediated aggregation in the solid state (46,47). Jiang and Schwendeman used the model protein antigen, for-maldehyde-treated BSA (i.e., formalinized BSA or f-BSA), to develop a rational approach to stabilize formalinized protein antigens when encapsulated in PLGA microsphere delivery systems, and the result was slow and continuous release of stable f-BSA for 2 months (48).

Formalinization is one of the most commonly used meth-ods to produce inactivated vaccine antigens. The formalinized tetanus toxoid and diphtheria toxoid are the two top antigens of the World Health Organization for the development of sin-gle-dose vaccine formulations (49). However, the antigen instability in polymeric systems is the primary obstacle pre-venting the successful development of controlled-release systems for these important antigens (18,50). As seen in Figure 8, when f-BSA or BSA was encapsulated in PLGA 50/50 microspheres by the w/o/w double emulsion-solvent evaporation method, these proteins could not be released from the microspheres after the initial burst release. The incomplete release was attributed to insoluble aggregation of both BSA and f-BSA in PLGA microspheres, as demonstrated in Figure 9. By 28 days, 40% of initially encapsulated BSA had become insoluble and unreleasable from the polymer. Insoluble f-BSA inside microspheres increased from 8% at 7 days to 31% at 28 days of incubation. However, as seen in Table 6, virtually none (<11%) of the f-BSA aggregates formed in PLGA microspheres was soluble in combined denaturing and reducing agents (8 M urea + 10 mM DDTT + 1 mM EDTA), after

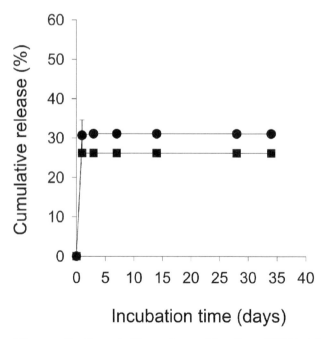

Figure 8 Cumulative release kinetics of BSA (●) and f-BSA (■) from 1.5% protein/PLGA 50/50 (i.v. = 0.20 dl/g) w/o/w microspheres at 37°C (average ± S.D., $n = 3$). (From Ref. 48.)

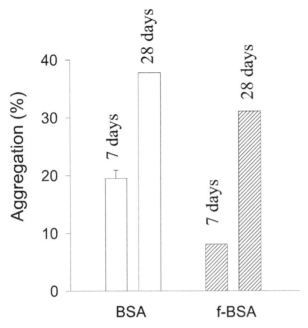

Figure 9   Aggregation of BSA (open bar) and f-BSA (solid bar) in 1.5% protein/PLGA 50/50 (i.v. = 0.20 dl/g) w/o/w microspheres when incubated at 37°C and in the release medium for 7 days and 28 days (average ± S.D., $n = 3$). (From Ref. 48.)

the microspheres were incubated either in the release medium or under humid conditions (80% RH) for 28 days. By contrast, aggregates of the encapsulated control antigen, BSA, were always soluble in the combined reducing and denaturing solvents irrespective of condition used, although the distribution of non-covalent to disulfide-bonded aggregates was largely dependent on the specific deleterious condition.

These data revealed that, unlike BSA which underwent primarily non-covalent aggregation or thiol-disulfide interchange, f-BSA aggregated primarily by the formaldehyde-mediated aggregation pathway (FMAP). As reported earlier, formalinized antigens possess special reactive species that trigger a unique formaldehyde-mediated aggregation (46,47). Briefly, labile antigen-formaldehyde linkages formed during formaldehyde treatment are converted to highly reactive elec-

Table 6 Solubility of Aggregates of Antigens in Denaturing and Reducing Solvents

Antigens formulations	Experimental conditions	Reconstituting solvent	Aggregate solubility (%) (Average ± S.D., $n=3$)[a]	Aggregation mechanism
f-BSA (1.5%f- BSA/w/o/w)	Release medium, 37°C, 28 days	8 M urea	11 ± 4	FMAP
f-BSA (1.5%f- BSA/w/o/w)	80% R.H., 37°C, 28 days	8 M urea +10 mM DTT / 8 M urea	11 ± 4 / 3.5 ± 0.2	FMAP
f-BSA no polymer	80% R.H., 37°C, 12 days	8 M urea +10 mM DTT / 8 M urea	3.1 ± 0.6 / 0	FMAP
BSA (1.5%f- BSA/w/o/w)	Release medium, 37°C, 28 days	8 M urea + 10 mM DTT / 8 M urea	8 ± 2 / 76 ± 4	Non-covalent, disulfide bonded
BSA (1.5%f- BSA/w/o/w)	80% R.H., 37°C, 28 days	8 M urea + 10 mM DTT / 8 M urea	99 ± 4 / 46 ± 5	Non-covalent, disulfide bonded
BSA no polymer	80% R.H., 37°C, 12 days	8 M urea + 10 mM DTT / 8 M urea	100 ± 20 / 0 / 92 ± 6	Disulfide bonded

[a]The percentage of aggregate solubility was calculated from the amount of soluble protein recovered in different solvents relative to the total amount of protein aggregates.
Source: From Ref. 48.

trophiles (e.g., Schiff bases) under appropriate conditions, such as intermediate levels of moisture in the solid state. Schiff bases may combine with nucleophiles on another protein molecule to form intermolecular crosslinks, eventually leading to the formation of non-disulfide-linked covalently bonded insoluble aggregates.

The same authors demonstrated that strongly formaldehyde-interacting free amino acids such as L-lysine and L-histidine could significantly inhibit or stop FMAP under the most severe moisture level for this reaction (47). Therefore, free histidine was incorporated together with f-BSA in the polymer to inhibit the FMAP. An o/o encapsulation was used to achieve high encapsulation efficiency and to avoid the exposure of protein to the detrimental w/o interface as occurs during the w/o/w encapsulation. A 1:5 (w/w) trehalose was also used to stabilize the protein by preferential hydration and to protect the protein during lyophilization. As seen in Table 7, without any stabilizer, f-BSA retained only 30% of its solubility in o/o microspheres after 6 days of incubation, whereas f-BSA co-encapsulated with histidine, histidine + trehalose resulted in 81% and 102% soluble protein after 28 days of incubation, respectively. These data strongly suggested that histidine successfully inhibits the FMAP in PLGA systems and the addition of trehalose adds to this stabilization effect.

However, when f-BSA, histidine and trehalose were encapsulated in PLGA 0.64 dl/g 50/50 by o/o microencapsulation, a large bust release was observed. A high 80% of f-BSA was released in 1 day and the larger burst effect could not be reduced even when the loading was decreased to 5%. Another problem associated with 50/50 PLGA microspheres was the expected acidic microenvironment, which may induce hydrolysis of proteins. In addition, most bacterial toxoids are generally unstable below pH 4.5, making it more desirable for protein antigens to be released from a neutral microenvironment. To solve the large burst release and the acidic microenvironment, slow-degrading PLA (i.v. 1.07) was used together with the pore-former PEG (35 KDa) (9) to encapsulate f-BSA and the stabilizing excipients. As demonstrated in Figure 10 and Table 8, in the absence of additives, 35% of f-BSA was

Table 7 Effect of Encapsulation Method, Histidine, and Trehalose on f-BSA Aggregation in Microspheres after Exposure to 80% R.H. and 37°C

Antigens formulation	Incubation condition	Soluble f-BSA (%)
Effect of encapsulation method		
f-BSA w/o/w microspheres[a]	97% R.H., 37°C, 16 days	50 ± 2
f-BSA o/o microspheres[a]	97% R.H., 37°C, 16 days	112 ± 1
Effect of histidine and trehalose		
f-BSA o/o microspheres[a]	80% R.H., 37°C, 6 days	30 ± 5
	80% R.H., 37°C, 6 days	106 ± 10
f-BSA+His o/o microspheres[b,c]	80% R.H., 37°C, 28 days	81 ± 5
	80% R.H., 37°C, 6 days	
	80% R.H., 37°C, 28 days	127 ± 7
f-BSA+His+Tre o/o microspheres[a,c]	80% R.H., 37 °C, 28 days	102 ± 14

[a]Theoretical protein loading was 8%; PLGA50/50 (i.v. = 0.64 dl/g) was used for microsphere preparation.
[b]Histidine was co-encapsulated in microspheres. The weight ratio of histidine to f-BSA was 1:5.
[c]Histidine and trehalose were co-encapsulated in microspheres. The weight ratio of histidine:trehalose:f-BSA was 0.5:0.5:5.0.
Source: From Ref. 48.

released from PLA microspheres after 60 days of incubation. After blending PLA with PEG, both release rate and total releasable amount of f-BSA was increased. After 60 days of incubation, formaldehyde-mediated aggregates formed in microspheres were reduced to <15%. With co-encapsulation of histidine and trehalose (the weight ratio of f-BSA/histidine/trehalose was 1:0.5:0.5), the FMAP was completely halted. For the stabilized formulation, 94% of f-BSA was continuously released out of the polymer device and the

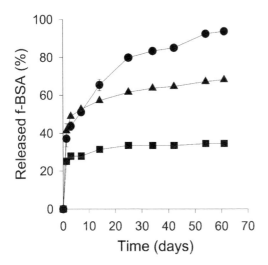

Figure 10   Cumulative release kinetics of f-BSA from f-BSA/PLA (i.v. = 1.07 dl/g) s/o/o microspheres (■), f-BSA/PLA (i.v. = 1.07 dl/g) /PEG 35,000 s/o/o microspheres (▲), f-BSA+histidine +trehalose /PLA (i.v. = 1.07 dl/g)/PEG 35,000 s/o/o microspheres (●) at 37°C (average ± S.D., $n = 3$). (From Ref. 48.)

unreleased protein fraction remained completely soluble in microspheres.

4.5. Stabilization of Therapeutic Proteins in PLGA Millicylinders

Therapeutic proteins have also been stabilized in PLGA milli-cylinders by neutralizing the acidic microenvironment with insoluble basic additives in combination with additional exci-pients that improve the intrinsic stability of the encapsulated protein. Two examples are provided below using t-PA and basic fibroblast growth factor.

Kang and Schwendeman reported a millicylindrical PLGA implant, which can release active t-PA completely and continuously over an one month period (42). The t-PA is a tissue type endogenous serine protease involved in thrombi dissolution (51). The US FDA has approved the use of recombinant t-PA for the treatment of myocardial infarction

Table 8 Inhibition of FMAP in f-BSA Microspheres after 60-Day Release in Phosphate Buffered Saline Containing 0.02% Tween 80 (PBST) at 37°C

Antigens formulations	Released f-BSA (%)[f]	Soluble f-BSA[d,g](%)	Non-covalent and disulfide-bonded aggregates[e,g] (%)	Recovery[h] (%)
f-BSA/PLA[a]	35 ± 0	27 ± 2	5 ± 2	65 ± 3[e]
f-BSA/PLA/ PEG blend[a,b]	68 ± 1	16 ± 1	3 ± 1	87 ± 1[e]
f-BSA+His +Tre PLA/ PEG blend[a-c]	94 ± 1	18 ± 1	0	111 ± 1

[a]Theoretical protein loading was 6%; PLA (i.v. = 1.07 dl/g) concentration (w/v) during encapsulation was 20% in the absence of PEG and 14% in the PLA/PEG blend.
[b]PEG (MW 35,000) concentration (w/v) was 6%.
[c]Histidine and trehalose were co-encapsulated into microspheres. The weight ratio of histidine:trehalose:f-BSA was 1:1:2.
[d]Soluble in PBST.
[e]Soluble in combined denaturing and reducing agents (6 M GnCl and 10 mM DTT)
[f]Insoluble aggregates were visible in combined denaturing and reducing agents, indicating the presence of the FMAP.
[g]Percentage was calculated relative to the original encapsulated protein amount.
Source: From Ref. 48.

(52,53). Controlled-release t-PA implants for local delivery have also been developed to control wound healing and for prevention of proliferative vitreoretinopathy (PVR) (54,55).

The t-PA was encapsulated in PLGA millicylinders together with arginine and other excipients existing in the t-PA lyophilized powder (2% t-PA, 75% L-arginine, 22% phosphoric acid, and 1% polysorbate 80), similarly as previously described (55). Excess free arginine serves as an excellent stabilizing ligand for t-PA. The loading of t-PA and total water-soluble solids (protein+excipients) was 10%, respectively. Arginine hydrochloride and BSA were added in the release medium to improve the stability of t-PA once released. To neutralize the acidic microenvironment in PLGA during release, 3% $Mg(OH)_2$ was added. The release profile and active residue of t-PA after release was evaluated by enzymatic colorimetric assay. As seen in Figure 11, upon addition of $Mg(OH)_2$ to the polymer, 1-month's release of t-PA was increased from $77 \pm 3\%$ to $98 \pm 0\%$, and recovery (active released part+active residue) was increased from $83 \pm 3\%$ to $100 \pm 1\%$, respectively.

Another example of improved stability of therapeutic proteins in PLGA delivery systems by addition of poorly soluble bases and protein-specific stabilization excipients is the delivery of basic fibroblast growth factor (bFGF) (8). The bFGF is a potent angiogenic factor, which can modulate both cell proliferation and differentiation in vitro and in vivo and is being studied for wound healing and tissue regeneration (56). bFGF is a very unstable protein, having an in vitro half life in pH 7 buffer of 24 h and an in vivo half life of 3 min (57).

To stabilize bFGF in PLGA millicylinders, several additives were encapsulated with bFGF in PLGA. Since bFGF is a very potent protein and loaded in the polymer at a very low level (i.e. <0.1%), a water-soluble carrier molecule such as BSA is needed to aid in the transport of the poorly soluble base throughout the polymer for uniform acid neutralization (40). Therefore, $Mg(OH)_2$ and BSA were used to neutralize

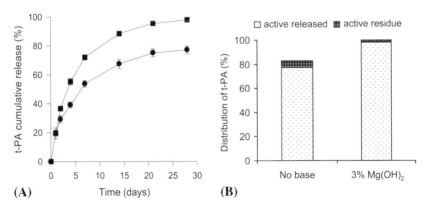

Figure 11 Release (A) and stability (B) of t-PA from PLGA millicylinders with (■) or without (•) 3% $Mg(OH)_2$. Active released and active residue indicate total t-PA release and recovery from the polymer after a 4-week incubation, respectively (average ± S.E., $n=3$). (From Ref. 42.)

the acidic microenvironment. Excess BSA could also potentially inhibit the deleterious adsorption of bFGF to PLGA. Heparin at a weight ratio of 1:1 (heparin to bFGF) was necessary to stabilize the heparin binding site of bFGF (37). EDTA was incorporated as a chelating reagent to inhibit the effect of trace metal ions on bFGF stability. Sucrose was used to help retain the bFGF structure in the solid state (56).

As shown in Figure 12 and Table 9, when bFGF was encapsulated ($\sim$0.0025%) in the $Mg(OH)_2$/BSA/PLGA millicylinders, the growth factor was released continuously. Over 28 days, 71% of immunoreactive bFGF was detected in the release medium and 21% remained in the polymer, accounting for approximately 92% of initially encapsulated bFGF. Both heparin and the $Mg(OH)_2$/BSA combination were necessary to retain bFGF immunoreactivity. For example, when heparin was removed from the stabilized formulation, only

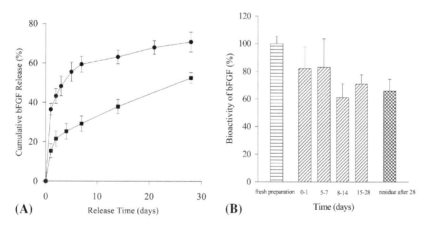

Figure 12 Controlled-release of bFGF. (A) PLGA millicylinders were loaded with 3% $Mg(OH)_2$/0.0025% bFGF/0.0025% heparin/0.01% EDTA/0.6% sucrose/14.4% BSA ($\bullet$) and 3% $Mg(OH)_2$/0.01% bFGF/0.01% heparin/0.01% EDTA/2.3% sucrose/12.7% BSA ($\blacksquare$) (average $\pm$ S.E., $n = 3$). (B) Evaluation of biological activity of bFGF samples from the slow-releasing millicylinders. Bioactivity (%) = concentration determined by bioassay/concentration determined by ELISA $\times$ 100%. (From Ref. 8.)

Table 9   Stabilization Effect of the Basic Salt on bFGF Delivery
from PLGA Millicylinders

Formulations		Released[b] (%)	Soluble residue[c] (%)	Recovery (%)
bFGF/PLGA millicylinders[a]	15%BSA/3%Mg (OH)$_2$/no heparin	1.9 ± 1.3	0	2
	20% gum arabic/heparin	32 ± 1	6 ± 3	38
	15%BSA/3% Mg(OH)$_2$/heparin	71 ± 5	21 ± 2	92

[a]bFGF loading was ~0.0025% and the release study was carried out for 28 days.
[b]All the data represent mean ± S.E., $n = 3$.
[c]Soluble in PBST/1% BSA/10 µg/ml heparin.
Source: From Ref. 8.

2% bFGF was released over 1 month with no immunoreactive
bFGF in the residue fraction. Similarly, when 20% arabic
gum was substituted for 3% Mg(OH)$_2$/BSA, no bFGF was
observed in the release medium after 4 days and only 38%
was accounted for in both the release and residual fraction.
It was also found that when bFGF and sucrose content in
the formulation was increased, the bFGF release could be
adjusted to a slow and continuous release for up to 4 weeks
and the protein retained $> 66\%$ bioactivity over the same
interval (8).

5.   CONCLUDING REMARKS

In the 1980s numerous impediments were overcome to provide controlled-release peptide dosage forms, which has
changed the paradigm of drug delivery for many of these
important drugs, i.e., particularly leuprolide for treatment
of prostate cancer. Over the last 10 years, numerous
impediments to controlled-release of proteins have been
similarly identified, and several solutions demonstrated, as
evidenced by the marketed Nutropin Depot formulation
for growth hormone, and some of the examples described

here. Strategies for microclimate pH control have proven very useful in the stabilization of PLGA-encapsulated proteins. The next several years may provide the answer as to whether these and other initiatives will bear fruit and many more therapeutic proteins will be successfully formulated into controlled-release dosage forms suitable for commercialization.

ACKNOWLEDGMENTS

The authors gratefully acknowledge Dr. Wenlei Jiang of Novartis Pharmaceutical Corp. and Dr. Gaozhong Zhu of Transkaryotic Therapies, Inc. for their help and whose Ph.D. dissertation work is described extensively herein. Much of the research described in this contribution was supported by NIH 68345.

REFERENCES

1. Eppstein DA, Longenecker JP. Alternative delivery systems for peptides and proteins as drugs. Crit Rev Ther Drug Carrier Syst 1988; 5:99–139.

2. Saffran M, Kumar GS, Savariar C, Burnham JC, Williams F, Neckers DC. A new approach to the oral administration of insulin and other peptide drugs. Science 1986; 233:1081–1084.

3. Schwendeman SP, Cardamone M, Brandon MR, Klibanov A, Langer R. Stability of proteins and their delivery from biodegradable polymer microspheres. In: Cohen S, Bernstein H, eds. Microparticulate Systems for the Delivery of Proteins and Vaccines. New York: Marcel Dekker, 1996:1–49.

4. Putney SD, Burke PA. Improving protein therapeutics with sustained-release formulations. Nat Biotechnol 1998; 16: 153–157.

5. Schwendeman SP. Recent advances in the stabilization of protein encapsulated in injectable PLGA delivery systems. Crit Rev Ther Drug Carrier Syst 2002; 19:1–16.

6. Fu K, Klibanov AM, Langer R. Protein stability in controlled-release systems. Nat Biotechnol 2000; 18:24–25.

7. Crotts G, Park TG. Protein delivery from poly(lactic-co-glycolic acid) biodegradable microspheres: release kinetics and stability issues. J Microencapsul 1998; 15:699–713.

8. Zhu G, Mallery SR, Schwendeman SP. Stabilization of proteins encapsulated in injectable poly(lactide-co-glycolide). Nat Biotechnol 2000; 18:52–57.

9. Jiang W, Schwendeman SP. Stabilization and controlled release of bovine serum albumin encapsulated in poly(D,L-lactide) and poly(ethylene glycol) microsphere blends. Pharm Res 2001; 18:878–885.

10. Van de Weert M, Hoechstetter J, Hennink WE, Crommelin DJ. The effect of a water/organic solvent interface on the structural stability of lysozyme. J Contr Rel 2000; 68:351–359.

11. Frangione-Beebe M, Albrecht B, Dakappagari N, Rose RT, Brooks CL, Schwendeman SP, Lairmore MD, Kaumaya PT. Enhanced immunogenicity of a conformational epitope of human T-lymphotropic virus type 1 using a novel chimeric peptide. Vaccine 2000; 19:1068–1081.

12. Johnson OL, Cleland JL, Lee HJ, Charnis M, Duenas E, Jaworowicz W, Shepard D, Shahzamani A, Jones AJS, Putney SD. A month-long effect from a single injection of microencapsulated human growth hormone. Nat Med 1996; 2:795–799.

13. Cleland JL, Mac A, Boyd B, Yang J, Duenas ET, Yeung D, Brooks D, Hsu C, Chu H, Mukku V, Jones AJ. The stability of recombinant human growth hormone in poly(lactic-co-glycolic acid) (PLGA) microspheres. Pharm Res 1997; 14:420–425.

14. Kim HK, Park TG. Microencapsulation of human growth hormone within biodegradable polyester microspheres: protein aggregation stability and incomplete release mechanism. Biotechnol Bioeng 1999; 65:659–667.

15. Kersten GF, Donders D, Akkermans A, Beuvery EC. Single shot with tetanus toxoid in biodegradable microspheres protects mice despite acid-induced denaturation of the antigen. Vaccine 1996; 14:1627–1632.

16. Schwendeman SP, Costantino HR, Gupta RK, Tobio M, Chang AC, Alonso MJ, Siber GR, Langer R. Strategies for stabilising tetanus toxoid towards the development of a single-dose tetanus vaccine. Dev Biol Stand 1996; 87:293–306.

17. Chang AC, Gupta RK. Stabilization of tetanus toxoid in poly(DL-lactic-co-glycolic acid) microspheres for the controlled release of antigen. J Pharm Sci 1996; 85:129–132.

18. Johansen P, Men Y, Audran R, Corradin G, Merkle H, Gander B. Improving stability and release kinetics of microencapsulated tetanus toxoid by co-encapsulation of additives. Pharm Res 1998; 15:1103–1110.

19. Tobio M, Schwendeman SP, Guo Y, McIver J, Langer R, Alonso MJ. Improved immunogenicity of a core-coated tetanus toxoid delivery system. Vaccine 1999; 18:618–622.

20. Sanchez A, Villamayor B, Guo Y, McIver J, Alonso MJ. Formulation strategies for the stabilization of tetanus toxoid in poly (lactide-co-glycolide) microspheres. Int J Pharm 1999; 185:255–266.

21. Morlock M, Kissel T, Li YX, Koll H, Winter G. Erythropoietin loaded microspheres prepared from biodegradable LPLG-PEO-LPLG triblock copolymers: protein stabilization and in-vitro release properties. J Contr Rel 1998; 56:105–115.

22. Lam XM, Duenas ET, Daugherty AL, Levin N, Cleland JL. Sustained release of recombinant human insulin-like growth factor-I for treatment of diabetes. J Contr Rel 2000; 67:281–292.

23. Cleland JL, Duenas ET, Park A, Daugherty A, Kahn J, Kowalski J, Cuthbertson A. Development of poly-(D,L-lactide–coglycolide) microsphere formulations containing recombinant human vascular endothelial growth factor to promote local angiogenesis. J Contr Rel 2001; 72:13–24.

24. Lu W, Park TG. Protein release from poly(lactic-co-glycolic acid) microspheres: protein stability problems. PDA J Pharm Sci Technol 1995; 49:13–19.

25. Sandor M, Riechel A, Kaplan I, Mathiowitz E. Effect of lecithin and $MgCO_3$ as additives on the enzymatic activity of carbonic anhydrase encapsulated in poly(lactide-co-glycolide) (PLGA) microspheres. Biochim Biophys Acta 2002; 1570:63–74.

26. Cleland JL, Jones AJ. Stable formulations of recombinant human growth hormone and interferon-gamma for microencapsulation in biodegradable microspheres. Pharm Res 1996; 13:1464–1475.

27. Pean JM, Boury F, Venier-Julienne MC, Menei P, Proust JE, Benoit JP. Why does PEG 400 co-encapsulation improve NGF stability and release from PLGA biodegradable microspheres? Pharm Res 1999; 16:1294–1299

28. Shao PG, Bailey LC. Stabilization of pH-induced degradation of porcine insulin in biodegradable polyester microspheres. Pharm Dev Technol 1999; 4:633–642.

29. Iwata M, Nakamura Y, McGinity JW. In vitro and in vivo release properties of brilliant blue and tumor necrosis factor-alpha (TNF-alpha) from poly(D,L-lactic-co-glycolic acid) multiphase microspheres. J Microencapsul 1999; 16:777–792.

30. Cohen S, Yoshioka T, Lucarelli M, Hwang LH, Langer R. Controlled delivery systems for proteins based on poly(lactic/glycolic acid) microspheres. Pharm Res 1991; 8:713–720.

31. Castellanos IJ, Cruz G, Crespo R, Griebenow K. Encapsulation-induced aggregation and loss in activity of gamma-chymotrypsin and their prevention. J Contr Rel 2002; 81:307–319.

32. Perez-Rodriguez C, Montano N, Gonzalez K, Griebenow K. Stabilization of alpha-chymotrypsin at the CH2Cl2/water interface and upon water-in-oil-in-water encapsulation in PLGA microspheres. J Contr Rel 2003; 89:71–85.

33. Volkin DB, Klibanov AM. Minimizing protein inactivation. In: Creighton TE, ed. Protein Function: A Practical Approach. Oxford: Oxford University Press, 1989:1–24.

34. Kissel T, Koneberg R. Injectable biodegradable microspheres for vaccine delivery. In: Cohen S, Bernstein H, eds. Microparticulate Systems for the Delivery of Proteins and Vaccines. New York: Marcel Dekker, 1996:51–87.

35. Manning MC, Patella L, Borchardt RT. Stability of protein pharmaceuticals. Pharm Res 1989; 6:903–918.

36. Peters T. Serum albumin. Adv Protein Chem 1985; 37:161–245.

37. Sommer A, Rifkin D. Interaction of heparin with human basic fibroblast growth factor: protection of the angiogenic protein from proteolytic degradation by a glycosaminoglycan. J Cell Physiol 1989; 138:215–220.

38. Hutchinson FG, Furr BJA. Biodegradable polymers for sustained release of peptides. Biochem Soc Trans 1985; 13:520–523.

39. Heller J. Development of poly(ortho esters): a historical overview. Biomaterials 1990; 11:659–665.

40. Zhu G, Schwendeman SP. Stabilization of proteins encapsulated in cylindrical poly(lactide-co-glycolide) implants: mechanism of stabilization by basic additives. Pharm Res 2000; 17:351–357.

41. Shenderova A, Burke TG, Schwendeman SP. The acidic microclimate in poly(lactide-co-glycolide) microspheres stabilizes camptothecins. Pharm Res 1999; 16:241–248.

42. Kang J, Schwendeman SP. Comparison of the effects of $Mg(OH)_2$ and sucrose on the stability of bovine serum albumin encapsulated in injectable poly(D,L-lactide-co-glycolide) implants. Biomaterials 2002; 23:239–245.

43. Fu K, Pack DW, Klibanov AM, Langer R. Visual evidence of acidic environment within degrading poly(lactic-co-glycolic acid) (PLGA) microspheres. Pharm Res 2000; 17:100–106.

44. Kang J, Schwendeman SP. Dynamics of the reversible transition between open and isolated pores in PLGA and its effect on the controlled release of proteins. Proc Int Symp Contr Rel Bioact Mater 2004; 31:32.

45. Magre EP, Sam AP. Hydrolytic degradation of PLGA, calculation of rate constants from various types of in vitro degradation curves. J Contr Rel 1997; 48:318–319.

46. Schwendeman SP, Costantino HR, Gupta RK, Siber GR, Klibanov AM, Langer R. Stabilization of tetanus and diphtheria toxoids against moisture-induced aggregation. Proc Natl Acad Sci USA 1995; 92:11234–11238.

47. Jiang W, Schwendeman SP. Formaldehyde-mediated aggregation of protein antigens: comparison of untreated and formalinized model antigens. Biotechnol Bioeng 2000; 70:507–517.

48. Jiang W, Schwendeman SP. Stabilization of a model formalinized protein antigen encapsulated in PLGA-based microspheres. J Pharm Sci 2001; 90:1558–1569.

49. Aguado MT. Future approaches to vaccine development: single-dose vaccines using controlled-release delivery systems. Vaccine 1993; 11:596–597.

50. Tobio M, Nolley J, Guo Y, McIver J, Alonso MJ. A novel system based on a poloxamer/PLGA blend as a tetanus toxoid delivery vehicle. Pharm Res 1999; 16:682–688.

51. Nugyen TH, Carole W. Stability characterization and formulation development of Alteplase, a recombined tissue plasminogen activator. In: Wang YJ, Pearlman R, eds. Stability and Characterization of Protein and Peptide Drugs: Case Histories. New York: Plenum Press, 1993:91–133.

52. Modi NB, Fox NL, Clow FW, Tanswell P, Cannon CP, Van de Werf F, Braunwald E. Pharmacokinetics and pharmacodynamics of tenecteplase: results from a phase II study in patients with acute myocardial infarction. J Clin Pharmacol 2000; 40:508–515.

53. Maksimenko AV, Tischenko EG. New thrombolytic strategy: bolus administration of tPA and urokinase-fibrinogen conjugate. J Thromb Thrombolysis 1999; 7:307–312.

54. Hubbell JA. Hydrogel systems for barriers and local drug delivery in the control of wound healing. J Contr Rel 1996; 39:305–313.

55. Zhou T, Lewis H, Foster RE, Schwendeman SP. Development of a multiple-drug delivery implant for intraocular management of proliferative vitreoretinopathy. J Contr Rel 1998; 55:281–295.

56. Wang YJ, Shahrokh Z, Vemuri S, Eberlein G, Beylin I, Busch M. Characterization, stability, and formulations of basic fibroblast growth factor. In: Pearlman R, Wang YJ, eds. Formulation, Characterization, and Stability of Protein Drugs. New York: Plenum Press, 1996:141–180.

57. Edelman ER, Nugent MA, Karnovsky MJ. Perivascular and intravenous administration of basic fibroblast growth factor: vascular and solid organ deposition. Proc Natl Acad Sci USA 1993; 90:1513–1517.

10

Recombinant Polymers for Drug Delivery

ZAKI MEGEED

Department of Pharmaceutical Sciences,
University of Maryland,
Baltimore, Maryland, U.S.A.

HAMIDREZA GHANDEHARI

Department of Pharmaceutical
Sciences and Greenebaum Cancer
Center, University of Maryland,
Baltimore, Maryland, U.S.A.

1. INTRODUCTION

Proteins are among the most exquisite components of nature's cellular machinery. Just 20 amino acids comprise the reservoir from which molecules can be constructed that perform such diverse functions as DNA replication, active transport of biological cargo, and structural scaffolding for the cell and organism. However, the amazing versatility of proteins comes at a cost, namely a requirement for exquisite fidelity in their synthesis. The deletion or misplacement of as few as one amino acid can render a protein dysfunctional and have catastrophic consequences for the cell and organism. In order to assure their functionality, elegant mechanisms have evolved for the synthesis of proteins. These include the various regulatory elements and checkpoints that make up the DNA replication and protein synthesis pathways, ensuring that genes are faithfully replicated during cell division, and that proteins with the correct sequences and post-translational modifications are produced from the corresponding genes.

From a polymeric materials science perspective, the properties of biologically synthesized proteins are interesting on three fundamental levels: *functionality, diversity,* and *fidelity.* The use of proteins or their subunits in polymeric materials offers the possibility to incorporate biofunctional and/or biorecognizable motifs that can interact with the physiological environment. Examples of biorecognizable materials include those containing recognition sites for proteolytic enzymes for controlled biodegradation, and cellular attachment motifs. As described in the previous paragraph, naturally occurring proteins exhibit tremendous diversity in their material properties and functionality. This diversity can be further enhanced by engineering slight alterations in the sequences of natural proteins, or combining natural proteins to produce artificial proteins with hybrid functionalities. The silk-elastinlike protein polymers, a hybrid class of materials based on silk fibroin and mammalian elastin, are one example (1). Finally, the fidelity with which biological systems synthesize proteins offers precise control over sequence, composition, stereochemistry, and molecular weight that is largely unattainable by traditional chemical methods. This fidelity can be exploited to synthesize macromolecules with complex structures and rational placement of functional motifs.

Taken together, all of these characteristics have stimulated significant interest in the use of protein-based biomaterials for drug delivery, gene delivery, and other biomedical applications. The purpose of this chapter is to present the current state of the art in the use of biologically synthesized, protein-based polymers for biomedical applications, with an emphasis on drug and gene delivery.

1.1. Genetically Engineered Polymers

Cappello has defined *genetically engineered protein-based polymers* (hereafter also referred to as *genetically engineered polymers* or *protein-based polymers*) as polypeptide chains composed of tandemly repeated amino acid monomer units, synthesized by recombinant techniques (2). These monomer units are ligated, at the DNA level, to form a template for the synthesis of a high molecular weight, repetitive polymer. In contrast, *sequential polypeptides* are synthesized by chemical polymerization of short peptides, which are usually obtained by solid-phase techniques. Random *poly(amino acid)* homo- or copolymers are synthesized by random polymerization of a homogeneous or heterogeneous pool of amino acids.

Both sequential polypeptides and poly(amino acid)s have heterogeneous molecular weights. Though sequential polypeptides offer some degree of control over sequence and composition, random poly(amino acid) copolymers have a random sequence and their composition depends on the method of synthesis and reactivity of the comonomers. By contrast genetically engineered polymers have homogeneous molecular weights, and recombinant techniques enable long-range and high fidelity control over polymer sequence.

2. SYNTHESIS AND CHARACTERIZATION

All strategies for the production of genetically engineered polymers rely on the basic principle of self-ligation (concatamerization) of DNA monomers to form concatameric DNA sequences (3,4). These concatamers are then cloned into the appropriate plasmid and introduced into a biological expression system, where they are transcribed and translated by the cellular machinery to produce a protein-based polymer.

Synthesis begins with the conceptual design of the desired polymer and the corresponding oligonucleotide sequence that encodes the monomer(s). When designing these sequences, several biological constraints are considered: First, the codon usage preference of the organism in which the polymers are to be synthesized (often *Escherichia coli*) must be considered. Repetitive usage of rare codons results in transfer RNA (tRNA) depletion, completely inhibiting the synthesis of some polymers and causing truncation of others. Second, while polymeric gene sequences require repetition by nature, the repetition of identical codons should be minimized. This can be achieved by utilizing the redundancy of the genetic code. The consequences of highly repetitive codon usage include tRNA depletion and genetic instability due to recombination and/or deletion (5–7). Third, the sequence should be designed to minimize complementarity that can cause messenger RNA (mRNA) secondary structure formation. Formation of mRNA secondary structures can cause pausing and disengagement of the ribosome during translation of the mRNA, resulting in truncated polymers.

The monomer oligonucleotide sequence is typically synthesized by automated chemical synthesizers, which are currently limited in their ability to produce oligonucleotides with a size greater than 100 bases. This size limitation may require the synthesis of multiple oligonucleotides, which can then be enzymatically ligated to form monomers with a length greater than 100 bases. After synthesis, the monomers are purified and annealed with their complementary strands to make double-stranded DNA that is suitable for cloning.

Synthesis of recombinant proteins (polymers) is often accomplished through the use of two types of plasmids (or vectors), the *cloning* vector and the *expression* vector. Cloning vectors are plasmids that lack the DNA sequences necessary for the transcription of an inserted gene. They are normally used to "store" genes in a stable form until expression and to limit the problems that can arise from basal levels of expression of toxic proteins. Expression vectors contain the DNA sequences necessary for gene transcription and, in combination with an appropriate biological host, can be used to produce a protein (polymer) of interest. To minimize genetic instability early in the process, a cloning vector is often used in the initial stages of polymer synthesis. The monomer gene is inserted into such a vector and transformed into *E. coli*, which can be grown to produce large quantities of the vector and hence the monomer gene within it. Before multimerization, the presence and identity of the monomer gene are confirmed by restriction digestion and DNA sequencing (8).

In order to produce polymers, at the DNA level, the monomers must be concatamerized. The first step in the synthesis of concatamers is the production of relatively large amounts of the monomer. This can be accomplished through large-scale plasmid preparations or the polymerase chain reaction (PCR), followed by digestion with the appropriate restriction enzymes and gel purification. The classic technique for concatamerization of DNA monomers involves the incubation of monomers with ligation enzymes such as T4 DNA ligase (1). This strategy produces an assortment of DNA concatamers in which the size may be roughly controlled by the reaction time and the concentration of the monomers and ligase. However, this technique suffers from three primary limitations: First, there is no guarantee that a concatamer of a desired size will be obtained. Second, as concatamer size increases, circularization occurs, preventing further manipulation. Third, this method requires the inclusion of unique restriction enzyme recognition sites that may require the insertion of extraneous codons (and hence amino acids) between monomers. These limitations have been addressed by the development of altered synthetic strategies. Recursive directional ligation is a method that can be used to synthesize concatameric DNA sequences of precisely defined lengths while avoiding circularization problems (9). Another technique, relying on the isolation and ligation of pre-ligated concatamers, has also been utilized to control length and limit circularization (10). Finally, a technique called *seamless cloning* has been used to produce DNA monomers and concatamers without any extraneous codons between monomers (11,12).

After synthesis, the DNA concatamers are ligated into either another cloning vector or an expression vector. The choice of the vector is partially based upon the availability of the necessary restriction sites and desired purification strategy. When the polymer gene is cloned into an expression vector, it may be transformed first into a strain of bacteria that is incapable of expressing it. As with the use of a cloning vector, this approach avoids potential problems related to genetic stability and/or cytotoxicity due to basal levels of polymer expression. However, the production of polymers eventually requires that an expression plasmid containing the polymer gene be transformed into an expression host.

Many *E. coli* expression systems contain an inducible promoter that can be used to turn on polymer production once the cells have grown to an optimal density. This strategy is intended to maximize the efficiency of polymer production. Production of recombinant proteins redirects metabolic resources that are normally used for cellular growth, maintenance, and division, often leading to a slowing or arrest of growth. By allowing the cells to reach a relatively high density prior to inducing polymer expression, the need for further division can be somewhat circumvented and the yield of the polymer may be increased.

Purification of polymers is performed by standard techniques that have previously been established for recombinant proteins. Frequently these include the use of affinity tags [e.g., poly(histidine), glutathione-*s*-transferase] that can be used to chromatographically purify recombinant proteins. If necessary, the tag can subsequently be removed by enzymatic cleavage.

Some polymers have been purified on the basis of their physicochemical properties. For example, silk-like polymers have been purified by taking advantage of their low solubility in aqueous medium (13). Elastin-like polymers (ELPs) have been purified by temperature cycling above and below their inverse temperature transition (T_t) (14). This technique has been extended to produce an ELP-tag that can be used to purify a number of recombinant proteins by temperature cycling, which may be faster and less expensive than affinity chromatography (15).

After purification, polymers are typically characterized by a standard set of techniques that can include amino acid content analysis, mass spectrometry, sodium dodecyl sulfate polyacrylamide gel electrophoresis (SDS-PAGE), and immunoblotting. These methods are intended to verify the identity of the polymer. Depending on the type of polymer being synthesized, a series of polymer-specific characterizations will then be performed.

While biological synthesis of polymers confers many advantages, it also imposes some limitations. As described previously, there are several issues that must be considered when designing the monomer and concatamer gene sequences. Furthermore, the potential toxicity of the genetically engineered polymer to the expression host, and the resultant effect on the integrity of the product must be considered.

While the 20 natural amino acids do provide significant structural diversity, the natural amino acid pool can also be viewed as a limitation. Kiick et al. have begun to address this limitation by incorporating artificial amino acid analogs into genetically engineered polymers (16). Incorporation of the amino acid azidohomoalanine into the medium of methionine-depleted bacterial cultures resulted in the replacement of natural methionine with azidohomoalanine. The significance of this substitution is that it allows chemical modification of the azide group by the Staudinger ligation, offering the ability for site-specific modification of the protein-based polymer.

3. DRUG DELIVERY APPLICATIONS

Drug delivery research is substantially focused on improving methods to deliver medications to the necessary location, in the correct amount, at the correct time. The relatively recent emergence of nucleic acid based therapies has enhanced the appreciation for protection and targeting at the subcellular level. Because of the potential to incorporate several functions into a single molecule, there is significant interest in polymeric biomaterials for drug and gene delivery applications requiring the circumvention of multiple biological barriers. The ability of some polymers to traverse these barriers and deliver bioactive agents without significant toxicity lies primarily in their chemical structure. By enabling more precise control over the macromolecular architecture, the structural factors influencing the effectiveness of drug delivery from polymeric biomaterials can be more clearly elucidated, and more efficient delivery systems can be designed.

Though still at an embryonic stage of development, the application of genetically engineered polymers to drug delivery thus far can be divided into two general approaches: polymers for systemic administration and gel-forming polymers for localized, controlled release applications. Both approaches may yield either a localized (i.e., targeted) or systemic effect, depending on the macromolecular structure and the properties of the therapeutic entity.

3.1. Polymers for Systemic Drug Delivery

3.1.1. Thermally Targeted ELP Carriers

Applications of genetically engineered polymers for targeted systemic drug delivery have primarily aimed at treating solid tumors. The targeting of anticancer medications to solid tumors is an extremely challenging problem, due to the vascular and structural heterogeneity, and elevated hydrostatic pressure within the tumor. Traditional methods of tumor targeting have relied on affinity approaches, selectively targeting specific epitopes on tumor cells, or the passive enhanced permeability and retention (EPR) effect. This refers to the enhanced permeability of the tumor vasculature and retention of fluids within the tumor, due to a locally dysfunctional lymphatic drainage system (17). Macromolecular carriers of appropriate molecular weight passively accumulate within tumors by transport across the leaky tumor vasculature. One approach to increase the efficiency of this accumulation involves using localized hyperthermia to induce aggregation of thermally responsive polymeric carriers within the tumor.

ELPs are macromolecules composed of the monomeric pentapeptide (VPGXG), where X can be any amino acid residue except proline (18,19). Through variation in the length of the polymer and the residue in the X position, numerous polymers have been synthesized that exhibit sharp phase transitions, due to hydrophobic collapse and aggregation, when the temperature is raised above a temperature termed the inverse temperature transition (T_t) (19). The responsiveness of these materials to temperature can be further modulated by combining monomers with different X residues, to construct block copolymers in which the constituent blocks undergo phase transitions at distinct temperatures (20,21).

Meyer et al. have pioneered the use of ELPs as thermally targeted carriers for the treatment of solid tumors (22). Using genetic engineering techniques, they constructed a library of ELPs with various sequences and molecular weights, both of which influence T_t. From this library, a polymer was selected with a T_t of approximately 40°C, intermediate between body temperature ($\approx 37°C$) and a temperature induced by localized heating of the tumor (≈ 42–$43°C$). This approach enhances localization by enabling thermally-induced aggregation of the ELP within the tumor. Furthermore, localized heating of tumors is known to increase macromolecular extravasation and sensitivity to some therapeutics (23,24).

Prior to in vivo characterization of the thermally targeted ELP, in vitro characterizations were performed to evaluate the effects of drug conjugation, solvent, and polymer concentration on T_t (22). Conjugation of reporter molecules to the ELPs was achieved by chemical coupling to a short N-terminus peptide leader sequence (e.g., SKGPG), engineered at the genetic level. Conjugation of the reporter molecules iodobenzoate and rhodamine reduced the T_t of the ELPs, possibly due to increased hydrophobic interactions between the polymers. Analysis of the phase transition in murine and mock (PBS + 0.9 mM BSA) plasma was found to decrease T_t by approximately 4°C. Although increasing the concentration of the polymer resulted in a decrease in T_t, the T_t of the selected polymer remained within the desired temperature range (37–42°C) over an approximately 10-fold range of concentration.

Intravenous administration of the thermally responsive ELP-rhodamine conjugate resulted in localized precipitation in tumor tissue heated to 42°C, while no precipitation was observed when a control ELP-rhodamine conjugate, with a T_t of 70°C, was administered. Localized hyperthermia resulted in a 2-fold increase in intratumoral accumulation and a two- to three-fold increase in cellular uptake when compared to unheated controls (22,25,26). ELP-doxorubicin conjugates induced cytotoxicity comparable to free doxorubicin in vitro, despite differences in the subcellular localization of the free and conjugated drug (27). Free doxorubicin accumulated in the nucleus, while the ELP-doxorubicin conjugates were dispersed throughout the cytoplasm, with significantly less nuclear accumulation, indicating the possibility of a difference in the mechanism of action of the two forms. These studies highlight the potential of temperature-responsive genetically engineered polymers as carriers for anticancer medications and have been reviewed in detail elsewhere (21,28).

3.1.2. Micelle-Forming Polymers

As mentioned previously, ELPs can be further customized by combining different pentapeptide blocks, to produce block copolymers with sensitivity to several different temperatures (20,21). Block copolymers are of great interest for drug and gene delivery applications, significantly because of their ability to form micellar structures with the ability to encapsulate drugs or nucleic acids within the core, and to display targeting moieties on the periphery of the shell (29).

Conticello and colleagues have studied the potential of amphiphilic diblock (**AB**) and triblock (**ABA**) elastin-like copolymers, where **A** is a hydrophilic and **B** a hydrophobic block, to reversibly self-assemble into well-defined micellar aggregates (20,30). Collapse of the hydrophobic block above T_t results in the formation of elastin-based nanoparticles. To provide diversity in the mechanical properties of the micellar structures, the amino acid sequence of the hydrophobic block was varied between plastic (VPAVG) and elastomeric (VPGVG) in nature. The hydrophilic block is designed to maintain solubility and form a protective core that prevents protein adsorption and clearance by the reticuloendothelial system.

Reversible self-assembly into monodisperse spherical micellar particles, 50–90 nm in diameter, was observed for diblock copolymers (**AB**) and triblock copolymers (**ABA**) (30). The temperature-dependent loading of solutes into these micelles was demonstrated with the fluorescent probe 1-anilinonaphthalene-8-sulfonic acid (1,8-ANS) (31). The emission intensity of the fluorophore was observed to undergo a sharp change at 15°C, which coincided with the phase transition temperature of the hydrophobic block, indicating that the probe was encapsulated within the hydrophobic core of the micelle.

Meyer and Chilkoti have synthesized elastin-like block copolymers composed of one block with a T_t of 35°C and another block with T_t of >90°C (9). This block copolymer was designed such that the block with the lower T_t should collapse and aggregate above 35°C, while that with the higher T_t should remain solvated below 90°C, thus forming a core-shell structure. The evolution of particle formation was followed, by dynamic light scattering (DLS) and turbidimetry, as a function of temperature (9). The results from both techniques indicated that the particle size of the elastin-like block copolymer solutions changed in four distinct steps as the temperature increased. The structures formed during each phase were hypothesized to be free elastin-like block copolymer (4.4 nm, ~35–40°C), micellar nanoparticle (20.4 nm, ~40–47.5°C), rearranged nanoparticle (54.5 nm, 47.5–50.8°C), and aggregate (1400 nm, >50.8°C). This work demonstrates the potential to precisely control nanoparticle size, and potential for drug loading and unloading, through variations in temperature. Biological fate of polymeric micelles, such as vascular extravasation and cellular uptake, are influenced by the particle size.

3.2. Hydrogel-Forming Polymers

Hydrogels are water-swollen polymer networks formed by chemical or physical crosslinking of the polymeric chains. Chemical crosslinking involves the formation of covalent bonds between crosslinking reagents and corresponding functional groups on the polymer chains. Physically crosslinked hydrogels can form by a number of mechanisms, including ionic interactions, crystallization, and hydrophobic interactions. For a thorough review of hydrogel crosslinking methods, the reader is referred to a recent comprehensive article by Hennink and van Nostrum (32).

Perhaps due to the fact that hydrogels can absorb large quantities of water, they tend to be relatively biocompatible. This biocompatibility, as well as the ability to control the swelling and deswelling by varying the crosslinking density, and the incorporation of stimuli-responsive elements, has made hydrogels the subject of intensive study in the controlled drug delivery field. The release of solutes from hydrogels is governed by the physicochemical properties of the polymer, the degree of crosslinking, and the properties of the solute. Recent developments in hydrogels and their biomedical applications, including controlled release, have been reviewed in Ref. 50.

3.2.1. Thermoreversible Hydrogels from ELPs

In addition to the characterizations of the micelle-forming diblock and triblock copolymers, previously described, Wright et al. have also synthesized and characterized triblock copolymers in which the hydrophobic blocks constitute the end units (**BAB**) (34). These copolymers undergo hydrophobic self-assembly to form physically crosslinked, thermoreversible hydrogels when the temperature is raised above the T_t of the hydrophobic **B** block. The polymers could be molded into various shapes by injecting them in the solution state and then incubating the mold above T_t (20). Scanning electron microscopy of flash-frozen gels showed networks of micellar aggregates, with individual micelles ranging from 20 to 30 nm in diameter. The complete reversibility of the sol to gel transition in these hydrogels is notable and distinct from hydrogels formed by chemical crosslinking of ELPs.

3.2.2. Silk-Elastinlike Hydrogels

The silk-elastinlike family of protein block copolymers (SELPs) are one class of genetically engineered biomaterials that has received considerable attention for applications in drug and gene delivery. These polymers are composed of tandemly repeated units of silk-like (GAGAGS) and elastin-like (GVGVP) peptide blocks (35). X-ray diffraction studies have shown that the silk-like blocks self-assemble via hydrogen bonding to form β-sheet crystals that impart thermal and chemical stability (36,37). Periodic inclusion of elastin-like blocks decreases the overall crystallinity of the material and increases its aqueous solubility (2). The biological, physicochemical, and material properties of the polymer chains can be specifically tailored, by varying the block lengths, sequence, and compositional ratio (2). For an in-depth review of the properties of SELPs, the reader is referred elsewhere (38).

Gelation of Silk-Elastinlike Polymers

Selected polymers from the SELP family (e.g., SELP-47K, Fig. 1) spontaneously irreversibly form hydrogels under physiological conditions, in the absence of solvents, crosslinking reagents, or reactive monomers (39). The kinetics of the sol to gel transition are dependent on the solution conditions and the structure and concentration of the polymer. Bioactive molecules can be incorporated homogeneously into the hydrogel matrix by simple mixing with the liquid polymer solution prior to gelation (39–41).

MDPVVLQRRDWENPGVTQLNRLAAHPPFASDPMGAGSGAGAGS

[(GVGVP)₄GKGVP(GVGVP)₃(GAGAGS)₄]₁₂

(GVGVP)₄GKGVP(GVGVP)₃(GAGAGS)₂GAGAMDPGRYQDLRSHHHHHH

Figure 1 The amino acid sequence of SELP-47K. The 884 amino acids have a molecular weight of 69,814 Da. It is composed of head and tail portions, and a series of silk-like (GAGAGS) and elastin-like (GVGVP) repeats (primary repetitive sequence in bold). Abbreviation key: A = alanine; D = aspartic acid; E = glutamic acid; F = phenylalanine; G = glycine; H = histidine; K = lysine; L = leucine; M = methionine; N = asparagine; R = arginine; P = proline; Q = glutamine; S = serine; T = threonine; V = valine; W = tryptophan; Y = tyrosine.

The formation of SELP hydrogels has been characterized by differential scanning calorimetry (DSC) and rotational viscometry (39). Consistent with x-ray diffraction results, DSC revealed an exothermic peak that indicates self-assembly via crystallization of the silk-like blocks. The rate of crystallization, and thus gelation, was enhanced by the addition of small quantities of the appropriate seed crystals and disrupted by the addition of 6 M urea, which disrupts hydrogen bonds. The rate of increase in the viscosity of SELP solutions was directly related to the number of silk-like blocks contained in the polymer. SELP-0K, containing two silk-like blocks, exhibited no increase in viscosity over the 125 min analysis, while SELP-5 and SELP-47K both exhibited significant increases in viscosity. SELP-5, containing eight silk-like blocks, was observed to gel approximately twice as fast as SELP-47K, with four silk-like blocks. This difference was attributed to a lag phase of approximately 1 h for SELP-47K, over which no increase in viscosity was observed. Given enough time (i.e., twice as much) for gelation, SELP-47K attained a viscosity equal to that of SELP-5. A positive correlation was observed between temperature and gelation rate, though SELP-5 gelled faster than SELP-47K at all temperatures.

Swelling of Silk-Elastinlike Hydrogels

The degree of swelling describes the amount of water imbibed by a hydrogel network. This parameter provides insight into the relative crosslinking density of the network and hence its porosity and solute transport characteristics. Swelling studies showed that physically crosslinked SELP-47K hydrogels were relatively insensitive to changes in pH and ionic strength (42). This is in contrast to chemically cross-linked elastin-like hydrogels without silk-like blocks (43). This insensitivity was explained by the irreversible crystallization of the silk-like blocks. However, decreasing the polymer concentration from 10 to 8 wt.% resulted in an increase of over 50% in the degree of swelling. A significant decrease in swelling was observed as the cure time increased from 1 h to 24 h. At reduced concentrations or cure times there is a lower probability that the polymer chains will interact, therefore decreasing the crosslinking density, which in turn leads to a higher degree of swelling. These results indicate that the swelling ratio of the SELP hydrogels studied is influenced more by polymer concentration and cure time, than by changes in environmental conditions such as temperature, pH, and ionic strength.

Increasing the concentration and/or cure time of the hydrogels also resulted in less polymer soluble fraction (42). In addition, the sample-to-sample variability in soluble fraction decreased at the longer cure time (24 h). As with decreased swelling, these results were also attributed to an increased density of physical crosslinks with increasing polymer concentration and cure time.

The soluble fraction data were used to calculate equilibrium polymer concentrations and equilibrium hydration of the hydrogels (42). The data indicate that, after release of the initial soluble fraction, no hydrogel degradation or dissolution occurred.

Drug Delivery from Silk-Elastinlike Hydrogels

The ability of SELPs to form hydrogels in situ, along with their biocompatibility and customizable structure, has stimulated interest in their use for the localized, controlled delivery of therapeutic agents. To evaluate the suitability of SELP hydrogels as controlled delivery vehicles, Cappello et al. studied the release of several fluorescently labeled probes, with molecular weights ranging from 380 to 70,000 Da (39). The release kinetics of these probes were all approximately first order, with a negative correlation between molecular weight and release rate. Interestingly, no significant difference in the time for release of 50% of the loaded compound (T_{50}) was observed between dansyl-lysine (MW = 380) and fluorescein-dextran (MW = 10,000). This indicates a relatively large pore size in these hydrogel networks, which do not significantly restrict the diffusion of solutes in this molecular weight range.

The delivery of proteins from SELP hydrogels was first investigated by Cappello et al., who studied the release of a recombinant mitotoxin (Pantarin) from SELP-47K hydrogels (39). This protein inhibits the proliferation of cells expressing the FGF receptor, making it an attractive candidate for localized delivery to tumors. Release of bioactive ^{125}I-labeled protein occurred over a period of 8 days, with a significant burst effect in the first 24 h.

The delivery of another protein, cytochrome c (MW = 12,384) and two smaller compounds, vitamin B_{12} (MW = 1,355) and theophylline (MW = 180) were investigated by Dinerman et al. (41). Release curves of equilibrium-loaded hydrogels fit a mathematical model for two-dimensional diffusion from a cylinder, in the axial and radial directions, indicating the occurrence of Fickian diffusion (Fig. 2) (44,45). Increasing the hydrogel cure time from 1 h to 24 h reduced the intra-gel diffusivity of theophylline and vitamin B_{12}, but not cytochrome c, which remained essentially constant despite a decrease in swelling with longer cure time. Calculation of normalized solute diffusivities indicated hindered transport in all three cases.

To quantify the influence of partitioning on solute transport, normalized diffusivity and dimensionless permeability were compared for each solute (41). Increasing the cure time of the hydrogel from 1 h to 24 h decreased the partition coefficient for all three compounds, though this decrease was not statistically significant for theophylline, indicating potentially size-dependent partitioning behavior (41). With a 1 h cure time, only the transport of vitamin B_{12} appeared to be affected by partitioning. In hydrogels cured for 24 h, solute partitioning was found to significantly affect the transport of both vitamin B_{12} and cytochrome c. The polymer volume fraction influenced both the effective diffusivity and partitioning of cytochrome c, though this effect was minimal in the volume fraction range of 0.04–0.09. A more pronounced effect was observed at greater polymer volume fractions (>0.1).

Direct incorporation of cytochrome c into 12 wt.% SELP-47K solutions prior to gelation yielded hydrogels that released nearly all of the loaded dose within 2 h (41). Cure time did not influence the effective diffusivity of cytochrome c over a range of 1–24 h, despite a decrease in the degree of hydration.

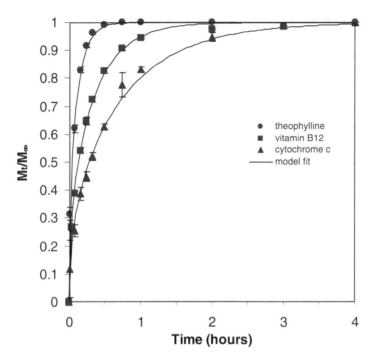

Figure 2 Experimental release profiles for solute diffusion from hydrogels prepared from 12 wt.% polymer 47K solutions incubated for 24 h at 37°C. Symbols represent mean value ± 1 standard deviation ($n=3$). The y-axis denotes the fraction of solute released from the hydrogel at time t. (From Ref. 50.)

Equilibrium uptake loading of pre-formed hydrogels with cytochrome c yielded effective diffusivities that were similar to those obtained by the direct incorporation method, indicating that either technique may be used to load the hydrogels.

The equilibrium swelling ratio and gel dimensions were determined before and after release, for the hydrogels in which cytochrome c was directly incorporated (41). Regardless of gelation time, release of cytochrome c was accompanied by an approximately two-fold increase in hydrogel swelling. Despite the change in swelling, the dimensions of the gels remained essentially constant, suggesting that the increased swelling is probably due to a decreased crosslinking density in the hydrogel network, resulting in the release of soluble fraction throughout the study.

3.2.2. Hybrid Hydrogels

Hybrid polymers made from genetically engineered motifs and chemically synthesized constructs have been made with the intent of expanding the diversity of materials that are accessible via only the chemical or biological route. Thus far, research has focused substantially on the incorporation of protein subunits, with well-defined physico-mechanical properties, into chemically synthesized polymers. The overall mechanical properties of proteins are determined by the mechanical properties of the individual, modular components. These modular "building blocks" are present in a number of proteins and underlie many important physiological processes. One example is the fibronectin type III (Fn3) module, found in proteins that exhibit responsiveness to mechanical force. In the giant muscle protein titin, the unfolding of Fn3 and immunoglobulin-like (Ig) modules allows the protein to alter its spring constant in response to force (46). These structural protein modules can serve as a resource for the development of new biomaterials with well-defined properties and unique responsiveness to environmental stimuli.

Wang et al. have utilized the Ig domain of titin to crosslink water-soluble N-(2-hydroxpropyl)methacrylamide copolymers (47). The hydrogels were formed by chelation between a 6X histidine tag, engineered on the ends of the Ig domain, with Ni^{2+}-chelated acrylamide copolymer. This approach should be widely applicable to a number of protein crosslinking domains, and allows crosslinking to occur in aqueous medium, at neutral pH, at ambient temperature, and in the absence of free radical generating reagents, thus preserving the structural integrity of the protein. These hybrid hydrogels swell substantially in response to increases in temperature, as a result of the thermally induced unfolding of the Ig domain. This positive temperature-volume response is unusual in thermosensitive hydrogels and highlights one potentially useful contribution of protein engineering to the biomaterials field.

In addition to the hybrid hydrogels crosslinked with Ig domains, Wang et al. have utilized proteins forming coiled-coils as crosslinking reagents for HPMA hydrogels (48). Coiled-coils are composed of α-helices that wrap around each other, forming a left- or right-handed superhelical bundle (49). The coiled-coil motif is found widely in nature and is known to occur in over 200 proteins. Increasing the temperature of the coiled-coil results in collapse due to unfolding of the rod-like structure.

Hydrogels were synthesized with crosslinkers that varied in the number of coiled-coil repeats (1–3 repeats) (48). Equilibrium swelling of the hydrogels was found to increase as the number of coiled-coil repeats, and hence cross-linker length, increased. On the other hand, increasing the crosslinking density led to a decrease in equilibrium swelling. All hydrogels underwent a thermally induced collapse with increased temperature, with the crosslinkers containing one and three coiled-coil repeats exhibiting a greater change in swelling than those containing two repeats. The temperature at which this transition occurred corresponded well with the known temperature at which the coiled-coil domains are known to unfold. This work highlights the potential of protein polymers in engineering hydrogels that undergo volume transitions at very specific temperatures.

4. GENE DELIVERY APPLICATIONS

It is widely recognized that the current limitations of gene delivery form a bottleneck that prevents safe and effective gene therapy. The major challenges in gene delivery, like drug delivery, can be distilled fundamentally to *spatio-temporal control* over release. Furthermore, since the delivery of nucleic acids may result in short- or long-term effects, a method by which the expression of the delivered gene can be regulated would be desirable.

We have investigated the potential of genetically engineered SELP-47K to act as an in situ gel-forming matrix for the controlled, long-term delivery of plasmid DNA and adeno-viral vectors to solid tumors (40,50). The use of polymeric matrices for gene delivery allows the manipulation of the release profile of the gene, potential protection against degradative enzymes such as nucleases and proteases, and precise spatial localization at the desired site of action.

The current limitations of actively targeted, systemic cancer gene delivery systems make intratumoral injection an attractive alternative. The primary advantages of intratumoral gene delivery stem from its very precise spatial localization. Direct injection into the tumor obviates the need for systemic targeting, though the efficiency of the therapy may be increased by attaching tumor-specific ligands to the vector (51). The primary disadvantage of intratumoral gene therapy is that, unless a significant bystander effect occurs, each tumor must be individually identified and injected. Hence, by virtue of the difficulty in their detection and distributed nature, small metastases could be impossible to treat by this method. Intratumoral cancer gene therapy may thus be best suited for the treatment of cancers that grow slowly (e.g., prostate cancer) and/or where potential complications or disability make tumor resection less desirable (e.g., prostate cancer, head and neck cancer).

4.1. Delivery of Plasmid DNA from Silk-Elastinlike Hydrogels

Our objective is to design gene delivery systems that can deliver plasmid DNA and viral vectors over a period of at least 28 days, thus ensuring that the expression of the transgene occurs for therapeutically relevant periods of time. As a first step toward this goal, the in vitro release of plasmid DNA from SELP-47K hydrogels has been evaluated over a 28 day period (40,50). DNA-containing hydrogels were fabricated by mixing aqueous DNA solution with aqueous SELP-47K solution. DNA/polymer solutions were incubated at 37°C, for 1 or 4 h to induce gelation.

DNA release was initially evaluated as a function of polymer concentration, DNA concentration, ionic strength, and cure time (40). Consistent with previous swelling studies, increasing the polymer concentration and/or cure time decreased the rate of DNA release from the hydrogels. Within a range of 50 to 250 μg/mL, the DNA concentration did not influence the rate of release. DNA release was strongly affected by the ionic strength of the medium, due to an ionic interaction between the negatively charged DNA and positively charged SELP (Fig. 3). Turbidimetric analysis confirmed an interaction between the polymer and DNA that was dependent on the ionic strength of the medium and consistent with the release data. Such ionic interactions between macromolecular solutes and hydrogel networks have previously been used to control solute release. Genetic

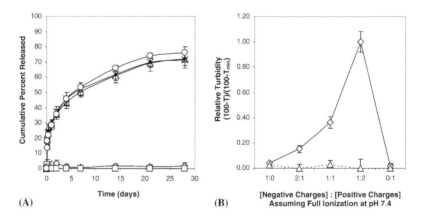

Figure 3 (A) Cumulative release of pRL-CMV from 12 wt.% SELP-47K hydrogels, in PBS with $\mu = 0.03$ M ($\Diamond$), 0.10 M ($\Box$), 0.17 M ($\triangle$), 0.25 M (X), and 0.50 M ($\bigcirc$). Hydrogels were cured for 1 h at 37°C before placement in the appropriate buffer. Each point represents average ± standard deviation ($n = 3$). (B) Effect of ionic strength on the formation of insoluble complexes between SELP-47K and pRL-CMV, in PBS with $\mu = 0.03$ M ($\Diamond$, solid line) and $\mu = 0.17$ M ($\Box$, dashed line). Ratios on the x-axis indicate the molar ratio of negative (DNA) charges to positive (polymer) charges, assuming 100% ionization of each at pH 7.4. The y-axis represents relative turbidity $(100-T)/(100-T_{min})$. Data points represent average ± standard deviation ($n = 3$). (From Ref. 40.)

engineering techniques offer the ability to precisely modulate the charge density and periodicity of the network, thereby facilitating the study of these interactions with a fidelity that is difficult to achieve by chemical methods alone.

DNA release was observed for more than 28 days, with apparent intra-gel diffusivities of approximately 10^{-9} to 10^{-10} cm^2/s, depending on the polymer concentration and cure time (40). Transfection studies performed on encapsulated DNA indicated that it retained in vitro bioactivity equivalent to stock DNA for at least 28 days (50).

The conformation of plasmid DNA, namely supercoiled, open circular, or linear has been hypothesized to play a role in its transfection efficiency. While conventional thinking has generally indicated a preference for supercoiled DNA, at least one study has shown that the delivery of linear DNA results in prolonged transgene expression in vivo, due to concatamerization (52). Other studies have shown that the conformation of DNA is irrelevant to the transfection efficiency in vitro and in vivo (53).

In order to evaluate the influence of plasmid conformation on release from SELP-47K hydrogels, plasmid DNA predominantly in the supercoiled, open circular, and linear conformations were prepared (50). The linear form of plasmid DNA was released most rapidly from the hydrogels, followed by the supercoiled form (Fig. 4). The open circular form was practically not released, probably due to its impalement on the polymer chains. The influence of plasmid size on release was also investigated, with size-dependent release observed for plasmids from 2.6 to 11 kilobases (kb) in size, from 10 wt.% SELP hydrogels (50). Since hydrogels form from SELPs with concentrations as low as 4 wt.%, it is possible that plasmids larger than 11 kb could be delivered.

The effect of hydrogel geometry on DNA release was also studied by fabricating hydrogels in the form of cylinders and flat discs (50). Disc-like hydrogels released DNA faster than their cylindrical counterparts. This was attributed to their larger surface to volume ratio and was described by fitting the release data to an equation that describes

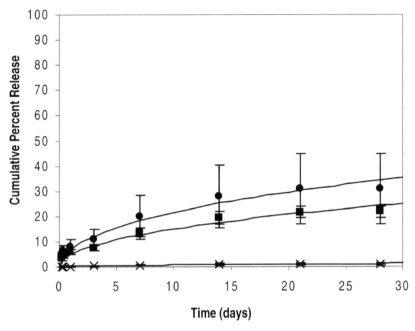

Figure 4 Effect of pRL-CMV conformation on in vitro release from SELP-47K hydrogels: (●) linear, (■) supercoiled, (×) open-circular; (—) Theoretical release based on model for two-dimensional diffusion from a cylinder (Ref. 44). Each data point represents the mean ± standard deviation for $n=3$ samples. (From Ref. 50.)

two-dimensional diffusion from a cylinder with geometric considerations (44,45).

While the applications of SELP-mediated controlled gene delivery are numerous, we have focused our efforts on controlled delivery to solid tumors. Our initial studies in this arena involved the intratumoral delivery of a reporter plasmid (*Renilla* luciferase) to solid tumors in a murine (athymic *nu/nu*) model of human breast cancer (MDA-MB-435 cell line) (50). Delivery of the *Renilla* luciferase plasmid from SELP-47K matrices resulted in significantly enhanced tumor transfection for up to 21 days compared to naked DNA (Fig. 5). In particular, delivery of the plasmid from matrices containing 4 or 8 wt.% polymer resulted in enhanced transfection

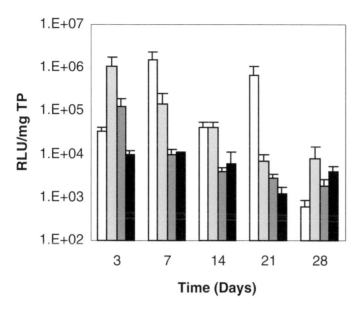

Figure 5 Expression of *Renilla* luciferase in MDA-MD-435 tumors grown subcutaneously in athymic *nu/nu* mice, after intratumoral injection. Bars represent 4 wt.% polymer (white), 8 wt.% polymer (light gray), 12 wt.% polymer (dark gray), and naked DNA without polymer (black). Each bar represents the mean ± standard error of the mean for $n = 4$ or $n = 5$ samples. (From Ref. 50.)

up to 21 days, while 12 wt.% matrices enhanced transfection up to 3 days. These results are consistent with sustained delivery from the 4 and 8 wt.% matrices and entrapment of the DNA within the 12 wt.% matrix.

The levels of tumor transfection mediated by the three concentrations of polymer were statistically equivalent until 7 days, when the 4 and 8 wt.% matrices were both more effective than 12 wt.%. The greater transfection persisted until 21 days for 4 wt.% polymer and 14 days for 8 wt.% polymer. Overall, the delivery of DNA from 4, 8, and 12 wt.% hydrogels resulted in a mean 142.4-fold, 28.7-fold, and 3.5-fold increase in tumor transfection, respectively, compared with naked DNA over the entire 28 day period.

In addition to the delivery of DNA that occurs within the tumor, it can be expected that some DNA will diffuse from the matrix into the surrounding tissue. In order to evaluate this, transfection of the skin approximately 1 cm around the tumor was evaluated (50). The enhancement of delivery to the tumor was compared to the levels of transfection in the tumors and skin at each time point and polymer concentration. While statistically significant differences were not detected between all compositions, the mean tumor transfection was 42.0, 27.2, and 4.6 times greater than skin transfection for 4, 8, and 12 wt.% hydrogels, respectively, over the entire 28 day period. This is in contrast to a 1.3-fold difference between tumor and skin transfection for naked DNA.

4.2. Delivery of Adenoviral Vectors from Silk-Elastinlike Hydrogels

Despite the promising results obtained from delivering plasmid DNA from SELP hydrogels, the delivery of naked DNA results in relatively low transfection efficiency and thus has limited applications. For this reason, we sought to explore the potential of SELP-47K to act as a matrix for the controlled delivery of viral vectors (50). The use of viruses that integrate into the host genome (e.g., retrovirus) can ensure long-term expression of a transgene. However, as was recently observed in a study of x-linked severe combined immunodeficiency, insertion of the viral DNA at or near an oncogenic regulation site can lead to the development of cancer (54). Thus, there is also an intense interest in non-integrating viral vectors, such as adenovirus, which may have a higher margin of safety than integrating viruses. However, without integration into the host genome the duration of transgene expression is limited. A controlled release approach may extend the duration of transgene expression by continuously delivering adenovirus at the site of action. In addition, it is conceivable that the immune response to the virus may be modulated by encapsulation in a polymeric matrix and allowing the release of very small quantities of virus in a given time. Current polymers have not provided the capability to deliver viable adenoviral vectors effectively over prolonged periods of time. Coacervate microspheres of gelatin and alginate have shown poor encapsulation efficiency, low virus bioactivity, and poor virus release kinetics (55). Collagen-based matrices have potential, but control over the rate of release is complicated by the limited control over polymeric structure and crosslinking density (56,57). Genetic engineering techniques may allow the design and synthesis of new polymers with precisely defined architecture and crosslinking density, where biodegradation can be controlled to release viable viral vectors at specific sites and rates.

As a first step toward this goal, we evaluated the potential of SELP-47K hydrogels to act as matrices for the controlled delivery of an adenovirus containing the green fluorescent protein (gfp) reporter gene (AdGFP) (50). SELP-47K/AdGFP solutions were prepared at 4, 8, and 11.3 wt.% polymer, allowed to gel, and placed in a PBS release medium. At predetermined time points, release medium was collected, entirely replaced, and used to transfect HEK-293 cells. These cells contain a relatively high density of adenoviral receptors and are thus a good screening tool for the presence and bioactivity of adenovirus in the release medium. Transduction was observed up to 22 days with the viruses released from the 4 wt.% hydrogel (Fig. 6). The number of transduced cells obtained with the viruses released from the 8 wt.% hydrogel was fewer than that obtained from the 4 wt.% hydrogel. The 11.3 wt.% hydrogel did not release any detectable adenovirus after the first day. These experiments demonstrate that adenoviral release can be controlled over a continuum by varying polymer composition from no release (11.3 wt.% gel) to greater release (4 wt.% gel). Control samples (viral particles without hydrogels in release medium) were bioactive until day 29 (with few gfp+ cells on day 29). However, as anticipated, bioactivity of the vectors decreased over time. Future research will include quantifying the amount of adenovirus released, the proportion that is bioactive, and the in vivo biodistribution, efficacy, and toxicity of adenoviral particles delivered from SELP-47K.

4.3. Soluble Nucleic Acid Carriers

A genetically engineered silk-like polymer derivatized with cationic functional groups, Pronectin F+, has recently been

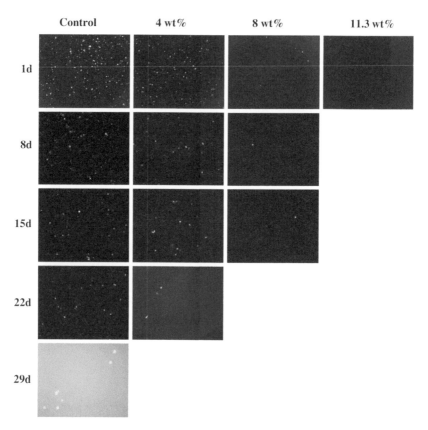

Figure 6 Adenovirus release from SELP-47K and the corresponding bioactivity results. The percentage of polymer increases from left to right. The time of release of virus from gels used to transfect cells or control increases from top to bottom. The images are from fluorescent microscopy at 40× magnification. Bright spots represent individual cells transfected with AdGFP. (From Ref. 50.)

studied as a non-viral vector for gene delivery (58). Pronectin F is a silk-like polymer with one fibronectin domain between every nine silk-like repeats. This polymer was originally synthesized by Cappello and Ferrari, and has been explored as a substrate to enhance cellular attachment to hydrophobic materials (35).

Hosseinkhani and Tabata cationized Pronectin F by reaction of ethylenediamine, spermidine, and spermine with the hydroxyl groups of the serine residue in the silk-like blocks (GAGAGS). For comparison, similarly cationized derivatives of gelatin were prepared. When complexed with plasmid DNA at a weight ratio of 50:1 protein:DNA, all three cationized Pronectins (Pronectin F+) formed particles with DNA with slightly positive ($\sim$10 mV) zeta potential and a size of approximately 200 nm.

All three Pronectin F+ derivatives increased the in vitro transfection of rat gastric mucosal cells by a reporter (luciferase) plasmid in comparison to naked plasmid. Due to its higher buffering capacity, the spermine derivative was found to be significantly more effective than the ethylenediamine and spermidine derivatives. Cellular attachment mediated by the Pronectin F+ derivatives, containing 13 RGD motifs, was found to be significantly greater than the cationized gelatin, which contained only one RGD motif. The amount of plasmid uptake mediated by the Pronectin F+ derivatives was found to be greater than that mediated by cationized gelatin. This study shows the potential of biorecognizable genetically engineered polymers as soluble non-viral vectors and surface coatings for gene delivery.

5. TISSUE REPAIR APPLICATIONS

There is significant interest in the synthesis of novel biomaterials for tissue repair or tissue engineering. Many of these materials are designed to include cellular recognition sequences (e.g., RGD) for the attachment of cells at specific sites. Genetic engineering techniques provide the ability to incorporate these biorecognition sequences at precise locations within the polymer backbone. Furthermore, protease recognition sites can be engineered into the polymer to provide controlled degradation in response to exposure to specific enzymes.

As previously mentioned, Cappello and coworkers have synthesized silk-like polymers that contain periodically spaced fibronectin sequences (35,59). These polymers, called SLP-F (also, Pronectin F), have been shown to support the attachment of a variety of cell types, including epithelial cells, fibroblasts, and cancer cells (59). In addition, silk-like polymers containing regions of human laminan have been synthesized (SLP-L) (2). The SLP-L polymers promoted greater attachment and spreading of fibrosarcoma and rhabdomyosarcoma cell lines than polylysine or laminan coatings. The inclusion of these attachment sequences in-between silk-like blocks allows them to withstand autoclaving without loss of activity (2).

Similarly, Panitch et al. have incorporated periodically spaced fibronectin CS5 domains into ELPs (60). These polymers are intended to function as surfaces for vascular grafts and enhance the attachment of endothelial cells to a glass substrate.

Halstenberg et al. have synthesized hybrid protein-*graft*-poly(ethylene glycol) polymers for tissue repair (61). The design of this biomaterial included motifs for rationally controlling both cellular attachment and biodegradation. Hydrogels prepared by crosslinking the polymer acted as scaffolds for three-dimensional outgrowth of fibroblasts, which attached to the polymer via specific interactions between cellular integrins and an RGD sequence on the polymer. Outgrowth involved serine protease degradation of the polymer at sites engineered for this purpose.

ELPs have been investigated as materials for cartilaginous tissue repair (62). Chondrocytes cultured in ELP matrices retained their phenotype and synthesized significant amounts of extracellular matrix molecules that are regarded as important for cartilage function. This work highlights the potential use of genetically engineered polymers as biocompatible extracellular matrices for tissue repair and engineering. It may also be possible to incorporate drugs and/or genes into these systems to guide the development of new tissue.

6. BIOCOMPATIBILITY AND BIODEGRADATION

The utility of genetically engineered polymers for biomedical applications will depend, to a large extent, on their biocompatibility and biodegradation. An obvious concern in the use of protein-based materials is the potential activation of an immune response toward the material. In terms of bio-degradation, protein-based polymers are thought to degrade to their amino acid constituents, which should be relatively non-toxic and may be used as nutrients. Unfortunately, the biocompatibility and biodegradation of a large number of pro-tein-based materials remains uncharacterized. In the follow-ing paragraphs, we highlight some of the characterizations that have been performed to date.

The rate of resorption of polymeric biomaterials can often be adjusted by controlling their composition. Interestingly, it has been shown that the resorption rate of protein-based block copolymers (SELPs) can be controlled by adjusting *both* the *sequence* and *composition* of the copolymer (2). Control of degradation through the adjustment of polymeric sequence is especially advantageous, because these alterations often do not impact the overall chemical properties of the material and may be made without significantly impacting its physico-mechanical properties. SELPs have been found to degrade primarily by enzymatic proteolysis (e.g., via elastase) and to produce amino acid degradation products that are largely non-toxic. Little is known about the degradation of the silk-like blocks.

Resorption of subcutaneously implanted SELP films has been evaluated in rats, over the course of 7 weeks (2). A collagen control and SELP-0, with a 4:1 elastin to silk ratio were both resorbed within 1 week. SELP-8 implants, with a 2:1 ratio of elastin to silk, retained 18% of their initial mass after 7 weeks, while SELP-3 implants, with a 1:1 elastin to silk ratio retained 58% of their initial mass. SELP-4 and SELP-5, each containing eight silk-like blocks and 3:2 and 2:1 elastin to silk ratios, respectively, showed no evidence of resorption after 7 weeks. These studies demonstrate that the resorption of SELPs is controlled more by the length of the silk-like blocks (i.e., sequence) than the elastin to silk ratio (i.e., composition).

Histological analysis of the implanted films generally showed a mild immune response up to 1 week, with some macrophage surveillance (2,39). Beyond 1 week, remaining films were observed to be surrounded by a zone of healing tissue, consisting of fibroblasts, collagen, and small numbers of macrophages. SELPs that were resorbed within 1 week left regions that were diffusely populated by macrophages. Experiments performed with SELP sponges yielded similar results, with no evidence of chronic inflammation or toxicity over 28 days (2).

To investigate their biocompatibility with wounded tissue, SELP-7 and SELP-5 fibrous meshes were applied to porcine dermal wounds. No adverse effects were observed and the wounds (2×2 cm partial and full dermal thickness) were completely epithelialized after 14 days. Histological evaluation revealed that some SELP filaments had been incorporated into the healing tissue. Similar results were observed with SELP sponges (2).

The immunogenicity of SLP-F and SELP copolymers has been evaluated in rabbits. Compared to hyperimmune positive control rabbit sera, the immunogenicity of all polymers was found to be relatively low. In cases where an elevated antibody titer was observed (e.g., against SLP-F), the antibody response was found to be directed only at the silk-like blocks of the polymer. No response was detected against the elastin-like blocks or fibronectin cellular attachment sites (2).

A broad characterization of the biocompatibility of ELPs was performed by Urry et al. (63). This study involved applying standard biocompatibility tests to chemically synthesized (VPGVG) with a degree of polymerization greater than 120, and a crosslinked gel formed by gamma irradiation. Results indicated that the two forms of the polymer were non-mutagenic, non-toxic, non-antigenic, non-sensitizing, non-pyrogenic, non-hemolytic, and exhibited favorable compatibility after muscle implantation. These studies indicate that ELPs exhibit good biocompatibility that is favorable for biomedical applications.

7. CONCLUSIONS

The diverse functionality of proteins and the fidelity with which they are biologically synthesized is a resource that is just starting to be investigated by material scientists. Although the application of genetically engineered polymers for the delivery of bioactive molecules is in its infancy, the level of interest is gaining momentum and it is likely that new materials and applications will emerge in the near future. A more thorough understanding of protein structure and function will undoubtedly encourage the use of more complicated, stimuli-responsive elements in these materials in drug delivery, gene delivery, and many other biomedically relevant applications.

ACKNOWLEDGMENT

Financial support was made possible by grants from DOD (DAMD17-03-0237 and DAMD17-03-1-0534) (HG), and a National Cancer Center Predoctoral Fellowship (ZM).

REFERENCES

1. Cappello J, Crissman J, Dorman M, Mikolajczak M, Textor G, Marquet M, Ferrari F. Genetic engineering of structural protein polymers. Biotechnol Prog 1990; 6:198–202.

2. Cappello J. In: Wiseman DM, ed. Handbook of Biodegradable Polymers. Amsterdam: Harwood Academic Publishers, 1997: 387–416.

3. Ferrari F, Richardson C, Chambers J, Causey SC, Pollock TJ, Cappello J, Crissman JW. US Patent 5,243,038.

4. Ferrari F, Cappello J. In: Kaplan DL, ed. Protein-Based Materials. Boston: Birkhauser, 1997:37–60.

5. Lohe AR, Brutlag DL. Multiplicity of satellite DNA sequences in *Drosophila melanogaster*. Proc Natl Acad Sci USA 1986; 83:696–700.

6. Sadler JR, Tecklenburg M, Betz JL. Plasmids containing many tandem copies of a synthetic lactose operator. Gene 1980; 8: 279–300.

7. Carlson M, Brutlag D. Cloning and characterization of a complex satellite DNA from *Drosophila melanogaster*. Cell 1977; 11:371–381.

8. Sambrook J, Russell SJ. Molecular Cloning: A Laboratory Manual, 3rd ed., Cold Spring Harbor Laboratory, 2001.

9. Meyer DE, Chilkoti A. Genetically encoded synthesis of protein-based polymers with precisely specified molecular weight and sequence by recursive directional ligation: examples from the elastin-like polypeptide system. Biomacromolecules 2002; 3: 357–367.

10. Won JI, Barron AE. A new cloning method for the preparation of long repetitive polypeptides without a sequence requirement. Macromolecules 2002; 35:8281–8287.

11. McMillan RA, Lee TAT, Conticello VP. Rapid assembly of synthetic genes encoding protein polymers. Macromolecules 1999; 32:3643–3648.

12. Goeden-Wood NL, Conticello VP, Muller SJ, Keasling JD. Improved assembly of multimeric genes for the biosynthetic production of protein polymers. Biomacromolecules 2002; 3: 874–879.

13. Cappello J, Ferrari FA, Buerkle TL, Textor G. Purification of structurally ordered recombinant protein polymers. US Patent No. 5,235,041 (1993).

14. McPherson DT, Xu J, Urry DW. Product purification by reversible phase transition following *Escherichia coli* expression of genes encoding up to 251 repeats of the elastomeric pentapeptide GVGVP. Protein Exp Purif 1996; 7:51–57.

15. Meyer DE, Chilkoti A. Purification of recombinant proteins by fusion with thermally-responsive polypeptides. Nat Biotechnol 1999; 17:1112–1115.

16. Kiick KL, Saxon E, Tirrell DA, Bertozzi CR. Incorporation of azides into recombinant proteins for chemoselective modification by the Staudinger ligation. Proc Natl Acad Sci USA 2002; 99:19–24.

17. Maeda H, Fang J, Inutsuka T, Kitamoto Y. Vascular permeability enhancement in solid tumor: various factors, mechanisms involved and its implications. Int Immunopharmacol 2003; 3:319–328.

18. Urry DW, Harris CM, Luan CX, Luan C-H, Channe Gowda D, Parker TM, Peng SQ, Xu J. In: Park K, ed. Controlled Drug

Delivery: Challenges and Strategies. Washington, DC: American Chemical Society, 1997, 405–438.

19. Urry DW. Physical chemistry of biological free energy trans- duction as demonstrated by elastic protein-based polymers. J Phys Chem B 1997; 101:11007–11028.

20. Wright ER, Conticello VP. Self-assembly of block copolymers derived from elastin-mimetic polypeptide sequences. Adv Drug Delivery Rev 2002; 54:1057–1073.

21. Chilkoti A, Dreher MR, Meyer DE. Design of thermally responsive, recombinant polypeptide carriers for targeted drug delivery. Adv Drug Delivery Rev 2002; 54:1093–1111.

22. Meyer DE, Kong GA, Dewhirst MW, Zalutsky MR, Chilkoti A. Targeting a genetically engineered elastin-like polypeptide to solid tumors by local hyperthermia. Cancer Res 2001; 61: 1548–1554.

23. Maeda H, Seymour LW, Miyamoto Y. Conjugates of anticancer agents and polymers: advantages of macromolecular therapeu- tics in vivo. Bioconjug Chem 1992; 3:351–362.

24. Feyerabend T, Steeves R, Wiedemann GJ, Richter E, Robins HI. Rationale and clinical status of local hyperthermia, radiation, and chemotherapy in locally advanced malignan- cies. Anticancer Res 1997; 17:2895–2897.

25. Raucher D, Chilkoti A. Enhanced uptake of a thermally respon- sive polypeptide by tumor cells in response to its hyperthermia- mediated phase transition. Cancer Res 2001; 61:7163–7170.

26. Meyer DE. Drug targeting using thermally responsive polymers and local hyperthermia. J Contr Rel 2001; 74: 213–224.

27. Dreher MR, Raucher D, Balu N, Colvin OM, Ludeman SM, Chilkoti A. Evaluation of an elastin-like polypeptide-doxorubicin conjugate for cancer therapy. J Contr Rel 2003; 91:31–43.

28. Chilkoti A, Dreher MR, Meyer DE, Raucher D. Targeted drug delivery by thermally responsive polymers. Adv Drug Delivery Rev 2002; 54:613–630.

29. Harada-Shiba M, Yamauchi K, Harada A, Takamisawa I, Shimokado K, Kataoka K. Polyion complex micelles as vectors

in gene therapy–pharmacokinetics and in vivo gene transfer. Gene Ther 2002; 9:407–414.

30. Lee TAT, Cooper RP, Apkarian VP, Conticello VP. Thermo-reversible self-assembly of nanoparticles derived from elastin-mimetic polypeptides. Adv Mater 2000; 12:1105–1110.

31. Zhou Y, Conticello VP. Thermally responsive amphiphilic block copolymers for drug delivery. Polym Preprints 2000; 41: 1643–1644.

32. Hennink WE, van Nostrum CF. Novel crosslinking methods to design hydrogels. Adv Drug Delivery Rev 2002; 54:13–36.

33. Park K, ed. Recent developments in hydrogels. Adv Drug Delivery Rev 2002; 54:1–165.

34. Wright ER, McMillan RA, Cooper A, Apkarian RP, Conticello VP. Thermoplastic elastomer hydrogels via self-assembly of an elatin-mimetic triblock polypeptide. Adv Funct Mater 2002; 2: 149–154.

35. Cappello J, Ferrari F. In: Mobley DP, ed. Plastics from Microbes. Munich: Hanser Publishers, 1994:35–92.

36. Anderson JP, Cappello J, Martin DC. Morphology and primary crystal structure of a silk-like protein polymer synthesized by genetically engineered *Escherichia coli* bacteria. Biopolymers 1994; 34:1049–1058.

37. Anderson JP. Morphology and crystal structure of a recombinant silk-like molecule, SLP4. Biopolymers 1998; 45: 307–321.

38. Megeed Z, Cappello J, Ghandehari H. Genetically engineered silk-elastinlike protein polymers for controlled drug delivery. Adv Drug Delivery Rev 2002; 54:1075–1091.

39. Cappello J, Crissman JW, Crissman M, Ferrari FA, Textor G, Wallis O, Whitledge JR, Zhou X, Burman D, Aukerman L, Stedronsky ER. In-situ self-assembling protein polymer gel systems for administration, delivery, and release of drugs. J Contr Rel 1998; 53:105–117.

40. Megeed Z, Cappello J, Ghandehari H. Controlled release of plasmid DNA from a genetically engineered silk-elastinlike hydrogel. Pharm Res 2002; 19:954–959.

41. Dinerman AA, Cappello J, Ghandehari H, Hoag SW. Solute diffusion in genetically engineered silk-elastinlike protein polymer hydrogels. J Contr Rel 2002; 82:277–287.

42. Dinerman AA, Cappello J, Ghandehari H, Hoag SW. Swelling behavior of a genetically engineered silk-elastinlike protein polymer hydrogel. Biomaterials 2002; 23:4203–4210.

43. Urry DW, Harris CM, Luan CX, Luan C-H, Channe D Gowda, Parker TM, Peng SQ, Xu J. I Park K, ed. Controlled Drug Delivery: Challenges and Strategies. Washington, DC: American Chemical Society, 1997:405–438.

44. Fu JC, Hagemeir C, Moyer DL. A unified mathematical model for diffusion from drug-polymer composite tablets. J Biomed Mater Res 1976; 10:743–758.

45. Siepmann J, Ainaoui A, Vergnaud JM, Bodmeier R. Calculation of the dimensions of drug-polymer devices based on diffusion parameters. J Pharm Sci 1998; 87:827–832.

46. Wang K, Forbes JG, Jin AJ. Single molecule measurements of titin elasticity. Prog Biophys Mol Biol 2001; 77:1–44.

47. Wang C, Stewart RJ, Kopecek J. Hybrid hydrogels assembled from synthetic polymers and coiled-coil protein domains. Nature 1999; 397:417–420.

48. Wang C, Kopecek J, Stewart RJ. Hybrid hydrogels cross-linked by genetically engineered coiled-coil block proteins. Biomacromolecules 2001; 2:912–920.

49. Yu YB. Coiled-coils: stability, specificity, and drug delivery potential. Adv Drug Delivery Rev 2002; 54:1113–1129.

50. Megeed ZE, Haider M, Li D, O'Malley BW, Cappello J, Ghandehari H. In vitro and in vivo evaluation of recombinant silk-elastinlike hydrogels for cancer gene therapy. J Contr Rel 2004; 94:433–445.

51. Xu LN, Pirollo KF, Chang EH. Transferrin-liposome-mediated p53 sensitization of squamous cell carcinoma of the head and neck to radiation in vitro. Hum Gene Ther 1997; 8:467–475.

52. Chen ZY, Yant SR, He CY, Meuse L, Shen S, Kay MA. Linear DNAs concatemerize in vivo and result in sustained transgene expression in mouse liver. Mol Ther 2001; 3:403–410.

53. Bergan D, Galbraith T, Sloane DL. Gene transfer in vitro and in vivo by cationic lipids is not significantly affected by levels of supercoiling of a reporter plasmid. Pharm Res 2000; 17: 967–973.

54. Hacein-Bey-Abina S, von Kalle C, Schmidt M, Le Deist F, Wulffraat N, McIntyre E, Radford I, Villeval JL, Fraser CC, Cavazzana-Calvo M, Fischer A. A serious adverse event after successful gene therapy for X-linked severe combined immunodeficiency. New Engl J Med 2003; 348:255–256.

55. Kalyanasundaram S, Feinstein S, Nicholson JP, Leong KW, Garver RI Jr. Coacervate microspheres as carriers of recombinant adenoviruses. Cancer Gene Ther 1999; 6:107–112.

56. Chandler LA, Doukas J, Gonzalez AM, Hoganson DK, Gu DL, Ma C, Nesbit M, Crombleholme TM, Herlyn M, Sosnowski BA, Pierce GF. FGF2-targeted adenovirus encoding platelet-derived growth factor-β enhances de novo tissue formation. Mol Ther 2000; 2:153–160.

57. Doukas J, Chandler LA, Gonzalez AM, Gu D, Hoganson DK, Ma C, Nguyen T, Printz MA, Nesbit M, Herlyn M, Crombleholme TM, Aukerman SL, Sosnowski BA, Pierce GF. Matrix immobilization enhances the tissue repair activity of growth factor gene therapy vectors. Hum Gene Ther 2001; 12: 783–798.

58. Hosseinkhani H, Tabata Y. In vitro gene expression by cationized derivatives of an artificial protein with repeated RGD sequences, Pronectin(R). J Contr Rel 2003; 86: 169–182.

59. Cappello J. Genetic production of synthetic protein polymers. MRS Bulletin 1992; 17:48–53.

60. Panitch A, Yamaoka T, Fournier MJ, Mason TL, Tirrell DA. Design and biosynthesis of elastin-like artificial extracellular matrix proteins containing periodically spaced fibronectin cs5 domains. Macromolecules 1999; 32:1701–1703.

61. Halstenberg S, Panitch A, Rizzi S, Hall H, Hubbell JA. Biologically engineered protein-graft-poly(ethylene glycol) hydrogels: A cell adhesive and plasmin-degradable biosynthetic material for tissue repair. Biomacromolecules 2002; 3: 710–723.

62.   Betre H, Setton LA, Meyer DE, Chilkoti A. Characterization of a genetically engineered elastin-like polypeptide for cartilaginous tissue repair. Biomacromolecules 2002; 3:910–916.

63.   Urry DW, Parker TM, Reid MC, Gowda DC. Biocompatibility of the bioelastic materials, poly (GVGVP) and its gamma-irradiation cross-linked matrix – summary of generic biological test results. J Bioact Compatible Polym 1991; 6:263–282.

11

Polymeric Nanoparticle Delivery of Cancer Vaccines

JOHN SAMUEL and GLEN S. KWON

Faculty of Pharmacy and Pharmaceutical
Sciences, University of Alberta, Edmonton,
Alberta, Canada

1. INTRODUCTION

Vaccines are the most effective means of prevention of diseases. Eradication of small pox worldwide attests to the power of this preventive strategy. The rarity of infectious diseases such as diphtheria, tetanus, pertussis, mumps, and measles in the Western world is largely due to successful immunizations. Development of effective vaccines is seen as the hope for prevention of other infectious diseases such as malaria, AIDS, and hepatitis B and C infections.

Vaccines may also be effective in treating diseases. Therapeutic vaccines against infections caused by human immunodeficiency virus, herpes simplex virus, hepatitis B virus,

455

and hepatitis C are in now in development (1). Therapeutic cancer vaccines based on tumor antigens is a rapidly expanding area of investigation (2–4). As many of the pathogenic bacteria become resistant to antibiotics, immune based strategies including therapeutic vaccinations will become important for treatment of bacterial diseases (5,6). Vaccines may also be used for treating degenerative diseases of nervous systems such as Alzheimer's diseases (7,8). It is reasonable to expect that the future application to vaccines will extend to autoimmune diseases such as multiple sclerosis, rheumatoid arthritis, or type I diabetes, where control of immunity through tolerance against "self antigens" is the goal (9–12). It is conceivable that "tolerogenic vaccines" designed to prevent graft rejection may be used in organ transplantation (13,14). Thus, vaccines have the potential for treating and managing a broad number of diseases.

Recent advances in immunology have clearly shown that the nature of immune responses against an antigen is largely influenced by the "context" in which the immune system encounters the antigen (15). For example, encounter of an antigen in an immunostimulatory microenvironment by T cells would lead to immune activation leading to elimination of the cells displaying the antigen, whereas the encounter of the same antigen in an immunosuppressive milieu can lead to tolerance, favoring the survival and acceptance of such cells. Thus it may be possible to induce radically opposing immune responses such as elimination or tolerance against a cell expressing the same antigen by appropriate control of the microenvironment of antigen encounter. Thus the delivery of antigens to appropriate immune cells and the control of the microenvironment of the antigen capture and presentation are emerging as significant issues in vaccinology (16,17). An effective vaccine not only requires the "right antigen," but also the "right delivery system." For therapeutic vaccines, the delivery system will largely determine the choice between immune responses that leads to protection or exacerbation of the diseases. Pharmaceutical vaccine delivery systems that can create the desired "context" in which the antigen is "seen" by the appropriate immune cells would advance the design of successful vaccines against a number of diseases. Clearly, the design of the delivery system should be in the context of the target antigen and the relevant disease. Although vaccine delivery research is still in its infancy, significant advances have been made in the past decade (18–25). This review discusses challenges in the design of therapeutic cancer vaccines and highlights the potential of polymeric nano-particulate delivery systems for meeting them.

2. CANCER VACCINES: PROMISE AND PROBLEMS

The potential of the immune system to recognize cancer antigens and mount anticancer responses is now well recognized (2–4,26–29). A broad range of cancer antigens have been characterized during the last two decades (26,30–37). Many of these are now being studied for immunotherapy of spontaneous tumors (38–40). Some examples are products of onco-genes generated by point mutations such as p21^{ras} or overexpressed due to gene amplification such as HER-2/*neu*, point mutated tumor suppresser genes such as *p*53, mela-noma associated antigens such as MAGE, and MUC1 mucin with aberrant glycosylation (41–46). Numerous animal model studies have validated the concept of cancer vaccination with antigen-specific rejection of cancer (47–50). However the results of most cancer vaccine clinical trials to date have been somewhat disappointing, with reports of only isolated cases of tumor regression and a limited number of disease stabiliza-tion, despite evidence for immune activation (51–54). Thus, in the majority of the patients, cancer cells seem to evade the immune responses elicited by vaccination. The immune evasion mechanisms proposed, many of which are interre-lated, include emergence of antigen-negative variants, loss of MHC class I molecules, defects in antigen processing, gen-eration of tolerogenic dendritic cells, activation of immuno-suppressive T regulatory (Tr) cells, induction of apoptosis or enargy of anti-tumor T cells, and defective death receptor signaling in tumors (55–62). Current efforts in cancer immunology are directed toward a better understanding of the relative significance of these issues, molecular mechanisms mediating them, and therapeutic approaches to counter them. Several of these issues deserve special attention in the context of antigen delivery.

3. CRITICAL ISSUES IN CANCER VACCINE DELIVERY

3.1. Antigen Delivery to Dendritic Cells

The central role of dendritic cells (DCs) in initiating and controlling immune responses is now well recognized (15,63–65). DCs are the key professional antigen presenting cells (APCs) capable of activation of naïve T cells. They originate from hemopoietic progenitors in the bone marrow. They are a heterogeneous population of cells (e.g., myeloid or lymphoid origin) serving as "sentries" in most peripheral tissues (66). Before antigen encounter, DCs are in the immature state. As they encounter microbes such as bacteria, fungi, and viruses, they engulf them through phagocytosis (67). This results in activation of complex signaling networks resulting in maturation and upregulation of a large number of genes including CD80, CD86, CD40, major histocompatibility complex class II (MHC-II), and IL-12. At this stage, DCs have switched from "antigen capture" mode to "antigen presentation" mode. Mature DCs migrate to draining lymph nodes, where they present processed antigens in association with MHC class I and class II molecules to the naïve T cells and activate them. The microenvironment of antigen capture and antigen presentation by DCs, controls the direction and magnitude of immune responses.

DCs distinguish the "non-self" from "self" by the recognition of pathogen-associated molecular patterns (PAMPs) on the invading microbes (68,69). Examples of PAMPs are lipopolysaccharide (LPS), peptidoglycans and unmethylated CpG motifs of bacteria, mannan of fungi, and double stranded RNA of viruses. These molecules are not produced by mammalian cells and are perceived by DCs as the "molecular signatures" of infection. A family of cell surface molecules known as toll-like receptors (TLRs) that recognize PAMPs have been extensively studied recently (70). So far, 10 TLRs have been characterized in humans and mice. Engagement of PAMPs by TLRs on DCs triggers signaling pathways responsible for DC maturation and activation. DCs also internalize self-antigens in circulation or in the apoptotic bodies derived from dead cells (71). However this does not result in their maturation. Antigen presentation by immature DCs to T cells is believed to be one of the mechanisms of activation of tolerogenic T cells (72–74). Some pathogens such as HIV have strategies that prevent maturation of DCs after their entry into DCs, which contribute to inappropriate immune responses that fail to provide protection and in some cases exacerbate the disease (75).

DCs loaded with cancer antigens ex vivo are now being studied for cancer vaccination (76–78). Approaches to generating such DC cancer vaccines include pulsing with peptides, tumor lysate, transfection with gene encoding tumor antigens, and generation of DCs fused with tumor cells (79,80). This approach allows antigen-loading of DCs in an immunostimulatory milieu and generation of DCs with the desired level of activation and maturation. Numerous animal model studies have demonstrated the efficacy of this approach, and several clinical trials are now in progress (81–86).

Delivery systems that can selectively target cancer antigens DCs in vivo would be of major significance to cancer vaccines (87). Particulate delivery systems with hydrophobic surfaces and below 5 μm in size would be suitable for uptake by DCs by phagocytosis. A cancer vaccine delivery system should have appropriate antigen retention/release characteristics. Since cancer antigens released into the systemic circulation can cause immune tolerance instead of immune activation, it is important that the delivery systems retain the antigen until the particles are taken up by the DCs. However once internalized by the DCs, the particles should release the antigens intracellularly for efficient antigen processing and presentation by both MHC class I and class II pathways. Co-delivery of cancer antigen and immunostimulatory molecules, especially ligands for TLR, would result in concurrent antigen processing and presentation as well as signaling of TLR pathway leading to generation of mature and activated DCs capable of induction of primary T cell responses. With better understanding of the "immune activating" signaling pathways of DCs, such delivery systems may be used to "turn on" the molecular switches most relevant to potent and lasting anticancer immunity. In this regard it is important that cancer vaccine delivery systems should have sufficient flexibility for incorporation of a broad number of such immunomodulatory molecules varying in physicochemical properties.

3.2. Induction of Protective T Cell Responses

Successful design of cancer vaccines requires definition of the effective immunological mechanisms of cancer rejection. Numerous animal model and several clinical studies highlight the important roles of T cells in mediating cancer rejection (30,32,88–94). Although CD8$^+$ cytotoxic T lymphocytes (CTLs) were initially considered as primary mediators of antitumor activity, more recent studies highlight the significance of CD4$^+$ T helper (Th) cells in controlling the antitumor responses (30,95). The two important subsets of Th cells are: Th1 cells that activate mainly the cell-mediated immune responses; Th2 cells, which provide help to the humoral responses (96). Each has a distinct pattern of cytokine secretion. Th1 cells secrete interleukin-2 (IL-2), interferon-γ (IFN-γ), and lymphotoxin (LT), but not IL-4, IL-5, IL-6, or IL-10, while Th2 cells produce IL-4, IL-5, IL-6, and IL-10, but not IL-2, IFN-γ, or LT. The cytokines produced by Th1 and Th2 subsets themselves reciprocally regulate the functions of each other. The choice of Th1 vs. Th2 pathway of activation of naïve CD4$^+$ T cells is influenced to a large extent by the microenvironment of activation of naïve T helper cells. Th1 rather than Th2 type of immune responses are considered to be mediators of cancer rejection. Activation of Th1 responses would also be expected to augment CTL responses, antibody responses that can mediate cytotoxicity against tumor, and provide stimulus to the natural immune mechanisms involving natural killer cells and macrophages.

The microenvironment of antigen capture by DCs, processing and presentation to T cells have a major influence on Th1/Th2 balance of the CD4+ T cell responses, activation of CD8+ cytotoxic responses, and establishment of robust T cell memory. For example, strong interaction of CD40 on DCs with CD40 ligand on the surface of T cells as well as the presence of IL-12 in the microenvironment of DC-T cell conjugation will favor Th1 response whereas the presence of IL-4 in the same milieu will favor the differentiation of the CD4$^+$ T cells towards a Th2 type (97,98). Therefore the delivery of immunomodulators to DCs, which can induce them to produce IL-12 during antigen presentation would be a useful strategy in cancer vaccination. Antigen delivery systems that promote escape of the antigens from phagosomes to cytoplasm would be beneficial in facilitating cross-presentation of exogenous antigens by MHC class I pathway and therefore induction of CTL responses (71,99). Cytokines such as IL-7 and IL-15 facilitate establishment of long-lived T cell memory (100,101). More detailed understanding of the immunological milieu of DC-T cell conjugation that favor the desired anticancer immune responses would provide additional clues to the design of appropriate delivery systems.

3.3. Overcoming Immune Tolerance

Immune tolerance is a major challenge in cancer vaccination at two levels. First, since the majority of cancer antigens so far characterized are "self-antigens," the immune system would be tolerant to these antigens (102). The high-affinity T cells capable of recognizing such antigens are usually deleted in the thymus (central tolerance) and the corresponding low-affinity T cells are kept under anergy by the peripheral tolerance mechanisms. The latter is believed to be largely through *natural* T regulatory cells (CD4$^+$ CD25+) (103,104). Several immunization strategies including particulate antigen delivery have been able to overcome the "self-tolerance" against cancer antigens and activate antigen-specific T cell responses, presumably recruiting low-affinity T cells, although the mechanism of "breaking" tolerance by these approaches have been inadequately characterized.

Second, there is abundant evidence that cancer cells actively induce tolerance against cancer antigens (105,106). While all the mechanisms of this "acquired" tolerance is not yet well characterized, the roles of "tolerogenic DCs" and *"acquired"* T regulatory cells are worthy of special attention. In some cancers there is a lack of adequate recruitment of DCs to the tumor site (107). Other studies have shown that DCs infiltrate cancer, but in many cases are prevented from maturation and activation. Thus when DCs take up dying cancer cells or soluble cancer antigens they may do so in the "self-antigen capture mode" rather than "pathogen capture mode." Further, many tumors produce "tolerogenic" factors such as vascular endothelial growth factor (VEGF), transforming growth factor-β (TGF-β and IL-10, all of which have been implicated in the generation of immature tolerogenic DCs. In addition certain cancer antigens such as MUC1 mucin also can induce DCs to become "tolerogenic" (108). Such DCs, upon migration to secondary lymphoid organs, would be expected to generate new cancer-antigen specific Tr cells, which will be recruited to tumors and inhibit the action of any effector T cells that are capable of killing tumor cells. In addition the immunosuppressive cytokines produced by the Tr cells may further facilitate generation of tolerogenic DCs. Current literature suggests that this reinforcing cycle of immune-tolerance may be a major hindrance to cancer immunotherapy and may explain the reason for the failure of the clinical responses in cancer patients despite activation of cancer-antigen-specific T cells.

Would it be possible to reverse this tumor-induced dominant immunosuppressive milieu? There is some preliminary evidence suggesting that appropriate engagement of TLR may offer an avenue for breaking tolerance. In vitro studies have shown that stimulation of TLR-4 or TLR-9 on DCs can block the immunosuppressive effect of Tr cells on T effector cells (109). More recently, animal model studies have further provided support to this concept, where sustained TLR stimulation was shown to reverse the immunosuppressive effects of the tumor-induced Tr on anticancer immune responses leading to cancer regression (110). Cancer vaccine delivery systems that facilitate a sustained TLR signaling in DCs concurrently with antigen presentation may be the key to translating this understanding to clinical applications.

4. CANCER VACCINE DELIVERY: LIVING VS. NON-LIVING SYSTEMS

Numerous live vectors based on attenuated viruses or bacteria are currently under investigation as vaccine delivery systems (111,112). Most of these are potent initiators of immune activation. However safety concerns have limited their human applications. They are also limited in terms of the type of immunomodulatory molecules that can be incorporated in them. Further, the unwanted immune responses against the proteins of the vector may also limit their use in booster immunization (e.g., adenovirus vectors). Pharmaceutical vaccine delivery systems based chemically defined materials such as phospholipids or biodegradable polymers have a better safety profile. They offer greater flexibility with respect to the manipulation of physicochemical properties of the vehicle and the range of antigens and immunomodulators that can be incorporated in them. A limitation of the non-living delivery systems is the need for repeat administration and relatively lower magnitude of the immune responses. However with appropriate engineering, it may be possible to develop particulate delivery systems that can give robust and long-lasting immunity comparable to the live vectors. It is conceivable that the living and the non-living systems may be used in sequence in a prime-boost immunization strategy for cancer vaccination so as to exploit their complementary strengths. The following discussions will highlight nanoparticles made of biodegradable polymer poly(D,L-lactic-co-glycolic acid) (PLGA) as a prototype of pharmaceutically acceptable cancer vaccine delivery system and its potential in meeting the challenges discussed above.

5. PLGA-BASED PHARMACEUTICAL DELIVERY SYSTEMS

PLGA is a biodegradable polymer well known for its long-term use in resorbable surgical sutures. This polyester is degraded into lactic and glycolic acid in vivo, both of which are normal metabolites, well tolerated in the body. PLGA is now widely studied as a biomaterial for controlled release drug delivery systems (113,114). It is suitable for encapsulation of a broad range of biologically active compounds varying in physicochemical properties. The release of the encapsulated material can be controlled by several parameters such as ratio of lactic to glycolic acid, molecular weight of the components, and particle size. Several pharmaceutical products based on this delivery system are already in clinical use.

PLGA is well suited as a biomaterial for vaccine delivery. A number of research groups have encapsulated proteins, peptides, lipopeptides, plasmid DNA, oligonucleotides, viruses, and a variety of immunomodulatory molecules in this biomaterial (19,115–121). Hydrophobic materials such as lipopeptides and immunomodulatory glycolipids may be incorporated in a single emulsion (oil/water) solvent evaporation method where as hydrophilic materials such as proteins, peptides, and nucleic acids may be incorporated by double emulsion (water/oil/water) solvent evaporation method. Alternatively proteins and nucleic acids may be absorbed on the surface of the cationic PLGA particles. PLGA vaccine formulations of proteins, peptides, and plasmid DNA induce potent and antigen-specific immune responses. PLGA particles may also be designed for slow or pulsed release of the antigen or for rapid uptake by antigen presenting cells. A controlled release vaccine delivery system incorporating PLGA particles of varying characteristics is under investigation as a single shot delivery system that can give three pulsed release of antigens mimicking three immunizations (122). Another major application of the PLGA delivery system is mucosal delivery of antigens (18). PLGA microparticles can protect the antigens from degradation in the gastrointestinal tract and can facilitate antigen delivery to the Peyer's patches to induce mucosal immune responses (123).

6. "PATHOGEN MIMICKING" NANOPARTICLES FOR CANCER VACCINES

In view of the need for targeting cancer antigens to DCs and activation of the TLR signaling pathway, we propose the design of "pathogen-mimicking" nanoparticles as a cancer vaccine delivery system. These particles are designed for rapid and efficient uptake by DCs, rapid release of the antigen in DCs for MHC class I and class II pathways, and mediation of "danger signal" through engagement of TLR by the ligand co-delivered with the antigen.

6.1. Uptake of PLGA Nanoparticles by Dendritic Cells

We and others have established the suitability of PLGA nano and microparticles for uptake by human and mouse DCs generated in vitro (124–127). We have used three DC cultures as in vitro models: human peripheral bood-monocyte derived DCs, human cord-blood CD34+ stem cell derived DCs, and murine bone-marrow derived DCs. All three systems conclusively demonstrated that > 90% of the cells of the DC phenotype phagocytose PLGA nanoparticles, with phagocytosis being complete within a 12 h incubation period. The uptake was dependent on actin polymerization as evidenced by its inhibition by cytochalasin B. Phenotypes of the cells taking the particles were established by multicolor flow cytometry, where surface-marker phagocytic cells with fluorescent particles were analyzed and intracellular location confirmed by confocal microscopy (Fig. 1). This was also further supported by electron microscopy studies where particles in the process of phagocytosis with membrane ruffling around them could be visualized (Fig. 2).

The uptake of PLGA nanoparticles containing ligands for TLR-4 (monophosphoryal lipid A) or TLR-9 (immunostimulatory CpG oligonucleotides) by DCs resulted in the upregulation

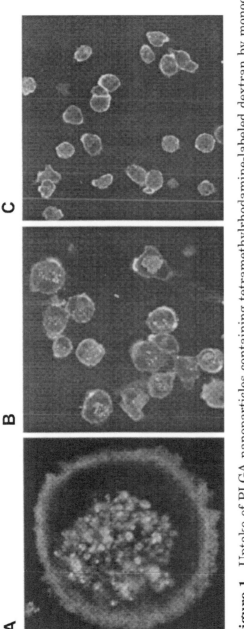

Figure 1 Uptake of PLGA nanoparticles containing tetramethylrhodamine-labeled dextran by monocyte-derived human immature DCs in culture. DC surface was labeled FITC-labeled concanavalin and shows veiling on the cell membrane (A). Pretreatment of cells with cytochalasin B resulted in inhibition of the nanoparticle uptake (B) whereas placebo treated cells showed strong uptake (C). (From Ref. 125.)

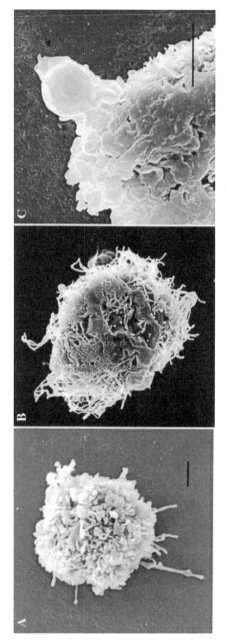

Figure 2 Uptake of PLGA nanoparticles by human cord blood $CD34^+$ cell-derived DCs (day 12) as seen under scanning electron microscope. Cells were incubated with nanoparticles for 4 h at 37°C. Photomicrographs show (A) unpulsed DC; (B,C) uptake of nanoparticles (as shown by arrows). Cell membrane ruffling around a bigger particle can be seen clearly in the latter. Bar = 1 μm. (From Ref. 127.)

of both MHC class II molecule as well as costimulatory molecules (Fig. 3). Such DCs showed higher capacity for allostimulation in a mixed lymphocyte response assay and secreted a number of pro-inflammatory cytokines including IL-12, TNF-α, and IL-6 (unpublished results). These activities were dependent on the presence of TLR ligands within the particles and could not be achieved by comparable quantities of TLR ligands in soluble form or simple mixing of the particles with TLR ligands in a solution. These results are consistent with the view that our "pathogen mimicking" nanoparticles provide a potent delivery system to stimulate DCs towards maturation and activation, critical steps in induction of anticancer immune responses.

Are PLGA nanoparticles administered in vivo internalized by DCs? Would the site of immunization significantly influence the type of cells internalizing the particles? We have demonstrated that PLGA nano and microparticles administered by intradermal route result in their uptake by DCs in mice (128). DCs carrying such particles were located in the draining lymph nodes of the immunized mice. The phenotype of the cells with intracellular fluorescent particles was established by dual color flow cytometric detection of two key DC cell surface markers DEC 205 and CD 86 (Fig. 4). In contrast, the intraperitoneal immunization resulted in the uptake of the particles largely by cells of the macrophage phenotype. These results suggest that the site of immunization may influence the type of cells internalizing the particles. However additional investigations are required to gain better understanding of the in vivo dynamics of cellular uptake of PLGA particles in vivo by different routes of immunization and their potential consequences in terms of immune responses elicited.

6.2. Antigen-loading of DCs by PLGA Nanoparticles Ex Vivo for Induction of T Cell Responses

Antigen delivery to DCs ex vivo is of particular significance to cancer immunotherapy. As already discussed, cancer may induce defects in DCs in cancer patients and may create an

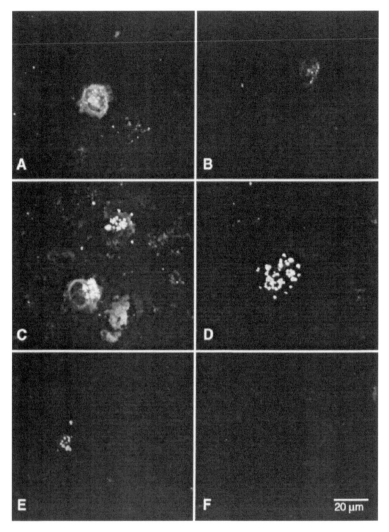

Figure 3 Uptake of PLGA nanoparticles by murine DCs in vivo. DCs were isolated from the draining lymph nodes of mice administered with nanoparticles containing TMR dextran. Cells were examined for double color immunofluorescence after staining with the following FITC-labeled antibodies: (A) DEC-205, (B) DEC-205 isotype control, (C) CD86, (D) CD86 isotype control. Negative controls consisted of unlabeled cells with (E) and without (F) TMR dextran-loaded nanoparticles. (From Ref. 128.)

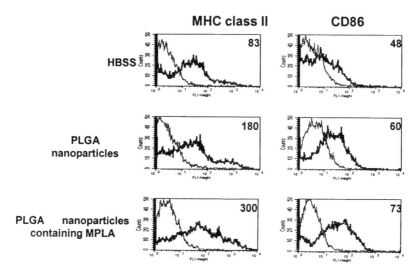

Figure 4   Enhanced expression of MHC class II and CD86 on murine bone-marrow derived DCs after uptake of PLGA nanoparticles. DCs were pulsed with nanoparticles (with or without MPLA) for 24 h and analyzed for expression levels of MHC class II and CD86. Control groups were treated with HBSS. Thin black line in the histogram plot represents the background staining with isotype controls. Soluble TMR-dextran induced no detectable change in the expression of the markers. Numbers in the histogram indicate MFI values for the respective markers. (From Ref. 126.)

immunosuppressive context for capture of tumor antigens. However, immunocompetent DCs may be generated from blood monocytes of cancer patients. They may be loaded with cancer antigens ex vivo in an immunostimulatory milieu and then administered to the cancer patients. Alternatively, such DCs may be used to activate cancer-antigen specific T cells ex vivo, which in turn can be infused into the cancer patients. Another advantage of this method is the generation of a larger number of DCs ex vivo as compared to the small number of DCs accessible in vivo by conventional routes of immunization. DCs loaded with cancer antigens ex vivo are now used in several clinical trials (129,130). Most of the current methods use soluble antigens or transfection approaches for loading DCs.

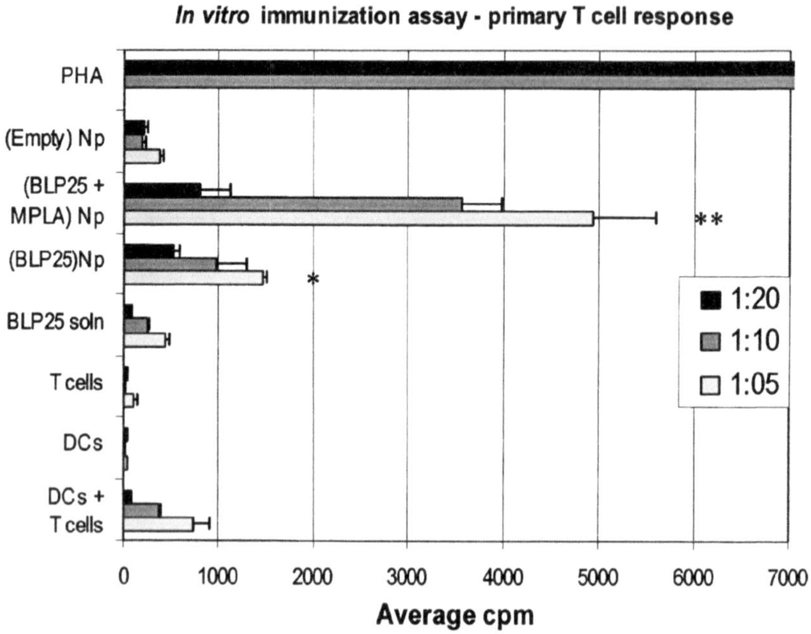

Figure 5 Induction of primary T cell response against an MUC1 lipopeptide (BLP25), following in vitro immunization using human cord blood derived DCs. DCs used were separately pulsed with MUC1 (BLP25) in nanoparticles or in solution prior to co-culture with T cells. Either MPLA or CpG ODN was used as immunomodulator. DC:T cell ratios used were 1:5 to 1:20 where the number of autologous naïve T cells were maintained at 1.5×10^5/well. Each data set comprises of triplicate samples. Standard deviation is shown as error bars. For clarity, student's unpaired t-test analysis is shown only for 1:5 ratios of DC:T cells to compare between the efficiencies of soluble vs. nanoparticulate antigen delivery; $^*P < 0.005$; $^{**}P < 0.0005$. Np = Nanoparticles. (From Ref. 127.)

Do PLGA nanoparticulate formulations of antigens have any advantage over the corresponding soluble formulations in antigen loading of DCs? Our preliminary results indicate that particulate delivery can achieve a 10- to 100-fold higher amount of antigen delivery as compared to soluble form (unpublished results). Particulate delivery also results in sustained antigen release, processing and presentation by DCs (131). The particulate formulations induced about a 1000-fold greater amount of proinflammatory cytokines such as IL-12, TNF-α and IL-6 by the DCs as compared to the soluble formulations containing the same dose of the TLR ligand (unpublished results). DCs loaded with particulate antigens have been more efficient in the activation of primary T cell responses as compared to the soluble antigens. This was demonstrated in both murine and human DC culture systems using a number of antigens including model antigens (tetanus toxoid and ovalbumin peptide $Ova_{323-339}$) and an MUC1 lipopeptide, an important cancer vaccine candidate (126,127). A significant observation in these studies was the efficiency of the delivery system in overcoming self-tolerance against MUC1 in DC/T cell co-cultures of human origin or from MUC1-transgenic mice. In these studies primary antigen-specific T cell activation required co-delivery of the antigen and TLR ligands in the same particles. Our preliminary results indicate the loading of the DCs with "pathogen mimicking" nanoparticles containing both antigen and TLR ligands induce more potent primary T cells responses as compared to DCs loaded with corresponding soluble formulations.

6.3. Immune Responses In Vivo: Th1/Th2 Balance and Anticancer Effects

Numerous studies have demonstrated the immunopotentiating effects of PLGA micro and nanoparticles for proteins and peptides (18). An important question in the context of cancer vaccines is whether the delivery system can be used to manipulate the type and magnitude of immune responses. Of particular interest are the effects of the delivery system on the Th1/Th2 balance and activation of potent CTL responses. We have shown that with co-delivery of the antigen, appropriate TLR ligands (e.g., lipid A analogs and CpG oligonucleotides) using PLGA nanoparticles can be used to bias the immune responses towards a Th1 type. This has been demonstrated for protein and peptide antigens (115,116,132). Even when a particular peptide has a tendency to elicit a Th2 type of immune response, nanoparticulate delivery may be used to elicit a Th1 response. Further, this system may be useful to qualitatively alter an ongoing Th2 response against a peptide into a Th1 response. Therapeutic vaccines for cancer and chronic viral infections are often administered to a patient who already may have an ongoing Th2 response against the vaccine antigen. It is often difficult to elicit a Th1 immune response specific for a peptide epitope, when there is an already established ongoing Th2 response. Using a hepatitis core peptide ($HBcAg_{129-140}$) which typically induces a Th2 response in C57BL/6 strain ($H-2^b$), we have demonstrated that delivery systems such as liposomes and PLGA nanoparticles can switch the response from a Th2 to a Th1 mode (133). In addition, such a Th1 response can be elicited using a booster immunization even after creating an "imprint" of Th2 response with a primary immune response (unpublished results). Whether such effects can be achieved in the context of an ongoing disease such as cancer or chronic hepatitis B infection remains to be seen.

PLGA nanoparticulate formulation of an MUC1 lipopeptide, BLP25, provided about 90% protection against subsequent challenge with MUC1-transfected Lewis lung carcinoma in MUC1-transgenic mice, with long-term survival (manuscript submitted). However therapeutic immunizations of mice with an already established lung metastasis were less effective.

6.4. Dose Sparing of TLR Ligands by Nanoparticle Delivery

Since vaccine delivery systems would protect the encapsulated molecules from the undesired in vivo degradation before their uptake by DCs, they would be expected to reduce the minimal effective dose of antigens as well as the immunomodulatory molecules significantly. Since the tolerated dose range of protein and peptides are usually very large, the dose sparing of antigens may not be a major issue in vaccination. On the other hand, the majority of the immunomodulatory molecules including TLR ligands are not well tolerated in vivo at higher doses, especially if administered repeatedly. Many of these compounds such as lipid A analogs and immunostimulatory oligodeoxynucleotides (ODN) may show serious undesired effects when administered at higher doses. A recent study reported that repeated daily injections of even modest doses of 60 µg of CpG ODN in mice caused splenomegaly within 7 days and drastic damages to lymphoid tissues and hepatic toxicity by day 14 of the treatment (134). Since overcoming tumor-specific immunosuppression mediated by T regulatory cells may require persistent signaling through TLRs (110), approaches to reduction of the effective doses of TLR ligands and their controlled delivery are emerging as important aspects of cancer vaccine delivery. We have recently shown that for an immunostimulatory CpG ODN, the effective dose in mice for antigen-specific T cell activation could be reduced by 10–100-fold by its delivery in PLGA nanoparticles (Fig. 6) (135). In this formulation, the CpG ODN, both CpG and the model antigen tetanus toxoid, were encapsulated within the nanoparticle. These nanoparticles would be expected to "hide" CpG ODN from other cells such as B cells and NK cells, while facilitating greater uptake by DCs than achievable by the soluble formulations. It is worth noting that TLR9 is predominantly localized in the endosomes or phagosomes of DCs rather than their cell surface, and its optimal signaling is dependent on acidic pH (136,137). Since the nanoparticles are internalized into phagosomes where CpG ODN will be released in an acidic environment, this approach of delivery is most suited for potent activity in the target cells. In addition, lactic and glycolic acids, the degradation products of PLGA, will also contribute towards further acidification of the phagosomes, providing the most favored microenvironment of TLR9 signaling.

6.5. Microparticles vs. Nanoparticles: Does Size Matter?

Particulate vaccine delivery systems seek to target DCs making use of their phagocytic properties. Since the upper size limit reported for phagocytosis by macrophages, closely

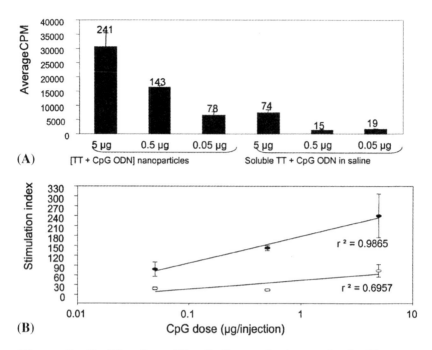

Figure 6 Proliferation of T cells from mice immunized with nano-particulate of soluble formulations in response to ex vivo stimulation with the recall antigen (tetanus toxoid). The T cells were derived from the mice immunized with TT and various doses of CpG ODN. The number of T cells and APCs incubated in the assay plate wells were 5×10^5 and 1×10^6 respectively. The recall proliferation response is shown on the y-axis as (A) CPM; for clarity SI values are also given on the top of each bar; (B) stimulation index for nanoparticulate (solid circles) and soluble (hollow circles) mode of vaccine delivery. Error bars indicate standard deviation. (From Ref. 135.)

related antigen-presenting cells, is in the range of 5–10 μm, it is reasonable to expect that both micro and nanoparticles will be efficiently internalized by DCs. However, a recent study has reported an inverse relationship between particle size and the extent of uptake by DCs (138). In our studies both nano and microparticles (below 5 μm) are taken up by DCs; however antigen loading by nanoparticles (>500 nm), rather than by microparticles (1–2 μm), was better suited for generation of DCs capable of primary activation of T cell responses in vitro (unpublished results). Due to the increased surface area, nanoparticles would be expected to degrade faster and release the antigens in the phagosomes more quickly than the larger microparticles. Although it is difficult to extrapolate these in vitro results to in vivo immunizations, it is worth noting another report showing that PLGA particles below 500 nm were more efficient in CTL activation in vivo against an encapsulated peptide antigen as compared to particles above 2 μm (139). More detailed studies on the effect of particle size, polymer characteristics, and surface properties of the delivery systems on uptake by DCs, antigen release in endosomes, antigen presentation by MHC class I and class II pathways, and maturation and activation of dendritic cells are necessary for the optimal design of effective cancer vaccine delivery systems.

7. CONCLUDING REMARKS

Do cancer vaccines offer a promising new approach that can potentially revolutionize cancer treatment? The answer appears to be a resounding "yes." However major hurdles remain to be overcome before this becomes a successful clinical modality. Several of these issues are related to vaccine delivery. Novel delivery systems that can target cancer antigens and immunomodulatory molecules to subsets of immune cells in a controlled release mode may be able to meet several of these challenges. The PLGA nanoparticulate system may serve as a prototype of such a delivery module.

ACKNOWLEDGMENTS

The authors thank the Canadian Institute of Health Research and Natural Sciences and Engineering Council of Canada for their ongoing support of this research program.

REFERENCES

1. Autran B, Carcelain G, Combadiere B, Debre P. Therapeutic vaccines for chronic infections. Science 2004; 305:205–208.

2. Acres B, Paul S, Haegel-Kronenberger H, Calmels B, Squiban P. Therapeutic cancer vaccines. Curr Opin Mol Ther 2004; 6:40–47.

3. Blattman JN, Greenberg PD. Cancer immunotherapy: A treatment for the masses. Science 2004; 305:200–205.

4. Gilboa E. The promise of cancer vaccines. Nat Rev Cancer 2004; 4:401–411.

5. Rossi G, Ruggiero P, Peppoloni S, Pancotto L, Fortuna D, Lauretti L, Volpini G, Mancianti S, Corazza M, Taccini E, Di Pisa F, Rappuoli R, Del Giudice G. Therapeutic vaccination against helicobacter pylori in the beagle dog experimental model: safety, immunogenicity, and efficacy. Infect Immun 2004; 72:3252–3259.

6. Nuermberger EL, Bishai WR. Antibiotic resistance in streptococcus pneumoniae: What does the future hold? Clin Infect Dis 2004; 38(suppl 4):s363–S371.

7. Heppner FL, Gandy S, McLaurin J. Current concepts and future prospects for alzheimer disease vaccines. Alzheimer Dis Assoc Disord 2004; 18:38–43.

8. Monsonego A, Weiner HL. Immunotherapeutic approaches to alzheimer's disease. Science 2003; 302:834–838.

9. Hackstein H, Morelli AE, Thomson AW. Designer dendritic cells for tolerance induction: guided not misguided missiles. Trends Immunol 2001; 22:437–442.

10. Santiago-Schwarz F. Dendritic cells: friend or foe in autoimmunity? Rheum Dis Clin North Am 2004; 30:115–134.

11. Xiao BG, Huang YM, Link H. Dendritic cell vaccine design: strategies for eliciting peripheral tolerance as therapy of autoimmune diseases. BioDrugs 2003; 17:103–111.

12. Friedman A, al-Sabbagh A, Santos LM, Fishman-Lobell J, Polanski M, Das MP, Khoury SJ, Weiner HL. Oral tolerance: a biologically relevant pathway to generate peripheral

tolerance against external and self antigens. Chem Immunol 1994; 58:259–290.

13. Oluwole SF, Oluwole OO, Adeyeri AO, DePaz HA. New strategies in immune tolerance induction. Cell Biochem Biophys 2004; 40:27–48.

14. Waldmann H, Cobbold S. Exploiting tolerance processes in transplantation. Science 2004; 305:209–212.

15. Steinman RM. The control of immunity and tolerance by dendritic cell. Pathol Biol (Paris) 2003; 51:59–60.

16. Beverley PC. Immunology of vaccination. Br Med Bull 2002; 62:15–28.

17. Schijns VE. Induction and direction of immune responses by vaccine adjuvants. Crit Rev Immunol 2001; 21:75–85.

18. Kersten G, Hirschberg H. Antigen delivery systems. Expert Rev Vaccines 2004; 3:453–462.

19. O'Hagan DT, Singh M, Ulmer JB. Microparticles for the delivery of DNA vaccines. Immunol Rev 2004; 199:191–200.

20. Starks H, Bruhn KW, Shen H, Barry RA, Dubensky TW, Brockstedt D, Hinrichs DJ, Higgins DE, Miller JF, Giedlin M, Bouwer HG. Listeria monocytogenes as a vaccine vector: virulence attenuation or existing antivector immunity does not diminish therapeutic efficacy. J Immunol 2004; 173: 420–427.

21. Horn ME, Pappu KM, Bailey MR, Clough RC, Barker M, Jilka JM, Howard JA, Streatfield SJ. Advantageous features of plant-based systems for the development of hiv vaccines. J Drug Target 2003; 11:539–545.

22. Alpar HO. Strategies for vaccine delivery. J Drug Target 2003; 11:459–461.

23. Morrow WJ, Sheikh NA. Summary: modern vaccine adjuvants and delivery systems meeting, Dublin, 4–6 june 2003. Vaccine 2004; 22:2361.

24. Lenarczyk A, Le TT, Drane D, Malliaros J, Pearse M, Hamilton R, Cox J, Luft T, Gardner J, Suhrbier A. Iscom based vaccines for cancer immunotherapy. Vaccine 2004; 22:963–974.

25. Stayton PS. Delivering the vaccination mail. Trends Biotechnol 2003; 21:465–467.

26. Coulie PG, Hanagiri T, Takenoyama M. From tumor antigens to immunotherapy. Int J Clin Oncol 2001; 6: 163–170.

27. Espinoza-Delgado I. Cancer vaccines. Oncologist 2002; 7: 20–33.

28. Kochenderfer JN, Molldrem JJ. Leukemia vaccines. Curr Oncol Rep 2001; 3:193–200.

29. Emens LA, Jaffee EM. Cancer vaccines: an old idea comes of age. Cancer Biol Ther 2003; 2:S161–S168.

30. Zeng G. Mhc class II-restricted tumor antigens recognized by CD4+ T cells: New strategies for cancer vaccine design. J Immunother 2001; 24:195–204.

31. Inaji H, Komoike Y, Motomura K, Koyama H. Tumor markers in breast cancer. Nippon Rinsho [Jap J Clin Med] 2000; 58:140–145.

32. Parmiani G. Melanoma antigens and their recognition by T cells. Keio J Med 2001; 50:86–90.

33. Van den Eynde BJ, Boon T. Tumor antigens recognized by t lymphocytes. Int J Clin Lab Res 1997; 27:81–86.

34. Robbins PF, Kawakami Y. Human tumor antigens recognized by T cells. Curr Opin Immunol 1996; 8:628–636.

35. Van den Eynde B, Brichard VG. New tumor antigens recognized by T cells. Curr Opin Immunol 1995; 7:674–681.

36. Boon T, Cerottini JC, Van den Eynde B, van der Bruggen P, Van Pel A. Tumor antigens recognized by t lymphocytes. Annu Rev Immunol 1994; 12:337–365.

37. Lewis JD, Reilly BD, Bright RK. Tumor-associated antigens: From discovery to immunity. Int Rev Immunol 2003; 22: 81–112.

38. Mitchell MS. Cancer vaccines, a critical review–part i. Curr Opin Invest Drugs 2002; 3:140–149.

39. Mitchell MS. Cancer vaccines, a critical review–part ii. Curr Opin Invest Drugs 2002; 3:150–158.

40. Stevanovic S. Identification of tumour-associated t-cell epitopes for vaccine development. Nat Rev Cancer 2002; 2:514–520.

41. Parajuli P, Pisarev V, Sublet J, Steffel A, Varney M, Singh R, LaFace D, Talmadge JE. Immunization with wild-type p53 gene sequences coadministered with flt3 ligand induces an antigen-specific type 1 t-cell response. Cancer Res 2001; 61:8227–8234.

42. Matsuda K, Masaki T, Watanabe T, Kitayama J, Nagawa H, Muto T, Ajioka Y. Clinical significance of muc1 and muc2 mucin and p53 protein expression in colorectal carcinoma. Jap J Clin Oncol 2000; 30:89–94.

43. Gjertsen MK, Buanes T, Rosseland AR, Bakka A, Gladhaug I, Soreide O, Eriksen JA, Moller M, Baksaas I, Lothe RA, Saeterdal I, Gaudernack G. Intradermal ras peptide vaccination with granulocyte-macrophage colony-stimulating factor as adjuvant: Clinical and immunological responses in patients with pancreatic adenocarcinoma. Int J Cancer 2001; 92:441–450.

44. Gjertsen MK, Gaudernack G. Mutated ras peptides as vaccines in immunotherapy of cancer. Vox Sang 1998; 74(suppl 2): 489–495.

45. Disis ML, Knutson KL, McNeel DG, Davis D, Schiffman K. Clinical translation of peptide-based vaccine trials: The her-2/neu model. Crit Rev Immunol 2001; 21:263–273.

46. Vlad AM, Kettel JC, Alajez NM, Carlos CA, Finn OJ. Muc1 immunobiology: from discovery to clinical applications. Adv Immunol 2004; 82:249–293.

47. Tatsumi T, Takehara T, Kanto T, Miyagi T, Kuzushita N, Sugimoto Y, Jinushi M, Kasahara A, Sasaki Y, Hori M, Hayashi N. Administration of interleukin-12 enhances the therapeutic efficacy of dendritic cell-based tumor vaccines in mouse hepatocellular carcinoma. Cancer Res 2001; 61: 7563–7567.

48. Akiyama Y, Maruyama K, Nara N, Hojo T, Cheng JY, Mori T, Wiltrout RH, Yamaguchi K. Antitumor effects induced by dendritic cell-based immunotherapy against established pancreatic cancer in hamsters. Cancer Lett 2002; 184:37–47.

49. Acres B, Apostolopoulos V, Balloul JM, Wreschner D, Xing PX, Ali-Hadji D, Bizouarne N, Kieny MP, McKenzie IF. Muc1-specific immune responses in human muc1 transgenic mice immunized with various human muc1 vaccines. Cancer Immunol Immunother 2000; 48:588–594.

50. Saeki A, Nakao K, Nagayama Y, Yanagi K, Matsumoto K, Hayashi T, Ishikawa H, Hamasaki K, Ishii N, Eguchi K. Diverse efficacy of vaccination therapy using the alpha-feto-protein gene against mouse hepatocellular carcinoma. Int J Mol Med 2004; 13:111–116.

51. Yu Z, Restifo NP. Cancer vaccines: progress reveals new complexities. J Clin Invest 2002; 110:289–294.

52. Drake CG, Pardoll DM. Tumor immunology—towards a paradigm of reciprocal research. Semin Cancer Biol 2002; 12:73–80.

53. Bodey B, Bodey B Jr, Siegel SE, Kaiser HE. Failure of cancer vaccines: the significant limitations of this approach to immunotherapy. Anticancer Res 2000; 20:2665–2676.

54. Bitton RJ. Cancer vaccines: a critical review on clinical impact. Curr Opin Mol Ther 2004; 6:17–26.

55. Vicari AP, Caux C, Trinchieri G. Tumour escape from immune surveillance through dendritic cell inactivation. Semin Cancer Biol 2002; 12:33–42.

56. French LE, Tschopp J. Defective death receptor signaling as a cause of tumor immune escape. Semin Cancer Biol 2002; 12:51–55.

57. Antony PA, Restifo NP. Do CD4+ CD25+ immunoregulatory T cells hinder tumor immunotherapy? J Immunother 2002; 25:202–206.

58. Ahmad M, Rees RC, Ali SA. Escape from immunotherapy: possible mechanisms that influence tumor regression/progression. Cancer Immunol Immunother 2004; 53:844–854.

59. Terabe M, Berzofsky JA. Immunoregulatory T cells in tumor immunity. Curr Opin Immunol 2004; 16:157–162.

60. Mapara MY, Sykes M. Tolerance and cancer: Mechanisms of tumor evasion and strategies for breaking tolerance. J Clin Oncol 2004; 22:1136–1151.

61. Strege RJ, Godt C, Stark AM, Hugo HH, Mehdorn HM. Protein expression of fas, fas ligand, Bcl-2 and TGFβ2 and correlation with survival in initial and recurrent human gliomas. J Neurooncol 2004; 67:29–39.

62. Mukherjee P, Ginardi AR, Madsen CS, Tinder TL, Jacobs F, Parker J, Agrawal B, Longenecker BM, Gendler SJ. Muc1-specific ctls are non-functional within a pancreatic tumor microenvironment. Glycoconj J 2001; 18:931–942.

63. Lanzavecchia A, Sallusto F. Regulation of T cell immunity by dendritic cells. Cell 2001; 106:263–266.

64. Banchereau J, Steinman RM. Dendritic cells and the control of immunity. Nature 1998; 392:245–252.

65. Stockwin LH, McGonagle D, Martin IG, Blair GE. Dendritic cells: immunological sentinels with a central role in health and disease. Immunol Cell Biol 2000; 78:91–102.

66. Shortman K, Liu YJ. Mouse and human dendritic cell subtypes. Nat Rev Immunol 2002; 2:151–161.

67. Underhill DM, Ozinsky A. Phagocytosis of microbes: complexity in action. Annu Rev Immunol 2002; 20:825–852.

68. Janeway CA Jr, Medzhitov R. Innate immune recognition. Annu Rev Immunol 2002; 20:197–216.

69. Barton GM, Medzhitov R. Control of adaptive immune responses by toll-like receptors. Curr Opin Immunol 2002; 14:380–383.

70. Alexopoulou L, Holt AC, Medzhitov R, Flavell RA. Recognition of double-stranded rna and activation of nf-kappab by toll-like receptor 3. Nature 2001; 413:732–738.

71. Guermonprez P, Valladeau J, Zitvogel L, Thery C, Amigorena S. Antigen presentation and T cell stimulation by dendritic cells. Annu Rev Immunol 2002; 20:621–667.

72. Steinman RM, Nussenzweig MC. Avoiding horror autotoxicus: The importance of dendritic cells in peripheral T cell tolerance. Proc Natl Acad Sci USA 2002; 99:351–358.

73. Roncarolo MG, Levings MK, Traversari C. Differentiation of t regulatory cells by immature dendritic cells. J Exp Med 2001; 193:F5–F9.

74. Steinman RM, Hawiger D, Nussenzweig MC. Tolerogenic dendritic cells. Annu Rev Immunol 2003; 21:685–711.

75. Niedergang F, Didierlaurent A, Kraehenbuhl JP, Sirard JC. Dendritic cells: the host achille's heel for mucosal pathogens? Trends Microbiol 2004; 12:79–88.

76. Stift A, Friedl J, Dubsky P, Bachleitner-Hofmann T, Schueller G, Zontsich T, Benkoe T, Radelbauer K, Brostjan C, Jakesz R, Gnant M. Dendritic cell-based vaccination in solid cancer. J Clin Oncol 2003; 21:135–142.

77. Zhou Y, Bosch ML, Salgaller ML. Current methods for loading dendritic cells with tumor antigen for the induction of antitumor immunity. J Immunother 2002; 25:289–303.

78. Reinhard G, Marten A, Kiske SM, Feil F, Bieber T, Schmidt-Wolf IG. Generation of dendritic cell-based vaccines for cancer therapy. Br J Cancer 2002; 86:1529–1533.

79. Terando A, Chang AE. Applications of gene transfer to cellular immunotherapy. Surg Oncol Clin North Am 2002; 11:621–643.

80. Morse MA, Lyerly HK. DNA and rna modified dendritic cell vaccines. World J Surg 2002; 26:819–825.

81. Yamanaka R, Zullo SA, Ramsey J, Yajima N, Tsuchiya N, Tanaka R, Blaese M, Xanthopoulos KG. Marked enhancement of antitumor immune responses in mouse brain tumor models by genetically modified dendritic cells producing semliki forest virus-mediated interleukin-12. J Neurosurg 2002; 97:611–618.

82. Shibagaki N, Udey MC. Dendritic cells transduced with protein antigens induce cytotoxic lymphocytes and elicit antitumor immunity. J Immunol 2002; 168:2393–2401.

83. Schott M, Feldkamp J, Klucken M, Kobbe G, Scherbaum WA, Seissler J. Calcitonin-specific antitumor immunity in medullary thyroid carcinoma following dendritic cell vaccination. Cancer Immunol Immunother 2002; 51:663–668.

84. Pecher G, Haring A, Kaiser L, Thiel E. Mucin gene (muc1) transfected dendritic cells as vaccine: results of a phase i/ii

clinical trial. Cancer Immunol Immunother 2002; 51: 669–673.

85. Soares MM, Mehta V, Finn OJ. Three different vaccines based on the 140-amino acid muc1 peptide with seven tandemly repeated tumor-specific epitopes elicit distinct immune effector mechanisms in wild-type versus muc1-transgenic mice with different potential for tumor rejection. J Immunol 2001; 166:6555–6563.

86. Sadanaga N, Nagashima H, Mashino K, Tahara K, Yamaguchi H, Ohta M, Fujie T, Tanaka F, Inoue H, Takesako K, Akiyoshi T, Mori M. Dendritic cell vaccination with mage peptide is a novel therapeutic approach for gastrointestinal carcinomas. Clin Cancer Res 2001; 7:2277–2284.

87. Foged C, Sundblad A, Hovgaard L. Targeting vaccines to dendritic cells. Pharm Res 2002; 19:229–238.

88. Foss FM. Immunologic mechanisms of antitumor activity. Semin Oncol 2002; 29:5–11.

89. Panelli MC, Wang E, Monsurro V, Jin P, Zavaglia K, Smith K, Ngalame Y, Marincola FM. Vaccination with T cell-defined antigens. Expert Opin Biol Ther 2004; 4:697–707.

90. Klebanoff CA, Finkelstein SE, Surman DR, Lichtman MK, Gattinoni L, Theoret MR, Grewal N, Spiess PJ, Antony PA, Palmer DC, Tagaya Y, Rosenberg SA, Waldmann TA, Restifo NP. Il-15 enhances the in vivo antitumor activity of tumor-reactive CD8+ T cells. Proc Natl Acad Sci USA 2004; 101:1969–1974.

91. Kalos M. Tumor antigen-specific T cells and cancer immunotherapy: current issues and future prospects. Vaccine 2003; 21:781–786.

92. Pardoll D. T cells take aim at cancer. Proc Natl Acad Sci USA 2002; 99:15840–15842.

93. Schirrmacher V. T-cell immunity in the induction and maintenance of a tumour dormant state. Semin Cancer Biol 2001; 11:285–295.

94. Kirk CJ, Hartigan-O'Connor D, Nickoloff BJ, Chamberlain JS, Giedlin M, Aukerman L, Mule JJ. T cell-dependent antitumor immunity mediated by secondary lymphoid tissue

chemokine: Augmentation of dendritic cell-based immunotherapy. Cancer Res 2001; 61:2062–2070.

95. Velders MP, Markiewicz MA, Eiben GL, Kast WM. Cd4+ T cell matters in tumor immunity. Int Rev Immunol 2003; 22:113–140.

96. Mosmann TR, Sad S. The expanding universe of t-cell subsets: Th1, th2 and more. Immunol Today 1996; 17:138–146.

97. Chen HW, Huang HI, Lee YP, Chen LL, Liu HK, Cheng ML, Tsai JP, Tao MH, Ting CC. Linkage of CD40l to a self-tumor antigen enhances the antitumor immune responses of dendritic cell-based treatment. Cancer Immunol Immunother 2002; 51:341–348.

98. Bianchi R, Grohmann U, Vacca C, Belladonna ML, Fioretti MC, Puccetti P. Autocrine il-12 is involved in dendritic cell modulation via CD40 ligation. J Immunol 1999; 163: 2517–2521.

99. Moron G, Dadaglio G, Leclerc C. New tools for antigen delivery to the mhc class i pathway. Trends Immunol 2004; 25:92–97.

100. Marrack P, Kappler J. Control of T cell viability. Annu Rev Immunol 2004; 22:765–787.

101. Carrio R, Bathe OF, Malek TR. Initial antigen encounter programs CD8+ T cells competent to develop into memory cells that are activated in an antigen-free, IL-7- and IL-15-rich environment. J Immunol 2004; 172:7315–7323.

102. Wei WZ, Morris GP, Kong YC. Anti-tumor immunity and autoimmunity: a balancing act of regulatory T cells. Cancer Immunol Immunother 2004; 53:73–78.

103. Hori S, Takahashi T, Sakaguchi S. Control of autoimmunity by naturally arising regulatory CD4+ T cells. Adv Immunol 2003; 81:331–371.

104. Takahashi T, Sakaguchi S. Naturally arising cd25+CD4+ regulatory T cells in maintaining immunologic self-tolerance and preventing autoimmune disease. Curr Mol Med 2003; 3:693–706.

105. Wang HY, Lee DA, Peng G, Guo Z, Li Y, Kiniwa Y, Shevach EM, Wang RF. Tumor-specific human CD4+ regulatory T

cells and their ligands: implications for immunotherapy. Immunity 2004; 20:107–118.

106. Wolf AM, Wolf D, Steurer M, Gastl G, Gunsilius E, Grubeck-Loebenstein B. Increase of regulatory T cells in the peripheral blood of cancer patients. Clin Cancer Res 2003; 9:606–612.

107. Dallal RM, Christakos P, Lee K, Egawa S, Son YI, Lotze MT. Paucity of dendritic cells in pancreatic cancer. Surgery 2002; 131:135–138.

108. Monti P, Leone BE, Zerbi A, Balzano G, Cainarca S, Sordi V, Pontillo M, Mercalli A, Di Carlo V, Allavena P, Piemonti L. Tumor-derived muc1 mucins interact with differentiating monocytes and induce IL-10highIL-12low regulatory dendritic cell. J Immunol 2004; 172:7341–7349.

109. Pasare C, Medzhitov R. Toll pathway-dependent blockade of CD4+CD25+ T cell-mediated suppression by dendritic cells. Science 2003; 299:1033–1036.

110. Yang Y, Huang CT, Huang X, Pardoll DM. Persistent toll-like receptor signals are required for reversal of regulatory T cell-mediated CD8 tolerance. Nat Immunol 2004; 5:508–515.

111. Bonnet MC, Tartaglia J, Verdier F, Kourilsky P, Lindberg A, Klein M, Moingeon P. Recombinant viruses as a tool for therapeutic vaccination against human cancers. Immunol Lett 2000; 74:11–25.

112. Cheadle EJ, Jackson AM. Bugs as drugs for cancer. Immunology 2002; 107:10–19.

113. Varde NK, Pack DW. Microspheres for controlled release drug delivery. Expert Opin Biol Ther 2004; 4:35–51.

114. Okada H, Toguchi H. Biodegradable microspheres in drug delivery. Crit Rev Ther Drug Carrier Syst 1995; 12:1–99.

115. Diwan M, Tafaghodi M, Samuel J. Enhancement of immune responses by co-delivery of a cpg oligodeoxynucleotide and tetanus toxoid in biodegradable nanospheres. J Contr Rel 2002; 85:247–262.

116. Newman KD, Samuel J, Kwon G. Ovalbumin peptide encapsulated in poy(d,l-lactic-co-glycolic acid) microspheres

is capable of inducing a T helper type 1 response. J Contr Rel 1998; 54:49–59.

117. Newman KD, Sosnowski DL, Kwon GS, Samuel J. Delivery of muc1 mucin peptide by poly(d,l-lactic-co-glycolic acid) microspheres induces type 1 T helper immune responses. J Pharm Sci 1998; 87:1421–1427.

118. Wang D, Robinson DR, Kwon GS, Samuel J. Encapsulation of plasmid DNA in biodegradable poly(d,l-lactic-co-glycolic acid) microspheres as a novel approach for immunogene delivery. J Contr Rel 1999; 57:9–18.

119. Alonso MJ, Gupta RK, Min C, Siber GR, Langer R. Biodegradable microspheres as controlled-release tetanus toxoid delivery systems. Vaccine 1994; 12:299–306.

120. Hunter SK, Andracki ME, Krieg AM. Biodegradable microspheres containing group b streptococcus vaccine: Immune response in mice. Am J Obstetr Gynecol 2001; 185:1174–1179.

121. Uchida T, Goto S. Oral delivery of poly(lactide-co-glycolide) microspheres containing ovalbumin as vaccine formulation: Particle size study. Biol Pharm Bull 1994; 17:1272–1276.

122. Cleland JL. Single-administration vaccines: Controlled-release technology to mimic repeated immunizations. Trends Biotechnol 1999; 17:25–29.

123. Damge C, Aprahamian M, Marchais H, Benoit JP, Pinget M. Intestinal absorption of plaga microspheres in the rat. J Anat 1996; 189:491–501.

124. Walter E, Dreher D, Kok M, Thiele L, Kiama SG, Gehr P, Merkle HP. Hydrophilic poly(dl-lactide-co-glycolide) microspheres for the delivery of DNA to human-derived macrophages and dendritic cells. J Contr Rel 2001; 76:149–168.

125. Lutsiak ME, Robinson DR, Coester C, Kwon GS, Samuel J. Analysis of poly(d,l-lactic-co-glycolic acid) nanosphere uptake by human dendritic cells and macrophages in vitro. Pharm Res 2002; 19:1480–1487.

126. Elamanchili P, Diwan M, Cao M, Samuel J. Characterization of poly(d,l-lactic-co-glycolic acid) based nanoparticulate

system for enhanced delivery of antigens to dendritic cells. Vaccine 2004; 22:2406–2412.

127. Diwan M, Elamanchili P, Lane H, Gainer A, Samuel J. Biodegradable nanoparticle mediated antigen delivery to human cord blood derived dendritic cells for induction of primary T cell responses. J Drug Target 2003; 11:495–507.

128. Newman KD, Elamanchili P, Kwon GS, Samuel J. Uptake of poly(d,l-lactic-co-glycolic acid) microspheres by antigen-presenting cells in vivo. J Biomed Mater Res 2002; 60:480–486.

129. Cranmer LD, Trevor KT, Hersh EM. Clinical applications of dendritic cell vaccination in the treatment of cancer. Cancer Immunol Immunother 2004; 53:275–306.

130. Cerundolo V, Hermans IF, Salio M. Dendritic cells: a journey from laboratory to clinic. Nat Immunol 2004; 5:7–10.

131. Audran R, Peter K, Dannull J, Men Y, Scandella E, Groettrup M, Gander B, Corradin G. Encapsulation of peptides in biodegradable microspheres prolongs their MHC class-I presentation by dendritic cells and macrophages in vitro. Vaccine 2003; 21:1250–1255.

132. Newman KD, Sosnowski D, Kwon GS, Samuel J. The delivery of muc1 mucin peptide by poly(d,l-lactic-co-glycolic acid) microspheres induces type 1 T helper immune responses. J Pharm Sci 1998; 87:1421–1427.

133. Lutsiak CM, Sosnowski DL, Wishart DS, Kwon GS, Samuel J. Use of a liposome antigen delivery system to alter immune responses in vivo. J Pharm Sci 1998; 87:1428–1432.

134. Heikenwalder M, Polymenidou M, Junt T, Sigurdson C, Wagner H, Akira S, Zinkernagel R, Aguzzi A. Lymphoid follicle destruction and immunosuppression after repeated cpg oligodeoxynucleotide administration. Nat Med 2004; 10:187–192.

135. Diwan M, Elamanchili P, Cao M, Samuel J. Dose sparing of cpg oligodeoxynucleotide vaccine adjuvants by nanoparticle delivery. Curr Drug Del, 2004; 1:405–412.

136. Ahmad-Nejad P, Hacker H, Rutz M, Bauer S, Vabulas RM, Wagner H. Bacterial cpg-DNA and lipopolysaccharides

activate toll-like receptors at distinct cellular compart-ments. Eur J Immunol 2002; 32:1958–1968.

137. Takeshita F, Gursel I, Ishii KJ, Suzuki K, Gursel M, Klinman DM. Signal transduction pathways mediated by the interac-tion of cpg DNA with toll-like receptor 9. Semin Immunol 2004; 16:17–22.

138. Reece JC, Vardaxis NJ, Marshall JA, Crowe SM, Cameron PU. Uptake of HIV and latex particles by fresh and cultured dendritic cells and monocytes. Immunol Cell Biol 2001; 79:255–263.

139. Nixon DF, Hioe C, Chen PD, Bian Z, Kuebler P, Li ML, Qiu H, Li XM, Singh M, Richardson J, McGee P, Zamb T, Koff W, Wang CY, O'Hagan D. Synthetic peptides entrapped in microparticles can elicit cytotoxic T cell activity. Vaccine 1996; 14:1523–1530.

12

Polymer Assemblies

Intelligent Block Copolymer Micelles for the Programmed Delivery of Drugs and Genes

YOUNSOO BAE and KAZUNORI KATAOKA

Department of Materials Science and Engineering,
Graduate School of Engineering,
The University of Tokyo, Tokyo, Japan

1. INTRODUCTION

Since the advent of new concepts for the exploiting of polymer conjugates as drug carriers in the 1970s, polymer science has played a pivotal role in both the design of novel drug carriers for cancer treatment and the enhancement of the efficacy and bioavailability of existing drugs (1–2). The major advantage of these polymer-based carriers is the easy modification (depending on the purpose at hand) of their structural and functional features, an advantage that may overcome the limits of natural carriers, like safety and production costs, by providing the appropriate properties for therapeutics to mass producible polymers and their assemblies (3). For this reason, interest has recently centered on the improved functionalities and the chemical structures of polymer assemblies, advances in which have led to new types of cancer chemotherapy; moreover, the requirements for the development of carriers whose properties are optimized for anticancer drugs, diagnostic agents, and genes are growing day by day. In this chapter, we will take a look at the recent studies that concern this issue and that advance suggestions for future design, development, and clinical applications of polymer assemblies in the delivery of drugs and genes, particularly focusing on one of the unique carriers, the polymeric micelle, that our research group has been investigating for the past decade.

2. RATIONALE FOR DELIVERY SYSTEM USING POLYMER ASSEMBLIES

2.1. Block Copolymers and Polymer Assemblies

Before introducing the recent studies of polymeric delivery systems using polymer assemblies, it might be helpful to review some general concepts about block copolymers and polymer assemblies in order to get a better understanding of their current approaches and achievements in the drug delivery system (DDS) field.

Most anticancer drugs are low molecular weight materials and are injected into the body through a vein. The intravenously injected drugs spread all over the body, and this rapid and non-specific distribution limits the range of their applications in spite of their great efficacy. Moreover, because many anticancer drugs are water insoluble, injection into human beings with an optimized aqueous solution formulation is not simple. In order to clarify these problems, many approaches have been carried out to increase the solubility of hydrophobic drugs by conjugating hydrophobic anticancer drugs to natural or artificial macromolecules; in this area, polymers have become some of the most promising drug carrier formulations (4–7).

A polymer is generally defined as a molecular complex from many repeating units, and the group of repeating units that methodically arrange in a single polymer strand is called a block or a segment. A block copolymer is a polymer that has two or more blocks in the main chain, and is characterized by forming nano- or micro-sized polymer assemblies according to the condition of the molecular affinity interaction affected by solvents, ionic strength, temperature, and so on. The capacity for forming polymer assemblies distinguishes block copolymers from random and graft copolymers. Like most molecules that aggregate or repel each other according to their physicochemical properties, the blocks in a polymer chain also undergo molecular interaction. However, their mobility is restricted for steric reasons, and the block copolymers self-assemble into polymer assemblies through a thermodynamic interaction between blocks, rearranging and forming domains in which blocks with the same physicochemical properties are segregated into the most entropically stabilized state (8).

In addition to such self-assembling features, the easily modifiable chemical structure offering new functionalities is another important advantage for carrier designs. Because block copolymers spontaneously form polymer assemblies, we can (when considering which block would be placed in the domains of the prepared assemblies) introduce functional groups to desired positions (9). Moreover, because modification between the inside and the surface of the assemblies is possible at the same time, the materials to be loaded can be widely selected by optimizing molecular interactions that are important to the forming of assemblies as a driving force based on the hydrophilicity, lipophilicity, and electrostatic ratio. In this regard, the functionalities of the assemblies are controlled by conjugation of those materials that selectively interact with charged groups or specific counterparts on the surface of the assemblies, inducing segregation between the inside and the outside of the assemblies. Therefore, polymer assemblies from block copolymers provide useful tools to design carriers.

2.2. Distribution of Polymer Assemblies in the Body

In order to develop polymeric drug carriers optimized for in vivo use, the fate of polymer assemblies and their constituent polymers in the body should be considered. An understanding of their behaviors in the body and an ability to control their structural and functional features when they are present in the body enable us to design carriers that safely protect materials being delivered from both external environments and the host defense system. This basic knowledge is particularly important for the delivery of materials such as drugs, proteins, genes, and imaging agents to a specific site in our bodies, like solid tumors.

Generally, the behavior of external materials after their injection into the body is influenced by their physicochemical properties. In comparison with low molecular weight materials, polymer assemblies with high molecular weights that hardly penetrate blood vessel walls will be placed in a vascular space after intravenous injection (Fig. 1). If the polymer assemblies are sufficiently water soluble, they move through the blood stream and reach the first gate to pass, the kidneys. The kidneys excrete into the urine external materials that have molecular weights and sizes of less than 50,000 and 6 nm, respectively (10). In addition, because the kidneys' capillaries are negatively charged, polymer assemblies with positive charges become filtered, a change that is evident when these polymer assemblies are compared with materials that have similar or the same molecular weights (11,12). Even though the polymer assemblies pass the kidneys, if they are not water soluble, the adsorption of proteins occurs, and the reticuloendothelial system (RES), a host defense system, becomes activated. The RES is composed of monocytes and macrophages located in the reticular connective tissues, like the liver and spleen, and is responsible for engulfing and removing cellular debris, old cells, and unwelcome external invaders from the bloodstream.

In the meantime, the polymeric assemblies that overcame the in vivo barriers still have fundamental problems

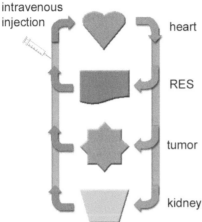

Figure 1 Dispositions of the injected polymer assemblies and their constituent polymers in the body.

related with the amount of used polymers that are needed to maintain the hydrophilicity of the polymer assemblies for clinical use. Although it is inevitable that the amount of polymers increases in line with the increase in the loaded drugs, this may cause potential side effects due to their accumulation in the body, and it becomes particularly serious when repeated injections are required. It is certain that, as certified by the FDA, the polymers used for biomedical science are generally biocompatible and non-toxic. However, it is preferable to reduce the amount of polymers; therefore, polymers after drug delivery should either be designed to be biodegradable or be safely excreted from the body without accumulation after delivery of the drugs and genes.

2.3. Targeting

Polymer assemblies distribute in the body depending on their physicochemical properties. This means that we can guide the polymer assemblies to a specific site in the body by changing their physicochemical properties, and we call this process "targeting."

The concept of targeting is grouped into the two categories of "passive" and "active" targeting. Passive targeting denotes a method by which to exploit the dispositions of the materials that naturally distribute in the body without any artificial operation and design, based on their intrinsic characteristics such as molecular weight, charge, hydrophilicity, and lipophilicity balance. Active targeting is a technique by which to control the natural distribution of materials more precisely through the use of selective interaction between ligands and receptors. This second technique has mostly been utilized in the piloting of materials to both the liver and solid tumors, which contain the corresponding receptors.

One might simply conclude that active targeting is more advanced than passive targeting, but active targeting cannot be realized until the material to be targeted has the capacity to be used for passive targeting. Therefore, even though newly designed polymer assemblies for the delivery of drugs and genes have functionality and target selectivity, their intrinsic characteristics play an important role in determining their distribution in the body; indeed, most existing carriers with the capacity to be used for passive targeting have a structural stability that remains over a long period of time and, for this reason, are not quickly eliminated from the body. In this regard, understanding the correlation between the physico-chemical properties of the assemblies and their behaviors in the body is very important, and the combination of these two features is required for the design of ultimate carriers.

Depending on the targeted level, targeting can be further grouped into three steps. First, carriers are limited in the targeted organs or tissues. Second, carriers are guided to the cell. And third, carriers are aimed to interact with organelles including the mitochondria and nucleus in each cell. When these levels of targeting are chosen, consideration is given to the nature of the materials being delivered. For example, the drugs that have low molecular weight can freely access all organs owing to the high diffusivity, and thereby, they can exert pharmaceutical activity irrespective of the targeted sites, either extracellular or intracellular space. However, in the case of materials like proteins or genes that show the efficacy by interacting with cells or DNA inside the cell nuclei, intracellular targeting is a prerequisite. Consequently, by understanding the targeting related to the organ, we can use cellular and subcellular levels in order to tailor a suitable delivery system based on the in vivo mechanism that the carriers will undergo.

3. BLOCK COPOLYMER MICELLES FOR DRUG DELIVERY

3.1. Polymeric Micelles

We have briefly reviewed polymer assemblies from block copolymers, their biodistribution, and the targeting technique that guides them to a specific site in the body. In this section, we will describe the practical efforts that have been made in the field of recent DDS using polymer assemblies.

As described, block copolymers form well-defined polymer assemblies according to given conditions. Among these assemblies are amphiphilic block copolymers having both hydrophilic and hydrophobic blocks in the same polymer chain, one that is built of spherical polymer assemblies in aqueous solutions, called "polymeric micelles," whose structures are characterized by nanosized and core-shell segregated domains similar to those of viruses in nature (13–17). As viruses protect the nucleic acid core with a protein-coated lipid envelope, polymeric micelles have a hydrophobic inner core that is surrounded by a hydrophilic outer shell, which provides a nano depot for loading materials, instead of viral DNA and RNA, protected by biocompatible shields (Fig. 2). In addition, heterogeneous functionalities can be introduced in each domain by the design of the block copolymers, which broadens the applications of the micelles in line with the changing materials and behaviors in the body, and which makes the polymeric micelles suited for the in vivo delivery of drugs.

In order to prepare the micelles using self-assembling amphiphilic block copolymers, it is very important to determine the hydrophilic/hydrophobic balance. This is because

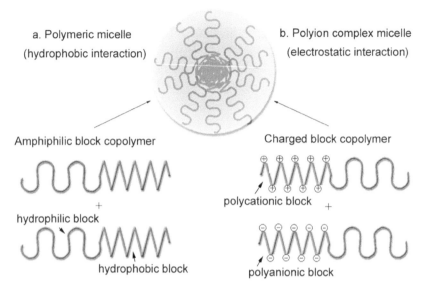

a. Polymeric micelle
(hydrophobic interaction)

b. Polyion complex micelle
(electrostatic interaction)

Amphiphilic block copolymer

Charged block copolymer

hydrophilic block

polycationic block

hydrophobic block

polyanionic block

Figure 2 Micellar polymer assemblies from self-assembling block copolymers: the polymeric micelles (a) and the polyion complex micelles (b).

amphiphilic block copolymers with hydrophilic and hydrophobic blocks are segregated into two phases in aqueous solutions: the hydrophilic blocks face toward the surface of water and the hydrophobic blocks face away from the water, and these rearrangements of the blocks are controlled by the hydrophilic/hydrophobic balance. Even though sedimentation occurs when hydrophilic and hydrophobic blocks exist alone, the amphiphilic block copolymers that bind these heterogeneous blocks together induce the formation of a structure in which the microsegregated hydrophobic domain tethers the hydrophilic domain; in turn, the hydrophilic domain surrounds the hydrophobic domain. This phenomenon depends on the priority of thermodynamic stability between solvents and the hydrophilic/hydrophobic blocks, which makes the structure of the micelles very stable (18). Micelle formation, in fact, was originally observed in low molecular weight surfactants (LMWS), and the typical way to show the stability of LMWS is the critical micelle concentration (CMC), which means the minimum concentration in which the micelle structure can be formed. A low CMC means that the micelles remain stable in aqueous solutions, and the CMC of LMWS is usually 10^{-3}–10^{-4} M. For polymeric micelles, however, the CMC becomes 10^{-6}–10^{-7} M, which means that the polymeric micelles are 1000-fold more stable than LMWS so that the former can maintain micelle formation under the diluted conditions in the blood after injection (19). Indeed, many experimental results have prompted reports that the polymeric micelles from hydrophilic-hydrophobic interactions remain stable in aqueous solutions for days. Another method to prepare the polymeric micelle involves the inducing of strong interactions between the core-forming blocks, such as electrostatic interaction (20), metal complexation (21), cross-linking (22), and crystallization (23). This method is based on the fact that the micelle structure is formed by the microsegregation of hetero-blocks of block copolymers; therefore, micelles are spontaneously formed if a strong interaction occurs between blocks that form stable cores. Among these, the electrostatic interaction is utilized in the forming of carriers that deliver charged materials, such as metal chealates, proteins, and genes, which will be described in detail later (Fig. 2b).

In addition to the stable micelle architecture with its core-shell structures, the small size of the micelles (on the order of nanometers) is another great advantage that it possesses as a drug carrier for clinical applications. The injecting of external materials into the body requires a sterilization process, and nano-sized micelles can be simply sterilized by filtering. Moreover, moving through the blood capillaries, the micelles have access to the entire body owing to their small size, which is particularly valuable for the delivery of drugs to solid tumors. Unlike normal tissues, solid tumors feature large vascular permeability and high interstitial diffusivity; and yet a lack of lymphatic drainage is observed. Therefore, macromolecular carriers, like polymeric micelles, are easily accumulated in a solid tumor, and the accumulated materials can hardly return to the blood stream, a fact that results in tumor-specific accumulation. This phenomenon is called the enhanced permeability and retention (EPR) effect, and has been accepted for passive targeting in many carrier systems (24). In order to achieve the maximum EPR effect, there are a few factors that carriers should clear in order to overcome the host defense systems in our bodies, such as rapid renal clearance and phagocytosis by RES. Moreover, while remaining stable in the blood over a long time period, the carriers should be small enough to pass through the small pores of blood vessels that have a size of less than 400 nm (25). From this point of view, the nano-sized polymeric micelle has structural features that make it the best application in terms of the EPR effect and, thus, in terms of overcoming in vivo barriers.

3.2. Delivery of Hydrophobic Anticancer Drugs

Polymeric micelles, with their various advantages (mentioned above) can be functionalized depending on the purpose or goal at hand, and their application as a carrier for the delivery of anticancer drugs is one such example, having crucial importance. Most anticancer drugs are limited in their clinical applications because of their high toxicity and low solubility in aqueous solutions. The toxicity of anticancer drugs results mainly from their non-specific systemic spread in the body. A significant problem stems from the fact that the range of drug concentrations having pharmaceutical activity without toxicity (a safely injectable effective dose) is extremely narrow. In addition, even though the formulation of drugs was decided in terms of achieving optimal therapeutics, the injecting of drugs into the body is difficult owing to the fact that the solubility of many drugs is too low. Meanwhile, rapid clearance of anticancer drugs from the body results in repeated administerings of the drug in order to maintain an effective concentration of it in blood; these repeated administerings create the potential for either chronic toxicity or the body's acquired resistance to the anticancer drug.

In order to solve these problems and realize a patient-friendly chemotherapy, researchers have developed many intelligent carriers using polymeric micelles (13,14). When designing drug-loaded carriers, we should consider all of the steps that, together, account for the binding, the delivery, and the release of drugs, particularly those drugs whose efficacy remains reliable before and after chemical modification; in addition, a clear pathological mechanism should be selected for the realization of such an appropriate system. Doxorubicin (DOX) has been used in many studies for that reason. Even though its clinical application has been limited owing to its low water-solubility and its side effects such as cardiotoxicity, myelosuppression, and nephrotoxicity, DOX has been utilized because its antitumor activity is effective in the treatment of leukemia, breast carcinoma, and other solid tumors. Moreover, functional groups can be easily introduced to the DOX molecules for chemical modification, but DOX does not lose its pharmaceutical activity as long as such modifications are reversible, and this is a great advantage in the design of drug carriers. In the early carriers were drug-conjugated systems from natural macromolecules like transferrin or from artificial hydrophilic polymers. Even though these systems achieved a reduction in the toxicity of the drugs, most of them ended in failure and frustration because they either failed to clear the host defense systems or deteriorated in the presence of conjugated drugs. In the 1970s, Ringsdorf and coworkers suggested that a DDS based on polymer science and supramolecular chemistry could overcome these limitations (1), and such recent approaches have been experimentally proven to enhance the efficacy of the drugs using a great many novel drug carrier systems. For example, the polymeric micelles from the DOX-conjugated amphiphilic block copolymers stably dispersed a large amount of hydrophobic anticancer drugs without sedimentation in aqueous solutions, showing effective tumor suppressing activity as well as reduced toxicity (26). These results occurred because the micelles can circulate in the blood stream for a long period of time and selectively accumulate in a solid tumor. These were the findings in animal experiments in which a high accumulation of the micelles, after their injection, was noted in solid tumors, and a low accumulation was noted in normal organs (27,28). It should also be noted that, in these experiments, there was a significant difference between drug-polymer conjugates and polymer assemblies from drug-bound block copolymers, which indicates an understanding that the design of polymers, as well as their assemblies, is important and should be considered together.

Preparation of these drug-incorporated micelles began with the synthesis of self-assembling amphiphilic block copolymers, poly(ethylene glycol)-poly(aspartic acid)[PEG-P(Asp)(DOX)] (Fig. 3). For the shell-forming segments, poly(ethylene glycol) (PEG) was used in this system. PEG is one of the widely used biocompatible polymers for in vivo use owing to its high water solubility and chain mobility. PEG is non-toxic and a polymer chain with a molecular weight of less than 30,000 can be cleared from the body through renal filtration (29). Because of the high flexibility and the large exclusion volume in water, PEG forms a shell that renders the micelles sterically stabilized, and this is useful for preventing the adsorption of proteins and the adhesion of cells. PEG shielding was used to minimize the non-specific

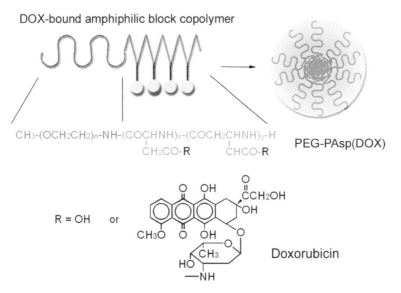

Figure 3 Preparation of the polymeric micelles loading hydrophobic drugs in the core.

interaction between plasma proteins and the surface of the micelles. DOX was covalently conjugated to the P(Asp) segments of the block copolymers, and the drug-bound blocks formed a stable core owing to the π-π interaction between the hydrophobic DOX molecules. In this regard, using carbodiimide compounds, we conjugated DOX to the side chain of the P(Asp) segment of the block copolymer between the carboxylic groups of the P(Asp) segment and the glycosidic primary amino group of the DOX molecule. Approximately 50% of the carboxylic moieties in the P(Asp) segment were conjugated with DOX, making the P(Asp) segment hydrophobic enough to form micelles in aqueous solutions (30–32). The high polymer content causes coagulation or gelation in most drug-polymer conjugates but no such phenomenon was shown in the micelles. On the contrary, the self-association between DOX molecules through the π-π interaction increases the cohesive force within the core, and this increase causes the additional entrapment of DOX molecules in a physical way, with yields improving from 10 to 65%. It must be mentioned that the micelle structure stabilized when the amount of physically entrapped DOX in the core increased. This increase reduced the systemic leakage of DOX and both enhanced DOX accumulation in a solid tumor and lessened toxic side effects from the drugs' non-specific distribution to normal organs. Because the drugs conjugated to polymers, and because the free drugs that were added later shared the same chemical structures, the free drugs that were incorporated later behaved as a filler molecule. These results enhanced the stability of the micelles, preventing dissociation of the micelles in the blood. This system is now under phase II testing and is showing effective activity, even when compared with the free DOX (33).

3.3. Delivery of Metal-Complex Anticancer Drugs

We saw that the matching molecular structures of drugs and polymers are important, and the design of the micelles slightly changes if the drugs are metal complexes. This is because functional groups are required to form a stable core and do so by substituting ligands of metal chealate. The well-known metal complex *cis*-diamminedichloroplatinum(II) (cisplatin, CDDP) is an anticancer agent, but its clinical use is limited owing to its low water solubility and its particular side effects of acute or chronic nephrotoxicity. Because, in a chloride-free condition, other reacting groups can substitute for the chloride ligands of the platinum(II) [Pt(II)] atom of CDDP, carboxylate-containing block copolymers were used to prepare the micelles (34). On the other hand, these carboxylate ligands can be exchanged with chloride ions to regenerate CDDP at physiological salt concentrations owing to their low nucleophilicity.

Block copolymers with carboxyl functional groups were prepared from PEG segments for the hydrophilic shell and polyamino acid blocks for the drug-bound core, which included poly(ethylene glycol)-poly(aspartic acid) [PEG-P(Asp)] and poly(ethylene glycol)-poly(glutamic acid) [PEG-P(Glu)] (Fig. 4). When CDDP was mixed with PEG-P(Asp) in distilled water, the micelles with a polymer-metal complexed core spontaneously formed, having a narrow distribution with a 20 nm diameter. To prepare a stable micelle structure, we determined the critical substitution molar ratio of CDDP to Asp residues in PEG-P(Asp) (CDDP) to be 0.5. Interestingly, the micelles dissociated within 10 h of the induction period as the molar ratio of CDDP to Asp residues in the micelles decreased to the critical value of 0.5 in physiological saline (0.15 M Nacl solution) at 37°C, followed by a sustained release over 50 h. This profile is of great advantage for the delivery of drugs because the micelles, after being administered, need to be stable during the circulation but degradable at the working site so as to release the loaded drugs. Indeed, CDDP-loaded micelles in tumors were revealed, first, to have a 14-fold higher plasma concentration than that of the free CDDP and, second, to reduce CDDP-induced nephrotoxicity without compromising the anticancer cytotoxicity of CDDP. To achieve a more effective tumor accumulation, PEG-P(Glu) was used instead of PEG-P(Asp). In other words, one methylene group at the side chain increased the hydrophobicity of

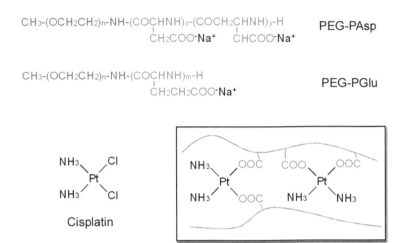

CH₃-(OCH₂CH₂)ₙ-NH-(COCHNH)ₓ-(COCH₂CHNH)ᵧ-H   PEG-PAsp

CH₃-(OCH₂CH₂)ₙ-NH-(COCHNH)ₘ-H   PEG-PGlu

Cisplatin

Figure 4 Interaction between metal-complex anticancer drugs, cisplatin (CDDP), and the core-forming segments of the block copolymers PEG-P(Asp) and PEG-P(Glu). The methane group in the side chain of PEG-P(Glu) was reported to play an important role in significantly increasing the stability of the micelles compared to PEG-P(Asp).

the micelle core. The prepared polymer-metal complex micelles PEG-P(Glu) (CDDP) had a 30 nm diameter, and in physiological saline, PEG-P(Glu) (CDDP) micelles showed a slower release rate (half-value period: >95 h) than the aforementioned PEG-P(Asp) (CDDP) micelles with a 30 h half-life in the blood circulation and a longer induction period (>20 h) for the micelle dissociation than the PEG-P(Asp) (CDDP) micelles (approximately 10 h). In particular, the biodistribution of the micelles were drastically changed. The PEG-P(Glu) (CDDP) micelles maintained a high plasma Pt level (11% at 24 h after injection), that was longer than the level associated with the PEG-P(Asp) (CDDP) micelles (1.5% at the same time period), while accumulations in the liver and spleen remained low. As a result, tumor accumulation of the PEG-P(Glu) (CDDP) micelles eventually reached a 20-fold higher value with respect to the free CDDP, inducing effective antitumor activity (35). Consequently, we can understand that the control of the drug's release and of micelle stability is possible through an optimization of the chemical structure of the block copolymers.

4. BLOCK COPOLYMER MICELLES FOR THE DELIVERY OF GENES AND CHARGED MATERIALS

4.1. Polyion Complex Micelles

The charged materials repulse and coagulate owing to an electrostatic interaction. In particular, the coagulated materials become sedimented by losing their charge, a phenomenon that also occurs in charged polymers and that can be utilized in the preparation of new types of polymeric micelles. Depending on their charge, polymers are called polycations and polyanions with positive and negative charges, respectively. If either the polycations or the polyanions are combined with hydrophilic polymers forming block copolymers, they can disperse without sedimentation (Fig. 2b).

Indeed, block copolymers with the oppositely charged polyelectrolyte segment, poly(ethylene glycol)-poly(L-lysine) [PEG-P(Lys)] and PEG-P(Asp), spontaneously assemble into the core-shell structures whose core is composed of the polyion complex of PEG-P(Lys) and PEG-P(Asp). This phenomenon is due to an electrostatic interaction, which is called a "polyion complex (PIC) micelle" and which thus distinguishes the PIC micelles from the polymeric micelles that utilize a hydrophobic interaction as the driving force for micelle formation (36). In comparison with the ion complexes that form a sediment at a mixing ratio to neutralize charges, the PIC micelles show high water-solubility and stability even in a 50% serum-containing medium without dissociation. The high stability of the micelles is owing to the structure in which the polyion complexed core is surrounded by a PEG shell, which means that the micelle can be used for delivering bioactive peptides and nucleotides in vivo (37). These structures are related to the association number of the block copolymers and the length of the charged segments, and therefore, the size of the core can be controlled by the degree of polymerization of the polycations and polyanion blocks (38). On the contrary, the thickness of the shell does not significantly change. PEG chains can stretch independently of the core size and protect the polyion complexed core from external environments, two features that make the nano-compartment shielded by the PEG shell more stable than the general ion complexes. These simple and well-defined PIC micelle structures are maintained by interactions between the charges and thermodynamical stability, which can be utilized in attempts to control intrinsic properties such as stability, solubility, and affinity in aqueous solutions. Because electrostatic interaction is the driving force behind the preparation of PIC micelles, varying PIC micelles can be prepared not only from charged block copolymers but also from the charged oligopeptides or oligonucleotides by mixing them in aqueous solutions (39,40). Moreover, the PIC micelle core is segregated from external environments, and the nano-system in which various functionalities are implemented can be built up in this regard.

4.2. Protein Delivery

Living creatures maintain life and proliferate by biochemical reactions. Every characteristic that cells show is caused by a biochemical reaction, which is catalyzed by the proteins that function with specific counterparts. For this reason, the characteristics expressed by a specific protein change, and biochemical reactions in the cell can be controlled using this mechanism. These features have encouraged scientists to develop carriers for protein delivery.

Over the years, delivery systems conjugating proteins to macromolecules or pilot molecules have been used to deliver proteins, and these systems have indeed improved the pharmaceutical, pharmacokinetic, and immunological properties of proteins (41). However, a protein is a large biological molecule in which amino acids are arranged in an order determined by genetic codes, and its activity is determined by its three-dimensional structure; therefore, proteins and the chemical structures of the macromolecules undergoing binding should be carefully considered in order to prevent a loss in the activity of the proteins. In this regard, when using charged groups, lipophilic groups, and saccharides for the modification of a specific part of proteins, scientists consider their physicochemical and biological properties, and the PIC micelles achieve protein delivery without chemical modification while safely protecting themselves from the in vivo environment.

Proteins have various sizes with a broad range of pK in different types, and when their net charge becomes zero, the solubility involved with the charge-charge interaction suddenly decreases. This change means that the proper selection of proteins and block copolymers enables the delivery of proteins using PIC micelles. Indeed, a protein-incorporated PIC micelle can be prepared by changing the mixing ratio between the lysozyme and PEG-P(Asp), and the prepared micelles are stable for over a month without sedimentation, and also show stability against the cationic counter charge accompanying the zeta-potential decrease below 10 mV (42). This is compared with ion colloids that have a 30 mV zeta potential, which means that the protein-loaded core is completely protected by the PEG shell when the lysozyme has a high isoelectric point (pI = 11) and is positively charged over a wide pH range.

The PIC micelles exhibit interesting properties against the salt concentration so that scientists can regulate the lytic activity (turn it on and off) by controlling the charge balance. A recent study revealed that lysozyme-incorporated PIC micelles showed no enzymatic activity against cells tested because the PEG shell effectively inhibits the cells from interacting with the lysozyme in the core (43). However, an increase in the ionic strength resulted in a dissociation of the PIC micelles, allowing the lysozyme to be exposed to the environment. This exposure caused the lysozyme to exert its inherent lytic activity against the cells. In the meantime, our research group has found that on-off switching of enzymatic reactivity can be achieved by applying a pulse electric field (44). These results demonstrate that the lysozyme-loaded PIC micelles acted as a nano-sized enzymatic reactor whose activity is controlled by the salt concentration in the environment or a pulse electric field. This reversible reaction in structural change can be the basis of other intelligent carrier systems.

4.3. Gene Delivery

The delivery of genes to a targeted site has an important meaning for cancer treatment and gene therapy. Gene therapy is a new type of therapeutics, the aim of which is to cure disease based on its causes and mechanisms by controlling gene expressions that activate or deactivate protein production. Oligonucleotides, plasmid DNA, and other varying nucleic acid-based drugs are widely used for this purpose (45). However, because genes are unstable against nuclease in the blood and have an anionic nature of repelling cells, the carriers for the safe and precise delivery of these genes to cells become required.

Generally, the carrier that delivers genes in order to induce a gene expression is called a "vector." Vectors are classified as viral and non-viral according to their origins. Viral vectors are considered the most effective system for the delivery of genes, but their clinical use is limited for safety reasons. On the other hand, over the past several decades, non-viral gene vectors, mostly prepared by polymer synthesis, have emerged as a safe and economical gene delivery system. Many synthetic vectors have been developed, and some of them are already commercially available. Nevertheless, many problems still remain to be solved not only because non-viral vectors are inferior to viral vectors in terms of transfection efficiency but also because polymeric non-viral vectors are generally cationic and form ion complexes with anionic genes so that the toxicity of the polymer is occasionally present. In addition, there are several important techniques that are necessary in the improvement of delivery effects; for example, smuggling genes into the cell (based on the organic synthesis of materials such as cationic polymers, cationic lipids, and cationic dendrimers), making the entered genes move smoothly within cytosol, guiding them into the cell's nucleus, and letting them produce proteins by gene expression. In this regard, the PIC micelle and the micelle-like polyplex have been drawing interest as alternative non-viral vectors to viral vectors because they provide a system that can safely deliver charged materials into the body. In particular, PIC micelles can be prepared from various sizes of genes because the driving force forming the gene-incor-porated micelle is an electrostatic interaction, which is a great advantage for the creation of gene carriers. Moreover, PIC micelles resemble viruses in nature owing to their size and functional aspects. A virus has a diameter ranging from 20 to 300 nm and its lipid shielding protects the genes from the outer environment. The micelle shows a particle size from 30 to 200 nm, depending on the loaded materials in its core, which is surrounded by a biocompatible hydrophilic PEG shell.

In preparing gene-incorporated PIC micelles, scientists select genes depending on the overall purpose. The essence of a gene being its DNA, many experimental reports have revealed that DNA dominates the characteristic properties of a living body and becomes a foundation for the synthesis of polypeptides, which modifies DNA and, thus, enables the mass production of a specific protein. Cell functions can be controlled when on (upregulation) and off (downregulation) by methods that modify the DNA's capacity for producing pro-teins. Among them, antisense DNA is oligonucleotide (ODN) with 20–30 bp used for the downregulation of the cell func-tions blocking the mRNA that delivers genetic information about protein synthesis for damaged genes. On the contrary, plasmid DNA from thousands of bp is generally used for gene therapy with the upregulation of cell functions. Actually, plasmid is an expression vector that delivers gene codes for the synthesis of proteins but is usually considered the DNA for gene transfection. In addition to these categories, the therapy that artificially controls the genetic codes or the intracellular signal cascade is called gene therapy. The action that carries antisense DNA, plasmid DNA, and siRNA is generally called "gene delivery."

In order to prepare PIC micelles, polycations are neces-sary because DNA has negative charges on its surface

(Fig. 5). Poly-L-lysine [P(Lys)] is, for this reason, a widely used polycation and provides a useful developmental tool for the development of numerous non-viral vectors that contribute to safer and more effective gene delivery systems. However, as shown in other ion complexes, DNA and the P(Lys) homopolymers aggregate and sediment, thus neutralizing the charge balance. To prevent this, poly(ethylene glycol)-poly(L-lysine) [PEG-P(Lys)] in which a hydrophilic PEG segment is implemented is utilized in the preparation of DNA-incorporated PIC micelles (46). Instead, the PEG-P(Lys) prepared micelles change their shapes depending on the size of the incorporated DNA. With ODN, the PIC micelle forms a spherical shape like that of the oligopeptide and the protein, but a toroidal shape with plasmid DNA. Recently, poly(ethylene glycol)-poly(dimethylaminoethyl methacrylate) [PEG-PAMA] (47), and poly(ethylene glycol)-poly(ethylenimine)[PEG-PEI] (48) have been reported as being effective polycations that form PIC micelles incorporating genes. These genes have been developed so that they give rise to better transfection effects concerning the intracellular behaviors of PIC micelles, as described below.

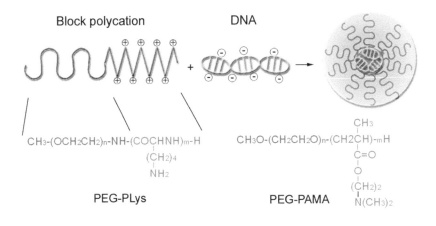

Figure 5 The PIC (polyion complex) micelles from varying block polycations and DNA for delivery of genes.

The structure of PIC micelles can be controlled by charge concentration because it becomes dissociated depending on the outer charges, and this plays an important role in the releasing of genes into cells. In order to transfect the cells, the genes incorporated in the vector should escape from the endosomes in the cell before the lysosomal enzymes bring about decomposition. To this end, a proton sponge effect (PSE) is generally exploited as an easy and definite method to make the vector liberate the incorporated genes in the intracellular compartments, such as endosomes and lysosomes (49). As described, PIC micelles are prepared by an electrostatic interaction and have a stable neutralizing charge balance between the polycations and the genes. When they were put under acidic conditions, however, the polycations became protonated, and the influx of counter ions maintained electroneutrality owing to PSE. In general, endosomes become swelled and burst so that the PIC micelle and its incorporated DNA leave the cytoplasm. This process is called endosomal escape, the effect of which changes depending on the buffering capacity of polycations determined by pKa. In particular, pKa is an index of the acid strengths of the polycations; therefore, pKa values are extremely useful for predicting whether a prepared polycation will be adequate for PSE. In other words, pKa values of polycations play an important role in achieving a better gene transfection owing to an enhanced endosomal escape of the genes. For this reason, polycations with varying pKa values have been synthesized and include poly(dimethylaminoethyl methacrylate) (PAMA) (50), poly,ethyl methacrylate) (PTMAEM) (51), and poly(ethyleneimine) (PEI) (52,53), and among these polycations, PEI is reported to show a significant endosomal escape effect.

4.4. Delivery of Photosensitizers

Among the materials that have charges and need to be targeted to a specific site in the body is a photosensitizer that is drawing interest for applications in photodynamic therapy (PDT). PDT is a new type of cancer therapy that works by transferring energy from a specific wavelength of light that treats cancer with a toxic form of oxygen, singlet oxygen, produced by a photosensitizer. Depending on the light, the photosensitizers remain inactive or create singlet oxygen, which destroys the membrane of cancer cells that take up photosensitizers in a non-invasive way. Therefore, if a system can deliver these photosensitizers to a specific site in the body, it would exert cytotoxic activity against the targets by selectively emitting light while producing the least damage to normal cells.

Porphyrin is a widely used photo-sensitive dye with a heterocyclic aromatic ring made from four pyrrole subunits joined on opposite sides through four methine links. However, the steric characteristics of the benzene rings induce stacking between the porphyrin molecules, thus decreasing their solubility. Consequently, a system that disperses porphyrins and maximizes their photochemical properties has been created, and the porphyrin-dendrimer conjugates constitute one of the most successful systems (54). A dendrimer is a polymer defined by regular, highly branched monomers leading to a monodispersed, tree-like, or generational structure. In order to realize this system, synthesizing monodisperse polymers demands a high level of synthetic control that is achieved through stepwise reactions; that is, the building up of dendrimers, one monomer layer, called "generation," at a time. Each dendrimer consists of a multifunctional core molecule with a dendritic wedge attached to each functional site. The core molecule is referred to as "generation 0." Each successive repeat unit along all branches forms the next generation, "generation 1," "generation 2," and so on until the terminating generation. As generation increases, the efficiency of the delivery of light energy to the inner core also increases, a fact that can be used in the preparation of strong photosensitizer-polymer conjugates that deliver absorbed light energy to isolated photosensitizers (55). In addition, functional groups can be attached to the surface of the dendrimers to produce charges. In spite of these significant functionalities, however, the size of the porphyrin-dendrimer conjugates, less than 5 nm, becomes a problem for their in vivo use. Namely, the conjugates spread non-specifically in the body and are excreted from the body through renal filtration before accessing the specific sites targeted for PDT. In order to overcome this limitation, a PIC micelle that is from porphyrin-dendrimer and block copolymers and that possesses its counter charges to maximize the PDT effect has recently been developed.

Depending on the type of charges, polycationic PEG-P(Lys) and polyanionic PEG-P(Asp) are used for the preparation of such PIC micelles (Fig. 6). Based on recent research, core-shell type PIC micelles with a 52 nm diameter were prepared from a Zn-porphyrin-dendrimer with 32 carboxylate groups on the periphery, 32(−)DPZn, and PEG-P(Lys) block copolymers (56,57). On the contrary, when the dendrimer structure is cationic, a third-generation porphyrin-dendrimer with 32 primary amine groups on the periphery, 32(+)DPZn, and PEG-P(Asp) were used to prepare a PIC micelle, and had a spherical structure with a 55 nm-sized diameter (58). Both

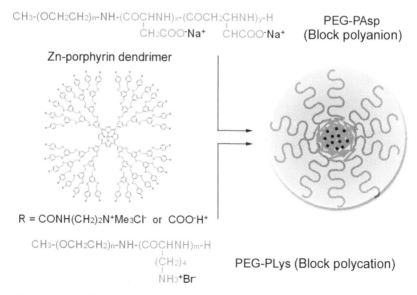

CH$_3$-(OCH$_2$CH$_2$)$_n$-NH-(COCHNH)$_x$-(COCH$_2$CHNH)$_y$-H   PEG-PAsp
                    | |
                 CH$_2$COO$^-$Na$^+$ CHCOO$^-$Na$^+$ (Block polyanion)

Zn-porphyrin dendrimer

R = CONH(CH$_2$)$_2$N$^+$Me$_3$Cl$^-$ or COO$^-$H$^+$

CH$_3$-(OCH$_2$CH$_2$)$_n$-NH-(COCHNH)$_m$-H
                              |
                          (CH$_2$)$_4$ PEG-PLys (Block polycation)
                              |
                          NH$_3$$^+Br^-$

Figure 6 The PIC micelles loading photosensitizers for photodynamic therapy. PEG-P(Asp) (block polyanions) and PEG-P(Lys) (block polycations) are selected according to the surface charge of Zn-porphyrin-dendrimers.

of the polyion dendrimer micelles showed a high stability upon dilution with 150 mM NaCl (the physiological salt condition). This system showed potential as a carrier for light-harvesting ionic Zn-porphyrin-dendrimer photosensitizers. The high salt stability is attributed to the hydrogen-bonding network formed in the micellar core among the amide groups of the dendrimer and the segments of the block copolymers, which could be destroyed by urea, a typical hydrogen-bonding cleaver. Interestingly, compared to the porphyrin-dendrimer, 32(+)DPZn, a relatively low cellular uptake of the PIC micelle was observed, yet the latter exhibited enhanced photodynamic efficacy on the cell line (59). In addition, the PIC micelle did not show any toxicity whereas the cationic porphyrin-dendrimer is toxic, which means that the biocompatible PEG shell completely neutralized the charge.

5. PROGRAMMED DELIVERY USING BLOCK COPOLYMER MICELLES

Up to now, we have reviewed the recent developments of block copolymer micelles, and described some prerequisites for carrier designs. We will see their applications using new types of block copolymer micelles, called intelligent or smart polymeric micelles, in which programmed functionalities are integrated to make the micelle interact with complex in vivo environments in the body. In order to design such polymeric micelles whose structural and functional features are controllable, as we require, a series of behaviors in the body (such as the delivery by blood circulation of materials that target specific sites and release loaded materials) should be considered a single event: we call the delivery systems constructed through this process "controlled delivery" or "programmed delivery." To this end, a great many systems have been developed and designed to optimize their functionalities, depending on the overall purpose, and the polymeric micelles introduced below will exemplify how these applications will be realized by carrier designs and their constituent block copolymers.

5.1. Design of Modified Amphiphilic Block Copolymers

Because the micelles are prepared in aqueous solutions that combine hydrophilic and hydrophobic block copolymers or charged materials and the block copolymers with their counter charges, carrier systems can be modified by designing amphiphilic block copolymers with different structures and functionalities for that purpose. In particular, block copolymer designs are closely related with characteristics such as polymer morphology, the selecting of assemblies' structures, the method for delivery, triggering signals, and so on.

As a hydrophilic block, N-(2-hydroxypropyl) methacrylamide (PHPMA) (including the PEG described above) has recently drawn the attention of researchers (60). The factors that are required for a hydrophilic block include hydrophilicity, flexibility, low toxicity, and low immunogenicity. In addition to these factors, PHPMA has the advantage of modifying the side chains and hydrophilicity. PEG and PHPMA have both advantages and disadvantages, and PEG that is characterized by its properties, such as a large molecular exclusion volume and strong hydration, is widely used as we saw throughout this chapter. Moreover, even the hetero-bifunctional PEG with different functional groups on each chain end can be easily synthesized so that it provides a useful tool for the synthesis of functional block copolymers; and the utility of PEG is still growing (61).

For hydrophobic and charged blocks, biodegradable polyamino acids such as P(Asp), P(Glu) and P(Lys) are generally used. During the synthesis of the polyamino acids, their charged moieties on the side chains are capped by protecting groups, a phenomenon that renders the synthesized polymers hydrophobic. However, the removal of protecting groups allows the polyamino acids to act as polyanions or polycations according to the end functional groups. Moreover, because polyamino acids (owing to amide bonds) are very stable, their side chains can be selectively substituted to the functional groups or linkers, depending on the purpose at hand, while the chemical structures of the polymer backbone are maintained.

As the linkers for the modification of the side chains, the enzymatically degradable polypeptides, the pH-sensitive Schiff base, and glutathion-sensitive disulfide bonds are used, the process of which is described in detail below. Even though other kinds of polymers can be also used for the design of functional block copolymers, it is not easy to satisfy the low toxicity and biocompatibility requirements, and limited polymers are allowed for practical use.

5.2. Surface-Modified Block Copolymer Micelles

In the case of the carriers for cell targeting, the fundamental problems such as intracellular trafficking and the release of incorporated materials are often encountered during the materials' intracellular movement. These problems are particularly crucial to the carriers in terms of the delivery of materials that become pharmaceutically effective after entering cells (Fig. 7).

Because macromolecular carriers normally cannot enter cells by passive diffusion across the plasma membrane, the general mechanism for passing the cell membrane is endocytosis. Endocytosis is a way in which the cells take up large materials like micelles by folding the cell membrane inward,

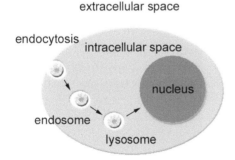

Figure 7 Endocytotic pathway for the intracellular trafficking of drugs and genes. Compared to the condition in extracellular space (pH 7.4), endosomes and lysosomes have acidic environments (pH 5–6), which can induce the pH-triggered cleavage of drug binding linkers or protonation of polycations.

forming small bubbles called endocytotic vesicles. A macromolecule dissolved in extracellular fluid can enter a cell with a medium but at a relatively slow rate. This process is called fluid-phase endocytosis. Macromolecular carriers without any specific affinity for the targeting of cells are considered to be "endocytosed" by this mechanism. If the macromolecule adheres to the cell's surface, it enters the cell through adsorptive endocytosis. During adsorptive endocytosis, macromolecules bound to the plasma membrane are internalized at a rate usually faster than those of fluid-phase endocytosis. The cells "endocytose" cationic macromolecules by this process because cationic materials are adsorptive on the anionic plasma membrane owing to an electrostatic interaction. In the meanwhile, macromolecular carriers for active targeting, such as glycoproteins, hormones, and lectins, are rapidly and effectively internalized via receptor-mediated endocytosis. The rate and extent of the endocytosis of drug-macromolecular carrier complexes are primary factors affecting their therapeutic efficacy. Therefore, surface modification can affect the internalization of the carriers so that a great many approaches can be carried out. After endocytosis, the carriers undergo a further transport process in the cell. Endocytotic vesicles then fuse with an endosome and are ultimately taken to lysosomes where the environment is acidic for digestion, thus generating metabolites used by the cell (pH 5–6). Although the process from endosomes to lysosomes is yet uncertain in the field of biology, this serial mechanism is a widespread process having to do with the transport of extracellular solutes into cells. Consequently, drug targeting that is related to these intracellular environments and material transports should consider the interaction between carriers and the cell membranes.

Recent experimental results using surface-modified polymeric micelles clearly demonstrate that the surface of the carriers plays an important role in these interactions. One of these examples for surface modification involves the binding of ligands on the surface of the micelles in order to internalize them inside the cells through receptor-mediated endocytosis. For this reason, researchers use the shell-forming PEG with acetal groups at the end of the amphiphilic block polymers. It was revealed that the system with a modified hydrophilic surface maintains its structural stability after the surface is functionalized with pilot molecules such as a ligands, peptides, and sugars. Among them, sugars are used owing to their selectivity on the epitopes of specific cells (62).

Asialoglycoprotein receptor-galactose, transferrin receptor-transferrin, folate receptor-folic acid, integrin receptor-RGD, and the triantennary N-glycan ligand (Tri) can be attached to the micelle's surface as ligands, while glycoproteins, transferrin, insulin, and the monoclonal antibodies are used for better recognition as targets. In order to synthesize these block copolymers with acetal groups, NCA is polymerized through ring-opening polymerization using heterobifunctional PEG prepared by initiating the polymerization of ethylene oxide (EO) with a series of potassium alkoxides possessing a protected functional group as a macroinitiator. After polymer synthesis, we can activate the functional groups by deprotecting the acetal groups into aldehydes under acidic conditions. Depending on the material to be loaded, the deprotection process changes in form from either block copolymers or micelles. This synthetic method was used in preparing α-acetal-poly(ethylene glycol)-block–poly(D,L-lactide) (α-acetal-PEG-PDLLA) block copolymers (63,64). With potassium 3,3-diethoxypropanolate as an initiator, EO and D,L-lactide are polymerized in a tandem manner, followed by deprotecting acetal groups into reactive aldehyde groups in order to attach sugars such as galactose, lactose, and mannose that selectively interact with lectin proteins. Galactose- and lactose-installed micelles reacted with RCA-1 whereas the mannose-installed micelles were sensitive to Con A, which was expected to have a wide utility in the field of drug delivery as glyco-receptor-directed carrier systems. Useful in the preparation of other polycations, this synthetic method allows for the surface modification of the PIC micelle gene vectors, which we described above. α-acetal-PEG-block-poly[2-(N,N-dimethylaminoethyl) methacrylate] is one of these examples, which was synthesized via the anion polymerization of 2-(N,N-dimethylaminoethyl) methacrylate

(AMA) in which acetal-PEG-OK was used as an initiator. Cationic PAMA blocks and plasmid DNA formed a PIC micelle with a 100–150 nm diameter. The surface of the micelle was then modified with lactose (65). In the same way, DNA-incorporated PIC micelles with a surface modification were prepared from the acetal-PEG/poly(ethylene imine) block copolymer (acetal-PEG/PEI). Acetal-PEG/PEI is synthesized by the cationic polymerization of 2-methyl-2-oxazoline initiated by acetal-PEG-OSO$_2$CH$_3$, followed by the hydrolysis of the pendant acyl groups (48). The usefulness of the surface modification for active targeting was revealed by animal experiments using polymeric micelles from the block copolymers, α-acetal-PEG-PDLLA, functionalized by attaching small peptidyl ligands such as tyrosine (Tyr) and tyrosyl-glutamic acid (Tyr-Glu). The micelles with Tyr and Tyr-Glu on their surface behave as nanocarriers with neutral and anionic charges, respectively. For the biodistribution of the micelles after injection into mice, both types of micelles exhibited long circulating times in the blood, and 25% of the total injection dose remained in the plasma after 24 h. However, the anionic Tyr-Glu-conjugated PEG-PDLLA micelles showed a lower accumulation in the liver and spleen than the neutral Tyr-conjugated micelles, which elucidated the micelles with an anionic surface that can prevent non-specific organ uptake (66). This research demonstrated that surface modification would drastically change the biodistribution and pharmacokinetic behavior of the particulated systems and play an important role in interacting with peptidyl receptors on the cell membrane.

5.3. Core-Modified Block Copolymer Micelles

We saw that a modification of the shell-forming segments in the block copolymers affects the internalization of the micelles inside the cells. On the other hand, modification of the core-forming segments of the block copolymers induces a change in the structural features of the micelles. Because a stable core is a prerequisite for the micelles' maintaining of their structure, the micelle structure can be dissociated if the core is designed to be time-dependently degradable or environmentally sensitive. For this approach, linkers that respond to a specific signal in the body (which include pH, glutathion, and other enzymes) are selected.

In the design of functional drug carriers, for instance, drugs are coupled via linkers to the functional groups of a macromolecular backbone, and these linkages are cleavable at an appropriate rate. However, when forming micelles, these linkers are protected within the polymer assemblies and become stable until selectively activated at the targeted site. Therefore, the chemical and/or biological stability of the linkage between the drug and the carriers should be considered because their pharmacological effectiveness requires the release of free drugs, pharmacologically active, from the carriers again. In particular, when the linkage is to be cleaved by enzymatic reactions, animal species' differences in relation to the type, and the activity, of the enzymes also need to be carefully considered. Among the functional groups in the drug and the carrier molecules used for chemical coupling are amino, carboxyl, hydroxyl, and free thiol groups. Most of the reactions are performed in aqueous solutions under mild conditions to avoid the denaturation of both the parent drug and the macromolecular carrier.

An example related with this approach is shown in a pH-sensitive polymeric micelle system that releases the incorporated drugs interacting with living cells (67). The micelles were prepared from self-assembling amphiphilic block copolymers, and the drugs were conjugated to the core-forming segments of the block copolymers through acid-sensitive linkers (Fig. 8). This system has the advantage of utilizing all the incorporated drugs without loss, a fact that distinguishes it from the drug-loaded micelles described above, which incorporated drugs as filler molecules for a structure that matches effects between drug molecules. This system is optimized for the intracellular pH that triggers drug release by cutting drug binding linkers; therefore, drug leakage in the circulation of blood can be suppressed so that the toxicity of the drugs is alleviated. As macromolecules normally cannot enter cells by passive diffusion across the plasma membrane

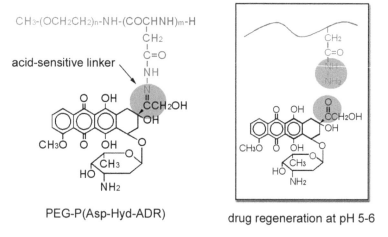

PEG-P(Asp-Hyd-ADR)   drug regeneration at pH 5-6

Figure 8 An example of the intracellular environment-sensitive polymeric micelles. The anticancer drugs known as Adriamycin are clustered to a core-forming segment of the block copolymers through acid-sensitive bonds. Prepared micelles can release the loaded drugs by responding to pH decrease within intracellular compartments such as endosomes and lysosomes.

but by the endocytosis described above, the micelle dissolved in the extracellular fluid enters cells with the medium and moves to endosomes (pH 6) and lysosomes (pH 5) in which the proton concentration is high enough to cleave the acid-sensitive linkers, a process that is followed by a release of drugs in a pharmaceutically active form (68). Although some drug-polymer conjugates exhibit activities in the conjugated form without liberating free drugs, most conjugates exhibit pharmacological efficacy after the release of free drugs at the site of action in the body. Therefore, drug selection, conjugation with polymer chains, assembly structure, and drug release at the working sites should be considered together. The pH-sensitive micelles were fabricated based on the rationale for the preparation of drug-incorporated polymeric micelles from the block copolymer composed of PEG and poly(b-benzyl-L-aspartate). The drug, adriamycin (ADR), was conjugated through acid-labile hydrazone bonds, which became cleaved for releasing the bound drugs as the pH of the environment decreases. This micelle showed the characteristic of delayed cytotoxicity and enhanced therapeutic efficacy owing to site-specific drug release.

By increasing the stability of the micelle core, core modification also improved the gene transfection effect using gene-incorporated PIC micelles. Even though PIC micelles are stable in aqueous solutions, they easily dissociate by releasing incorporated genes as the salt concentration in the environment increases because the driving force for the PIC micelle formation is electrostatic interaction, as we saw above. For this reason, crosslinking and crystallization have recently been drawing the attention of researchers in the field who are concerned with stabilizing the gene-incorporated core of the PIC micelles. Among these approaches is a system stably incorporating genes through disulfide crosslinking between the core-forming segments (22). Disulfide bonds are selectively cleaved by glutathion (GSH) in the cell, which, under physiological conditions, increased the stability of the plasmid DNA. This PIC micelle was prepared from the thiolated PEG-block-poly(L-lysine) (PEG-thioPLL) and antisense oligonucleotide (ODN), and had a 40 nm diameter in aqueous solutions. In contrast to the PIC micelles that dissociated depending on the salt concentration, the PIC micelle system with a disulfide cross-linkage showed stability against the salt concentration, protecting genes safely from nuclease. However, incorporated oligonucleotides were released in an active form when reductive agents like GSH were added, an outcome suggesting the creation of a system that is responsive to the release of incorporated oligonucleotides in the living cells. Furthermore, the glutathion concentration is approximately 3 mM in the cells, which is 300-fold higher than under physiological conditions. What should be of importance in designing crosslinked systems is the fact that incorporated genes cannot be released from the PIC micelle because the core becomes too stable. Recent research data revealed that the DNA-incorporated PIC micelles showed the best transfection effect with a 28% thiolation degree (69). Therefore, the balance between the cationic charge and the disulfide-crosslinking density is important for this system.

Another method for core-stabilization is the core-crystallized micelle system consisting of calcium phosphate, oligonucleotide, and poly(ethylene glycol)-block-poly(aspartic acid) block copolymers [PEG-PAA] (23). This characteristic PIC micelle was prepared from calcium/DNA and phosphate/PEG-PAA. Calcium phosphate crystals were confirmed to stably exist in a 100 nm-sized assembly incorporating DNA inside (70).

6. FUTURE PROSPECTS OF BLOCK COPOLYMER MICELLES

The ultimate goal for intelligent polymeric micelles might be conjectured from the recent collaboration between the National Aeronautics and Space Administration (NASA) and the National Cancer Institute (NCI) (71). NASA and NCI are trying to construct a biodevice that, for prolonged periods of time, can patrol in the body, sense abnormalities like the presence of tumors, diagnose them, and take suitable action to alleviate the problem. Such a device would, in turn, incorporate a process for the integration of all the successful achievements of novel carriers at present.

In this regard, a drug delivery technique using a polymer assembled from functional block copolymers would hold promise for the development of versatile drug carriers implemented with various functions into a single molecular complex. In particular, because segregated core-shell structures can incorporate various kinds of materials like hydrophilic anticancer drugs, peptides, genes, and imaging agents into the inner core while a dense PEG shell, acting as a biocompatible shield, protects the core, it is expected that the polymer assemblies including the micelles will produce new types of cancer therapy in the future. In particular, we saw in this chapter that, by modifying the chemical structures of the block copolymers, the micelles have given rise to four important results: the creation of carriers that exhibit prolonged stability in the blood stream, site-specific targeting with pilot molecules on the surface, sensitivity to external signals, and effective activity. Consequently, polymer assemblies may broaden the modulated delivery of drugs and genes in the field of DDS. The wide breadth of roles in drug delivery is evident, and due to their unique ability to form nanoscopic supramolecular structures, more roles in drug delivery and fields such as tissue regeneration by gene expression will likely emerge. In particular, the keys to the clinical success of targeted DDSs are involved with the chemical and physicochemical considerations associated with the design of drug carriers. Furthermore, it becomes more important to understand the assessment of the pharmacokinetic properties of drug carriers in humans, the elucidation of their intracellular pharmaco-kinetic mechanisms, and the reasons for side effects related to toxicity and antigenicity. Therefore, the development of drug carriers depends on the progress that is being made in a variety of research fields involving biochemistry, immunology, cell/molecular biology, pharmacokinetics, and pharmacology. Of no less importance is the integration of progress that is being made simultaneously in multidisciplinary research, for such integration will enhance applications using polymer assemblies as drug carriers in the DDS field.

REFERENCES

1. Ringsdorf H. Structure and properties of pharmacologically active polymers. J Polym Sci Pol Sym 1975; 51:135–153.

2. Maeda H, Takeshita J, Kanamaru R. A lipophilic derivative of neocarzinostatin. A polymer conjugation of an antitumor protein antibiotic. Int J Pet Protein Res 1979; 14:81–87.

3. Duncan R. The dawning era of polymer therapeutics. Nat Rev Drug Discov 2003; 2:347–360.

4. Putnam D, Kopecek J. Polymer conjugates with anticancer activity. Adv Polym Sci 1995; 122:55–123.

5. Kabanov AV, Alakhov V. Pluronic block copolymers in drug delivery: from micellar nanocontainers to biological response modifiers. Crt Rev Ther Drug Target 2002; 19:1–73.

6. Cammas-Marion S, Okano T, Kataoka K. Functional and site-specific macromolecular micelles as high potential drug carriers. Colloid Surf B 1999; 16:207–215.

7. Torchilin VP. PEG-based micelles as carriers of contrast agents for different imaging modalities. Adv Drug Deliv Rev 1995; 16:295–309.

8. Allen C, Maysinger D, Eisenberg A. Nano-engineering block copolymer aggregates for drug delivery. Colloid Surfaces B 1999; 16:3–27.

9. Scholz C, Iijima M, Nagasaki Y, Kataoka K. A novel reactive polymeric micelle. Polymeric micelle with aldehyde groups on its surface. Macromolecules 1995; 28:7295–7297.

10. Seymour LW, Duncan R, Strohalm J, Kopecek J. Effect of molecular weight (Mw) of N-(2-hydroxypropyl)methacrylamide copolymers on body distribution and rate of excretion after subcutaneous, intraperitoneal, and intravenous administration to rats. J Biomed Mater Res 1987; 21: 1341–1358.

11. Takakura Y, Hashida M. Macromolecular carrier systems for targeted drug delivery: pharmacokinetic considerations on biodistribution. Pharm Res 1996; 13:820–831.

12. Takakura Y, Fujita T, Hashida M, Sezaki H. Distribution characteristics of macromolecules in tumor-bearing mice. Pharm Res 1990; 7:339–346.

13. Lavasanifar A, Samuel J, Kwon G. Poly(ethylene oxide)-block-poly(L-amino acid) micelles for drug delivery. Adv Drug Deliv Rev 2002; 54:169–190.

14. Kataoka K, Harada A, Nagasaki Y. Block copolymer micelles for drug delivery: design, characterization and biological significance. Adv Drug Deliv Rev 2001; 47:113–131.

15. Kwon G, Kataoka K. Block copolymer micelles as long-circulating drug vehicles. Adv Drug Deliv Rev 1995; 16: 295–309.

16. Kataoka K, Kwon G, Yokoyama M, Okano T, Sakurai Y. Block copolymer micelles as vehicles for drug delivery. J Contr Rel 1993; 24:119–132.

17. Kabanov A, Batrakova EV, Melik-Nubarov NS, Fedoseev NA, Dorodnich TY, Alakhov VY, Chekhonin VP, Nazarova IR, Kabanov VA. A new class of drug carriers: micelles of poly-(oxyethylene)-poly(oxypropylene) block copolymers as micro-containers for drug targeting from blood in brain. J Contr Rel 1992; 22:141–158.

18. Webber, SE, Munk P, Tuzar Z, eds. Solvents and Self-Organization of Polymers. Vol. 327. Dordrecht: Kluwer Academic Publishers, 1996.

19. Kwon G, Naito M, Yokoyama M, Okano T, Sakurai Y, Kataoka K. Micelles based on ab block copolymers of poly-(ethylene oxide) and poly(β-benzyl L-aspartate). Langmuir 1993; 9:945–949.

20. Harada A, Kataoka K. Formation of polyion complex micelles in aqueous milieu from a pair of oppositely-charged block copolymers with poly(ethylene glycol) segments. Macromolecules 1995; 28:5294–5299.

21. Nishiyama N, Yokoyama M, Aoyagi T, Okano T, Sakurai Y, Kataoka K. Preparation and characterization of self-assembled polymer-metal complex micelle from cis-dichloro-diammineplatinum (II) and poly(ethylene glycol)-poly(aspartic acid) block copolymer in an aqueous medium. Langmuir 1999; 15:377–383.

22. Kakizawa Y, Harada A, Kataoka K. Environment-sensitive stabilization of core-shell structured polyion complex micelle by reversible cross-linking of the core through disulfide bond. J Am Chem Soc 1999; 121:11247–11248.

23. Kakizawa Y, Kataoka K. Block copolymer self-assembly into monodispersive nanoparticles with hybrid core of antisense DNA and calcium phosphate. Langmuir 2002; 18:4539–4543.

24. Matsumura Y, Maeda H. A new concept for macromolecular therapeutics in cancer chemotherapy: mechanism of tumori-tropic accumulation of proteins and the antitumor agent SMANCS. Cancer Res 1986; 46:6387–6392.

25. Jain RK, Munn LL, Fukumura D. Dissecting tumour patho-physiology using intravital microscopy. Nat Rev Cancer 2002; 2:266–276.

26. Yokoyama M, Okano T, Sakurai Y, Ekimoto H, Shibazaki C, Kataoka K. Toxicity and antitumor activity against solid tumors of micelle-forming polymeric anticancer drug and its extremely long circulation in blood. Cancer Res 1991; 51:3229–3236.

27. Yokoyama M, Okano T, Sakurai Y, Fukushima S, Okamoto K, Kataoka K. Selective delivery of adriamycin to a solid tumor using a polymeric micelle carrier system. J Drug Target 1999; 7:171–186.

28. Kwon G, Suwa S, Yokoyama M, Okano T, Sakurai Y, Kataoka K. Enhanced tumor accumulation and prolonged circulation times of micelle-forming poly(ethylene oxide-aspartate) block copolymers-adriamycin conjugates. J Contr Rel 1994; 29:17–23.

29. Yamaoka T, Tabata Y, Ikada Y. Distribution and tissue uptake of poly(ethylene glycol) with different molecular weights after intravenous administration to mice. J Pharm Sci 1994; 83:601–606.

30. Yokoyama M, Satoh A, Sakurai Y, Okano T, Matsumura Y, Kakizoe T, Kataoka K. Incorporation of water-insoluble anti-cancer drug into polymeric micelles and control of their particle size. J Contr Rel 1998; 55:219–229.

31. Kwon G, Naito M, Yokoyama M, Okano T, Sakurai Y, Kataoka K. Block copolymer micelles for drug delivery: loading and release of doxorubicin. J Contr Rel 1997; 48:195–201.

32. Yokoyama M, Okano T, Sakurai Y, Kataoka K. Improved synthesis of adriamycin-conjugated poly(ethylene glycol)-poly(aspartic acid) block copolymer and formation of unimodal micellar structure with controlled amount of physically entrapped adriamycin. J Contr Rel 1994; 32:269–277.

33. Nakanishi T, Fukushima S, Okamoto K, Suzuki M, Matsumura Y, Yokoyama M, Okano T, Sakurai Y, Kataoka K. Development of the polymer micelle carrier system for doxorubicin. J Contr Rel 2001; 74:295–302.

34. Nishiyama N, Kato Y, Sugiyama Y, Kataoka K. Cisplatin-loaded polymer-metal complex micelle with time-modulated decaying property as a novel drug delivery system. Pharm Res 2001; 18:1035–1041.

35. Nishiyama N, Okazaki S, Cabral H, Miyamoto M, Kato Y, Sugiyama Y, Nishio K, Matsumura Y, Kataoka K. Novel cisplatin-incorporated polymeric micelles can eradicate solid tumors in mice. Cancer Res 2003; 63:8977–8983.

36. Kataoka K, Togawa H, Harada A, Yasugi K, Matsumoto T, Katayose S. Spontaneous formation of polyion complex micelles with narrow distribution from antisense oligonucleotide and cationic block copolymer in physiological saline. Macromolecules 1996; 29:8556–8557.

37. Aoyagi T, Sugi K, Sakurai Y, Okano T, Kataoka K. Peptide drug carrier: studies on incorporation of vasopressin into nano-associates comprising poly(ethylene glycol)-poly(L-aspartic acid) block copolymer. Colloid Surf B 1999; 16: 237–242.

38. Harada A, Kataoka K. Chain length recognition: core-shell supramolecular assembly from oppositely charged block copolymers. Science 1999; 283:65–67.

39. Kabanov AV, Vinogradov SV, Suzdaltseva YG, Alakhov VY. Water-soluble block polycations as carriers for oligonucleotide delivery. Bioconj Chem 1995; 6:369–643.

40. Kataoka K, Togawa H, Harada A, Yasugi K, Matsumoto T, Katayose S. Spontaneous formation of polyion complex micelles with narrow distribution from antisense oligonucleotide and cationic block copolymer in physiological saline. Macromolecules 1996; 29:8556–8557.

41. Weissig V, Whiteman KR, Torchilin VP. Accumulation of protein-loaded long-circulating micelles and liposomes in subcutaneous Lewis lung carcinoma in mice. Pharm Res 1998; 15:1552–1556.

42. Harada A, Kataoka K. Novel polyion complex micelles entrapping enzyme molecules in the core: preparation of narrowly-distributed micelles from lysozyme and poly(ethylene glycol)-poly(aspartic acid) block copolymer in aqueous medium. Macromolecules 1998; 31:4208–4212.

43. Harada A, Kataoka K. On-off control of enzymatic activity synchronizing with reversible formation of supramolecular assembly from enzyme and charged block copolymer. J Am Chem Soc 1999; 121:9241–9242.

44. Harada A, Kataoka K. Switching by pulse electric field of the elevated enzymatic reaction in the core of polyion complex micelles. J Am Chem Soc 2003; 125:15306–15307.

45. Luo D, Saltzman WM. Synthetic DNA delivery systems. Nat Biotechnol 2000; 18:33–37.

46. Katayose S, Kataoka K. Water-soluble polyion complex associates of DNA and poly(ethylene glycol)-poly(L-lysine) block copolymer. Bioconj Chem 1997; 8:702–707.

47. Kataoka K, Harada A, Wakebayashi D, Nagasaki Y. Polyion complex micelles with reactive aldehyde groups on their surface from plasmid DNA and end-functionalized charged block copolymers. Macromolecules 1999; 32:6892–6894.

48. Akiyama Y, Harada A, Nagasaki Y, Kataoka K. Synthesis of poly(ethylene glycol)-block-poly(ethylenimine) possessing an acetal group at the PEG end. Macromolecules 2000; 33: 5841–5845.

49. Behr JP. Synthetic gene transfer vectors. Acc Chem Res 1993; 26:274–278.

50. Laus M, Sparnacci K, Ensoli B, Butto S, Caputo A, Mantovani I, Zuccheri G, Samori B, Tondelli L. Complex associates of plasmid DNA and a novel class of block copolymers with PEG and cationic segments as new vectors for gene delivery. J Biomater Sci Polymer Ed 2001; 12:209–228.

51. Konak C, Mrkvickova L, Nazarova O, Ulbrich K, Seymour LW. Formation of DNA complexes with diblock co-polymers of poly(N-(2-hydroxypropyl)methacrylamide), polycations. Supramol Sci 1998; 5:67–74.

52. Boussif O, Lezoualc'h F, Zanta MA, Mergny MD, Scherman D, Demeneix B, Behr JP. A versatile vector for gene and oligonucleotide transfer into cells in culture and in vivo: polyethyleneimine. Proc Natl Acad Sci 1995; 92:7297–7301.

53. Ogris M, Brunner S, Schuller S, Kircheis R, Wagner E. PEGylated DNA/transferrin-PEI complexes: reduced interaction with blood components, extended circulation in blood and potential for systemic gene delivery. Gene Ther 1999; 6: 595–605.

54. Jian DL, Aida T. Morphology-dependent photochemical events in aryl ether dendrimer porphyrins: cooperation of dendron subunits for singlet energy transduction. J Am Chem Soc 1998; 120:10895–10901.

55. Sadamoto R, Tomioka N, Aida T. Photoinduced electron transfer reaction through dendrimer architecture. J Am Chem Soc 1996; 118:3978–3979.

56. Nishiyama N, Stapert HR, Zhang GD, Takasu D, Jiang DL, Nagano T, Aida T, Kataoka K. Light-harvesting ionic dendrimer porphyrins as new photosensitizers for photodynamic therapy. Bioconj Chem 2003; 14:58–66.

57. Stapert HR, Nishiyama N, Kataoka K, Jiang DL, Aida T. Polyion complex micelles encapsulating light-harvesting dendrimer porphyrins. Langmuir 2000; 16:8182–8188.

58. Zhang GD, Nishiyama N, Harada A, Jiang DL, Aida T, Kataoka K. pH-Sensitive assembly of light-harvesting dendrimer Zinc porphyrin bearing peripheral groups of primary amine with poly(ethylene glycol)-b-poly(aspartic acid) in aqueous solution. Macromolecules 2003; 36:1304–1309.

59. Zhang GD, Harada A, Nishiyama N, Jiang DL, Koyama H, Aida T, Kataoka K. Polyion complex micelles entrapping cationic dendrimer porphyrin: effective photosensitizer for photodynamic therapy of cancer. J Contr Rel 2003; 93:141–150.

60. Kopecek J, Bazilova H. Poly(N-(2-hydroxypropyl)methacrylamide). I. Radical polymerization and copolymerization. Eur Polym J 1973; 9:7–14.

61. Akiyama Y, Otsuka H, Nagasaki Y, Kato M, Kataoka K. Selective synthesis of heterobifunctional poly(ethylene glycol) derivatives containing both mercapto and acetal terminals. Bioconj Chem 2000; 11:947–950.

62. Nagasaki Y, Yasugi K, Yamamoto Y, Harada A, Kataoka K. Sugar-installed block copolymer micelles: their preparation and specific interaction with lectin molecules. Biomacromolecules 2001; 2:1067–1070.

63. Yasugi K, Nakamura T, Nagasaki Y, Kato M, Kataoka K. Sugar-installed polymer micelles: synthesis and micellization of poly(ethylene glycol)-poly(D,L-lactide) block copolymers

having sugar groups at the PEG chain end. Macromolecules 1999; 32:8024–8032.

64. Nagasaki Y, Okada T, Scholz C, Iijima M, Kato M, Kataoka K. The reactive polymeric micelle based on an aldehyde-ended poly(ethylene glycol)-poly(lactide) block copolymer. Macromolecules 1998; 31:1473–1479.

65. Yamamoto Y, Nagasaki Y, Kato M, Kataoka K. Surface modification of poly(ethylene glycol)-poly(D,L-lactide) block copolymer micelle: conjugation of charged peptide. Colloid Surf B 1999; 16:135–146.

66. Yamamoto Y, Nagasaki Y, Kato Y, Sugiyama Y, Kataoka K. Long-circulating poly(ethylene glycol)-poly(D,L-lactide) block copolymer micelles with modulated surface charge. J Contr Rel 2001; 77:27–38.

67. Bae YS, Fukushima S, Harada A, Kataoka, K. Design of environment-sensitive supramolecular assemblies for intracellular drug delivery: polymeric micelles that are responsive to intracellular pH change. Angew Chem Int Edit 2003; 42:4640–4643.

68. Jones AT, Gumbleton M, Duncan R. Understanding endocytic pathways and intracellular trafficking: a prerequisite for effective design of advanced drug delivery systems. Adv Drug Deliv Rev 2003; 55:1353–1357.

69. Miyata K, Kakizawa Y, Nishiyama N, Harada A, Yamasaki Y, Koyama H, Kataoka K. Block catiomer polyplexes with regulated densities of charge and disulfide cross-linking directed to enhance gene expression. J Am Chem Soc 2004; 126: 2355–2361.

70. Kakizawa Y, Miyata K, Furukawa S, Kataoka K. Size-controlled formation of a calcium phosphate-based organic-inorganic hybrid vector for gene delivery using poly(ethylene glycol)-block-poly(aspartic acid). Adv Mater 2004; 16:699–702.

71. Goldin DS, Dahl CA, Olsen KL, Ostrach LH, Klausner RD. The NASA-NCI collaboration on biomolecular sensors. Science 2001; 292:443–444.

13

Polymeric Micelles for the Targeting of Hydrophobic Drugs

MASAYUKI YOKOYAMA

Kanagawa Academy of Science and Technology,
Takatsu-ku, Kawasaki-shi, Kanagawa-ken, Japan

1. POLYMERIC MICELLES FOR DRUG CARRIERS

1.1. What Is a Polymeric Micelle for Drug Targeting?

A polymeric micelle is a macromolecular assembly that forms from block copolymers or graft copolymers, and has a spherical inner core and an outer shell (1). As shown in Fig. 1 in which an AB type block copolymer is used, a micellar structure forms if one segment of the block copolymer can provide enough interchain cohesive interactions in a solvent. Most studies of polymeric micelles both in basic and applied aspects have been done with AB or ABA type block copolymers

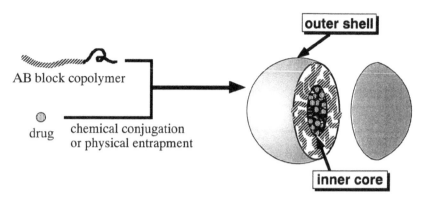

Figure 1   Polymeric micelles for drug carriers.

because the close relationship between micelle forming behavior and the structure of polymers can be evaluated more easily with AB or ABA type block copolymers than with graft or multi-segmented block copolymers.

The cohesive interactions in the inner core utilized as the driving force of micelle formation include hydrophobic, electrostatic, and π-π interactions, as well as hydrogen bonding. Because most drug molecules possess a hydrophobic character, hydrophobic interactions are used most commonly for drug targeting. Actually, most reported examples of polymeric micelle drug carrier systems were done by this combination (2–14) with a few exceptions (15). The second interaction, electrostatic interaction, may be applied to macromolecules with electric charges at a high density. DNA and RNA are appropriate candidates for this because of their anionic phosphodiester bonds that exist per each repeating unit (16). In another chapter, Kataoka describes polymeric micelles containing DNA. Proteins with a large number of charged groups such as aspartic acid and lysine residues are also included for this combination (17). Hydrogen bonding and π-π interactions may work cooperatively with other cohesive interactions. For drugs with aromatic rings, π-π interactions are considered to work cooperatively with hydrophobic interactions.

Drugs can be incorporated into polymeric micelles both by chemical conjugation and physical entrapment. Drugs can be incorporated into both the inner core and the outer shell; however, the inner core is considered an appropriate site for drug incorporation for the following two reasons. The first reason is that possible interactions between the incorporated drug molecules and the outer shell segment of the block copolymer may lead to intermicellar aggregation. Intermicellar aggregation should be avoided when the goal is to achieve polymeric micelle delivery, not the delivery of aggregates. The second reason concerns a shielding function of the outer shell for drug targeting. The minimization of hydrophobic interactions between drug carriers and bio-components such as proteins and cells is an important key to targeting, especially for passive tumor targeting. By incorporating hydrophobic drugs in the inner core, hydro-phobic interactions arising from the incorporated drugs can be effectively inhibited owing to the shielding effects of the hydrophilic outer shell. This subject is explained also in Sec. 1.3.

1.2. The Advantages of Polymeric Micelles for Drug Targeting

Polymeric micelles possess strong and unique advantages as drug carriers. Table 1 summarizes these advantages. The first advantage is their very small size. Polymeric micelles are formed typically in a diameter range from 10 to 100 nm with a substantial narrow distribution. Because it evades the reti-culoendothelial system's uptake and renal excretion, this size range is considered ideal for the attainment of stable, long-term circulation of the carrier system in the blood-stream. Alternatively, the small size of polymeric micelles is

Table 1 Advantages of Polymeric Micelles as Drug Carriers

1. Very small diameter (10–100 nm)
2. High structural stability
3. High water solubility
4. Low toxicity
5. Separated functionality

a big benefit in the sterilization processes associated with pharmaceutical production. Polymeric micelles are easily and inexpensively sterilized by filtration using typical sterilization filters with 0.45 or 0.22 μm pores.

The second advantage that polymeric micelles offer drug carriers is the former's high structural stability. It is known that polymeric micelles possess high structural stability provided by the entanglement of polymer chains in the inner core. This stability has two aspects: static and dynamic (18–21). Static stability is described by a critical micelle concentration (CMC). Generally, polymeric micelles show very low CMC values in a range form 1 to 10 μg/mL. These values are much smaller than typical CMC values of micelles forming from low molecular weight surfactants. The second aspect, dynamic stability, is described by the low dissociation rates of micelles, and this aspect may be more important than the former for in vivo drug delivery in physiological environments that are in non-equilibrium conditions. The high structural stability of polymeric micelles stated above is an important key to in vivo delivery in micellar forms and simultaneously eliminates the possible contribution of single polymer chains to drug delivery.

The third advantage is the high water solubility of the polymeric micelle drug carrier system incorporating hydrophobic drugs. Generally, in conventional polymeric drug carrier systems, a loss of the water solubility of the polymeric carrier resulting from the introduction of a hydrophobic drug creates a serious problem. For polymer-drug conjugates, the conjugation of drugs to a homopolymer easily leads to precipitation because of the high, localized concentration of hydrophobic drug molecules bound along the polymer chain. Several research groups reported this problem of the drug-polymer conjugates in syntheses (22–24) and in their intravenous injections (25). Therefore, conventional drug-polymer conjugates must be designed with a considerably low drug content so that the risk of precipitation is avoided or lessened. Alternatively, polymeric micelles can maintain their water solubility by inhibiting the intermicellar aggregation of the hydrophobic cores by utilizing a hydrophilic outer shell layer that works as a barrier against the intermicellar aggregation. In fact, it was reported that micellar systems exhibited a much larger hydrophobic drug content than did the conventional polymeric carrier systems. For conventional polymer-drug conjugates, the maximum weight of the contents of an anticancer drug like adriamycin (ADR) or daunomycin (an ADR derivative) was reported to range from 10 to 35 wt% (22,23,26,27). A polymeric micelle system was reported to contain 60 wt% ADR (28).

The beneficial character of low toxicity may be described as the fourth advantage. Generally, polymeric surfactants are known to be less toxic than low molecular weight surfactants such as sodium dodecyl sulfate. In fact, some Pluronic block copolymers [poly(propylene oxide)-poly(ethylene oxide)-poly-(propylene oxide)] have been approved for intravenous injection. Further information about Pluronic block copolymers can be obtained in another chapter written by Bronich and Kabanov.

Furthermore, polymeric micelles are considered very safe in relation to chronic toxicity. Possessing a much larger size than critical filtration values in the kidney, polymeric micelles can evade renal filtration, even if the molecular weight of the constituting block copolymer is lower than the critical molecular weight for renal filtration. Additionally, polymeric micelles are formed by intermolecular non-covalent interactions, and therefore all polymer chains can be released (as single polymer chains) from the micelles during a long time-period. This phenomenon results in the complete excretion of the block copolymers from the renal route if the polymer chains are designed with a lower molecular weight than the critical value for the renal filtration. Such a result constitutes an advantage of polymeric micelles over the conventional (non-micelle forming) and non-biodegradable polymeric drug carrier systems.

The fifth advantage is separated functionality. Polymeric micelles are composed of two phases: inner core and outer shell. Various functions required for drug delivery systems can be shared by these structurally separated phases. Each phase can play different roles in drug delivery. As shown in

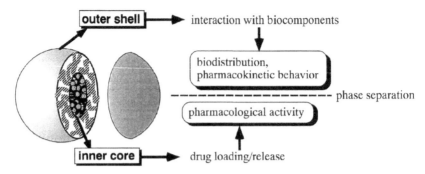

Figure 2 Separated functionality of polymeric micelles as drug carriers.

Figure 2, the outer shell is responsible for interactions with biocomponents such as proteins and cells. These interactions determine pharmacokinetic behavior and the biodistribution of drugs; therefore, the in vivo delivery of drugs may be controlled by the outer shell segment independently of the inner core, which is responsible for pharmacological activities through drug loading and release. This heterogeneous structure is more favorable in the constructing of highly functionalized carrier systems than in the conventional (non-micelle-forming) polymeric carrier systems, since properties of both phases are freely and independently controlled through a selection of the polymer chains that are appropriate for each segment of block copolymers.

1.3. The Purposes of the Incorporation of Drugs into Polymeric Micelles

For what purposes are drugs incorporated into polymeric micelle drug carriers? Table 2 summarizes the several purposes underlying the incorporation of drugs into polymeric micelles. These purposes are drug targeting, the controlled or sustained release of a drug, and the solubilization of hydrophobic or water-insoluble drugs into water or blood.

The first purpose, drug targeting, is defined as selective drug delivery to specific physiological sites—organs, tissues, or cells—where a drug's pharmacological activities are

Table 2   Purposes of Incorporation of Drugs into Polymeric Micelle Carriers

1. Drug targeting
2. Controlled or sustained release of drug
3. Solubilization of drug

required. The drug targeting could be classified into two methods: active and passive targeting (29,30). A discussion of the passive targeting of polymeric micelles will, at the outset, prove useful.

Passive targeting is defined as a method whereby the physical and chemical properties of carrier systems increase the target/non-target ratio of a quantity of a delivered drug. Biodistribution and the pharmacokinetic behaviors of polymeric micelles are determined by the micelles' size and surface properties (e.g., charge, hydrophilicity, or hydrophobicity).

The passive targeting of polymeric micelles to solid tumors can be achieved by the enhanced permeability and retention effect (EPR effect). Maeda and his coworkers presented this new drug targeting strategy in 1986 (31,32). As illustrated in Fig. 3, the vascular permeability of tumor tissues is enhanced by the actions of secreted factors such as kinin. As a result of this increased vascular permeability, macromolecules selectively increase their transport from blood vessels to tumor tissues. Furthermore, the lymphatic drainage system does not operate effectively in tumor tissues.

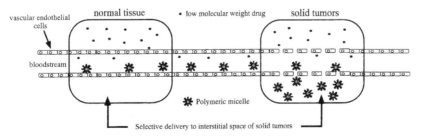

Figure 3   Passive targeting of macromolecules to solid tumors by EPR effect.

Therefore, macromolecules are selectively retained for a prolonged time in the tumor interstitium. In the EPR effect, specific targeting moieties such as antibodies are not necessary. As a result of this effect, macromolecules including polymeric micelles can selectively accumulate at solid tumor sites. However, carrier polymers must fulfill the following two requirements to avoid non-specific capture at non-tumor sites:

1. They must possess an appropriate size or molecular weight. The diameter of carriers must be smaller than ca. 200 nm if the reticuloendothelial system's uptake is to be evaded (33). Additionally, molecular weights larger than a critical value (approximately 40,000) are favorable for evading renal filtration.

2. They must not possess the character of strong interactions or uptake with/by normal organs (especially the reticuloendothelial systems). This character is typically seen for cationic (34) and hydrophobic polymers (35). Therefore, carrier polymers must be hydrophilic, have a neutral or weakly negative charge, and exhibit no other chemical structures that would be biologically recognizable to normal tissues.

Because polymeric micelles are formed in a diameter range from 10 to 100 nm, the size requirement for the EPR effect is inherently fulfilled by the polymeric micelle drug carrier systems. Additionally, the second requirement can be easily fulfilled by the selecting of hydrophilic and neutral/weakly negatively charged polymers for the outer shell forming block.

The other targeting method, active targeting, aims at an increase in the delivery of drugs to the target by utilizing biologically specific interactions such as antigen-antibody binding or by utilizing locally applied signals such as heating and sonication. Carriers classified in this method include specific antibodies, transferrin, and thermo-responsive liposomes and polymeric micelles.

As stated with regard to passive tumor targeting, polymeric micelles can circulate in the bloodstream for a long time-period by evading non-specific capture if the outer shell is hydrophilic and neutral/weakly negatively charged. The polymeric micelle carrier systems with this long-circulating property are an important base for an active targeting system because they either add specific ligands to the micelle's surface or endow it with a stimuli-responsive character. The reasons for this importance are two-fold:

1. The greater part of a living body consists of non-target sites. Even the liver, which is one of the largest targets in the body, occupies only 2% in weight of the whole body. That is, non-target sites in this case account for 98% of the body's weight. Drug carrier systems cannot access the target sites once they are captured by non-target sites. Therefore, the minimization of non-specific capture at non-target sites is important if the quantity of drugs to be delivered using active targeting systems is to be maximized.

2. Passive transfer phenomena precede biologically specific interactions for most active targeting systems. (Exceptions are cases for intravascular targets such as lymphocytes and vascular endothelial cells.) Most targets are located in extravascular space. To reach these targets through the bloodstream, the first step must be translocation through vascular endothelium, followed by permeation through interstitial space. Even for active targeting systems based on cells' biologically specific receptors such as tumor-specific antigens, the passive transendothelial step is both a necessary and anterior one.

The second purpose underlying the incorporation of drugs into polymeric micelles concerns controlled or sustained release. As the hydrophobic inner core works as a drug reservoir, the incorporated drug is released based on physical parameters such as the diffusion and partition coefficient of the drug, hydrophobicity, and the viscosity of the hydrophobic inner core. Up to now, polymeric micelles have been utilized for the purposes of targeting and solubilization only.

However, the controlled release of drugs from polymeric micelles has been potentially and effectively utilized in drug targeting. In most cases, a drug can express its pharmacological activity after its release from the carrier. If the drug is released quickly during circulation in the bloodstream, then the released drug lowers the targeting efficiency of the carrier system. Therefore, release of the drug must be set at a certain rate to maximize targeting efficiency. The optimized drug release rate varies depending on the time periods required for delivery to the target. Drug release must be slower for carrier systems requiring a longer time to reach targets. In the examples of successful drug targeting attempts that are described later, drug release from the micelle was set at an appropriate rate to achieve selective drug release at the target.

The third purpose for a drug's incorporation into micelles is solubilization. The field of chemotherapy has voiced a strong interest in the solubilization of water-insoluble drugs, since many newly developed and very potent drugs such as camptothecin and taxol are water-insoluble or hardly water-soluble. For the intravenous injection of these very hydrophobic or water-insoluble drugs, organic solvents and/or surfactants are used to make these drugs soluble in aqueous media. These substances often show substantial toxic side effects, an outcome that has been exemplified in the use of cremophor EL and ethanol for the solubilization of taxol. Because polymeric micelles are generally much less toxic than these surfactants, a drug's solubilization that is achieved by the drug's incorporation into polymeric micelles can decrease such toxic side effects. Furthermore, drug incorporation can improve injection methods. An anticancer agent KRN-5500 was injected into the central circulation using a catheter in order to evade severe peripheral vascular toxic side effects. These severe vascular side effects were observed in peripheral intravenous injections owing to the toxicity of organic solvents and surfactants used for solubilization. Therefore, incorporation of KRN-5500 into polymeric micelles not only diminishes the toxicity of solvents and surfactants, but also renders a clinically easier injection (peripheral intravenous injection) possible. In the case of the taxol micelle, two further clinical merits should be expected. In the conventional injection of taxol, a patient must be premedicated with an anti-inflammatory drug, and then the injection of the taxol takes place over the course of several hours. However, the polymeric micelle formulation containing taxol may be injected in a short time-period and without any premedication.

Toxic solubilizing substances can be used only for anticancer drugs because toxic side effects can be allowed in anticancer drugs to a much larger degree than in other categories of drugs. This indicates that many water-insoluble drugs cannot be intravenously injected even if they are very potent. Therefore, polymeric micelles may make it possible to inject these potential drugs along an intravenous route that is very useful in terms of quick delivery to blood circulation and easy dose control.

2. THE HISTORY OF POLYMERIC MICELLE DRUG CARRIERS

Studies on polymeric micelles were initiated in the 1960s and focused on their basic physicochemical aspects. The first attempt to utilize polymeric micelles as drug carriers was reported in 1984 by Ringsdorf and colleagues (36,37). A sulfido-derivative of cyclophosphamide (an anticancer drug analog) was conjugated to the lysine residues of a poly(ethylene glycol) (PEG)-poly(L-lysine) block copolymer. Because the sulfido-derivative of cyclophosphamide was hydrophilic, palmitic acid was also introduced as a hydrophobic component to the lysine residues of the block copolymer. The researchers reported a sustained release of the conjugated drug from the block copolymer, and attributed this decrease in the release rate to a depot effect of the hydrophobic micelle's inner core. Although the micelle formation of this polymer-drug conjugate was suggested from the data of dye-solubilization experiments, the formation of the micellar structure was not confirmed by more direct methods such as laser light scattering or gel-permeation chromatography. And although their study was limited to the in vitro stage, Ringsdorf and his colleagues suggested that the polymeric micelle—because it mimicked the functions of natural lipoproteins—could play a role in living bodies as a drug carrier.

Kabanov et al. (38) reported an increase in the activity in vivo of a neuroleptic drug (haloperidol) when it was physically associated with a polymeric amphiphile [Pluronic® P-85: a poly(propylene oxide)-poly(ethylene oxide) block copolymer] that was, itself, coupled to a specific antibody. This is the first example of polymeric micelles enhancing of drug activity in vivo. The reason for this enhancement of drug efficacy may concern either an increase in the amount of the drug delivered to targets due to drug targeting or an increase in permeability (through biological membranes) that the polymeric amphiphile bestows on such a drug. Later, Kabanov and Alakhov showed that the uptake of a model drug by the brain's microvessel endothelial cells increased in the presence of Pluronic P-85 at concentrations lower than its CMC (39). This indicates that Pluronic polymers could increase the permeability of biological membranes by working as a single-chain (non-micelle-forming) surfactant. They also reported that Pluronic polymers could circumvent multi-drug resistance against anticancer drugs in vitro by changing uptake amounts and the subcellular distribution of a drug inside the resistant cells or by inhibiting cellular functions for multi-drug resistance expression (40). This is an innovative application of a synthetic polymer to an anticancer drug therapy—namely, chemotherapy—that is not based on the selective delivery of a drug to the target. Please refer to another chapter written by Bronich and Kabanov concerning Pluronic polymers for drug and gene delivery.

With a clear focus on in vivo selective delivery to a target, Yokoyama and coworkers designed polymeric micelle systems using PEG-poly(amino acid) block copolymers in the late 1980s. They showed the first detailed characterization of the micelle formation of the drug carrier system (41–43), long-circulation in the bloodstream (44), improved in vivo anticancer activity against a murine leukemia, largely enhanced in vivo anticancer activity against solid tumors

(44), and selective delivery to solid tumors (45,46). Details of their examples are explained in Secs. 3 and 4.

In the 1980s, only the three research groups stated above conducted studies of polymeric micelle drug carrier systems. In the 1990s, a growing number of research groups brought about a substantial increase in the field's research activity, giving rise to a significant number of publications (47–56). In some papers (57–59), polymeric micelles that were studied had names like "nanospheres," "nanoparticles," or "nanoparticulates." In the original definition of the term "micelle," a polymeric micelle form exists in an equilibrium with polymer single chains. This means that micelle structures disappear through their repeated exposure to dilution procedures. This behavior is typically observed in micelles forming from low molecular weight surfactants. In contrast, polymeric micelles with a rigid (solid) inner core can keep their structure even after infinite dilution. The preservation of this structural integrity means that the kinetic constant stretching from the micellar structure into the single polymer chain is zero owing to a kinetically frozen inner core (60). These polymeric micelles with a rigid inner core exhibit exactly the same behavior as nanospheres and nanoparticles. Insofar as the two—phase structure of the inner core and the outer shell—are formed from block or graft copolymers, the drug delivery field must classify this structure as a polymeric micelle, just as is done in basic polymer sciences.

3. THE MOLECULAR DESIGN OF POLYMERIC MICELLE DRUG CARRIER SYSTEMS

3.1. Choice of Polymerization for Block Copolymers

Synthetic polymers are obtained by the polymerization reaction of monomers, as shown in Figure 4(a). By sequential propagation reaction, polymers are produced. A monomer (A) reacts with a polymer chain that has a reactive terminal (denoted by A*). Because of this reaction, one extra monomer unit—while possessing the same reactive terminal—adds to a

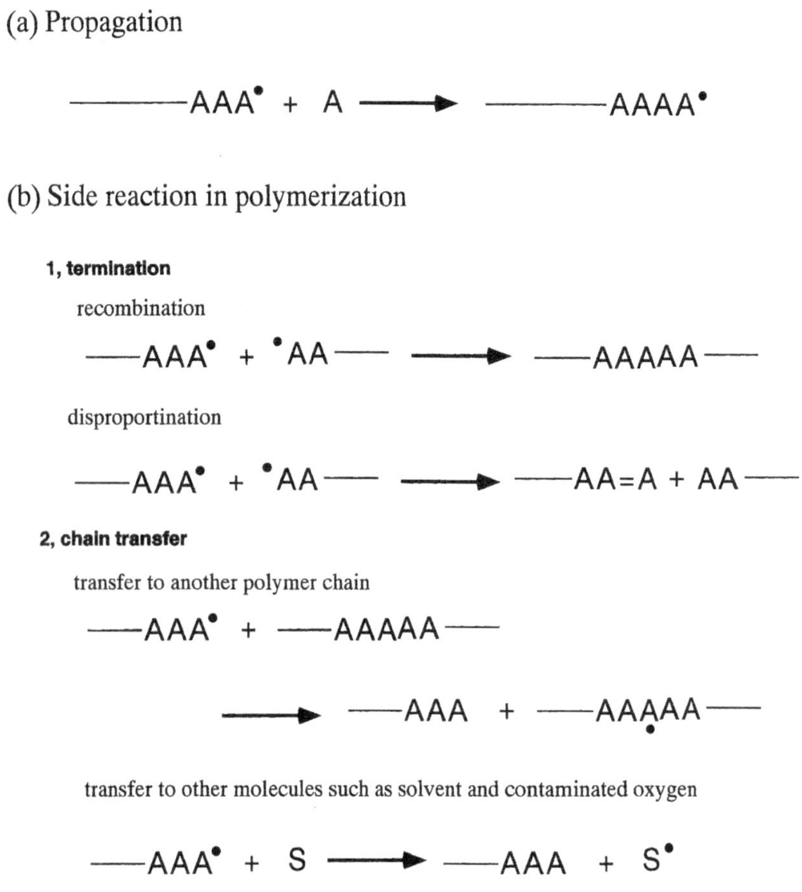

(a) Propagation

$$\text{———AAA}^\bullet + \text{A} \longrightarrow \text{———AAAA}^\bullet$$

(b) Side reaction in polymerization

1, termination

recombination

$$\text{——AAA}^\bullet + {}^\bullet\text{AA——} \longrightarrow \text{——AAAAA——}$$

disproportination

$$\text{——AAA}^\bullet + {}^\bullet\text{AA——} \longrightarrow \text{——AA=A + AA——}$$

2, chain transfer

transfer to another polymer chain

$$\text{——AAA}^\bullet + \text{——AAAAA——}$$

$$\longrightarrow \text{——AAA} + \text{——AA}\underset{\bullet}{\text{A}}\text{AA——}$$

transfer to other molecules such as solvent and contaminated oxygen

$$\text{——AAA}^\bullet + \text{S} \longrightarrow \text{——AAA} + \text{S}^\bullet$$

Figure 4   Reactions included in polymerization.

polymer terminus. In this scheme, block copolymers seem to be easily obtained: the second monomer is put in the reaction mixture after all the first monomers are consumed for polymerization. However, block copolymers cannot be obtained by the most common polymerization method—conventional radical polymerization—owing largely to the presence of side reactions. These side reactions are recombination, disproportionation, and chain transfer, as shown in Figure 4(b). By recombination and disproportionation reactions, polymerization stops. By chain transfer to another polymer chain, branched polymers are produced. By chain transfer to other molecules, molecular weights of an obtained polymer become smaller. As a consequence of these side reactions, reaction mixtures include short polymer chains lacking reactive ends and having branched structures. If the second monomers are added to this reaction mixture, the products contain considerable amounts of poly(A) and poly(B) homopolymers and branched polymers. On the contrary, block copolymers can be prepared by living polymerization, which diminishes the above-mentioned side reactions. In living polymerization, a reactive terminal reacts only with a monomer to generate a new reactive terminal that has one additional monomer unit. (The term "livening polymerization" comes from the fact that reaction terminals are living after all monomer molecules are consumed in polymerization.) Living polymerization is typically seen in anionic polymerization. Even if the side reactions are not completely inhibited, polymerization methods diminishing most side reactions may be classified as living-like polymerization or as polymerization with living character. These polymerization methods are also utilized for block copolymer preparations, as exemplified later.

An important factor relating to the polymerization method is molecular weight distribution. Synthetic polymers must possess distributions of molecular weight owing to the statistical propagation mechanisms in polymer syntheses. Therefore, their molecular weights are shown only in terms of average values. Exceptions are gene-engineered polymers and polymers obtained by step-wise syntheses such as solid phase peptide (or DNA) synthesis and dendrimers. The width of molecular weight distribution is evaluated according to a ratio of weight-average molecular weight (M_w) to number-average molecular weight (M_n). This ratio (M_w/M_n) is larger for polymers that exhibit a wider molecular weight distribution. For macromolecules with a single molecular weight such as a natural protein, this value is exactly 1. For synthetic polymers, this M_w/M_n ratio is significantly dependent on polymerization types. Applicable polymerization types are largely dependent on the chemical structures of monomers. (Polymers are obtained by the polymerization of a number of monomers.) Therefore, the width of molecular weight distribution is considerably determined by monomers.

PEG-[poly(ethylene oxide)] has one of the lowest ratios (M_w/M_n) because this substance is synthesized by anionic polymerization, which can produce polymers that have narrow molecular weight distributions. The ratios of PEG usually range from 1.05 to 1.20. In contrast, conventional radical polymerization, which is the most common way to produce synthetic polymers, produces polymers that have a wide distribution of molecular weight. Typically, polymers obtained in this way possess M_w/M_n ratios over 2.0. It is believed that block copolymers with narrow molecular distributions on each block are preferable for polymeric micelle studies because the close relationships between the molecular architectures of block copolymers and the properties of micelles are easily analyzed using narrowly distributed block copolymers.

In this perspective, polymers obtained by anionic polymerization are favorably used in basic physico-chemical studies of polymeric micelles. However, anionic polymerization can be applied only to vinyl monomers and to cyclic ether monomers when these monomers do not possess other reactive functional groups. Other polymerization methods, which—in terms of both the distribution width and the degree of side reactions—are placed between anionic polymerization and conventional radical polymerization, have been used in block copolymer preparations for polymeric micelle drug carriers. The ring opening polymerization of an amino acid derivative (*N*-carboxy anhydride) has been used for preparations of block copolymers containing poly(amino acid)s. Next to anionic polymerization, this polymerization method can produce polymers that have a narrow weight distribution owing to the low probability of side reactions in the polymerization process. In this case, typical M_w/M_n values range from 1.3 to 1.7. Another polymer frequently used for this purpose is the block copolymers-composed of PEG and poly(lactide). Lactide is a cyclic anhydride. The polymerization of cyclic anhydrides can proceed while giving rise to side reactions that are far less significant than those associated with conventional radical polymerization. This form of polymerization generates polymers with M_w/M_n values ranging from 1.5 to 2.0. Recently, special radical polymerization methods such as RAFT polymerization (61) have been developed and feature substantial reductions in the side reactions that typically correspond to polymerization. By utilizing this technique, scientists can obtain block copolymers for vinyl monomers that contain various functional groups.

3.2. Means of Drug Incorporation: Chemical Conjugation and Physical Entrapment

Drugs are incorporated into the inner core of the-polymeric micelles by means of both chemical conjugation and physical entrapment. Physical entrapment utilizing hydrophobic interactions can be applied to many drugs, since most drug molecules possess a hydrophobic moiety(ies) (even in the case of water-soluble drugs), and since functional groups that are required for chemical conjugation are not necessary for physical entrapment. The optimization of physical entrapment is described in the next section.

For chemical conjugation, drug release by cleavage of a chemical bond is an important issue. One common reaction for this cleavage is hydrolysis. Owing to the phase separation of the inner core from both the outer shell and an outer environment, the drug's access to water molecules, hydrogen ions, hydroxyl ions, and hydrolytic enzymes is considerably inhibited. Therefore, the cleavage rate is expected to be much lower than occurs with conventional polymer-drug conjugates. In turn, micelle structures can be utilized for the quick release of a drug if the hydrophobicity of the bound drug contributes to inner core association for micelle formation. As the drug molecules are released, drug release can be accelerated owing to a decrease in inner core hydrophobicity. From this point of view, Kwon and his colleagues have designed methotrexate-conjugated block copolymer micelles, which are introduced in Sec. 4.6. The other possible strategy for chemical conjugation is an alternative action mechanism for the drugs. In a physically entrapped system, only released drugs are expected to express pharmacological activity, even though the delivery systems change both the whole body's distribution and the intracellular distribution. In the case of chemical conjugation, a drug may exhibit activity in both a released form and a conjugated form. If the conjugated form can express activity, they may overcome the multidrug resistance induced by P-glycoprotein, since the P-glycoprotein is not expected to efflux the polymer-drug conjugates to the cells' exteriors.

3.3. Physical Factors for Targeting

Here, several physical factors of polymeric micelles that determine drug targeting and therapeutic effects of carrier systems are discussed. Table 3 summarizes six physical factors of polymeric micelles.

The chain length of each block is known to be an important factor that determines micelle-forming characters such as CMC. In biological evaluations of the drug delivery of polymeric micelles, the chain length has been found to greatly influence the biodistribution and the antitumor efficacy of ADR-incorporating polymeric micelles. As shown in Fig. 5, concentrations in the bloodstream of micelles forming from PEG-poly(ADR-conjugated aspartic acid) block copolymers were revealed to be completely dependent on their chain length of the block copolymers (62). Compositions of the four polymeric micelles are summarized in Table 4. The molecular weight of the (ethylene glycol) chain varied among 1000, 5000, and 12,000, while three unit numbers of aspartic acid (20, 40, and 80) were used. The micelle formation of these four samples was confirmed by gel-permeation chromatography. A

Table 3 Physical Factors of Polymeric Micelles for Targeting

1. Chain length
2. Drug content
3. Stability
4. Size
5. Secondary aggregation
6. Inner core segment

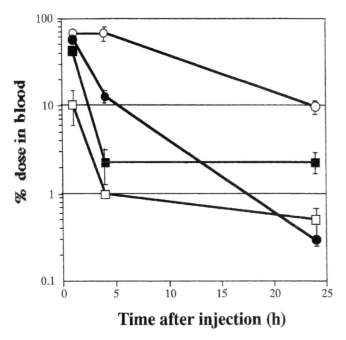

Time after injection (h)

Figure 5 The level of polymeric micelles in blood after intra-venous injection. ○: 5–20, ●: 12–20, □: 5–80, and ■: 1–40. (From Ref. 62.)

sample coded 12–20 was observed circulating in the blood-stream very stably; a dose of over 10% was still circulating 24 h post intravenous injection. During the first 4 h post-intravenous injection, a sample coded 5–20 followed the

Table 4 Compositions of ADR-Conjugated Polymeric Micelles

| Composition | Molecular weight | | ADR substitution (%)[a] | Diameter (nm) |
	PEG	P(Asp)		
1–40	1,000	4,800	12	ND[b]
5–20	5,000	2,100	30	30
5–80	5,000	8,700	46	36
12–20	12,000	2,100	104	40

[a]With respect to Asp residue.
[b]ND: not detected above 10 nm.

long-circulating character of sample 12–20. In contrast, the blood concentrations corresponding to the 5–80 and the 1–40 micelles quickly decreased after the injection. Antitumor activity was also revealed to be dependent on the chain length, as shown in Figure 6 (63). Four samples (5-09, 12–21, 12–41, and 12–81) largely enhanced antitumor activity by incorporating drugs into micelles. The enhanced antitumor activity was shown by greater tumor growth inhibition than occurred with free ADR at the maximum tolerated dose. These four polymeric micelles provide 100% long-term survivals at their maximum tolerated doses. Among these four samples, 12–21 exhibited the highest antitumor activity, since only this sample exhibited long-term survival results at the 100% level in the second-highest dose (the dose one step below the maximum tolerated). In contrast, one sample (1–48)

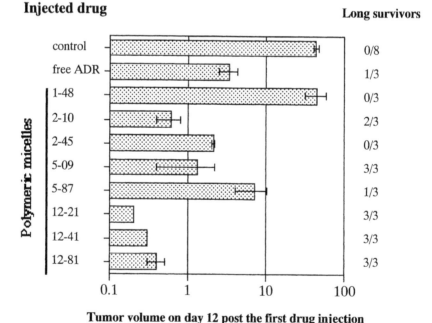

Figure 6 In vivo antitumor activity of polymeric micelles and free ADR at their maximum tolerated doses against C 26 murine tumor. (Modified from Ref. 63.)

showed lower antitumor activity than free ADR, and the other three samples (2–10, 2–45, and 5–87) showed antitumor activity that was either the same as or higher than free ADR. All these results indicate both that the incorporation of a drug into polymeric micelle is not enough to obtain the enhanced efficacy of the drug, and that the optimization of the block chain length is important.

The second physical factor, drug content, has been known to be influential on the targeting efficacy of the ADR-incorporating polymeric micelles (76). Figure 7 compares the in vivo antitumor activities of two polymeric micelle samples composed of identical chain lengths of both the PEG and the poly(aspartic acid) chains. The amounts of chemically conjugated and physically entrapped ADR differed between the two samples, as summarized in Table 5. The antitumor activities of these two runs are shown in relation to the dose of the physically entrapped amount, since the physically entrapped ADR was revealed to be responsible for in vivo antitumor activity. Run 1 exhibited substantial activity, which can be seen in the significant decreases in tumor volumes in all three doses. Run 2 did not show any antitumor

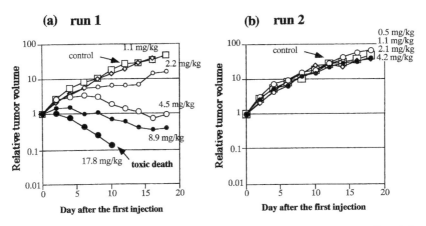

Figure 7 In vivo antitumor activity of two polymeric micelles differing in amounts of incorporated adriamycin. (a) run 1 and (b) run 2 of Table 5. Dose is shown by the physically encapsulated ADR. (From Ref. 76.)

Table 5 Compositions of ADR-Containing Polymeric Micelles

Molecular weight			ADR content[a]	
Run	PEG	P(Asp)	Conjugated ADR %	Conjugated ADR %
1	12,000	2,100	28.5	6.2
2	12,000	2,100	21.5	2.0

[a]Wt% in micelle.

activity. This difference is very stark when the two runs are compared with each other in terms of identical doses. Run 1 showed a considerable reduction in tumor volume at doses of 2.2 and 4.5 mg/kg, while run 2 did not at doses of 2.1 and 4.2 mg/kg.

As stated above, two physical factors, chain length and drug content, are shown to influence the efficacy of a drug. Then, how do these two factors in micelle composition affect drug efficacy? It has been postulated that an optimization of the above-mentioned two factors would give the polymeric micelles satisfactory stability for long-term circulation in the blood. Micelle stability is listed as the third physical factor in Table 3. Figure 8 shows plots of in vitro micelle stability (evaluated by gel-permeation chromatography) vs average tumor volume on day 12 of the post-drug injection (64). Smaller tumor volumes mean stronger antitumor activity. In this figure, a discreet tendency can be observed. Polymeric micelles with higher in vitro stability brought about higher in vivo antitumor activity. Furthermore, Figure 9 shows the gel-filtration chromatographs run 1 and run 2 in Figure 7 (74). Both of the runs eluted at much earlier volumes than did their corresponding molecular weights (20,000 for run 1 and 18,000 for run 2) indicating elution as polymeric micelles. There was, however, a little difference in the exact elution volume. Run 1 eluted at 4.4 mL, and its peak shape was very sharp. Run 2 had a slightly delayed elution, measuring 4.8 mL and exhibiting a wider peak shape. This indicates that the run 2 micelle was delayed in its elution by hydrophobic interactions with gels in the column. It was thought that

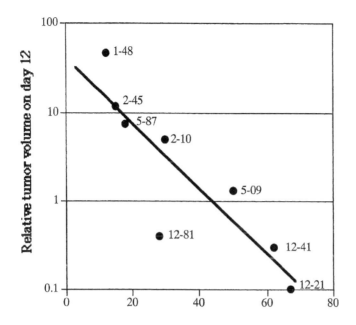

**Relative peak area of polymeric micelle peak
after incubation in serum**

Figure 8 Correlation between in vitro micelle stability and in vivo antitumor activity. In vitro micelle stability is plotted by area of micelle peak in gel-filtration chromatogram after incubation in serum. In vivo antitumor activity is shown by relative C26 tumor volume on day 12. Volume 1 means the volume at the first drug injection. Codes of plots are from Figure 6. (Modified from Ref. 64.)

run 2, owing to the incorporation of a small amount of hydrophobic ADR molecules (which, in turn, loosened the packing of the hydrophobic inner core), gave rise to these hydrophobic interactions.

What does the stability of polymeric micelles mean? The stability is classified according to two aspects. The first aspect is how stably polymeric micelle carriers circulate in the bloodstream. The stability of the other aspect is how slowly a physically entrapped drug is released from the carrier. The data shown in Figures 6 and 7 do not specify which aspect was essential. Further studies are required to answer this

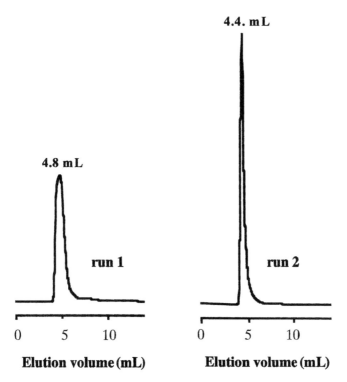

Figure 9 Gel-filtration chromatograms of polymeric micelles of run 1 and run 2 from Fig. 7. (From Ref. 76.)

question if the strategic molecular design of polymeric micelle drug carrier systems is to be established.

The fourth physical factor concerns the size of polymeric micelles. As explained before, polymeric micelles form along a narrow range of diameter; from 10 to 100 nm. Within this diameter range, no data have been reported about the effects of size on targeting or antitumor activity, since not only the size of polymeric micelles but also their stability undergo change in the presence of both block copolymer composition and drug content. It would be very interesting and valuable to know the effect of size on selective delivery in a polymeric micelle that ranges in diameter from 10 to 100 nm.

The formation of a polymeric micelle structure having a single inner core is sometimes accompanied by secondary

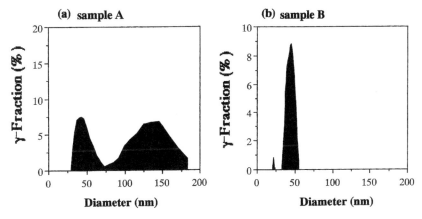

Figure 10 Diameter distribution measured by dynamic light scattering. (From Ref. 74.)

aggregates. These are aggregates of several micelles that range in diameter from 100 to 300 nm. As shown in Fig. 10, the formation of these aggregates was dependent on the composition of polymeric micelles. Sample B with a longer PEG chain was free of these secondary aggregates, whereas sample A was revealed to contain considerable proportions of the aggregates. It is considered that polymeric micelle formation without these aggregates is preferable because the diameter of aggregates may exceed 200 nm, which is a limit on size for non-selective capture by the reticuloendothelial system.

The last factor of polymeric micelles is the hydrophobic inner core segment. The hydrophobicity and the rigidity of the inner core segment are considered to determine drug incorporation stability and micelle stability. Polymeric micelles forming from Pluronic block copolymers can incorporate ADR into their inner core (65), but they instantly release the incorporated ADR upon dilution, since equilibrium shifts quickly and considerably from the incorporated status to the released status owing to the liquid and the low hydrophobic properties of the inner core, respectively. Therefore, Pluronic micelles cannot target ADR. In contrast, polymeric micelles with ADR-conjugated poly(aspartic acid) inner cores were found to maintain both drug incorporation and micelle

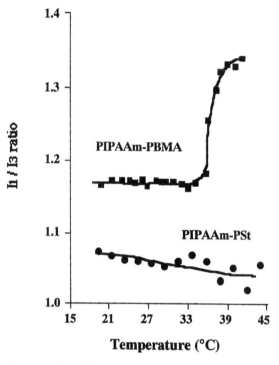

Figure 11   Measurements of hydrophobicity of micelle inner core using pyrene as a probe. The ratio (I_1/I_3) of two vibrational bands is smaller, hydrophobicity of the inner core is higher. (From Ref. 71.)

formation upon dilution and resulted in the drug's targeting of solid tumors. Another example of the effects of the hydrophobic inner core segment is shown in Fig. 11 (71). These are the hydrophobicity measurements of the inner core that were taken with a fluorescence probe pyrene. The intensity ratio (I_1/I_3) is smaller when the pyrene is in a more hydrophobic atmosphere. Figure 11 shows the change of the I_1/I_3 ratio of the P(IPAAm)-*b*-P(butyl methacrylate) micelle exposed to a heating process. The micelles' inner cores were more hydrophilic above the transition temperature. This change was considered to result from a mixing of the inner core with the dehydrated outer shell or from the invasion of water molecules into the inner core. The second possible cause came about when the aggregation of the dehydrated outer shell segment induced a mechanical distortion. It was observed that the drug was observed to be released more rapidly from a less hydrophobic polymer inner core above the LCST.

On the other hand, the fluorescence probe did not show any change for polymeric micelles forming from poly(IPAAm) - polystyrene block copolymers, as shown in the lower plots of Fig. 11. Turbidity measurements confirmed that this micelle having the polystyrene inner core underwent the phase transition at 32°C. Possible reasons for the difference between this micelle and the previous case are strong hydrophobicity and/or the solid character of the polystyrene core. In fact, its lower I_1/I_3 ratio revealed the polystyrene inner core to be more hydrophobic than the PBMA inner core. Additionally, fluorescence measurements of incorporated 1,3-*bis*(l-pyrenyl) propane that were used as a flexibility/rigidity marker revealed the polystyrene inner core to be more rigid than the P(butyl methacrylate) inner core. Although it is not known which physical property (hydrophobic degree or rigidity) contributed—upon the phase transition—to the change of the inner core, the choice of the inner core's chemical structure was revealed to be very important for drug release and thermo-responsive characters.

3.4. Additional Features

In the last three sections, discussion focused chiefly on passive drug targeting and its relation to polymeric micelles that have hydrophilic and inert outer shells such as PEG. Here, some additional functions of these micelles are discussed.

The first function is ligand attachment to the outer shell. In most AB-type block copolymer syntheses that have the PEG chain for polymeric micelles, a chemically unreactive functional group such as methoxy is used as the PEG terminal. Scholz et al. reported a synthesis of aldehyde group-terminated (on the PEG side) PEG-PLA block copolymers (66). By utilizing this aldehyde group, they succeeded in preparing polymeric micelles that had targeting ligands, such as antibodies on their surface. Such micelles actively target tumors and do so because of the ligands.

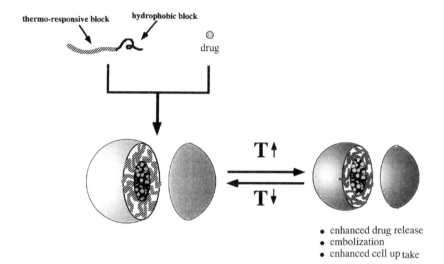

Figure 12   Thermo-responsive polymeric micelle.

On the other hand, Okano and coworkers have studied thermo-responsive polymeric micelles using poly(*N*-isopropylacrylamide) [P(IPAAm)] as a thermo-responsive and outer shell-forming block (67–72). As shown in Figure 12, below the LCST, micelles with the hydrated outer shell and the hydrophobic inner core are formed. Upon heating above the LCST, the outer shell shrinks and becomes hydrophobic. Micelle diameter decreases owing to the shrunk outer shell. Intermicellar aggregates may form dependent on a concentration of micelles and on the strength of the hydrophobic interactions of the shrunk outer shell layers. Drug release can be enhanced as the temperature rises. For hydrophobic polymer chains, which form the hydrophobic inner cores, poly(D,L-lactide) (PLA) and poly(butyl methacrylate) (PBLA) were used.

4. EXAMPLES OF DRUG TARGETING WITH POLYMERIC MICELLE CARRIERS

4.1. Adriamycin (Doxorubicin)

Yokoyama and colleagues succeeded in getting an anticancer drug, ADR (=doxorubicin), with a polymeric micelle system,

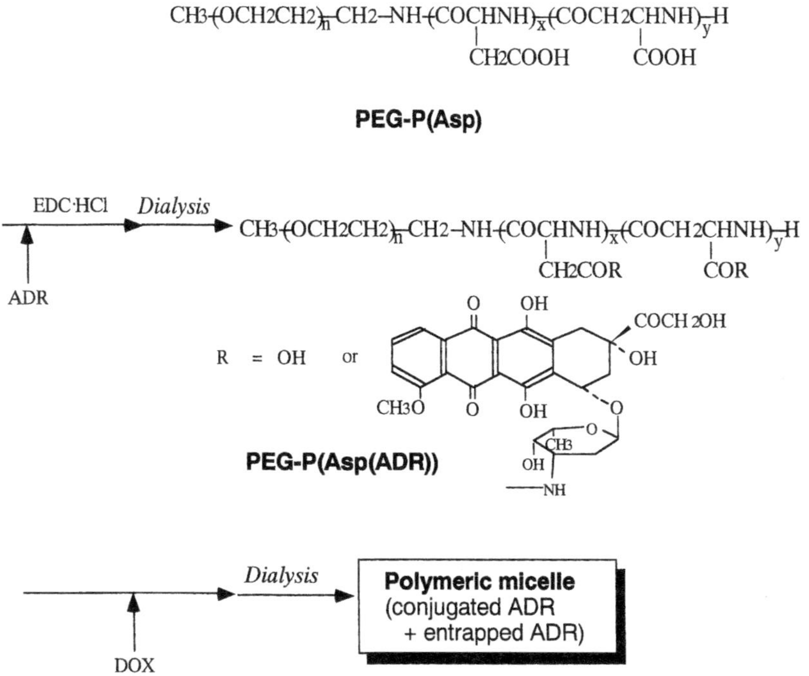

Figure 13 Chemical structure of ADR-containing polymeric micelle.

to passively target solid tumors (41–46,73–76). The molecular design of their system is shown in Figure 13. ADR was chemically conjugated to aspartic acid residues of PEG-poly(aspartic acid) block copolymers [PEG-P(Asp)] by amide bond formation. The PEG segment was hydrophilic, whereas the ADR-substituted poly(aspartic acid) chain was hydrophobic. Therefore, the obtained drug-block copolymer conjugate (PEG-P[Asp(ADR)]) formed micellar structures owing to its amphiphilic character. On the second step, ADR was incorporated into the inner core by physical entrapment utilizing hydrophobic and π-π interactions with the chemically conjugated ADR molecules. As a result, polymeric micelles containing both the chemically conjugated and the physically entrapped ADR in the inner core were obtained with the PEG outer shell.

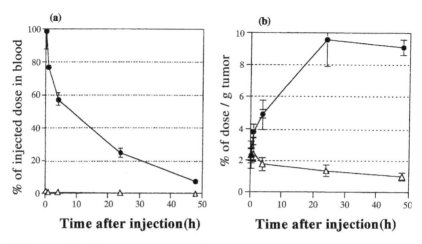

Figure 14 Concentrations in blood (a) and c26 tumor (b) after intravenous injection. (▲) free ADR and (●) polymeric micelles containing ADR. (From Ref. 46.)

As shown in Figure 14, the physically entrapped ADR circulated in the bloodstream for a long time-period (Fig. 14a) and was delivered to the solid tumor site at much higher concentrations than free ADR (Fig. 14b) (46). Furthermore, the observed time profile with a peak concentration at 24 h post-intravenous injection, as well as retention of this high concentration in longer time-periods post-injection, matched well to passive delivery by the EPR effect (31). On the other hand, accumulation of the physically entrapped ADR in the polymeric micelles at normal organs and tissues was the same or lower than free ADR.

In accordance with this highly selective delivery to solid tumor sites, a dramatic enhancement of antitumor activity was observed (46). Figure 15 shows in vivo antitumor activity against murine colon adenocarcinoma 26. For free ADR, only the maximum tolerated dose (10 mg/kg body weight) provided considerable inhibition effects on tumor growth; however, a decrease in tumor volume was never seen from the day of the first injection. For the polymeric micelles, the tumor completely disappeared in two doses (20 and 10 mg physically entrapped ADR/kg of body weight). All the polymeric

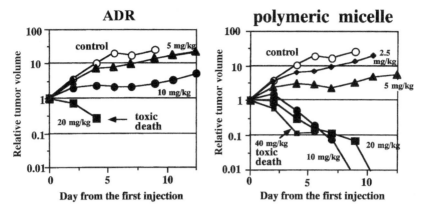

Figure 15 In vivo antitumor activity of free ADR and polymeric micelle containing ADR. (From Ref. 46.)

micelle-treated mice in these two doses survived over 60 days, whereas the control mice were dead by about day 20. Such a significant enhancement of antitumor activity has been scarcely seen in the many studies using various types of drug carriers. All these results clearly demonstrate the successful passive targeting of an anticancer drug using the polymeric micelle carrier system to solids, and this ADR-containing system finished the phase I clinical test, and entered the phase II test in autumn of 2003 at the National Cancer Hospital, Japan.

4.2. KRN-5500

Although ADR shows some hydrophobic properties, it is still water-soluble. In the field of cancer chemotherapy, a strong demand for the solubilization of water-insoluble drugs has been voiced particularly since many newly developed and very potent anticancer drugs such as camptothecin and taxol are water-insoluble or hardly water-soluble. In this perspective, the incorporation of a water-insoluble anticancer agent, KRN-5500 (KRN), into polymeric micelles was studied (77–79). By choosing an appropriate structure of a block copolymer, the water-insoluble KRN was successfully incorporated into polymeric micelles (77). This polymeric micelle showed higher antitumor activity than KRN injected into a conventional formulation. This polymeric micelle did not exhibit the severe vascular and pulmonary toxic side effects that surfaced in the KRN in the conventional formulation, a result that is due to the toxicity of the organic solvents and the surfactants used to dissolve KRN (78,79). This fact indicates that polymeric micelle carrier systems are not only a strong strategy for drug targeting but also very potent in dissolving water-insoluble drugs for safe intravenous injection.

4.3. Cisplatin

Cisplatin is a platinum chelate complex with two chloride and two ammonium ligands. This compound is water-soluble but not highly water-soluble. Nishiyama and his colleagues introduced by a ligand exchange reaction between chloride and carboxylate the cisplatin into the aspartic acid residues of PEG-poly(aspartic acid) block copolymers (80). Polymeric micelles were formed from these block copolymers. Because the micelle structures were not broken by the addition of a surfactant, the sodium dodecyl sulfate that broke the ADR-incorporating polymeric micelle structures, the crosslinking of the poly(aspartic acid) blocks through platinum atoms contributed to micelle formation. In physiological saline, cisplatin could be released from the micelles by a ligand exchange reaction between chloride and carboxylate, and simultaneously micelle structures would be broken. This cisplatin delivery system is now on an optimization stage of targeting tumors, as well as on an optimization stage of antitumor-activity.

4.4. Taxol

Taxol is a water-insoluble anticancer drug and is recognized as one of the most potent anticancer drugs among the approved anticancer drugs to appear in the 1990s. For intravenous injection, an organic surfactant cremophor EL and ethanol are used to solubilize taxol. Because cremophor EL is considerably toxic, injection without the use of cremophor EL is required. Samyoung Company, Daejeon, Korea developed PEG-PLA based polymeric micelles containing taxol. By this incorporation, water solubility was significantly increased as compared with the conventional cremophor EL formulation, and toxicity of cremophor EL was successfully diminished. This system is now in phase I clinical trial. This system, however, does not show targeting effects because all of the incorporated drug is released immediately after the intravenous injection. Alternatively, Nippon Kayaku Co., Ltd. (Tokyo, Japan) and Matsumura successfully incorporate taxol into polymeric micelles with exhibiting targeting ability to solid tumors. A phase I clinical trial of this targetable taxol system started in 2004 at the National Cancer hospital, Japan.

4.5. Amphotericin B

Amphotericin B is a potent antibiotic widely used in the treatment of systemic fungal infections. This drug is hardly water-soluble, and therefore, its intravenous injection is not possible. Kwon and his associates successfully incorporated amphotericin B into polymeric micelles forming from PEG-poly(*N*-hexyl aspartamide)-based block copolymers (81–83). They obtained enhancement of water-solubility and a decreased toxic side effect. It is known that amphotericin B exists in an intact (monomer) form and an aggregated form in physiological conditions, and that the aggregated form is mainly responsible for the toxic side effects. They speculated that the sustained release of amphotericin B and a larger proportion of monomers, which form owing to encapsulation in the hydrophobic environment, are the reasons for this decreased hemolysis toxicity. If the latter reason is true, this is the first case in which polymeric micelle carriers raised drug efficacy by changing the drug's existing forms owing to the hydrophobic environment for drug storage.

4.6. Methotrexate

Methotrexate (MTX) is a widely used anticancer drug, and recently this drug has also been used to treat rheumatoid arthritis. Li and Kwon conjugated MTX to a block copolymer, and obtained polymeric micelles, with the hydrophobicity of the conjugated MTX serving as the driving force behind the micelle formation (84). By controlling the amount of the conjugated MTX, micelle stability was successfully controlled. Micelle stability that is, the equilibrium between the micelle form and the single polymer chain, is correlated with the drug release rate, since hydrolysis between the polymer and MTX can proceed more rapidly in a single polymer chain form owing to both its easier access to water and other small molecules responsible for hydrolysis. This is the first example of a release control of the chemically conjugated drugs.

REFERENCES

1. Tuzar Z, Kratochvil P. Block and graft copolymer micelles in solution. Adv Colloid Interf Sci 1976; 6:201–232.

2. Yokoyama M, Inoue S, Kataoka K, Yui N, Okano T, Sakurai Y. Molecular design for missile drug: Synthesis of adriamycin conjugated with IgG using poly(ethylene glycoel)-poly(aspartic acid) block copolymer as intermediate carrier. Die Makromolekulare Chemie 1989; 190:2041–2054.

3. Kabanov AV, Chekhonin VP, Alakhov VY, Batrakova EV, Lebedev AS, Melik-Nubarov NS, Arzhakov SA, Levashov AV, Morozov GV, Severin ES, Kabanov VA. The neuroleptic activity of haloperidol increases after its solubilization in surfactant micelles: Micelles as microcontainers for drug targeting. FEBS Lett 1989; 258:343–345.

4. Yokoyama M, Fukushima S, Uehara R, Okamoto K, Kataoka K, Sakurai Y, Okano T. Characterization of physical entrapment and chemical conjugation of adriamycin in polymeric micelles and their design for in vivo delivery to a solid tumor. J Controlled Release 1998; 50:79–92.

5. Yokoyama M, Satoh A, Sakurai Y, Okano T, Matsumura Y, Kakizoe T, Kataoka K. Incorporation of water-insoluble anticancer drug into polymeric micelles and control of their particle size. J Controlled Release 1998; 55:219–229.

6. Li Y, Kwon GS. Methotrexate esters of poly(ethylene oxide)-block-poly(2-hydroxyethyl-L-aspartamide). Part 1: Effects of the level of methotrexate conjugation on the stability of micelles and on drug release. Pharm Res 2000; 17:607–611.

7. Lavasanifar A, Samuel I, Kwon GS. The effect of fatty acid substitution on the in vitro release of amphotericin B from micelles compoased of poly(ethylene oxide)-block-poly(N-hexyl stearate-L-aspartamide). J Contr Rel 2002; 79:165–172.

8. Allen C, Han J, Yu Y, Maysinger D, Eisenberg A. Polycaprolactone-b-poly(ethylene oxide) copolymer micelles as a delivery vehicle for dihydrotestosterone. J Contr Rel 2000; 51:275–286.

9. Trubetskoy VS, Torchilin VP. Polyethyleneglycol based micelles as carriers of the therapeutic and diagnostic agents. STP Pharma Sci 1996; 6:79–86.

10. Zhang X, Burt H-M, Von Hoff D, Dexter D, Mangold G, Degen D, Oktaba AM, Hunter W-L. An investigation of the antitumor activity and biodistribution of polymeric micellar paclitaxel. Cancer Chemother Pharmacol 1997; 40:81–86.

11. Rolland A, O'Mullane J, Goddard P, Brookman L, Petrak K. New macromolecular carriers for drugs. I. Preparation and characterization of poly(oxyethylene-b-isoprene-b-oxyethylene) block copolymer aggregates. J Appl Polym Sci 1992; 44: 1195–1203.

12. Inoue T, Chen G, Nakamae K, Hoffman AS. An AB block co polymer of oligo(methyl methacrylate) and poly(acrylic acid) for micellar delivery of hydrophobic drugs. J Contr Rel 1998; 41: 21–229.

13. Rapoport NY, Herron JN, Pitt WG, Pitina L. Micellar delivery of doxorubicin and its paramagnetic analog, ruboxyl, to HL-60 cells: Effect of micelle structure and ultrasound on the intracellular drug uptake. J Contr Rel 1999; 58:153–162.

14. Benahmed A, Ranger M, Leroux J-C. Novel polymeric micelles based on the amphiphilic diblock copolymer poly(N-vinyl-2-pyrrolidone)-block-poly(D,L-lactide). Pharm Res 2001; 18:323–328.

15. Nishiyama N, Yokoyama M, Aoyagi T, Okano T, Sakurai Y, Kataoka K. Preparation and characterization of self-assembled polymer-metal complex micelle from cis-dichlorodiammineplatinum (II) and poly(ethylene glycol)-poly(a, p- aspartic acid) block copolymer in an aqueous medium. Langmuir 1999; 15:377–383.

16. Kataoka K, Togawa H, Harada A, Yasugi K, Matsumoto T, Katayose S. Spontaneous formation of polyion complex micelles with narrow distribution from antisense oligonucleotide and cationic block copolymer in physiological saline. Macromolecules 1996; 29:8556–8557.

17. Harada A, Kataoka K. Novel polyion complex micelles entrapping enzyme molecules in the core: Preparation of narrowly-distributed micelles from lysozyme and poly(ethylene glycol)-poly(aspartic acid) block copolymer in aqueous medium. Macromolecules 1998; 31:288–294.

18. Calderara F, Hruska Z, Hurtrez G, Lerch J-P, Nugay T, Riess G. Investigation of olystyrene-poly(ethylene oxide)block copolymer micelle formation in organic and aqueous solutions by nonradiative energy transfer experiments. Macromolecules 1994; 27:1210–1215.

19. Wang Y, Kausch C-M, Chun M, Quirk R-P, Mattice WL. Exchange of chains between micelles of labeled polysty rene-block-poly(oxyethylene) as monitored by nonradiative singlet energy transfer. Macromolecules 1995; 28:904–911.

20. Wilhelm M, Zhao C-L, Wang Y, Xu R, Winnik R-A. Poly(styrene-ethylene oxide) block copolymer micelle formation in water: A fluorescence probe study. Macromolecules 1991; 24:1033–1040.

21. Desjardins A, Eisenberg A. Colloidal properties of block ionomers. 1.4 Characterization of reverse micelles of styrene-b-metal methacrylate diblocks by size-exclusion chromatography. Macromolecules 1991; 24:5779–5790.

22. Hoes CJT, Potman W, van Heeswijk WAR, Mud J, de Grooth B-G, Grave I, Feijen J. Optimization of macromolecular prodrugs of the antitumor antibiotic adriamycin. J Contr Rel 1985; 2:205–213.

23. Duncan R, Kopeckova-Rejmanova P, Strohalm J, Hume I, Cable HC, Pohl J, Lloyd JB, Kopecek J. Anticancer agents coupled to N-(2-hydroxypropyl)methacrylamide copolymers I. Evaluation of daunomycin and puromycin conjugates in vitro. Br J Cancer 1987; 55:165–174.

24. Endo N, Umemoto N, Kato Y, Takeda Y, Hara T. A novel covalent modification of antibodies at their amino groups with

retention of antigen-binding activity. J Immunol Meth 1987; 104:253–258.

25. Zunino F, Pratesi G, Micheloni A. Poly(carboxylic acid) polymers as carriers for anthracyclines. J Contr Rel 1989; 10: 65–73.

26. Hirano T, Ohashi S, Morimoto S, Tsukada K, Kobayashi T, Tsukagoshi S. Synthesis of antitumor-active conjugates of adriamycin or daunomycin with the copolymer of divinyl ether maleic anhydride. Makromol Chem 1986; 187:2815–2824.

27. Tsukada Y, Kato Y, Umemoto N, Takeda Y, Hara T, Hirai H. An anti-oc-fetoprotein antibody-daunorubicin conjugate with a novel poly-L-glutamic acid derivative as intermediate drug carrier. J Natl Cancer Inst 73; 1984:721–729.

28. Yokoyama M, Kwon GS, Okano T, Sakurai Y, Seto T, Kataoka K. Preparation of micelle-forming polymer-drug conjugates. Bioconj Chem 1992; 3:295–301.

29. Yokoyama M, Okano T. Targetable drug carriers: Present status and a future perspective. Adv Drug Delivery Rev 1996; 21:77–80.

30. Sugiyama Y. Importance of pharmacokinetic considerations in the development of drug delivery systems. Adv Drug Delivery Rev 1996; 19:333–334.

31. Matsumura Y, Maeda H. A new concept for macromolecular therapeutics in cancer chemotherapy: Mechanism of turn or tropic accumulation of proteins and the antitumor agent smancs. Cancer Res 1986; 46:6387–6392.

32. Maeda H, Seymour L-W, Miyamoto Y. Conjugates of anticancer agents and polymers: Advantages of macromolecular therapeutics in vivo. Bioconj Chem 1992; 3:351–361.

33. Litzinger DC, Buiting A-M-J, van Rooijen N, Huang L. Effect of liposome size on the circulation time and intraorgan distribution of amphipathic poly(ethylene glycol)-containing liposomes. Biochim Biophys Acta 1994; 1190:99–107.

34. Takakura Y, Hashida M. Macromolecular carrier systems for targeted drug delivery: Pharmacokinetic consideration on biodistribution. Pharm Res 1996; 13:820–831.

35. Ilium L, Davis SS, Miller R-H, Mak E, West P. The organ distribution and circulation time of intravenously injected colloidal carriers sterically stabilized with a block copolymer– Poloxamine 908. Life Sci 1987; 40:367–374.

36. Bader H, Ringsdorf H, Schmidt B. Water soluble polymers in medicine. Angew Chem 1984; 123/124:457–485.

37. Pratten MK, Lloyd JB, Horpel G, Ringsdorf H. Micelle-forming block copolymers: Pinocytosis by macrophages and interaction with model membranes. Makromol Chem 1985; 186:725–733.

38. Kabanov AV, et al. The neuroleptic activity of haloperidol increases after its solubilization in surfactant micelles: Micelles as microcontainers for drug targeting. FEBS Lett 1989; 258:343–345.

39. Kabanov A-V, Alakhov VY. Micelles of amphiphilic block copolymers as vehicles for drug delivery. Alexandridis P, Lindman B, eds. Amphiphilic Block Copolymers: Self Assembly and Applications. Netherlands: Elsevier, 1997:1–31.

40. Kabanov AV. Anthracycline antibiotics non-covalently incorporated into the block copolymer micelles: In vivo evaluation of anti-cancer activity. Br J Cancer 1996; 74:1545–1552.

41. Yokoyama M, Inoue S, Kataoka K, Yui N, Sakurai Y. Preparation of adriamycin-conjugated poly(ethylene glycol)-poly-(aspartic acid) block copolymer. A new type of polymeric anticancer agent. Makromolekul Chemi Rapid Commun 1987; 8:431–435.

42. Yokoyama M, Inoue S, Kataoka K, Yui N, Sakurai Y. Molecular design for missile drug: Synthesis of adriamycin conjugated with IgG using poly(ethylene glycol)-poly(aspartic acid) block copolymer as intermediate carrier. Makromol Chem 1989; 190:2041–2054.

43. Yokoyama M, Miyauchi M, Yamada N, Okano T, Sakurai Y, Kataoka K, Inoue S. Characterization and anti-cancer activity of micelle-forming polymeric anti-cancer drug, adriamycin-conjugated poly(ethylene glycol)-poly(aspartic acid) block copolymer. Cancer Res 1990; 50:1693–1700.

44. Yokoyama M, Yamada N, Okano X, Sakurai Y, Kataoka K, Inoue S. Toxicity and antitumor activity against solid tumors

of micelle-forming polymeric drug and its extremely long circulation in blood. Cancer Res 1991; 51:3229–3236.

45. Kwon GS, Suwa S, Yokoyama M, Okano T, Sakurai Y, Kataoka K. Enhanced tumor accumulation and prolonged circulation times of micelle-forming poly(ethylene oxide-aspartate) block copolymer-adriamycin conjugates. J Contr Rel 1994; 29:17–23.

46. Yokoyama M, Okano T, Sakurai Y, Fukushima S, Okamoto K, Kataoka K. Selective delivery of adriamycin to a solid tumor using a polymeric micelle carrier system. J Drug Target 1999; 7:171–186.

47. Newman KD, Samuel J, Kwon GS. Ovalbumin peptide encapsulated in Poly(d,l lactic-co-glycolic acid) microspheres is capable of inducing a T helper type 1 immune response. J Contr Rel 1998; 54:49–59.

48. Yu BG, Okano T, Kataoka K, Kwon GS. Polymeric micelles for drug delivery: Solubilization and haemolytic activity of amphotericin B. J Contr Rel 1998; 53:131–136.

49. Yu BG, Okano T, Kataoka K, Sardari S, Kwon GS. In vitro dissociation of antifungal efficacy and toxicity for amphotericin B-loaded poly(ethylene oxide)-block-poly((3-benzyl-L-aspartate) micelles. J Contr Rel 1998; 56:285–291.

50. Rolland A, O'Mullane J, Goddard L, Brookman L, Petrak K. New macromolecular carriers for drugs. 1. Preparation and characterization of poly(oxyethylene-b-isoprene-b-oxyethylene)block copolymer aggregates. J Appl Polym Sci 1992; 44:1195–1203.

51. Shin IL, Kim SY, Lee YM, Cho CS, Sung YK. Methoxy polyethylene glycol)/e-caprolactone amphiphilic block copolymeric micelle containing indomethacin 1. Preparation and characterization. J Contr Rel 1998; 51:1–11.

52. Allen C, Han J, Yu Y, Maysinger D, Eisenberg A. Polycaprolactone-b-poly(ethylene oxide) copolymer micelles as a delivery vehicle for dihydrotestosterone. J Contr Rel 2000; 63:275–286.

53. Inoue T, Chen G, Nakamae K, Hoffman AS. An AB block copolymer of oligo(methyl methacrylate) and poly(acrylic acid) for micellar delivery of hydrophobic drugs. J Contr Rel 1998; 51:221–229.

54. Hagan SA, Coombes AGA, Garnett MC, Dunn SE, Davies MC, Ilium L, Davies SS. Polylactide-poly(ethylene glycol) copolymers as drug delivery systems. 1. Characterization of water dispersible micelle-forming systems. Langmuir 1996; 12:2153–2161.

55. Trubetskoy VS, Torchilin VP. Polyethyleneglycol based micelles as carriers of therapeutic and diagnostic agents. STP Pharma Sci 1996; 6:79–86.

56. Piskin E, Kaitian X, Denkbas EB, Kucukyavuz Z. Novel PDLLA/PEG copolymer micelles as drug carriers. J Biomater Sci Polymer Edn 1995; 7:359–373.

57. Gref R, Minamitake Y, Peracchia MT, Torchilin V, Trubetskoy V, Langer R. Biodegradable long-circulating nanospheres. Science 1994; 28:1600.

58. Ha JC, Kim SY, Lee YM. Poly(ethylene oxide)-poly(propylene oxide)-poly(ethylene oxide)(plueronic/poly(e-caprolactone) (PCL) amphiphilic block copolymeric nanospheres I. Preparation and characterization. J Contr Rel 1995; 62:381–392.

59. Verrecchia T, Spenlehauer G, Bazile DV, Murry-Brelier A, Archimbaud Y, Veillard M. Non-stealth (poly(lactic acid/ albumin)) and stealth (poly(lactic acid-polyethylene glycol)) nanoparticles as injectable drug carriers. J Contr Rel 1995; 36:49–61.

60. Prochazka K, Kiserow D, Ramireddy C, Tuzar Z, Munk P, Webber SE. Time-resolved fluorescence studies on the chain dynamics of naphthalene-labeled polystyrene-block-poly-(methacrylic acid) micelles in aqueous media. Macromolecules 1992; 25:454–460.

61. Ganachaud E, Monteiro MJ, Gilbert RG, Dourges MA, Thang SH, Rizzaedo E. Molecular weight characterization of poly(N-isopropylacrylamide) prepared by living free-radical polymerization. Macromolecules 2000; 33:6738–6745.

62. Kwon GS, Yokohama M, Okano T, Sakurai Y, Kataoka K. Biodistribution of micelle-forming polymer-drug conjugates. Pharm Res 1993; 10:970–974.

63. Yokoyama M, Kwon GS, Okano T, Sakurai Y, Ekimoto H, Okamoto K, Mashiba H, Seto T, Kataoka K. Composition-dependent in vivo antitumor activity of adriamycin-conjugated

polymeric micelle against murine colon adenocarcinoma 26. Drug Delivery 1993; 1:11–19.

64. Yokoyama M, Kwon GS, Okano T, Sakurai Y, Kataoka K. Influencing factors on in vitro micelle stability of adriamycin-block copolymer conjugates. J Contr Rel 1994; 28:59–65.

65. Kabanov AV, Alakhov VY. Micelles of amphiphilic block copolymers as vehicles for drug delivery. In: Alexandridis P, Lindman B, eds. Amphiphilic Block Copolymers: Self Assembly and Applications. Netherlands: Elsevier, 1997: 1–31.

66. Scholz C, Iijima M, Nagasaki Y, Kataoka K. A novel reactive polymeric micelle with aldehyde groups on its surface. Macromolecules 1995; 28:7295–7297.

67. Chung J-E, Yokoyama M, Aoyagi T, Sakurai Y, Okano T. Effect of molecular architecture of hydrophobically modified poly(N-isopropylacrylamide) on the formation of thermoresponsive core-shell micellar drug carriers. J Contr Rel 1998; 53:119–130.

68. Kohori F, Sakai K, Aoyagi T, Yokoyama M, Sakurai Y, Okano T. Preparation and characterization of thermally responsive block copolymer micelles comprising poly(N-isopropylacrylamide-b-DL-lactide). J Contr Rel 1998; 55: 87–98.

69. Chung JE, Yokoyama M, Yamato M, Aoyagi T, Sakurai Y, Okano T. Thermo-responsive drug delivery from polymeric micelles constructed using block copolymers of poly(N-isopropylacrylamide) and poly(butylmethacrylate). J Contr Rel 1999; 62:115–127.

70. Kohori F, Sakai K, Aoyagi T, Yokoyama M, Yamato M, Sakurai Y, Okano T. Control of adriamycin cytotoxic activity using thermally responsive polymeric micelles composed of poly(N-isopropylacrylamide-co-N,N-dimethylacrylamide)-b-poly(DL-lactide). Colloid Surf B 1999; 16:195–205.

71. Chung JE, Yokoyama M, Okano T. Inner core segment design for drug delivery control of thermo-responsive polymeric micelles. J Contr Rel 2000; 65:93–103.

72. Kohori F, Yokoyama M, Sakai K, Okano T. Process design for efficient and controlled drug incorporation into polymeric micelle carrier systems. J Contr Rel 2002; 78:155–163.

73. Kwon GS, Naito M, Kataoka K, Yokoyama M, Sakurai Y, Okano T. Block copolymer micelles as vehicles for hydrophobic drugs. Colloid Surf B 1994; 2:429–434.

74. Yokoyama M, Okano T, Sakurai Y, Kataoka K. Improved synthesis of adriamycin-conjugated poly(ethylene oxide)-poly-(aspartic acid) block copolymer and formation of unimodal micellar structure with controlled amount of physically entrapped adriamycin. J Contr Rel 1994; 32:269–277.

75. Kwon GS, Naito M, Yokoyama M, Okano T, Sakurai Y, Kataoka K. Physical entrapment of adriamycin in AB block copolymer micelles. Pharm Res 1995; 12:192–195.

76. Yokoyama M, Fukushima S, Uehara R, Okamoto K, Kataoka K, Sakurai Y, Okano T. Characterization of physically entrapment and chemically conjugation of adriamycin in polymeric micelles and their design for in vivo delivery to a solid tumor. J Contr Rel 1998; 50:79–92.

77. Yokoyama M, Satoh A, Sakurai Y, Okano T, Matsumura Y, Kakizoe T, Kataoka K. Incorporation of water-insoluble anticancer drug into polymeric micelles and control of their particle size. J Contr Rel 1998; 55:219–229.

78. Matsumura Y, Yokoyama M, Kataoka K, Okano T, Sakurai Y, Kawaguchi T, Kakizoe T. Reduction of the adverse effects of an antitumor agent, KRN 5500 by incorporation of the drug into polymeric micelles. Jap J Cancer Res 1999; 90:122–128.

79. Mizumura Y, Matsumura Y, Yokoyama M, Okano T, Kawaguchi T, Moriyasu F, Kakizoe T. Incorporation of the anticancer agent KRN 5500 into polymeric micelles diminishes the pulmonary toxicity. Jap J Cancer Res 2002; 93:1237–1243.

80. Nishiyama N, Yokoyama M, Aoyagi T, Okano T, Sakurai Y, Kataoka K. Preparation and characterization of self-assembled polymer-metal complex micelle from cis-dichloro-diammineplatinum (II) and poly(ethylene glycol)-poly(a, p- aspartic acid) block copolymer in an aqueous medium. Langmuir 1999; 15:377–383.

81. Lavasanifar A, Samuel J, Satari S, Kwon GS. Block copolymer micelles for the encapsulation and delivery of amphotericin B. Pharm Res 2002; 19:418–422.

82. Lavasanifar A, Samuel J, Kwon GS. Micelles self-assembled from poly(ethylene oxide)-block-poly(N-hexyl stearate L-aspartamide) by a solvent evaporation method; effect on the solubilization and hemolytic activity of amphotericin B. J Contr Rel 2001; 77:155–160.

83. Adams ML, Kwon GS. Relative aggregation state and hemolytic activity of amphotericin B encapsulated by poly(ethylene oxide)-block-poly(N-hexyl-L-aspartamide)-acyl conjugate micelles: effects of acyl chain length. J Contr Rel 2003; 87:23–32.

84. Li Y, Kwon GS. Methotrexate ester of poly(ethylene oxide)-block-poly(2-hydroxyethyl-L-aspartamide). Part 1: Effects of the level of methotrexate conjugation on the stability of micelles and on drug release. Pharm Res 2000; 157:607–611.

14

Pluronic® Block Copolymers for Drug and Gene Delivery

ALEXANDER V. KABANOV and JIAN ZHU
College of Pharmacy, Department of
Pharmaceutical Sciences, Nebraska Medical
Center, Omaha, Nebraska, U.S.A.

1. INTRODUCTION

In the recent two decades, development of efficient delivery systems for biological agents, such as low molecular mass drugs and biomacromolecues (DNA, proteins, etc.), has attracted tremendous attention. It is believed that such systems will help to overcome enormous barriers encountered by these agents on target sites, as well as increase solubility and stability of these agents. Synthetic polymers are among the major materials used in drug and gene delivery systems (1). One important and promising example of such materials is Pluronic® block copolymers. Pluronic block copolymers have

577

been used in experimental medicine and pharmaceutical sciences for a long time (see for review Refs. 2–11). The current chapter focuses on the relatively dilute isotropic solutions of block copolymers in aqueous media. These solutions exist in the form of either molecular dispersion or as micelles of the block copolymers. They can be used as solubilizers for insoluble drugs as well as nanocontainers for site-specific drug delivery in body. The distinct properties of Pluronic block copolymers can also enhance drug performance by acting as biological response-modifying agents, which act directly upon the target cells. Very recently, Pluronic block copolymers have shown promising potential for the non-viral gene delivery.

2. PLURONIC STRUCTURE AND SYNTHESIS

Pluronic block copolymers (also known by their non-proprietary name "poloxamers") consist of ethylene oxide (EO) and propylene oxide (PO) blocks arranged in a basic A-B-A structure: EO_x-PO_y-EO_x. This arrangement results in an amphiphilic copolymer, in which the number of hydrophilic $EO(x)$ and hydrophobic $PO(y)$ units can be altered. The structural formula of Pluronic block copolymers is shown in Figure 1. Table 1 presents a list of selected Pluronic copolymers

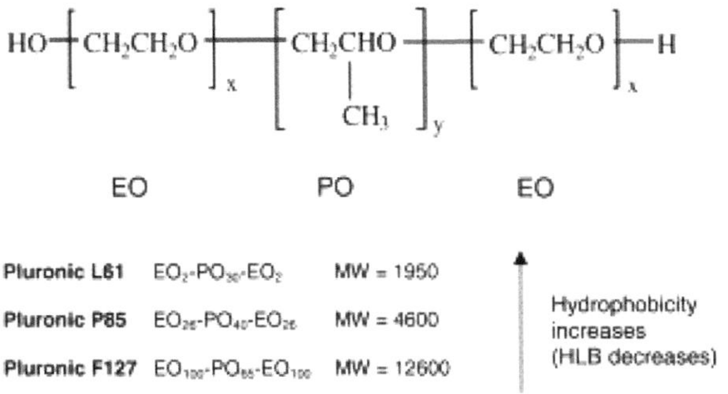

| EO | PO | EO |

Pluronic L61	EO_2-PO_{30}-EO_2	MW = 1950
Pluronic P85	EO_{26}-PO_{40}-EO_{26}	MW = 4600
Pluronic F127	EO_{100}-PO_{65}-EO_{100}	MW = 12600

Hydrophobicity increases (HLB decreases)

Figure 1 Pluronic block copolymers available from BASF Co (Wyandotte, MI), contain two hydrophilic EO blocks and a hydrophobic PO block. (From Ref. 93.)

Table 1 Physicochemical Characteristics of Pluronic Block Copolymers

Copolymer	MW[a]	Average number of EO units (x)[b]	Average number of PO units (y)[b]	HLB[c]	Cloud point in 1% aqueous solution (°C)[c]	CMC (M)[d]
L35	1,900	21.59	16.38	19	73	5.3×10^{-3}
L43	1,850	12.61	22.33	12	42	2.2×10^{-3}
L44	2,200	20.00	22.76	16	65	3.6×10^{-3}
L61	2,000	4.55	31.03	3	24	1.1×10^{-4}
L62	2,500	11.36	34.48	7	32	4.0×10^{-4}
L64	2,900	26.36	30.00	15	58	4.8×10^{-4}
F68	8,400	152.73	28.97	29	>100	4.8×10^{-4}
L81	2,750	6.25	42.67	2	20	2.3×10^{-5}
P84	4,200	38.18	43.45	14	74	7.1×10^{-5}
P85	4,600	52.27	39.66	16	85	6.5×10^{-5}
F87	7,700	122.50	39.83	24	>100	9.1×10^{-5}
F88	11,400	207.27	39.31	28	>100	2.5×10^{-4}
L92	3,650	16.59	50.34	6	26	8.8×10^{-5}
F98	13,000	236.36	44.83	28	>100	7.7×10^{-5}
L101	3,800	8.64	58.97	1	15	2.1×10^{-6}
P103	4,950	33.75	59.74	9	86	6.1×10^{-6}
P104	5,900	53.64	61.03	13	81	3.4×10^{-6}
P105	6,500	73.86	56.03	15	91	6.2×10^{-6}
F108	14,600	265.45	50.34	27	>100	2.2×10^{-5}
L121	4,400	10.00	68.28	1	14	1.0×10^{-6}
P123	5,750	39.20	69.40	8	90	4.4×10^{-6}
F127	12,600	200.45	65.17	22	>100	2.8×10^{-6}

[a] The average molecular weights provided by the manufacturer (BASF Co., Wyandotte, MI).
[b] The average numbers of EO and PO units were calculated using the average molecular weights.
[c] HLB values of the copolymers and the cloud points were determined by the manufacturer.
[d] CMC values were determined previously using the pyrene probe discussed in Ref. 96.
Source: From Ref. 95.

available from BASF Corp. Copolymers with various x and y values are characterized by distinct hydrophilic-lipophilic balance (HLB).

Pluronic block copolymers are synthesized by sequential addition of PO and EO monomers in the presence of an alkaline catalyst, such as sodium or potassium hydroxide (3). The reaction is initiated by polymerization of the PO block followed by the growth of EO chains at both ends of the PO block. Anionic polymerization usually produces polymers with a relatively low polydispersity index (M_n/M_w). Further chromatographic fractionation was employed in procedures for the manufacture of highly purified block copolymers (12,13).

3. SELF-ASSEMBLY OF PLURONIC BLOCK COPOLYMERS

A defining property of Pluronic is the ability of individual block copolymer molecules, termed "unimers," to self-assemble into micelles in aqueous solutions. These unimers form a molecular dispersion in water at block copolymer concentrations below the critical micelle concentration (CMC). At concentrations of the block copolymer above the CMC, the unimer molecules aggregate, forming micelles through a process called "micellization." The driving force for the micellization is the hydrophobic interactions of the PO blocks. The PO blocks self-assemble into the inner core of the micelles covered by the hydrophilic corona from EO blocks. Pluronic micelles are commonly pictured as spheres composed of a PO core and an EO corona. This portrayal is correct for most block copolymers, which have EO content above 30%, especially in relatively dilute solutions at body temperature. However, additional micelle morphologies, including lamella and rods (cylinders), can also form in Pluronic systems (14) (Fig. 2). When spherical micelles are formed, depending on the Pluronic type, the micelles commonly have an average hydrodynamic diameter ranging from about 20 to about 80 nm (14). The number of block copolymer unimers forming one micelle

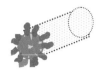 

Figure 2 The different micelle morphologies formed in Pluronic block copolymer systems.

is referred to as the "aggregation number." Usually this number ranges from several to over a hundred.

The process of transfer of water-insoluble compounds into the PO core of the micellar solution is referred to as "solubilization." Pluronic micelles containing solubilized low molecular mass drugs and polypeptides are being actively investigated as potential drug delivery systems (15–23). The core-shell architecture of polymeric micelles is essential for their utility for these applications. The core formed by the PO chains is a water-incompatible compartment that is segregated from the aqueous exterior by the hydrophilic chains of the EO corona, thereby forming, within the core, a "cargo hold" for the incorporation of various biological agents. As a result, polymeric micelles can be used as efficient carriers for compounds, which alone exhibit poor solubility, undesired pharmacokinetics, and low stability in a physiological environment. The hydrophilic shell contributes greatly to the pharmaceutical behavior of block copolymer formulations by maintaining the micelles in a dispersed state, as well as by decreasing undesirable drug interactions with cells and proteins through steric-stabilization effects.

4. PLURONIC® FORMULATIONS FOR TREATMENT OF DRUG-RESISTANT TUMORS

4.1. Sensitization of Drug-Resistant Cancers by Pluronic Block Copolymers

Tumors with the multidrug-resistant (MDR) phenotype are among some of the most difficult types to treat. MDR cells overexpress drug efflux proteins belonging to a superfamily of ATP binding cassette (ABC), such as P-glycoprotein (Pgp), multidrug resistance-associated proteins (MRP), and brain cancer resistance proteins that pump drugs out of cell (24,25). Furthermore, several other proteins (e.g., glutathione S-transferase, metallothionein, thioredoxin, topoisomerase I, II, O^6-alkylguanine-DNA alkyltransferase, etc.) are believed to contribute to the resistant phenotype as well (24). Alakhov and coworkers (26–28) demonstrated that Pluronic block copolymers sensitize resistant cells, resulting in an increase in the cytotoxic activity of the drug by 2–3 orders of magnitude. By addition of P85 or L61, the cytotoxic effects of doxorubicin in the resistant lines significantly surpassed those observed in the sensitive lines. Similar effects of Pluronic block copolymers have also been reported in vivo (19,29), and these studies indicate improved treatment of drug-resistant cancers with Pluronic block copolymers.

4.2. Effects of Pluronic Block Copolymers on Drug Resistance Systems

4.2.1. Inhibition of Pgp and Other Drug Transport Systems

Enhanced cytotoxicity observed with Pluronic block copolymers in drug-resistant cancer cells appears to be related to the effects of these block copolymers on the Pgp drug efflux transport system. Evidence are the observations that defects in the intracellular accumulation of doxorubicin in resistant cancer cells expressing Pgp can be overcome by treatment with Pluronic (28,30), while no alteration in drug uptake in the presence of Pluronic was observed with non-Pgp expressing parental cancer cells. This conclusion has been reinforced by the recent studies by Evers et al. (31) and Batrakova et al. (32), demonstrating that Pluronic block copolymers (L61, P85) have pronounced effects, increasing accumulation and permeability of various Pgp-dependent drugs in MDR1-transfected cells that overexpress Pgp. An additional support for the Pgp-mediated mechanism of Pluronic in Pgp expressing cells is that the block copolymer has no or little effect on the accumulation of non-Pgp dependent compounds in both resistant and parental cells (32–34). Therefore, the increased absorption of the Pgp substrates in Pgp-expressing cells is attributable to the effects of the block copolymer on the Pgp efflux system, rather than to non-specific alterations in membrane permeability of the substrates. Recently, however, evidence has begun to mount suggesting that the effects of Pluronic block copolymers might stem beyond solely inhibition of the Pgp efflux pump. Studies by Miller et al. (35), using the human pancreatic adenocarcinoma cell line, Panc-1, that expresses the MRP efflux pump, suggested that P85 inhibits efflux and increases cellular accumulation of the MRP-dependent probe, fluorescein, in these cells. Batrakova et al. (36) showed that decrease of MRP ATPase activity by P85 results in sensitization effects in MRP1 and MRP2-overexpressing cells toward anticancer agents, vinblastine, and doxorubicin, while the magnitude of these effects is less than sensitization effects in Pgp-overexpressing cells. Generally, the evaluation of the effects of Pluronic block copolymers on MRP is far from completion.

4.2.2. Effects on Drug Sequestration Within Cytoplasmic Vesicles

Another impediment to treatment, which is present in MDR cells, involves the sequestration of drugs within cytoplasmic vesicles, followed by extrusion of the drug from the cell (37–41). Drug sequestration in MDR cells is achieved through the maintenance of abnormally elevated pH gradients across organelle membranes–by the activity of H^+-ATPase, an ATP-dependent pump (42). Studies by Venne et al. (28) examined the effects of Pluronic block copolymers on intracellular localization of doxorubicin in the MDR cancer cell line, MCF-7/ADR. In these cells, free doxorubicin is sequestered in cytoplasmic vesicles, which might further diminish the amount of the drug available for interaction with the nucleus (41). Following incubation of the cells with doxorubicin and Pluronic, the drug was released from the vesicles and accumulated, primarily, in the nucleus (28).

4.2.3.   Effect on GSH/GST System

The activity of MRP drug efflux transporter with respect to many substrates is closely tied to the GSH/GST detoxification system in MDR cells (41). Therefore, studies of the effects of Pluronic block copolymers on GSH/GST system have begun. For example, significant decreases in both intracellular levels of GSH and GST activity were observed in MDCK cells expressing MRP, following exposure of these cells to P85 (36). GST activity, as assessed using a colorimetric substrate 1-chloro-2,4-dinitrobenzene, was significantly inhibited immediately following 2-h exposure of the MDCKII, MDCKII–MRP1 and MDCKII–MRP2 cells to P85. The levels of GSH did not decrease during 2-h exposure of these cells to P85. However, the GSH pools in these cells were significantly depleted when the cells were first exposed to Pluronic and then incubated in Pluronic-free media during 22 h. Inhibition of the GST/GSH detoxification system should result in the decrease of GSH conjugation of select substrates, such as doxorubicin, which can additionally decrease the extent of elimination of these substrates from the cells.

4.3.   ATP Depletion Induced by Pluronic Block Copolymers in MDR Cells

Various drug resistance systems, including drug transport and detoxification systems, require consumption of energy to sustain their function in MDR cells. Therefore, mechanistic studies have focused on the effects of Pluronic block copolymers on metabolism and energy conservation in drug-resistant cells. It was found that the ATP depletion induced by Pluronic block copolymers is one major reason for their effects on drug resistance systems. The key observation was the study by Batrakova et al. (43,44) that compared the effects of the P85 on ATP levels in several cell types that either exhibit MDR phenotype or do not. In this study, exposure of MDR and non-MDR cells to different doses of P85 resulted in a transient energy depletion, which was reversed following removal of the block copolymer. The MDR cells were much more responsive to P85, exhibiting a profound decrease in ATP levels at substantially lower concentrations of the block copolymer, compared to the non-MDR cells. The effective concentrations of P85 that induced a 50% decrease in ATP levels in the cells (EC_{50}) are presented in Table 2. This table also presents the relative responsiveness of the MDR cells compared to the non-MDR counterparts. These data suggested that the responsiveness of the cells to P85 correlate with the appearance of MDR in these cells, rather than the amount of ATP available in the cells prior to treatment. Based on the results of this study, the presence of the MDR phenotype is one factor that renders cellular metabolism responsive to treatment with Pluronic block copolymers.

The mitochondria are responsible for carrying out much of the metabolic activities of the cell, and might be a potential

Table 2   Effects of P85 on ATP Levels in Pgp and MRP Overexpressing Cells and Cells That Do Not Overexpress Pgp or MRP

Cells	Pgp or MRP overexpression	Initial ATP levels (nmol/mg protein)[a]	EC_{50} (%)	Relative responsiveness to P85[b]
MCF-7	No	30 ± 1.5	2.25	–
MCF-7/ADR	Pgp, MRP	300 ± 20	0.009	250[c]
KB	No	1 ± 0.01	0.675	–
KBv	Pgp	4 ± 0.1	0.036	19[d]
C2C12	No	15 ± 1.4	4.5	–
HUVEC	No	40 ± 4.9	0.0675	–
Caco-2	Pgp, MRP	5.5 ± 0.4	0.00067	6670[e]
BBMEC	Pgp, MRP	1.6 ± 0.04	0.018	250[e]
LLC-PK1	No	61 ± 6.9	0.45	–
LLC-MDR1	Pgp	79 ± 1.7	0.0045	100[f]

[a]Mean ± SEM ($n=4$).
[b]Calculated as the ratio of EC_{50} of non-Pgp non-MRP cells to EC_{50} of corresponding Pgp expressing cells.
[c]Compared to MCF-7 cells.
[d]Compared to KB cells.
[e]Compared to C2C12 cells.
[f]Compared to LLC-PK1 cells.
Source: From Ref. 93 and the data presented by Batrakova et al (43). In this study cells were exposed to P85 for 120 min prior to determining the intracellular ATP levels.

site of action for Pluronic block copolymers. It has long been known that non-ionic polymeric detergents, such as Tween 80 and Pluronic, can decrease oxidative metabolism of tissues, cells, and isolated mitochondria (45–47). There could be multiple reasons for the inhibitory activity of these compounds in mitochondria (45,48–52).

However, the reasons for the remarkable selectivity of Pluronic block copolymers with respect to MDR cells are not completely understood. One hypothesis is that the high rates of energy consumption by the drug efflux pumps, combined with Pluronic-induced inhibition of respiration, determine the responsiveness of the resistant cells to the block copolymer. Such a hypothesis is in line with the earlier observation that resistant cells have an increased glucose utilization rate compared to sensitive cells (53,54). Furthermore, the toxicity of the inhibitor of glycolysis, 2-deoxyglucose, was found to be consistently higher in MDR cells than in the non-MDR lines (54). According to Batrakova et al. (44), the extent of the selectivity of P85 with respect to MDR cells is the same or even greater than that of 2-deoxyglucose. Most recently, evidence has begun to mount suggesting that there might be an alternative explanation for the high selectivity of the block copolymers with respect to MDR cells. Our study suggested that P85 significantly inhibits ATPase activity of Pgp in isolated cell membranes (44). Since Pgp is one of the major ATPases overexpressed in MDR cells, the fact that Pluronic inhibits this activity makes the high energy consumption in MDR cells a less likely cause for the block copolymer selectivity in these cells. An alternative hypothesis would be that, for some reason, the metabolic processes in MDR cells are more sensitive to inhibition with Pluronic than the metabolic processes in non-MDR cells. This could result in the more pronounced ATP depletion observed following exposure of MDR cells to the block copolymer.

The ATP depletion phenomenon is obviously extremely important in order to understand the reasons for the elevated anticancer activity of Pluronic-based formulations in drug-resistant tumors. Transient energy depletion, as a result of the exposure of the cells to Pluronic in the absence of the chemotherapeutic agent, does not induce a cytotoxic effect in either MDR or sensitive cells (43,44). However, the exposure of MDR cells to the combination of both the chemotherapeutic agent and block copolymer results in a pronounced amplification of the cytotoxic activity, i.e., the sensitization effect (43,44). The apparent relationship between energy depletion and enhanced cytotoxicity in cancer cells is also supported by some other studies (55,56).

4.4. Membrane Interaction of Pluronic and Inhibition of Pgp ATPase Activity

Pluronic block copolymers are known to induce changes in the microviscosity of cell membranes (57–59). These changes can be attributed to the alterations in the structure of the lipid bilayers as a result of adsorption of the block copolymer molecules on the membranes. Interestingly, it appears that Pluronic block copolymers have different effects with respect to the membranes of some normal and cancer cells. Melik-Nubarov et al. (57) reported that the treatment with the same doses of P85 or L61 increased microviscosity ("fluidized") of the membranes of cancerous cells, while, in contrast, it decreased microviscosity ("solidified") of the membranes of normal blood cells. Membrane fluidization by various agents including nonionic surfactants, such as Tween 20, Nonidet P-40, and Triton X-100, is known to contribute to inhibition of Pgp efflux function (60). MDR modulation by membrane fluidizers occurs by abolishment of Pgp ATPase activity that results in the loss of Pgp-mediated drug efflux (60). Furthermore, recent studies demonstrated that P85 inhibits Pgp ATPase activity and inhibition of this activity is observed with the same doses of the block copolymer as those that inhibit Pgp efflux in Pgp expressing cells (44,59).

Figure 3 schematically presents the multiple effects of Pluronic block copolymers displayed in MDR cells. It is likely that Pluronic block copolymers have a "double-punch" effect in Pgp expressing cells: through ATP depletion and membrane interaction, which both have a combined result of potent inhibition of Pgp (44,59). It is possible that different

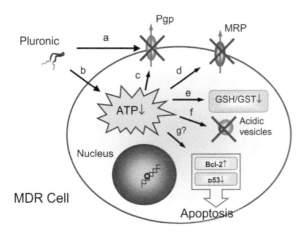

Figure 3 Schematic presenting multiple effects of Pluronic block copolymers displayed in MDR cell. These effects include (a) decrease in membrane viscosity ("fluidization"); (b) ATP depletion; (c,d) inhibition of drug efflux transport systems; (e) reduction in GSH/GST detoxification activity; and (f) drug release from acidic vesicles in the cell. Effects of Pluronic block copolymers on apoptosis (g) are not sufficiently studied at present. (From Ref. 94.)

drug transporters have different energy requirements in MDR cells and/or different sensitivity to changes in membrane microviscosity. As a result, the potency of Pluronic block copolymers with respect to modulation of these drug transport mechanisms may also vary.

4.5. Clinical Trials of Doxorubicin-Pluronic Formulation (SP1049C)

Pluronic block copolymers, L61 and F127, were selected for further preclinical development based on the in vitro and in vivo efficacy evaluation studies of anthracyclines (19,28). The final block copolymer system used in the clinical studies contains 0.25% L61 and 2% F127 formulated in isotonic buffered saline and is called SP1049C (Supratek Pharma, Montreal, PQ, Canada) (29). This system contains mixed micelles of L61 and F127 with an effective diameter of approximately. 22–27 nm, which do not aggregate in the presence of the serum proteins. Prior to administration, doxorubicin is mixed with this system, resulting in the spontaneous incorporation of the drug into micelles. The formulation is safe following systemic administration based on acute and subacute toxicity studies in animals (29). An open labeled, phase I dose escalation and pharmacokinetics clinical trial of SP1049C has been completed under the sponsorship of the UK Cancer Research Campaign in two sites: Christie Hospital in Manchester (UK) and Queen Elisabeth Hospital, Birmingham (UK) (61) (a full report is in preparation). The primary goal of this trial was to determine the maximal tolerable dose (MTD), toxicity, and the pharmacokinetics profile of SP1049C after intravenous administration. A total of 26 patients entered the trial. The dose limiting toxicity of myelosuppression was observed at MTD $90\,mg/m^2$. Other toxicities observed were reversible: alopecia, nausea, and lethargy. It was established that a dose level of $70\,mg/m^2$ is suitable for evaluation in phase II trials. Plasma pharmacokinetics showed that SP1049C exhibited slightly longer half-life compared to conventional doxorubicin (50 h vs. 30 h). One patient with relapsed previously treated esophageal sarcoma showed temporarily a 50% reduction in tumor size, which progressed rapidly on completion of treatment. Furthermore, antitumor activity was seen in relapsed previously treated soft tissue sarcoma. Phase II clinical trials are in progress.

5. PLURONIC EFFECTS ON BRAIN AND ORAL BIOAVAILABILITY OF DRUGS

5.1. Effects of Pluronic on Brain Accumulation of Drugs: In Vitro and In Vivo Evaluation

The transporters Pgp and MRP are also expressed in "normal" cells in the body, including brain microvessel endothelial cells (BMVEC), intestinal epithelial cells, and hepatocytes (62,63). The roles and effects of these transporters on the tissue distribution of drugs have attracted great attention. For example, immunohistochemical and functional studies indicate that Pgp is expressed at the luminal (apical) plasma membrane of BVMEC that forms the blood-brain barrier (BBB) (62,63). Less is known regarding the role of the MRP drug efflux transport system in the BBB.

Similar to the model of drug-resistant tumor cells, the studies by Batrakova et al. (59) suggested that Pluronic can enhance the delivery of Pgp substrate to the brain through the inhibition of the Pgp-mediated efflux system in brain endothelial cells by inducing ATP depletion and membrane fluidization which are the two main mechanisms. This group evaluated the effects of Pluronic block copolymers on the permeability of a broad panel of drugs using bovine BMEC monolayers as an in vitro model of BBB (64). The panel of compounds used in this study included known substrates for Pgp such as rhodamine 123, doxorubicin, digoxin, ritonavir, Taxol, and vinblastine. In addition, this panel included substrates of organic anion transporters: fluorescein, ziduvidin, and methotrexate, as well as compounds with less studied specificity for the efflux pumps, such as loperamide, valproic acid, and L-DOPA. Table 3 presents the apparent permeability coefficients (P_{app}) of the compounds in bovine brain microvessel endothelial cells (BBMEC) monolayers in the apical (AP) to basolateral (BL) direction. As is seen from the data, P85 increased AP to BL permeability in BBMEC monolayers with respect to a broad panel of structurally diverse compounds. The enhancement effects ranged from 1.3 times, in the case of methotrexate, to almost 20 times, in the case of vinblastine. These observations were reinforced by a study by Witt et al. using opioid peptides (65). These investigators demonstrated that P85 enhances transport of Pgp substrates, morphine, and [D-Pen2,D-Pen5]-enkephalin in BBMEC monolayers but not the transport of biphalin that is Pgp-independent.

The studies in our group also indicated Pluronic block copolymers can enhance the brain penetration of selected drugs in vivo. Kabanov et al. (15,66) demonstrated increased delivery of the neuroleptic drug, haloperidol, to the brain in mice with the use of insulin or antibody conjugated Pluronic micelles. In addition, haloperidol delivery to the brain was

Table 3 Effects of P85 on the Permeability of Various Solutes in Bovine BMEC Monolayers in AP to BL Direction

Solute	Brain to blood drug transporter	$P_{app} \times 10^6$ (cm/s) Assay buffer	P85	Effect[a]
Mannitol	None	5.7 ± 0.4	5.7 ± 0.4	n.s.
Loperamide	Unknown	25.2 ± 2.8	24.5 ± 1.1	n.s.
Valproic acid	Unknown	25.8 ± 1.4	22.8 ± 0.3	n.s.
Methotrexate[b]	MRP	8.2 ± 0.8	11.1 ± 0.4	1.3(*)
Fluorescein	MRP	16.5 ± 0.6	24.4 ± 0.2	1.5(*)
Rhodamine 123	Pgp	2.8 ± 0.2	4.5 ± 0.3	1.6(*)
Ziduvidin	MRP	16.5 ± 1.8	31.1 ± 9.1	2.0(*)
Doxorubicin	Pgp, MRP	13.2 ± 0.2	31.5 ± 7.3	2.4(*)
Digoxin[c]	Pgp	3.8 ± 0.4	15.5 ± 1.0	4.1(*)
Ritonavir[b]	Pgp	0.9 ± 0.1	6.5 ± 0.6	7.7(*)
Taxol	Pgp	1.5 ± 0.1	16.5 ± 2.1	11.2(*)
Vinblastine[b]	Pgp, MRP	0.6 ± 0.1	12.2 ± 2.0	19.0(*)

[a]Asterisk (*) shows statistically significant effects of the copolymer on the permeability of the drug.
[b]Data in preparation, not included in Ref. 64.
[c]Calculated using data reported by Batrakova et al. (32).
Source: From Ref. 93 and data reported by Batrakova et al. (64) except those stated to be different.

further increased in the presence of P85 (15,66). Batrakova et al. (32) examined the brain accumulation of digoxin in wild type mice, MDR1a knockout mice, and wild type mice treated with P85. The co-administration of 1% P85 with radiolabeled digoxin in wild-type mice increased the brain penetration of digoxin 3-fold and the digoxin level in the P85-treated wild-type mice was similar to that observed in the Pgp-deficient animals. This demonstrated that delivery to the CNS of a prototypical Pgp substrate could be significantly enhanced by co-administration of Pluronic. The already mentioned study by Witt et al. (65) reported that P85 significantly enhances the opioid peptide analgesia in mice. Interestingly, in this study the effects of the block copolymer on analgesia were observed not only with the Pgp substrates, morphine, and [D-Pen2, D-Pen5]-enkephalin, but also with biphalin, which was unaffected in in vitro transport studies. Therefore, in addition to

Pgp inhibition some other effect(s) of P85 in vivo could contribute to enhanced analgesia. Nevertheless, it has been concluded that the effects of the Pluronic block copolymers on drug efflux transporters technology will prove to be a valuable tool for enhancing drug delivery to selected organs.

5.2. Enhancement of Oral Bioavailability of Drugs by Pluronic Block Copolymers

There is increasing evidence that outwardly directed drug efflux systems hinder oral bioavailability of selected drugs. Intestinal epithelial cells are known to express a functionally active Pgp as well as various isoforms of MRP (67). By inhibiting the function of these drug efflux systems, it is possible to increase efficiency of delivery of these drugs through the oral route. The potential effectiveness of inhibition of the Pgp efflux system in enhancing oral bioavailability can be found in the studies by Nerurkar et al. (68,69) examining the effects of non-ionic detergents on peptide permeability in monolayers of the human colon epithelium cell line, Caco-2. The effects of Pluronic block copolymers on drug absorption and permeability in Caco-2 monolayers were also examined (34,70). The results obtained using rhodamine 123 were very similar to those observed during studies of drug transport in BBMEC monolayers. Specifically, the unimers of Pluronic block copolymers inhibited the Pgp efflux system in Caco-2 monolayers, resulting in a significant enhancement of absorption and permeability of a broad panel of drug and probe molecules. This suggested that Pluronic can be useful for increasing oral bioavailability of these compounds. This approach has been further validated by Alakhov and colleagues in an in vivo study in which C57 Bl/6 mice were given *p.o.* rhodamine 123. When the compound was formulated with a lipophilic Pluronic block copolymer its oral absorption rate increased by more than 3-fold (V. Alakhov, personal communication). Additionally the studies by Jannath et al. (71,72) demonstrated the increased oral uptake of two drugs, amikacin and tobramicin, which could be Pgp substrates, following oral administration to mice in the presence of poloxamer

CRL-1605. Furthermore, studies by Johnson et al. (73) using rat jejunal tissue as a model demonstrated that P85 not only increases intestinal uptake of digoxin and verapamil through inhibition of Pgp efflux, but also inhibits cytochrome P450 3A (CYP3A) metabolism in excised rat intestine. All of these studies provide substantial evidence that Pluronic block copolymers can be useful in increasing oral absorption of select drugs.

6. OPTIMIZATION OF PLURONIC COMPOSITION FOR PGP INHIBITION

The effects of Pluronic block copolymers in inhibiting the Pgp efflux system depend on the molecular composition of the block copolymer, and such relationship was evaluated in MDR cancer cell lines and BBMEC (30,58). These studies used a wide range of Pluronic block copolymer with various lengths of the EO and PO segments. As shown in Fig. 4A, the maximal rhodamine 123 accumulation levels observed with the most effective doses of each Pluronic were plotted as a function of the length of the hydrophobic PO block (N_{PO}). The hydrophilic copolymers (those having HLB from 20 to 29) were practically inactive in enhancing drug accumulation (empty circles). In contrast, hydrophobic copolymers (those having HLB < 20) enhanced rhodamine 123 accumulation to different extents, which depended on the molecular structure of the copolymers (filled circles). The most efficacious hydrophobic copolymers were those with intermediate lengths of the hydrophobic PO block (ranging from ~30 to ~60 PO repeating units). Based on the results of these studies, the copolymers were subdivided into four basic groups to describe their effects on Pgp. These groups include: hydrophilic copolymers with HLB 20–29 (group I); hydrophobic copolymers with intermediate PO block length of 30–60 PO units (group II); hydrophobic copolymers with PO block lengths less than 30 units (group IIIa); and hydrophobic copolymers with PO block lengths greater than 60 units (group IIIb). The relationships between the number

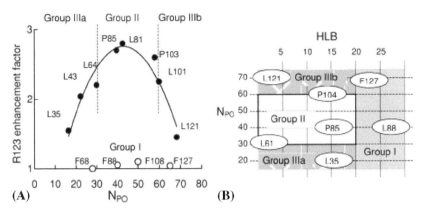

Figure 4 Efficacy of Pluronic block copolymer compositions displayed in inhibiting drug efflux function in BBMEC monolayers. (A) Rhodamine 123 (R123) enhancement factors are defined here as the ratios of rhodamine 123 accumulation in the cells in the presence of the block copolymer to rhodamine 123 accumulation in the assay buffer. (B) A grid of Pluronic indicating four groups determined based on the activity of these copolymers displayed in BBMEC monolayers as shown in A. (From Ref. 95.)

of PO units and HLB for the various groups, along with representative copolymers from each group, are shown in Figure 4B.

Pluronics in group I (Fig. 5) have extended hydrophilic EO block. These molecules adhere to the surface plasma membrane of the cells and limit the lateral mobility of membrane lipids, causing membrane solidification. As a result, they showed no or little inhibition of the Pgp efflux system in bovine BMEC monolayers. Pluronics in group II (Fig. 5) consist of lipophilic copolymers with intermediate length of PO block. Members of this group rapidly adhered to the cell membranes including mitochondria membranes and incorporated into them, resulting in membrane fluidization. These polymers has a two-fold effect; causing (i) a decrease of Pgp ATPase activity due to changes in the lipid microenvironment of Pgp, and (ii) an inhibition of ATP synthesis due to changes in the electron transport in the mitochondria membranes. Lipophilic Pluronic copolymers with short PO blocks in group

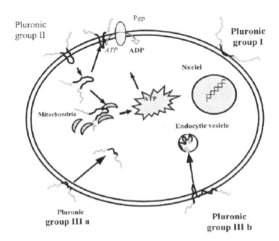

Figure 5 Scheme describing interactions between the BBB cells and four major groups of Puronics®. (From Ref. 58.)

IIIa (Fig. 5) could be placed between the hydrophilic Pluronics (group I) and the intermediate lipophilic Pluronics (group II) with respect to their effect on Pgp activity in bovine BMEC monolayers. They practically do not inhibit the Pgp efflux system due to their adhesion on the surface membranes of bovine BMEC causing the membrane solidification like the hydrophilic block copolymers. On the other hand, similar to the intermediate lipophilic Pluronics, they easily spread into the cytoplasm of the cell and reach intracellular compartments, including nuclei. However, the lack of effect on ATP intracellular levels is likely due to the absence of membrane fluidization, particularly, in mitochondria membranes. Extremely lipophilic Pluronic copolymers in group IIIb with long PO blocks (Fig. 5) are the most membranotropic block copolymers. They cause the highest fluidization effect on plasma membranes and the most efficient inhibition of Pgp ATPase activity in Pgp-containing membranes. Because of such high membranotropic properties these block copolymers anchor in the plasma membranes and remain there for an extended period of time. As a result, they are less efficiently transported into the intracellular compartments than intermediate

Pluronics (group II). Therefore, the extremely lipophilic Pluro-nic cause less energy depletion and, consequently, have less effect on Pgp efflux system in BBB cells than the inter-mediate block copolymers. An additional consideration with the very lipophilic Pluronic compositions is the low CMC. It has been shown previously that the effect of Pluronic is mediated by the copolymer single chain unimers, rather than by the micelles (74). Extremely lipophilic Pluronic tend to form micelles at low concentrations of the copolymer in water solutions. Thus, the micelle formation decreases the ability of Pluronic molecules to enter the cells and reduces the influ-ence of the copolymer on all systems in the barrier cells. All in all, a delicate balance between hydrophilic and lipophilic components in the Pluronic molecule should be accomplished to provide the best interactions and the most significant impact of the block copolymer on the endothelial cell transport.

The pattern of efficacy of Pluronic block copolymers observed in bovine BBMEC is quite similar to that previously reported in the study of Pgp expressing MDR cancer cell line, KBv (30). As is seen in Figure 6, the most efficacious block copolymers are those with intermediate lengths of the hydro-phobic blocks ranging from 30 to 60 PO repeating units, while the block copolymers with shorter or longer PO blocks are less efficient.

7. PLURONIC BLOCK COPOLYMERS FOR GENE THERAPY

The field of non-viral gene therapy has recently gained increased interest (75). It is widely believed that non-viral gene therapy can overcome some problems inherent to cur-rent viral-based therapies, including immune and toxic reac-tions as well as the potential for viral recombination (76). The Pluronic block copolymers, as one class of non-ionic polymers, have proven to be useful elements to the polyplexes for gene therapy applications. Furthermore, Pluronic block copoly-mers themselves appear to be very valuable for gene delivery

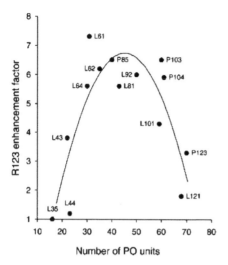

Figure 6 Optimization of Pluronic block copolymer composition in MDR cells. The entire set of hydrophilic copolymers with HLB varying from 20 to 29 had no or little effect on drug transport and is not presented in this figure. Only the copolymers with HLB less than 20 exhibiting varying degrees of activity in MDR cells are presented. (From Ref. 30.)

in skeletal muscle and other tissues. Both types of applications of the Pluronic block copolymers for gene delivery are considered along with some major aspects of self-assembly and biological properties of these block copolymers, which are essential for the development of such gene delivery systems.

7.1. Pluronic-Containing Polyplexes for Gene Delivery

Some recent reports suggest that Pluronic block copolymers can be used as the components of novel self-assembling gene delivery systems. Astafieva et al. (77) have demonstrated that Pluronic block copolymers can enhance polycation-mediated gene transfer in vitro. A synthetic polycation, poly(*N*-ethyl-4-vinylpyridinium bromide) (PEVP), and plasmid DNA, were mixed with 1% P85 to treat the cells. Both the DNA uptake

and transgene expression were significantly increased compared to the cells treated with the PEVP and DNA complex
alone (77). This study was reinforced by a recent report that
another block copolymer F127 enhances the receptor-
mediated gene delivery of plasmid DNA complexed with conjugates of poly(L-lysine) and ligands to hepatic and cancer cell
lines (78). Pluronic block copolymers were also able to prevent
the reduction of serum-mediated inhibition of gene transfer of
polyethyleneimine (PEI)-DNA complexes in NIH/3T3 cells
(79). Pluronics with higher HLB showed marked improvement of gene-expression levels in serum media. It was suggested that Pluronic block copolymers might bind with the
polyplexes and prevent their aggregation (79). However,
further studies are needed to demonstrate and characterize
self-assembly of Pluronic block copolymers with these
polyplexes.

Block-graft copolymers synthesized by covalent conjugation of Pluronic and branched PEI were used as materials for
preparation of polyplexes. Such polyplexes usually contain
three components: (i) DNA; (ii) Pluronic-PEI conjugate; and
(iii) free Pluronic (Fig. 7). The formulations were prepared
with both plasmid DNA and oligonucleotides (ODN), resulting
in stable polyplex dispersion with the particle size in the
ranges of 100–200 nm. These polyplexes were used successfully for delivery of plasmid DNA and antisense ODN in vitro

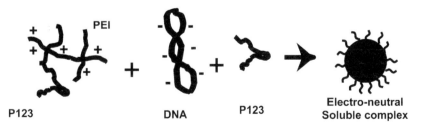

P123   DNA   P123   Electro-neutral
                                                          Soluble complex

Figure 7   Preparation of polyplexes with P123-g-PEI (2K). With
DNA and polycation alone the particle size was found to be
600 nm but when free pluronic P123 is added, an electroneutral
soluble complex of diameter 160 nm is obtained which are stable
even in presence of serum.

and in vivo (80–82). One such polyplex system, based on a Pluronic 123 polycation graft, namely P123-g-PEI(2K), has shown effective gene delivery in vitro compared to PEI and other polycationic vectors (80,83) (Fig. 8). One major advantage of this system demonstrated in vitro is that the complexes formed by P123-g-PEI(2K) and DNA with the free Pluronic are stable in a variety of conditions particularly when the serum proteins are present. The P123-g-PEI(2K)-based system also works well in vivo, which exhibits more uniform biodistribution and significant gene expression in the liver compared to unmodified PEI (80). This system was subsequently used for intravenous (I.V.) delivery of the gene encoding the murine ICAM-1 molecule in the liver of the transgenic ICAM-1 deficient mice (82). Recently similar delivery systems have been developed, such as those based on conjugates of Pluronic block copolymers with polylysine (84).

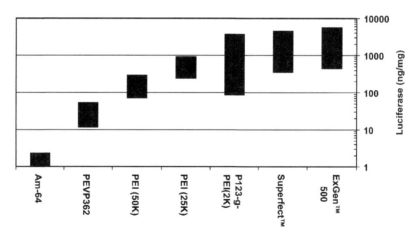

Figure 8 Relative transfection activity of various polycations in Cos-7 cells. Luciferase reporter gene along with polycations at their particular optimum ratio was used. Data represents values of luciferase expression (ng/mg) obtained following with a polyplex over a series of experiments. PEI, polyethyleneimine; PEVP, poly(n-ethyl-4-vinylpyridinium bromide); polyamidoamine dendrimer (Superfect™); AM-64, Astramol™. (From Ref. 83.)

7.2. Pluronic Block Copolymers Enhance Gene Expression in Stably Transformed Cell Models

Traditional wisdom in polymer-based gene delivery is that polymers are needed as structural and functional components to enhance delivery of DNA into the cells (the "artificial virus" or polyplex paradigm). However, it is also possible that some of these polymer components can affect the machinery for gene expression of delivered DNA. To test this hypothesis, we recently developed the models of NIH 3T3 cells stably transformed with the luciferase or green fluorescent protein (GFP) reporter gene (both under control of the CMV promoter). These cells were exposed to various free Pluronic block copolymers; also P123-g-PEI(2K) and gene expression levels were examined. As is shown in Figure 9 selected copolymers, such as P85 and P123, substantially increased gene expression by over ten times (in preparation). Similar results were observed with cells treated with P123-g-PEI (2K) (not shown), This study is important for understanding the effects that Pluronic block copolymers (as well as other water soluble

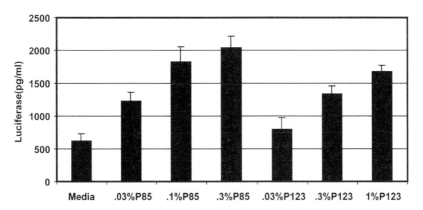

Figure 9 Effect of Pluronics on gene expression: NIH3T3 cells were stably transfected with gWIZluc. Cells were treated with the polymers for 4 h followed by 24 h incubation in fresh media, then harvested and the effect of polymer components on the reporter gene expression was evaluated (unpublished data).

polymers used in polyplexes) have on gene expression. These effects could also contribute to alterations in transgene expression observed with some viral gene delivery systems after formulation with Pluronic block copolymers (85–87).

7.3. Effect of Pluronic Block Copolymers on Gene Delivery In Vivo

Plasmid DNA can be injected into skeletal muscle and generate therapeutically meaningful levels of gene expression (88). However, in many cases the relatively low level of gene expression limits the applicability of naked DNA as a therapeutic agent. One of the alternative approaches to solve this problem is to enhance gene expression of the transfected DNA into muscles by using a delivery system. It was recently reported that certain Pluronic block copolymers significantly increase the expression of plasmid DNA in skeletal muscles of mice (89). The SP1017 formulation based on the mixture of the Pluronic block copolymers, L61 and F127, increased the expression of both reporter and therapeutic genes by 5–20-fold compared to naked DNA (Fig. 10). Unlike cationic DNA carriers, SP1017 does not condense DNA. SP1017 exhibited maximal activity at concentrations that are close to the CMC values of the Pluronic components in SP1017. This suggested that the unimers, not the micelles, are involved in this activity. With such low doses required for enhanced gene expression, the formulation provides at least a 500-fold safety margin in animals. Compared to PVP, SP1017 was found to be more efficient and required less DNA to produce the same level of transgene expression (89). Furthermore, the histology study using β-galactosidase as reporter gene demonstrated that SP1017 significantly enhanced distribution of transgene throughout the muscles and thus increased its "bioavailability." Another Pluronic block copolymer, L64 (PE6400), was also found to achieve gene expression in up to 35% fibers of mouse tibial cranial muscle, which is comparable to the effects obtained by electro-transfer techniques (90). More recently, SP1017 and Pluronic P123 were shown to enhance and prolong transgene expression by administrating DNA

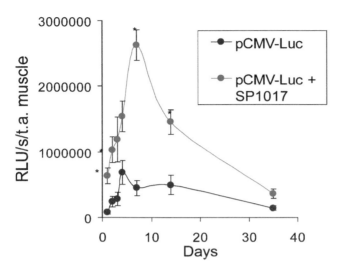

Figure 10 Time-course of gene expression in muscles: C57Bl/6 mice (6–8 week old females) were injected with 5 μg of pCMV-luc alone or formulated with SP1017 in tibialis anterior muscles. The muscles were harvested from day 1 to day 35 after injection and extracted with a lysis buffer to measure the luciferase activity. (From Ref. 89.)

into solid tumors (91). Furthermore, it was also reported that Pluronic F68 (Poloxamer 188) was used as a carrier for topical gene delivery in eye drops formulation (92).

The mechanism by which the Pluronic block copolymers enhance gene delivery in vivo is still unknown. Pluronic formulations have no apparent effect on delivery and expression of naked DNA in vitro in the murine muscle cell line C2C12 (89), isolated cardiomyocytes, or established muscle cell lines (90). However, when DNA was microinjected into the cell cytoplasm, Pluronic L64 promoted DNA translocation into the nucleus and induced gene expression (90). These results indicated that the delivery of naked DNA across the cell membrane, a limiting step for transgene expression, might not be affected by Pluronic, and that Pluronic might play critical roles in other processes consistent with the observations described in the previous sections.

Overall, the Pluronic block copolymers are currently among the most intriguing drug and gene delivery systems. They display abilities to self-assemble into micelles, incorporate drugs and genes, and transport them within the body. Pluronic block copolymers also have a broad spectrum of biological response modifying activities. All these properties encourage further investigations regarding the application of Pluronic formulations as promising drug and gene delivery systems.

ACKNOWLEDGMENTS

We would like to thank Dr. Elena Batrakova (UNMC) and Dr. Valery Alakhov (Supratek Pharma Inc.) for the discussions and critical advice. AVK acknowledges support from National Institute of Health related to the studies of drug transport to the brain and treatment of MDR cancer using Pluronic block copolymers as well as support from Nebraska research Initiative for his studies in gene delivery. AVK has been a co-founder, shareholder and scientific advisor of Supratek Pharma Inc.

REFERENCES

1. Langer R. Drug delivery and targeting. Nature 1998; 392:5–10.

2. Stolnik S, Illum L, Davis SS. Long circulating microparticulate drug carriers. Adv Drug Delivery Rev 1995; 16:195–214.

3. Schmolka IR. A review of block polymer surfactants. J Am Oil Chem Soc 1977; 54:110–116.

4. Geyer RP. Perfluorochemicals as oxygen transport vehicles. Biomater Artif Cells Artif Organs 1988; 16:31–49.

5. Allison AC, Byars NE. Adjuvant formulations and their mode of action. Semin Immunol 1990; 2:369–374.

6. Schmolka IR. Physical basis for poloxamer interactions. Ann NY Acad Sci 1994; 720:92–97.

7.   Hawley AE, Davis SS, Illum L. Targeting of colloids to lymph nodes: influence of lymphatic physiology and colloidal characteristics. Adv Drug Delivery Rev 1995; 17:129–148.

8.   Newman MJ, Actor JK, Balusubramanian M, Jagannath C. Use of nonionic block copolymers in vaccines and therapeutics. Crit Rev Ther Drug Carrier Syst 1998; 15:89–42.

9.   Malmsten M. Block copolymers in pharmaceutics. In: Alexandridis P, Lindman B, eds. Amphiphilic Block Copolymers: Self Assembly and Applications. Amsterdam, New York: Elesvier, 2000:319–346.

10. Moghimi SM, Hunter AC. Poloxamers and poloxamines in nanoparticle engineering and experimental medicine. Trends Biotechnol 2000; 18:412–420.

11. Anderson BC, Pandit NK, Mallapragada SK. Understanding drug release from poly(ethylene oxide)-b-poly(propylene oxide)-b-poly(ethylene oxide) gels. J Contr Rel 2001; 70: 157–167.

12. Emanuele MR, Balasubramanian M, Alludeen HS. Polyoxypropylene/polyoxyethylene copolymers with improved biological activity. Norcross, GA: CytRx Corporation, 1996.

13. Emanuele MR, Hunter RL, Culbreth PH. Polyoxypropylene/ polyoxyethylene copolymers with improved biological activity. US, CYTRX Corporation: Norcross, GA, 1996.

14. Nagarajan R. Solubilization of hydrocarbons and resulting aggregate shape transitions in aqueous solutions of Pluronic (PEO-PPO-PEO) block copolymers. Coll Surf B 1999; 16:55–72.

15. Kabanov AV, Chekhonin VP, Alakhov VY, et al. The neuroleptic activity of haloperidol increases after its solubilization in surfactant micelles. Micelles as microcontainers for drug targeting. FEBS Lett 1989; 258:343–345.

16. Kabanov AV, Batrakova EV, Melik-Nubarov NS, et al. A new class of drug carriers: micelles of poly(oxyethylene)-poly (oxypropylene) block copolymers as microcontainers for drug targeting from blood in brain. J Contr Rel 1992; 22:141–157.

17. Wang PL, Johnston TP. Enhanced stability of two model proteins in an agitated solution environment using poloxamer 407. J Parenter Sci Technol 1993; 47:183–189.

18. Wang PL, Johnston TP. Thermal-induced denaturation of two model proteins: effect of poloxamer 407 on solution stability. Int J Pharm 1993; 96:41–49.

19. Batrakova EV, Dorodnych TY, Klinskii EY, et al. Anthracycline antibiotics non-covalently incorporated into the block copolymer micelles: in vivo evaluation of anti-cancer activity. Br J Cancer 1996; 74:1545–1552.

20. Alakhov VY, Kabanov AV. Block copolymeric biotransport carriers as versatile vehicles for drug delivery. Exp Op Invest Drugs 1998; 7:1453–1473.

21. Rapoport N. Stabilization and activation of Pluronic micelles for tumor-targeted drug delivery. Colloids Surf B 1999; 16: 93–111.

22. Rapoport NY, Herron JN, Pitt WG, Pitina L. Micellar delivery of doxorubicin and its paramagnetic analog, ruboxyl, to HL-60 cells: effect of micelle structure and ultrasound on the intracellular drug uptake. J Contr Rel 1999; 58:153–162.

23. Kabanov AV, Alakhov VY. Pluronic block copolymers in drug delivery: from micellar nanocontainers to biological response modifiers. Crit Rev Ther Drug Carrier Syst 2002; 19:1–72.

24. Naito S, Yokomizo A, Koga H. Mechanisms of drug resistance in chemotherapy for urogenital carcinoma. Int J Urol 1999; 6:427–439.

25. Kuwano M, Toh S, Uchiumi T, Takano H, Kohno K, Wada M. Multidrug resistance-associated protein subfamily transporters and drug resistance. Anticancer Drug Des 1999; 14:123–131.

26. Alakhov VY, Moskaleva EY, Batrakova EV, Kabanov AV. Hypersensitization of multidrug resistant human ovarian carcinoma cells by pluronic P85 block copolymer. Bioconjug Chem 1996; 7:209–216.

27. Page M, Alakhov VY. Elimination of P-gp-mediated multidrug resistance by solubilization in Pluronic micelles. Proceedings of the Annual Meeting of the American Association for Cancer Research (AACR), Philadelphia, PA, USA. Vol. 33, 1992.

28. Venne A, Li S, Mandeville R, Kabanov A, Alakhov V. Hypersensitizing effect of pluronic L61 on cytotoxic activity,

transport, and subcellular distribution of doxorubicin in multiple drug- resistant cells. Cancer Res 1996; 56:3626–3629.

29. Alakhov V, Klinski E, Li S, et al. Block copolymer-based formulation of doxorubicin. From cell screen to clinical trials. Colloids Surf B 1999; 16:113–134.

30. Batrakova EV, Lee S, Li S, Venne A, Alakhov V, Kabanov A. Fundamental relationships between the composition of pluronic block copolymers and their hypersensitization effect in MDR cancer cells. Pharm Res 1999; 16:1373–1379.

31. Evers R, Kool M, Smith AJ, van Deemter L, de Haas M, Borst P. Inhibitory effect of the reversal agents V-104, GF120918 and pluronic L61 on MDR1 Pgp-, MRP1- and MRP2-mediated transport. Br J Cancer 2000; 83:366–374.

32. Batrakova EV, Miller DW, Li S, Alakhov VY, Kabanov AV, Elmquist WF. Pluronic P85 enhances the delivery of digoxin to the brain: in vitro and in vivo studies. J Pharmacol Exp Ther 2001; 296:551–557.

33. Nagarajan R, Ganesh K. Comparison of solubilization of hydrocarbons in (PEO-PPO) diblock versus (PEO-PPO-PEO) triblock copolymer micelles. J Colloid Interface Sci 1996; 184:489–499.

34. Batrakova EV, Han HY, Alakhov V, Miller DW, Kabanov AV. Effects of pluronic block copolymers on drug absorption in Caco-2 cell monolayers. Pharm Res 1998; 15:850–855.

35. Miller DW, Batrakova EV, Kabanov AV. Inhibition of multidrug resistance-associated protein (MRP) functional activity with pluronic block copolymers. Pharm Res 1999; 16:396–401.

36. Batrakova EV, Li S, Alakhov VY, Elmquist WF, Miller DW, Kabanov AV. Sensitization of cells overexpressing multidrug resistant protein by Pluronic P85. Pharm Res 2003; 20: 1581–1590.

37. Breuninger LM, Paul S, Gaughan K, et al. Expression of multidrug resistance-associated protein in NIH/3T3 cells confers multidrug resistance associated with increased drug efflux and altered intracellular drug distribution. Cancer Res 1995; 55:5342–5347.

38. Nooter K, Stoter G. Molecular mechanisms of multidrug resistance in cancer chemotherapy. Pathol Res Pract 1996; 192:768–780.

39. Cleary I, Doherty G, Moran E, Clynes M. The multidrug-resistant human lung tumour cell line, DLKP-A10, expresses novel drug accumulation and sequestration systems. Biochem Pharmacol 1997; 53:1493–1502.

40. Shapiro AB, Fox K, Lee P, Yang YD, Ling V. Functional intracellular P-glycoprotein. Int J Cancer 1998; 76:857–864.

41. Altan N, Chen Y, Schindler M, Simon SM. Defective acidification in human breast tumor cells and implications for chemotherapy. J Exp Med 1998; 187:1583–1598.

42. Benderra Z, Morjani H, Trussardi A, Manfait M. Role of the vacuolar H+-ATPase in daunorubicin distribution in etoposide-resistant MCF7 cells overexpressing the multidrug-resistance associated protein. Int J Oncol 1998; 12:711–715.

43. Batrakova EV, Li S, Alakhov VY, Kabanov AV. Selective energy depletion and sensitization of multiple drug resistant cancer cells by pluronic block copolymers. Polym Prep 2000; 41:1639–1640.

44. Batrakova EV, Li S, Elmquist WF, Miller DW, Alakhov VY, Kabanov AV. Mechanism of sensitization of MDR cancer cells by pluronic block copolymers: selective energy depletion. Br J Cancer 2001; 85:1987–1997.

45. Kirillova GP, Mokhova EN, Dedukhova VI, et al. The influence of pluronics and their conjugates with proteins on the rate of oxygen consumption by liver mitochondria and thymus lymphocytes. Biotechnol Appl Biochem 1993; 18:329–339.

46. Chiu P, Karler R, Craven C, Olsen DM, Turkanis SA. The influence of delta9-tetrahydrocannabinol, cannabinol and cannabidiol on tissue oxygen consumption. Res Commun Chem Pathol Pharmacol 1975; 12:267–286.

47. Rapoport N, Marin AP, Timoshin AA. Effect of a polymeric surfactant on electron transport in HL-60 cells. Arch Biochem Biophys 2000; 384:100–108.

48. van Zutphen H, Merola AJ, Brierley GP, Cornwell DG. The interaction of nonionic detergents with lipid bilayer membranes. Arch Biochem Biophys 1972; 152:755–766.

49. Atkinson TP, Smith TF, Hunter RL. Histamine release from human basophils by synthetic block co-polymers composed of polyoxyethylene and polyoxypropylene and synergy with immunologic and non-immunologic stimuli. J Immunol 1988; 141:1307–1310.

50. Atkinson TP, Smith TF, Hunter RL. In vitro release of histamine from murine mast cells by block co-polymers composed of polyoxyethylene polyoxypropylene. J Immunol 1988; 141:1302–1306.

51. Brierley GP, Jurkowitz M, Merola AJ, Scott KM. Ion transport by heart mitochondria. XXV. Activation of energy-linked K + uptake by non-ionic detergents. Arch Biochem Biophys 1972; 152:744–754.

52. Brustovetskii NN, Dedukhova VN, Egorova MV, Mokhova EN, Skulachev VP. Uncoupling of oxidative phosphorylation by fatty acids and detergents suppressed by ATP/ADP antiporter inhibitors. Biochem (Moscow) 1991; 56:1042–1048.

53. Kaplan O, Jaroszewski JW, Clarke R, et al. The multidrug resistance phenotype: 31P nuclear magnetic resonance characterization and 2-deoxyglucose toxicity. Cancer Res 1991; 51:1638–1644.

54. Kaplan O, Navon G, Lyon RC, Faustino PJ, Straka EJ, Cohen JS. Effects of 2-deoxyglucose on drug-sensitive and drug-resistant human breast cancer cells: toxicity and magnetic resonance spectroscopy studies of metabolism. Cancer Res 1990; 50:544–551.

55. Marton A, Mihalik R, Bratincsak A, et al. Apoptotic cell death induced by inhibitors of energy conservation–Bcl-2 inhibits apoptosis downstream of a fall of ATP level. Eur J Biochem 1997; 250:467–475.

56. Martin DS, Stolfi RL, Colofiore JR, Nord LD, Sternberg S. Biochemical modulation of tumor cell energy in vivo: II. A lower dose of adriamycin is required and a greater antitumor activity is induced when cellular energy is depressed. Cancer Invest 1994; 12:296–307.

57. Melik-Nubarov NS, Pomaz OO, Dorodnych T, et al. Interaction of tumor and normal blood cells with ethylene oxide and propylene oxide block copolymers. FEBS Lett 1999; 446:194–198.

58. Batrakova EV, Li S, Alakhov VY, Miller DW, Kabanov AV. Optimal structure requirements for pluronic block copolymers in modifying P-glycoprotein drug efflux transporter activity in bovine brain microvessel endothelial cells. J Pharmacol Exp Ther 2003; 304:845–854.

59. Batrakova EV, Li S, Vinogradov SV, Alakhov VY, Miller DW, Kabanov AV. Mechanism of pluronic effect on P-glycoprotein efflux system in blood-brain barrier: contributions of energy depletion and membrane fluidization. J Pharmacol Exp Ther 2001; 299:483–493.

60. Regev R, Assaraf YG, Eytan GD. Membrane fluidization by ether, other anesthetics, and certain agents abolishes P-glycoprotein ATPase activity and modulates efflux from multidrug-resistant cells. Eur J Biochem 1999; 259:18–24.

61. Ranson M, Ferry D, Kerr D, et al. Results of a cancer research campaighn phase I dose escalation trial of SP1049C in patients with advanced cancer. Proceedings of the Fifth International Symposium on Polymer Therapeutics: From Laboratory to Clinical Practice, Cardiff, UK, 3–5 January 2002. The Welsh School of Pharmacy, Cardiff University.

62. Kostarelos K, Tadros TF, Luckham PF. Physical conjugation of (tri-) block copolymers to liposomes toward the construction of sterically stabilized vesicle systems. Langmuir 1999; 15:369–376.

63. Cordon-Cardo C, O'Brien JP, Casals D, et al. Multidrug-resistance gene (P-glycoprotein) is expressed by endothelial cells at blood-brain barrier sites. Proc Natl Acad Sci USA 1989; 86:695–698.

64. Batrakova EV, Li S, Miller DW, Kabanov AV. Pluronic P85 increases permeability of a broad spectrum of drugs in polarized BBMEC and Caco-2 cell monolayers. Pharm Res 1999; 16:1366–1372.

65. Witt KA, Huber JD, Egleton RD, Davis TP. Pluronic p85 block copolymer enhances opioid peptide analgesia. J Pharmacol Exp Ther 2002; 303:760–767.

66. Kabanov AV, Slepnev VI, Kuznetsova LE, et al. Pluronic micelles as a tool for low-molecular compound vector delivery into a cell: effect of Staphylococcus aureus enterotoxin B on cell loading with micelle incorporated fluorescent dye. Biochem Int 1992; 26:1035–1042.

67. Hirohashi T, Suzuki H, Chu XY, Tamai I, Tsuji A, Sugiyama Y. Function and expression of multidrug resistance-associated protein family in human colon adenocarcinoma cells (Caco-2). J Pharmacol Exp Ther 2000; 292:265–270.

68. Nerurkar MM, Burton PS, Borchardt RT. The use of surfactants to enhance the permeability of peptides through Caco-2 cells by inhibition of an apically polarized efflux system. Pharm Res 1996; 13:528–534.

69. Nerurkar MM, Ho NF, Burton PS, Vidmar TJ, Borchardt RT. Mechanistic roles of neutral surfactants on concurrent polarized and passive membrane transport of a model peptide in Caco-2 cells. J Pharm Sci 1997; 86:813–821.

70. Batrakova EV, Han HY, Miller DW, Kabanov AV. Effects of pluronic P85 unimers and micelles on drug permeability in polarized BBMEC and Caco-2 cells. Pharm Res 1998; 15:1525–1532.

71. Banerjee SK, Jagannath C, Hunter RL, Dasgupta A. Bioavailability of tobramycin after oral delivery in FVB mice using CRL-1605 copolymer, an inhibitor of P-glycoprotein. Life Sci 2000; 67:2011–2016.

72. Jagannath C, Wells A, Mshvildadze M, et al. Significantly improved oral uptake of amikacin in FVB mice in the presence of CRL-1605 copolymer. Life Sci 1999; 64:1733–1738.

73. Johnson BM, Charman WN, Porter CJ. An in vitro examination of the impact of polyethylene glycol 400, Pluronic P85, and vitamin E d-alpha-tocopheryl polyethylene glycol 1000 succinate on P-glycoprotein efflux and enterocyte-based metabolism in excised rat intestine. AAPS Pharm Sci 2002; 4:40.

74. Miller DW, Batrakova EV, Waltner TO, Alakhov V, Kabanov AV. Interactions of pluronic block copolymers with

brain microvessel endothelial cells: evidence of two potential pathways for drug absorption. Bioconjug Chem 1997; 8: 649–657.

75. Felgner P, Barenholz Y, Behr J, et al. Nomenclature for synthetic gene delivery systems. Hum Gene Ther 1997; 8:511–512.

76. Hengge UR, Taichman LB, Kaur P, et al. How realistic is cutaneous gene therapy? Exp Dermatol 1999; 8:419–431.

77. Astafieva I, Maksimova I, Lukanidin E, Alakhov V, Kabanov A. Enhancement of the polycation-mediated DNA uptake and cell transfection with Pluronic P85 block copolymer. FEBS Lett 1996; 389:278–280.

78. Cho CW, Cho YS, Lee HK, Yeom YI, Park SN, Yoon DY. Improvement of receptor-mediated gene delivery to HepG2 cells using an amphiphilic gelling agent. Biotechnol Appl Biochem 2000; 32:21–26.

79. Kuo JH. Effect of Pluronic-block copolymers on the reduction of serum-mediated inhibition of gene transfer of polyethyleneimine-DNA complexes. Biotechnol Appl Biochem 2003; 37:267–271.

80. Nguyen HK, Lemieux P, Vinogradov SV, et al. Evaluation of polyether-polyethyleneimine graft copolymers as gene transfer agents. Gene Ther 2000; 7:126–138.

81. Ochietti B, Guerin N, Vinogradov SV, et al. Altered organ accumulation of oligonucleotides using polyethyleneimine grafted with poly(ethylene oxide) or pluronic as carriers. J Drug Target 2002; 10:113–121.

82. Ochietti B, Lemieux P, Kabanov AV, Vinogradov S, St-Pierre Y, Alakhov V. Inducing neutrophil recruitment in the liver of ICAM-1-deficient mice using polyethyleneimine grafted with Pluronic P123 as an organ-specific carrier for transgenic ICAM-1. Gene Ther 2002; 9:939–945.

83. Gebhart CL, Sriadibhatla S, Vinogradov S, Lemieux P, Alakhov V, Kabanov AV. Design and formulation of polyplexes based on pluronic- polyethyleneimine conjugates for gene transfer. Bioconjug Chem 2002; 13:937–944.

84. Jeon E, Kim HD, Kim JS. Pluronic-grafted poly-(L)-lysine as a new synthetic gene carrier. J Biomed Mater Res 2003; 66A:854–859.

85. March KL, Madison JE, Trapnell BC. Pharmacokinetics of adenoviral vector-mediated gene delivery to vascular smooth muscle cells: modulation by poloxamer 407 and implications for cardiovascular gene therapy. Hum Gene Ther 1995; 6:41–53.

86. Van Belle E, Maillard L, Rivard A, et al. Effects of poloxamer 407 on transfection time and percutaneous adenovirus-mediated gene transfer in native and stented vessels. Hum Gene Ther 1998; 9:1013–1024.

87. Croyle MA, Cheng X, Sandhu A, Wilson JM. Development of novel formulations that enhance adenoviral-mediated gene expression in the lung in vitro and in vivo. Mol Ther 2001; 4:22–28.

88. Wolff JA, Malone RW, Williams P, et al. Direct gene transfer into mouse muscle in vivo. Science 1990; 247:1465–1468.

89. Lemieux P, Guerin N, Paradis G, et al. A combination of polox-amers increases gene expression of plasmid DNA in skeletal muscle. Gene Ther 2000; 7:986–991.

90. Pitard B, Pollard H, Agbulut O, et al. A nonionic amphiphile agent promotes gene delivery in vivo to skeletal and cardiac muscles. Hum Gene Ther 2002; 13:1767–1775.

91. Gebhart C, Alakhov VY, Kabanov AV. Pluronic block copoly-mers enhance local transgene expression in skeletal muscle and solid tumor. Glasgow, Scotland, UK: Controlled Release Society, 2003.

92. Liaw J, Chang SF, Hsiao FC. In vivo gene delivery into ocular tissues by eye drops of poly(ethylene oxide)-poly(propylene oxide)-poly(ethylene oxide) (PEO-PPO-PEO) polymeric micelles. Gene Ther 2001; 8:999–1004.

93. Kabanov AV, Batrakova EV, Alakhov VY. Pluronic block copolymers as novel polymer therapeutics for drug and gene delivery. J Contr Rel 2002; 82:189–212.

94. Kabanov AV, Batrakova EV, Alakhov VY. Pluronic block copolymers for overcoming drug resistance in cancer. Adv Drug Delivery Rev 2002; 54:759–779.

95. Kabanov AV, Batrakova EV, Miller DW. Pluronic(R) block copolymers as modulators of drug efflux transporter activity in the blood-brain barrier. Adv Drug Delivery Rev 2003; 55:151–164.

96. Kozlov MY, Melik-Nubarov NS, Batrakova EV, Kabanov AV. Relationship between pluronic block copolymer structure, critical micellization concentration and partitioning coefficients of low molecular mass solutes. Macromolecules 2000; 33:3305–3313.

15

Graft Copolymers for Therapeutic Gene Delivery

ANURAG MAHESHWARI, JAMES W.
YOCKMAN, and SUNG WAN KIM
Center for Controlled Chemical Delivery,
University of Utah, Salt Lake City, Utah, U.S.A.

1. NON-VIRAL GENE THERAPY

The year 2003 was celebrated as the culmination of a half-century of discoveries regarding the structural aspects of DNA. Since the modern era of genetics got underway in 1953, the field of molecular therapy and recombinant DNA technology has made great strides and offers humanity immense potential in both the treatment of numerous diseases that are currently untreatable and the more effective treatment—free of side effects—of treatable diseases whose treatments, nevertheless, give rise to secondary and adverse outcomes.

615

   Gene therapy is defined as any procedure in which
genetic materials are transferred into patients so that the
resulting circumvention of genetic abnormalities can bring
out therapeutic effects in the patients. With gene therapy
approach, the treatment of human diseases takes place at a
molecular level and derives from either the restoration of
defective biological functions or the re-constitution of homeos-
tasis in the body. Both pre-clinical and clinical gene therapy
research have been progressing rapidly during the past 15
years. Gene therapy now appears to be a highly promising
new modality in, and a clinical reality for, the treatment of
numerous human disorders. Since 1989, when the first clinical test of gene therapy took place, more than 600 gene therapy protocols have been approved, and more than 3000
patients have received gene therapy. However, at the time
of writing this chapter, no gene therapy products have been
approved for clinical use owing to safety concerns. Additionally, there seems to be a huge performance gap between the
success of viral vectors and non-viral delivery mechanisms.
Whereas viral vectors constitute approximately 75% of all
clinical trails, non-viral methods constitute less than 25%
among clinical trials. Even more stupefying is the fact that,
to date, only 2 out of 147 on-going or completed (or abandoned) trials based on non-viral vectors are currently in phase
III clinical trials (1).
   This enormous imbalance has been challenged by way of
investigative research into not only comparisons of non-viral
vectors with viral vectors but also the current potential and
the current limitations of non-viral vectors, in themselves.
It has become common knowledge among pharmaceutical
scientists and gene therapy investigators that, although
viral vectors—having evolved over millions of years so as to
transfer genetic material in a most efficient way—hold stu-
pendous promises in the treatment of disease, we have not
yet discovered an example of one that is free from the potential to invoke toxic effects and counterproductive immune
responses. Although several excellent reviews enumerating
the promises and the challenges of viral vectors can be found,
very briefly, clinicians use gutted viruses as a vector with

which to transfer corrective genes into patients' cells. However, if a viral vector integrates itself into the host genome, that vector can cause the cells to mutate and ultimately become cancerous. Furthermore, viral vectors that are derived from adeno-associated viruses, which were thought to be one of the most promising viruses for therapeutic purposes, have been found to integrate themselves more often into genes than into non-coding regions of DNA. Additionally, the viral vectors inside the host can still invoke counterproductive immune responses thereby rendering the repeated administration of a treatment more difficult (2).

On the other hand non-viral vectors, which have only been used for the past two decades, have yet to become a marketing reality because they still do not meet adequate safety profiles, the standards for efficient delivery, or the rigorous testing of clinical trials. Non-viral vectors involving cationic liposomes have been most widely used in pre-clinical and clinical trials; however, apart from the non-viral delivery of genetic material, which is inefficient when compared to the delivery associated with viral vectors, non-viral vectors still suffer from some of the same toxicity issues from which viral vectors suffer. Some of the delivery issues such as lipoplex instability in the blood stream, interaction with plasma proteins, and non-specific cellular interactions persist, despite the extensive investigations and corrective maneuvers that have been conducted over the past decade. Therefore, currently, the only advantage with cationic lipids seems to be their inability to transform host cells into cancerous cells (3,4).

Another not so widely examined area concerns both the use of synthetic polymers and peptides (as well as their combinations) for gene delivery and, especially, their pre-clinical and clinical challenges. In fact, to date, not a single clinical trial has been undertaken so that the potential of polymer-based gene therapy can be examined. This research gap has infused in much of the scientific community a sense of urgency to take up these challenges. Although polymeric gene therapy suffers from some of the same challenges that are found among cationic liposomes, the advantage with polymers is the infinite number of ways in which their chemical structure can be modified and augmented not only to circumvent inefficient transfection, instability in the blood stream, non-specific uptake, and interactions with cells and plasma molecules but also to bypass the immune system. These synthetic modifications can lead to polymers with superior delivery dynamics tailored specifically to the delivery of genetic material in a site-specific and in some cases even time-controlled manner. Therefore, the limiting factor could be the human endeavor rather than shortcomings of research potential in this stagnating but promising field (5,6). Although polymeric gene delivery vectors come in various configurations, this chapter as is evident from the title, will focus exclusively on graft-copolymers as carriers of nucleic acids.

2. NOMENCLATURE OF GRAFT COPOLYMERS

A graft polymer is a polymer which comprises of molecules with one or more species of blocks connected to the main chain as side chains, the side chains having constitutional or configurational features that differ from those in the main chain. A constitutional unit is that whose repetition describes a regular polymer whereas a configurational unit is a constitutional unit having one or more sites of defined stereoisomerism. A graft copolymer is then defined as one whose distinguishing feature of the side chains is constitutional; i.e., the side chains comprise units derived from at least one species of monomer different from those which supply the units of the main chain (7). A running definition- somewhat less intricate is that a graft copolymer is comprised of a high molecular weight backbone to which a second polymer is attached at intervals along the chain. The backbone may be homopolymeric or copolymeric with pendant groups of either type (8). The nomenclature of the graft copolymers can be derived from a simple structure represented by A_k-*graft*-B_m where the monomer A is that which supplies the backbone or the main chain units and B is that which supplies the side chain(s). Thus for this simplest case the graft copolymer would be named polyA-*graft*-polyB. For example, chains of poly(ethylene glycol) (PEG) linked to a poly(L-lysine) (PLL) backbone would be called PLL-*graft*-PEG. An even more restricted definition of a graft copolymer with emphasis on "polymer" is provided by the IUPAC committee where block A = 14 and B = 15 are the minimum number of units constituting the backbone and the side chains respectively. There can be increasingly complex configurations such as a block copolymer grafted to another polymer resulting in a configuration such as (polyA-*block*-polyB)-*graft*-polyC. If more than one type of graft chain is attached to the backbone, semicolons are used to separate the names of the grafts; for instance, polyA-*graft*-(polyB; polyC) (7). An example of this can be a PLL backbone to which are attached two different side chains such as poly(L-histidine) and PEG resulting in PLL-*graft*-(PLH;PEG). Even more complex configurations require correspondingly detailed nomenclature; however, as the reader will note such configurations are rarely pursued in the laboratory and especially so in the case of synthesizing polymeric gene delivery vectors.

It must also be noted that the aforementioned nomenclature though recommended by IUPAC may or may not be followed strictly by research laboratories around the world. For example a PEG chain grafted to a PLL backbone may be written simply as PEG-graft-PLL or PEG-grafted-PLL, thereby reversing the IUPAC prescribed PLL-graft-PEG nomenclature. For these reasons, as far as possible, the author will try to indicate the backbone in the subsequent sections of this chapter.

3. CATIONIC GRAFT COPOLYMERS

Figure 1 represents the chemical structures of cationic polymers that can be used as a backbone for synthesizing graft copolymers to be used as gene delivery carriers. This list is by no means exhaustive; however, most of the work published so far has involved one of these polycations. Among these, PLL and poly(ethyleneimine) (PEI) are the most widely used backbones with more recent work leaning toward PEI as the main chain or backbone of choice. These cationic polymers have a common chemical characteristic, the presence of amine functional group, which is positively charged at physiological

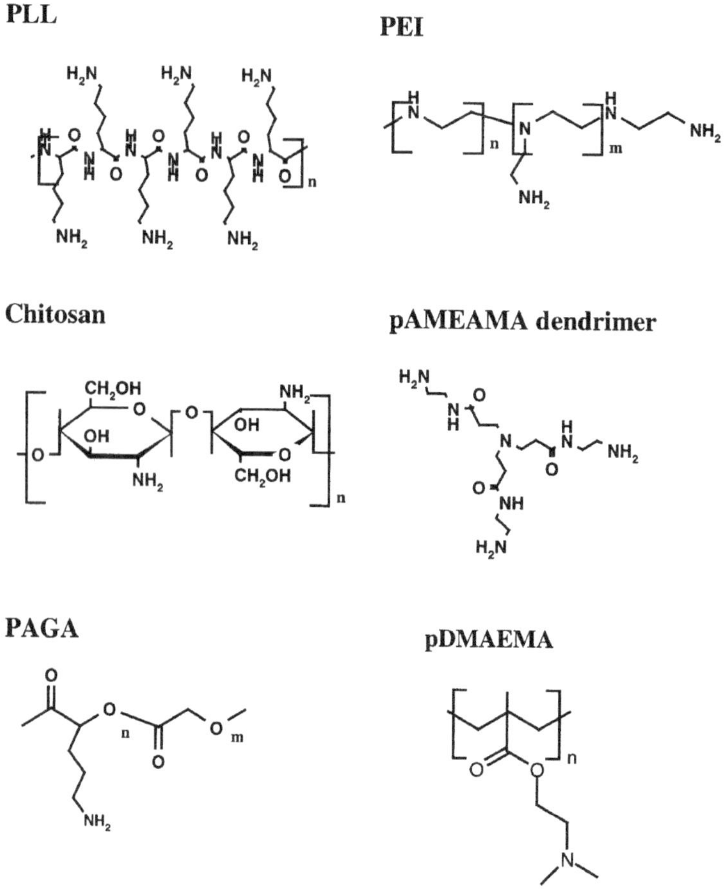

PLL

PEI

Chitosan

pAMEAMA dendrimer

PAGA

pDMAEMA

Figure 1 Chemical structures of cationic polymers used for gene delivery as well as potential grafting with semitelechelic macromolecules.

pH due to its pKa. The positive charges on the polymer neutralize and condense negatively charged nucleic acids such as DNA and RNA into the appropriate form for delivery into the cell. A slightly net over all positive charge is generally—though its not a rule—desired for efficient delivery of nucleic acids into the cells or tissues (9–14). Ideal gene delivery vectors should be non-toxic, biodegradable, biocompatible, non-immunogenic, and stable during synthesis, formulation, storage and post-administration, and be able to reach their target optimally for efficient gene transfer. This ideal vision was not formed in one sitting but has evolved to become more comprehensive during time and in the following sections we will see how through experimentation, systems based on graft copolymers evolved in terms of choice of materials as well as their structural complexity to their current state of art.

3.1. Poly(ʟ-lysine)

PLL is one of the oldest and, by today's standards, primitive cationic polymers under investigation for nucleic acid delivery. Although researchers had observed its DNA condensing properties at least as far back as 1972, its use as a routine gene delivery vector or transfection agent only became evident after 1976 (15,15a,16,16a). PLL has ε-amine functional groups on its side chains which condense DNA into toroidal or globular-shaped complexes (17,18). Polylysine mediated DNA condensation depends on various factors, chief among them being chain length of PLL, the pH and ionic strength of the medium, and the molar ratio of PLL to DNA. It was also discovered that high molecular weight (MW) PLL had greater avidity for DNA compared to small MW PLL in solution but tended to form more disordered aggregates with increasing concentration of DNA. Afterwards, it was also discovered that though PLL condensed DNA into small size and facilitated transfection it was very inefficient, formed frequently insoluble or large aggregates >250 nm with DNA depending upon its molecular weight under physiological conditions, and was toxic (19–21).

For these reasons there were various strategies undertaken to increase its solubility, decrease the toxicity in vitro and in vivo and enhance the cellular uptake of PLL and plasmid DNA (pDNA) complexes. One of the key strategies was to graft PLL with other polymers/moieties, which would impart as many aforementioned characteristics as possible without sacrificing its lone virtue, which was to condense DNA via electrostatic interactions. Kim et al. pioneered these developments (22). To repress the expression of glutamic acid decarboxylase (GAD) in transgenic nonobese diabetic (NOD) mice for treatment of type I diabetes, PEG-*g*-PLL was synthesized as a gene carrier. PEG-*g*-PLL was complexed with antisense GAD expression plasmid pRIP-AS-GAD which resulted in enhanced protection from DNAse I as well as high transfection efficiency at polymer/pDNA ratio of 3/1 (w/w). It was also found that as the mol% of PLL (MW ~ 25,000) backbone grafted to PEG (MW ~ 2000) through its carboxylic group increased from 10% to 15% and 20% the ability to protect against DNAse decreased. This confirmed the results from another report previously published, where 5, 10 and 25 mol% of PLL (MW ~ 25,000) were grafted with PEG (MW = 550). The 10 mol% grafted PEG-*g*-PLL was found to facilitate the highest transfection efficiency (23,24). PEG-*g*-PLL/pDNA complexes were found to possess 5–30-fold higher transfection efficiency compared to PLL/pDNA complexes in a human carcinoma cell line. Compared to commercially available cationic lipid transfection reagent Lipofectin™, PEG-*g*-PLL demonstrated lower cytotoxicity (23). In vivo experiments with PEG-*g*-PLL/pRIP-AS-GAD suggested that after intravenous administration in mice, antisense GAD mRNA persisted till after 3 days of injection (24). PLL-*g*-PEG comb-like graft copolymers were also extensively studied in terms of MW of PEG and PLL used. PEG size was varied as MW = 2000, 3000 and 5000 and PLL size was varied as MW = 10,000, 26,000 and 38,000. Copolymers with as little as 2 mol% grafted PEG chains stabilized the copolymer/pDNA complexes in physiological conditions even at neutral charge ratio. However it is not mentioned how the charge ratio was calculated properly. All polyplexes, formed with various PEGs and PLLs, remained relatively small (~100 nm) in saline. Higher degree of grafting led to significantly diminished binding with the morphology changing from spherical to worm-like structures. Enhanced levels of luciferase expression were observed with these polyplexes; however, the difference in gene expression among all these polymers with varying MW did not show any conclusive trends. The authors also did not explain the rationale behind using PEGs with not so widely varying molecular weights (25).

The rationale behind grafting PEG to PLL has been multifaceted. The polymer/pDNA complexes after administration enter the bloodstream, interact with a number of blood components particularly, blood borne immune cells, as well as the complement system. This leads to activation of the complement as well as secondary immune responses, leading to flushing out of the complexes sooner than expected by these host defense mechanisms. The complexes are also readily phagocytosed by the reticuloendothelial cells and finally cleared from blood circulation leading not only to poor bioavailability but also leading to antibody generation against the therapy itself as well as extensive damage to erythrocytes leading to serious side effects. These triple problems of non-specific immune interactions, alteration of blood chemistry, and their physical instability led to the investigation of PEG-grafted polymers and their impact in solving the aforementioned problems (26).

PEG has been widely used in a variety of research fields during the past three decades. Polyethylene glycol, when covalently conjugated to other molecules, acts as a non-immunogenic, hydrophilic surface shield which reduces the interactions with blood components and pacifies the immunogenic potential of shielded molecule or supramolecular complexes. In this manner, the circulation time in the bloodstream is amplified and renal clearance is reduced owing to increased molecular weight as well as molecular shielding effect. Thus pharmacokinetic parameters such as bioavailability, serum half life as well as biodistribution to various disease sites tend to improve. PEG conjugation also leads to improved solubility and formulation stability at physiologically relevant ionic strengths. Low molecular weight drugs, affinity ligands, proteins, oligonucleotides, liposomes, the surfaces of biomaterials, the synthesis of highly hydrated hydrogels have been extensively conjugated with PEG and its has long been approved by the FDA for human and preclinical use. A widely known explanation for such a beneficial effect of PEG is that owing to its chemical structure, it is highly mobile and fully hydrated in aqueous solutions. Its hydrodynamic diameter is much larger than those of other molecules of similar molecular weight. This leads to PEG acting as a cloak keeping other proteins flowing in the bloodstream at arm's length. Being structurally flexible and well hydrated, PEG chains can expose targeting moeties attached to other molecules in a more orderly fashion and molecules positioned on a PEG acting as a spacer can have relatively easy access to targeting sites (27,28).

Based on these observations, PEG was grafted to PLL's ε-amine functional groups as side chains to improve its solubility, physiological salt, and serum stability. Later on, as has already been mentioned, PEG-*g*-PLL was also thought to be superior in terms of long circulation time due to high gene expression with no noticeable side effects. However, it was later realized that further improvements are necessary to increase the delivery in vivo for prolonged transgene expression (23–25). In order to increase the effectiveness of PLL, poly(L-histidine) as side chains were grafted to it. The resulting comb shaped graft copolymer, *N*-Ac-poly(L-histidine)-graft-poly(L-lysine) or PLH-*g*-PLL had 25% of PLL's ε-amine functional groups tethered to the histidine polymer and the overall MW of the copolymer was approximately 45 KDa. The particle size of PLH-*g*-PLL based polymer/pDNA complexes were in the range of 117–306 nm and mediated higher transfection efficiency compared to naked pDNA as well as PLL/pDNA complexes (29). The imidazole ring of histidine has a pKa of 6.5 and has been found to be pH sensitive between pH 6 and 7 (30). This phenomenon results in what is termed as a proton sponge effect. This proton sponge effect leads to escape of polymer/pDNA complexes from endosomes and/or endolysosomes, which pose one of the main barriers to DNA delivery inside the nucleus, thereby leading to higher transfection efficiency. The protonation of the imidazole ring has been studied by 1H NMR spectroscopy by monitoring C2-H peak position versus the pH medium reflecting the charge state of the imidazole ring (31). It has also been found that even when histidine or poly(L-histidine) are linked to other polymers, over 95% of them are deprotonated at pH 7.2 and may not participate in DNA condensation, thus leaving them entirely in charge of contributing to the pH buffering effect (32,33). PLL has also been grafted with Pluronic block copolymers to enhance its solubility as well as transfection efficiency. Compared with PLL based polymer/pDNA complexes, PLL-*g*-Pluronic based complexes demonstrated 2-fold increase in transfection efficiency with similar cytotoxicity at polymer to DNA ratio of 1:1 (w/w). However, the size of these complexes was >300 nm and the PLL MW~50,000 used as backbone was too high and potentially toxic in vivo, all of which are unfavorable for gene transfer (34). Pluronics are believed to possess several physicochemical and biological properties, which may directly or indirectly enhance transgene expression. Pluronic block copolymers or poloxamers consist of a triblock structure, polyethylene oxide-polypropylene oxide-polyethylene oxide (PEO-PPO-PEO), which is amphiphilic in nature. Size as well as hydrophobic PPO to hydrophilic PEO balance can alter their amphiphilic properties. Poloxamers have the ability to interact with biological membranes and have been used as surfactants, immunoadjuvants as well as adhesives for postoperative wound healing since they enhance the sealing of damaged cell membranes, but improve the passage of molecules by enhanced membrane interactions in normal cells. The PPO in Pluronic can interact with cellular membranes due to hydrophobic-hydrophobic interactions leading to structural rearrangements of lipid membranes, reminiscent of fusogenic peptides derived from viruses designed by nature for similar functions (35,36). Polymers linked with Pluronic are believed to form self-assembled micelle like complexes with pDNA which are relatively more well-defined and less polydisperse compared to complexes formed between cationic polymers and pDNA. It is also believed that certain poloxamers act as synthetic biological response modifiers by altering the transcriptional control of transgene expression. In addition Kabanov et al. have also reported that Pluronic block copolymers are very effective in the treatment of multi-drug resistant tumors by inhibiting the ATP dependent pathways which in turn lead to inhibition of drug efflux pumps so that an increase in the cytotoxic activity of the drug or genes could be amplified (35,37). There have been other strategies where the PLL backbone has been attached with various types of polymers and biopolymers to generate species which is more effective in terms of transfection efficiency compared to PLL alone. High molecular weight PLL (MW = 134,000) grafted with short poly[*N*-(hydroxypropyl) methacrylamide] chains (MW = 7000) were studied for oligonucleotide (36 base pairs) delivery purposes. The polyplexes prepared with these graft copolymers were soluble at any charge ratios in aqueous medium. However, the overall molecular weight of graft copolymer was so high (MW = 314,000) that it was rendered impractical for in vivo administration (38).

PEG-*g*-PLL with PLL as backbone has also been used in tandem with positively charged fusogenic peptide KALA (39). KALA is a cationic amphipathic peptide that binds to DNA, destabilizes the endosomal membrane due to a conformational change from an α-helical form at physiological pH to a mixture of α-helical form and random coil at endosomal pH. Although KALA on its own was found to be capable of DNA delivery it was not as effective as promised earlier (40). PEG-*g*-PLL was complexed with DNA to form negatively charged polyplexes. KALA was then serially added and through electrostatic interactions the overall complex was made positively charged. The resultant complex did not aggregate due to steric hindrance between positively charged complexes although PLL/DNA/KALA complexes did aggregate. It was also found that increasing the content of KALA in PEG-*g*-PLL/DNA/KALA complexes led to higher transfection efficiency yet exhibited low cytotoxicity (39).

In a famously candid and comprehensive study, structure activity relationships of PLLs and their derivatives were studied in the context of gene delivery. Effects of molecular shape, weight, and extent of PEGylation on linear, grafted, dendritic, and branched polylysines with respect to their gene delivery properties were investigated. Ability to condense DNA as well as surface morphology, particle size, and transfection efficiency were studied for various polyplexes. A few trends were seen in terms of their physicochemical properties. Low molecular weight polymers had less avidity for DNA compared to large molecular weight polymers and were consequently less successful in condensing DNA. Linear polymers were more efficient than dendritic polymers in condensing DNA. As PEGylation was increased per PLL backbone, ability to condense DNA became less. It was also observed that although PEGylation increased cellular uptake and transfection efficiency the differences across all polyplexes were not that substantial. Overall transfection efficiency remained low despite noticeable differences in physicochemical properties. It could be said that although there was a correlation between structure-physicochemical activities, it did not extend to the biological activity of polyplexes (41). PLL has also been grafted to polysaccharides like hydrophilic dextran as side chains. These comb type graft copolymers, PLL-*g*-dextran, were successfully prepared with varying degrees of grafting as well as molecular weights of dextran. All preparations of PLL-*g*-dextran formed soluble complexes with DNA. Circular dichroism studies suggested that the copolymer interacted with DNA without altering its native structure or physicochemical properties. However, these polymers were not reported to be tested for their transfection efficiency in cultured cell or in vivo (42). PLL-*g*-PEG/DNA complexes have also been studied for their biocompatibility and pharmacokinetic parameters. PLL-*g*-PEG was compared with poly(trimethylammonioethyl methacrylate chloride)-poly[*N*-(2-hydroxypropyl) methacrylamide] (pHPMA-b-pTMAEM) block copolymers for its ability to condense oligonucelotides where the latter was found to be more efficient. In addition pHPMA-b-pTMAEM produced oligonucleotide complexes which were compact in size (~40 nm), low zeta potential, as well as enhanced stability in physiological salts and could be formed at a DNA concentration of 0.5 μg/μl. These characteristics are ideally suited for in vivo applications since most polyplexes are found to precipitate at DNA concentrations >0.2 μg/μl in laboratory conditions. Complexes formed with both copolymers reduced the renal clearance of oligonucleotides after intravenous administration in mice. However, the complexes were not persistent in the blood stream and within 30 min they were cleared. It was therefore envisaged that these complexes would be more suitable for localized in vivo delivery (43). PLL grafted with a range of hydrophilic polymers including PEG, dextran, and phPMA were simultaneously evaluated for their pharmaceutical characteristics. The particle size of complexes was typically around 100 nm viewed by atomic force microscopy. Hydrophilic side chains were able to shield the positive charge as depicted by zeta potential measurements and copolymer/DNA complexes were less toxic to cultured cells compared to PLL/DNA complexes. All complexes were readily soluble in aqueous solutions, and PLL-*g*-PEG demonstrated relatively high transfection efficiency in cultured cells (44). However, it is now a settled fact that PLL-based gene delivery systems have significant limitations that may not be overcome, and too many complicated modifications may be required for PLL to be a viable gene delivery carrier.

3.2. Poly[*N*-(2-hydroxypropyl)methacrylamide]

The majority of the polymeric anti-cancer drug conjugates that have reached clinical trials have used [*N*-(2- hyroxypropyl)methacrylamide] (HPMA) copolymers as the carrier. HPMA homopolymers as well as its copolymers were pioneered by Kopecek and co-workers (45,46). HPMA when grafted with oligonucleotides using a lysosomally cleavable linker led to efficient internalization compared to unconjugated copolymer into the cells. Oligonucleotides escaped from lysosomes and were able to enter the cytoplasm and nucleus. Linker type was critical in this successful approach since non-lysosomally cleavable linker was not able to facilitate escape of oligonucleotides from lysosomes. The HPMA-*g*-oligonucleotides possessed antiviral activity but free oligonucleotides did not (47). Block and graft copolymers of 2-(trimethylammonioethyl methacrylate) (TMAEM) and HPMA were synthesized for the preparation of polymer/DNA complexes and further characterized in terms of their physicochemical properties. The percentage of phPMA in the copolymers did not significantly affect the ability of pTMAEM parts to form complexes with DNA, but had an effect on the molecular parameters and aggregation of complexes. Size of the complexes was directly proportional to the increasing pHPMA content, and so was stability in the physiological salt conditions. HPMA content $> 40\%$ led to most stable formulations. With increasing HPMA content the shape of the complexes changed from spherical to coil-like. All the complexes showed resistance to nucleases. Transfection efficiency was greater than of the corresponding cationic homopolymer, but the complexes could not achieve DNA concentration higher than $40\,\mu g/ml$ so that their in vivo applications are doubtful (48,49). Stated in these terms, water-soluble HPMA copolymers also face serious challenges to become viable for in vivo applications as far as delivery of plasmids and oligonucleotides is concerned.

3.3. Polyethyleneimine

Polyethyleneimine (PEI) is a cationic polymer with molecular weight of 43 and empirical formula $(C_2H_5N)_n$.PEI comes in two forms, a linear and a branched polymer. While the linear polymer consists of mostly secondary amines with primary amines at the terminal ends, the branched polymer consists of 25% primary amines, 50% secondary amines, and 25% tertiary amines as determined by ^{13}C NMR spectroscopy (50,51). PEI has been shown to condense plasmids into colloidal particles that possess high transfection efficiency into a variety of cells in vitro and in vivo. These complexes are of spherical shape and under controlled physiological conditions can have narrow size distribution. This is one of the factors which leads to high transfection efficiency (14,52). The main factor however is the widely known "proton sponge" effect which is due to an inherent high buffering capacity of PEI. The overall protonation level of PEI increases reasonably from 20 to 45% at pH 7 to pH 5, respectively. When the complexes undergo endocytosis, the endosomes involved start maturing which leads to progressive lowering of their pH by ATP proton pumps. In response the amines in the PEI start absorbing those protons, become protonated and due to increase in structural positive charge, start swelling. This leads to influx of more negatively charged Cl^- counter ions, and this cascade leads to further swelling of the polymer leading to premature rupture of endosomes due to osmotic pressure. These escaped complexes then present themselves in the cytoplasm to be further introduced into the nucleus. In this manner, DNA does not undergo the harsh lysosomal treatment and consequently higher amount of DNA is available for transfection inside the cells leading to higher transfection efficiencies (53,54). PEI is known for some of the highest transfection efficiencies and has become the default starting point for further modifications to increase its safety, stability, specificity, and efficiency for gene transfer. In order to increase the safety and efficiency profile of PEI (MW~25,000 for branched or PEI-25K and ~22,000 for linear PEI or PEI-22K), it was grafted to dextran (MW~1500) resulting in PEI-*g*-dextran. Various degrees of grafting were investigated in an effort to explore how the conjugation affected the transgene expression. Grafting of dextran to PEI's primary amines was found to lessen the cytotoxicity, buffering capacity, DNA integrity as well as the cellular uptake as shown by ethidium monoazide labeled plasmids complexed with PEI-*g*-dextran. As the degree of grafting increased from 0 to 1.84%, all the aforementioned trends were more pronounced. However, at an optimal degree of grafting (~64%), the PEI-*g*-dextran enhanced the percentages of GFP positive cells to a level as high as three times compared to complexes formed with PEI alone in certain cell types (55). In an earlier study by the same group, low MW dextrans (MW~1500) and high MW dextrans (MW~10,000) were used to produce various degrees of grafting on linear and branched PEIs and these grafted polymers were used to prepare polymer/DNA complexes. Salt and serum stability of these complexes in the presence of bovine serum albumin was found to be higher than PEI based complexes. Although stability with high MW dextran was more in terms of absence of any precipitates, the percentage of transfected cells were much higher for linear PEI grafted low MW dextran. Branched PEI grafted low MW dextran was even more effective and transfected 15% or more of the cells as seen by green fluorescence protein measurements using flow cytometry (56).

As was the case with PLL, PEI-*g*-PEG has also been studied extensively and its merits and demerits somewhat follow the same trends as PLL-*g*-PEG. However, due to PEI's high efficiency compared to PLL in terms of facilitating transgene expression, it has been the subject of much more concerted effort in terms of finding that silver bullet which will have characteristics fit for successful in vivo applications as well as clinical trials. In order to investigate the possibility of prolonged gene expression over a period of weeks and perhaps months, PEI-25K was grafted with PEG (MW~2000), resulting in PEI-*g*-PEG. PEI-*g*-PEG/pDNA complexes were able to prolong the transgene expression for up to a week and no attenuation of gene expression was detected after repeated intrathecal injections, even in those rats receiving three doses administered 2 weeks apart. In comparison, PEI/pDNA complexes were found to attenuate gene expression by as much as 70% after re-injection thus clearly demonstrating the superiority of PEI-*g*-PEG for further in vivo studies (57).

In another study, PEI-*g*-PEG was investigated at various degrees of PEGylation and then tested in vivo. Linear PEG (~2000) was grafted to branched PEI-25K from the average number of PEG per PEI macromolecule at 1–14.5. As the degree of PEGylation increased, no benefits in terms of transgene expression were found, but the lowest degree of PEGylation led to 11-fold higher expression than PEI/PDNA complexes after intrathecal administration into the lumbar spinal cord subarachnoid space. In vitro studies had earlier corroborated the fact that the PEI conjugate with a low degree of PEG grafting was able to reduce the size of polymer DNA complexes, prevent the aggregation of complexes, decrease the interactions of the complexes with serum proteins, counter the inhibition of serum to gene transfer, and enhance transfection efficiency, although it was not significant in affecting complex formation and reducing in vitro cell toxicity of PEI (58). The effect of PEG MW has also been investigated. PEG of various MW (350, 750, and 1900 Da) were grafted to PEI in order to lower the cytotoxicity as well as improve the transfection efficiency and salt and serum stability. Although large MW PEG (~1900) grafted PEI showed better cytotoxicity profiles, the transgene expression was seriously hampered. It was also noticed that as the degree of PEGylation increased, the cell viability and serum stability increased due to charge shielding, but transfection efficiency was compromised. This article favored PEG ($\sim$350) with 4% by weight PEI chains grafted so that transfection efficiency remained as good as PEI/pDNA complexes alone with better cell viability. However, it remains to be seen if similar trends could be observed in vivo (59).

PEI-25K has been grafted with an even larger MW range of PEGs (550–25,000 Da) as well as varying degrees of substitution for a low MW PEG ($\sim$5000 Da). Atomic force microscopy showed that PEI-g-PEG (PEG, $\sim$5000 Da) significantly reduced the diameter of the spherical complexes from 142 ± 59 to 61 ± 28 nm. On the other hand, increased degree of PEG substitution reduced the zeta potential, reduced the ability to condense pDNA as well as retain the ability to maintain spherical shape. As the PEG MW was reduced the size of the complexes became larger, and the tendency to cause hemolysis and erythrocyte aggregation increased. High degree of PEG substitution led to low toxicity but toxicity was independent of PEG MW. Transfection efficiency was greatest with PEG ($\sim$550 Da) (60). In yet another study with a complimentary strategy, PEI with various MW (0.8–800 kDa) were grafted to low MW PEG ($\sim$550 Da). Almost all PEI-g-PEG were able to protect the oligonucleotides from serum and nucleases, except 0.8 kDa PEI grafted PEGs (61). Further in vitro characterization with ribozymes and oligonucleotides as drugs was carried out and it was found that PEI-g-PEG with PEG ($\sim$5000) did not influence complex size and just 2 PEGs per PEI were enough to achieve neutral charge ratio for the complexes. As PEG substitution increased, protection against complement in blood was higher. Ribozyme complexes with PEI-g-PEG achieved a 50% down-regulation of the target mRNA and the results were comparable with PEI/ribozyme complexes (62). Biodistribution of dual labeled PEI-g-PEG (PEI-25K, PEG $\sim$550 Da) and DNA complexes as well as controls such as PEI-25K and PEI-2.7K followed at time points 15 min, 2 h, and 12 h and showed that complexes were mainly distributed in liver and spleen. Organ profiles for DNA as well as polymers were similar till 2 h, indicating that they had not yet separated. Naked DNA after injection was rapidly degraded and eliminated within 15 min. At 12 h, the organ concentrations of labeled polymer remained high but labeled DNA levels decreased in a time-dependent manner, likely due to separation of the complexes and degradation of the DNA. Although PEI-*g*-PEG demonstrated a slower uptake into the RES organs compared with PEI-25K due to the shielding effect of PEG, it was not able to better stabilize the complexes in the circulation or protect DNA from degradation (63).

Kabanov et al. have extensively studied PEI-*g*-Pluronics which form self-assembled interpolyelectrolyte complexes and in some of their characteristics are superior to unmodified PEI-25K (35–37). When PEI-2K is modified with polyethylene oxide (~8 kDa) resulting in PEI-*g*-PEO and complexed with DNA, it results in the formation of species containing hydrophobic sites from neutralized DNA polycation complexes and hydrophilic sites from PEO. These complexes have improved solubility and stability and a compact size (~40 nm). When these polymers are complexed with plasmid DNA they form larger (70–200 nm) complexes, which are poor in transfection efficiency. This is because PEO provides too much of a stealth effect. However when PEO is replaced with Pluronics such as P123 ($EO_{20}PO_{70}EO_{20}$,MW~5820), the resultant copolymer resulted in high transfection efficiency. These complexes exhibited elevated levels of transgene expression after 24 h in spleen, heart, lungs, and especially liver following systemic administration in mice (64,65). These copolymers have also been found to be more effective than cationic lipid transfection reagents and they do not show any serum dependence for transfection. In addition, they perform even better when P123 is used in low concentrations as an excipient. It is believed that this better performance is due to optimization of size of the polyplex. However, one downside of these systems is that they do not protect from DNAse for long periods of time, so that their in vivo applications pose a challenge (66,67). PEI grafted with Pluronics of lower molecular weight such as P85 ($EO_{26}PO_{40}EO_{26}$, MW ~4600 Da) also facilitate accumulation in liver (68,69). PEI-*g*-P85 when formulated with an anti-tumor antisense oligonucleotide, also resulted in significant inhibition of tumor growth (70).

3.4. Poly(2-dimethylamino)ethyl Methacrylate

Poly(2-dimethylamino)ethyl methacrylate (pDMAEMA) is a water-soluble cationic polymer that contains tertiary amines and shows the "proton sponge" effect to some extent due to pKa=7.5. Approximately 50% of the amines are protonated at pH 7.2 implying that the other 50% participate in buffering. The absence of primary and secondary amines indicates that pDMAEMA binds relatively loosely to pDNA compared to polycations such as PEI-25K and other high MW PEI. It has been shown to condense pDNA into 150–180 nm size particles and mediate transfection in different cell types. However, complexes with pDMAEMA are not very stable in physiological salt and serum and since they do not possess attributes other than DNA condensation and modest endosomal buffering, attempts have been made to modify them (71)

4. BIODEGRADABLE AND TARGETED POLYMERIC GENE CARRIERS

Biodegradability is a virtue, which cannot be ignored especially by those laboratories where the focus is on finding novel, non-toxic, and biocompatible drug and gene delivery agents. Biodegradable thermosensitive tri-block copolymers based on PEG-PLGA-PEG blocks of various molecular weights have been used since more than a decade (72,73). Recently polymers of such kind have been investigated for gene delivery. PEG-PLGA-PEG polymers with sol-gel characteristics were used to entrap radioactively labeled pDNA and their release phenomena was investigated. The in vitro degradation of PEG-PLGA-PEG lasted for more than 30 days. The release profile of supercoiled pDNA from the polymer followed the zero-order kinetics up to 12 days. Peak gene expression of luciferase was seen at 24 h in the skin. The expression had dropped to almost baseline by 72 h. These kinds of release characteristics can be useful for long term gene expression if further investigation is made to prolong the release profile as well as resultant gene expression (74). Biodegradable multi-block co-polymers of PLL and PEG were synthesized using biodegradable linkers. These synthesized copolymers showed negligible toxicity and transfection efficiency similar to non-degradable PLL homopolymers. The copolymer degraded rapidly so that its MW was reduced to almost 20% of original within 3 days. Transfection efficiencies of copolymers were not affected by the presence of serum, while that of PLL homopolymers decreased to the level of naked DNA in the presence of serum. Based on the results, the new copolymers are believed to be potentially efficient carriers for the delivery of bioactive agents (75). In a similar manner crosslinked copolymers of PEI attached to PEG via biodegradable linkers were investigated. These biodegradable PEI were stable in physiological salt and serum and were able to complex with DNA resulting in high transfection compared to other commercially available polymers. Transfection efficiency, though not as high as PEI, was several-fold higher than naked plasmid. No toxicity was seen even at elevated ratios of biodegradable PEI to plasmid DNA (76). Biodegradable nanoparticles, which combine the functions of both degradability as well as targeting on their surfaces, were synthesized as a novel carrier for nucleic acids. The nanoparticles were obtained from poly(D,L-lactic acid) and PLL-graft-polysaccharide copolymers by using either a solvent evaporation method or a diafiltration method. Nanoparticles as small as 60 nm in diameter were successfully obtained from the graft copolymers with high polysaccharide content but not from the non-degradable PLL. Polysaccharide moieties on the surface of the nanoparticles were found to interact specifically with lectin. The polymer DNA complexes from nanoparticles were resistant to aggregation possibly due to polysaccharide moieties on the surface. These results suggested that the nanoparticles prepared from PLL-graft-polysaccharide copolymer and poly(D,L-lactic acid) were viable alternatives for gene delivery in vivo (77).

There are numerous targeting ligands that can be utilized for cell-specific gene transfer, including asialoglycoprotein,

Table 1   Cationic Graft Polymers and Their Properties

Graft polymeric gene carriers	Properties
PEG-g-polyspermine/oligonucleotide PEG-g-PLL/plasmid DNA	Enhance solubility Decrease cytotoxicity; increase transfection efficiency
PEG-g-PLL/plasmid DNA	Enhance solubility; average size: 50 nm; narrow size distribution
Phpma-g-ptmaem/plasmid DNA	Enhance solubility; average size: 50–120 nm
PEG-g-PEI/plasmid DNA	Poor DNA condensation; lower surface charge; decrease transfection efficiency
PEG-g-PEI/plasmid DNA	Decrease transfection efficiency; enhance solubility; reduce size
Dextran-PLL/plasmid DNA	Reduce cytotoxicity; increase transfection efficiency; average size: 100 nm
PDEAEMA-g-PLL dextran-PEI	pH-dependent DNA condensation; lower zeta potential; decrease transfection efficiency; average size: 500 nm at $N/P = 2$
PEG coating on DNA/PLL- transferrin complexes	Lower zeta potential and cytotoxicity; better biodistribution; enhance stability of complexes in vivo
Phpma coating on DNA/PLL complexes	Lower the phagocytosis of RES; decrease stability of DNA complexes; no difference in biodistribution

lactose, low-density lipoproteins (LDL), folic acid, antibodies, and so on. Asialoglycoprotein was first employed as a targeting ligand for cell-specific gene delivery, in which asialoglycoprotein-attached PLL selectively transfected HepG2 cells (hepatocytes containing 2.5×10^5 asialoglycoprotein receptor/cells). Further in vivo experiments showed that its systemic administration resulted in a liver-specific expression of the reporter gene (78).

LDL has been used as a targeting moiety in the "terplex" system, in which LDL molecules were incorporated into stearyl PLL via a hydrophobic interaction with stearyl groups. When these terplex systems were applied to HepG2 and A7R5 murine vascular smooth muscle cells, the terplex/DNA system exhibited higher transfection efficiency and lower cytotoxicity than either the lipofectamine/DNA complex or the stearyl PLL/DNA complex. It was also demonstrated that when LDL was incorporated in terplex, it enhanced gene transfer efficiency by facilitating cellular uptake by the LDL receptor, otherwise known as receptor mediated endocytosis. The terplex system was also used for administration in rabbit myocardium, in which the terplex system yielded a 20–100-fold increase in transgene expression over naked DNA and prolonged the gene expression up to 30 days. In an in vivo pharmacokinetic study, the terplex system was found to prolong the half-life in blood circulation and to stay intact for up to 3 days (79,80).

RGD-based peptide sequences have been used to target integrin receptors on endothelial cells in an effort to create an antiangiogenic therapy. When RGD peptide was conjugated to PEI-PEG small sized polyplexes (100–200 nm) were obtained. Their surface charge was also significantly reduced due to the charge shielding effect of PEG. When angiogenic endothelial cells were transfected with PEI-*g*-PEG-RGD there was a 5-fold increase in transfection efficiency over PEI which was cell specific (81).

Similarly, lactose was attached to the PLL-PEG copolymer so that PEG served as a spacer and lactose moiety served as a targeting ligand to lactose receptors on HepG2 cells. In the transfection with HepG2 cells, lactose-PEG-grafted PLL exhibited a 7-fold increase in the reporter gene expression over DNA/PLL complexes, whereas on a control cell line, the transfection efficiency was no better than naked pDNA. In addition, lactose-PEG-grafted PLL increased the solubility of their DNA complexes and, owing to the grafted PEG effect, significantly reduced the cytotoxicity induced by PLL (82).

Vitamin folate also has been used to promote gene transfer into malignant cells that over-express the folate receptors for the rapid DNA replication, such as ovarian carcinoma cells; and proteins or macromolecules containing folate have been proven to undergo internalization via folate receptor-mediated endocytosis. Therefore, folate-mediated gene transfer has been efficient and selective in delivering genes into cancer cells. For example, folate-PEG-folate-grafted PEI showed a high specificity to CT-26 colon adeno-carcinoma cells, compared with folate receptor-deficient smooth muscle cells. In addition, reduced cytotoxicity was seen with these targeted copolymers (83).

Although non-viral gene delivery has been practiced for nearly two decades, we still seek success in clinical trials. With a wealth of information in our hands and some of the barriers to gene delivery clearly elucidated, novel ideas must be explored in in vivo situations. Since there is poor correlation between these molecules tested in vitro for their therapeutic potential and their success in vivo (84), better models for predicting preclinical outcomes as well as a close relationship between investigators involved in drug discovery and pharmaceutical engineers is the key for a successful non-viral gene therapy application.

REFERENCES

1. Russell SJ, Peng KW. Primer on medical genomics. Part X: Gene Therapy. Mayo Clin Proc 2003; 78:1370–1383.

2. Check E. Harmful potential of viral vectors fuels doubts over gene therapy. Nature 2003; 423:573–574.

3. Tan Y, Huang L. Overcoming the inflammatory toxicity of cationic gene vectors. J Drug Target 2002; 10:153–160.

4. Nishikawa M, Huang L. Nonviral vectors in the new millennium: delivery barriers in gene transfer. Hum Gene Ther 2001; 12:861–870.

5. Thomas M, Klibanov AM. Non-viral gene therapy: polycation-mediated DNA delivery. Appl Microbiol Biotechnol 2003; 62:27–34.

6. Schmidt-Wolf GD, Schmidt-Wolf IG. Non-viral and hybrid vectors in human gene therapy: an update. Trends Mol Med 2003; 9:67–72.

7. Metanomski WV. Compendium of Macromolecular Nomenclature. Oxford: Blackwell Scientific Publications, 1991.

8. Burlant WJ, Hoffman AS. Block and Graft Polymers. New York: Reinhold Publishing Corporation, 1960.

9. De Smedt SC, Demeester J, Hennink WE. Cationic polymer based gene delivery systems. Pharm Res 2000; 17(2):113–126.

10. Lim YB, Han SO, Kong HU, Lee Y, Park JS, Jeong B, Kim SW. Biodegradable polyester, poly[α-(4-aminobutyl)-L-glycolic acid] as a non-toxic gene carrier. Pharm Res 2000; 17:811–816.

11. Cherng JY, van de Wetering P, Talsma H, Crommelin DJ, Hennink WE. Effect of size and serum proteins on transfection efficiency of poly[2-(dimethylamino)ethyl methacrylate]-plasmid nanoparticles. Pharm Res 1996; 13:1038–1042.

12. Roy K, Mao HQ, Huang SK, Leong KW. Oral delivery with chitosan-DNA nanoparticles generates immunologic protection in murine model of peanut allergy. Nat Med 1999; 5:387–391.

13. Wagner E, Plank C, Zatloukal K, Cotton M, Birnstiel ML. Influenza virus hemagglutinin HA-2 N-terminal fusogenic peptides augment gene transfer by transferrin-polylysine-DNA complexes: toward a synthetic virus-like gene-transfer vehicle. Proc Natl Acad Sci USA 1992; 89:7934–7938.

14. Boussif O, Lezoualc'h F, Zanta MA, Mergny MD, Scherman D, Demeneix B, Behr JP. A versatile vector for gene and oligonucleotide transfer into cells in culture and in vivo: polyethylenimine. Proc Natl Acad Sci USA 1995; 92:7297–301.

15. Carroll D. Complexes of polylysine with polyuridylic acid and other polynucleotides. Biochemistry 1972; 11:426–433.

15a. Wadhwa MS, Knoell DL, Young AP, Rice KG. Targeted gene delivery with a low molecular weight glycopeptide carrier. Bioconjug Chem 1995; 6:283–291.

16. Ehrlich M, Sarafyan LP, Myers DJ. Interaction of microbial DNA with cultured mammalian cells. Binding of the donor DNA to the cell surface. Biochim Biophys Acta 1976; 454:397–409.

16a. Maruyama A, Ishihara T, Kim JS, Kim SW, Akaike T. Nanoparticle DNA carrier with poly(L-lysine) grafted olysaccharide copolymer and poly(D,L-lactic acid). Bioconjug Chem 1997; 8:735–742.

17. Tang MX, Szoka FC. The influence of polymer structure on the interactions of cationic polymers with DNA and morphology of the resulting complexes. Gene Ther 1997; 4:823–832.

18. Dunlop DD, Maggi A, Soria MR, Monaco L. Nanoscopic structure of DNA condensed for gene delivery. Nucl Acids Res 1997; 25:3095–3101.

19. Wolfert MA, Seymour LW. Atomic force microscopic analysis of the influence of the molecular weight of poly(L)lysine on the size of polyelectrolyte complexes formed with DNA. Gene Ther 1996; 3:269–273.

20. Wagner E, Cotton M, Foisner R, Birnstiel ML. Transferrin-polycation-DNA complexes: the effect of polycations on the structure of the complex and DNA delivery to cells. Proc Natl Acad Sci USA 1991; 88:4255–4259.

21. Read ML, Etrych T, Ulbrich K, Seymour LW. Characterization of the binding interaction between poly(L-lysine) and DNA using the fluorescamine assay in the preparation of non-viral gene delivery vectors. FEBS Lett 1999; 461:96–100.

22. Han SO, Mahato RI, Sung YK, Kim SW. Development of biomaterials for gene therapy. Mol Ther 2000; 2:302–317.

23. Lee M, Han SO, Ko KS, Koh JJ, Park JS, Yoon JW, Kim SW. Repression of GAD autoantigen expression in pancreas beta-cells by delivery of antisense plasmid/PEG-*g*-PLL complex. Mol Ther 2001; 4:339–346.

24. Choi YH, Liu F, Kim JS, Choi YK, Park JS, Kim SW. Polyethylene glycol-grafted poly-L-lysine as polymeric gene carrier. J Contr Rel 1998; 54:39–48.

25. Banaszczyk M, Lollo C, Kwoh D, Phillips A, Amini A, Wu, D, Mullen P, Coffin C, Brostoff S. Poly-L-lysine-graft-PEG comb-like polycation copolymers for gene delivery. J Macromol Science 1999; 36:1061–1084.

26. Ogris M, Brunner S, Schuller S, Kircheis R, Wagner E. PEGylated DNA/transferrin-PEI complexes: reduced interaction with blood components, extended circulation in blood and

potential for systemic gene delivery. Gene Ther 1999; 6:595–605.

27. Harris JM, Zalipsky S. Poly(ethylene glycol): chemistry and biological applications. ACS Symp Ser No 680; 1997:1–15.

28. Harris JM. Poly(ethylene glycol) chemistry: Biotechnical and biomedical applications. New York: Plenum Press, 1992, 1–12.

29. Benns JM, Choi JS, Mahato RI, Park JS, Kim SW. pH-sensitive cationic polymer gene delivery vehicle: N-Ac-poly(L-histidine)-graft-poly(L-lysine) comb shaped polymer. Bioconjug Chem 2000; 11:637–645.

30. Midoux P, Monsigny M. Efficient gene transfer by histidylated polylysine/pDNA complexes. Bioconjug Chem 1999; 10: 406–411.

31. Meadows DH, Jardetzky O, Epand RM, Ruterjans HH, Scheraga HA. Assignment of the histidine peaks in the nuclear magnetic resonance spectrum of ribonuclease. Proc Natl Acad Sci USA 1968; 60:766–772.

32. Wang CY, Huang L. Polyhistidine mediates an acid-dependent fusion of negatively charged liposomes. Biochemistry 1984; 23(19):4409–4416.

33. Putnam D, Gentry CA, Pack DW, Langer R. Polymer-based gene delivery with low cytotoxicity by a unique balance of ide-chain termini. Proc Natl Acad Sci USA 2001; 98: 1200–1205.

34. Jeon E, Kim HD, Kim JS. Pluronic-grafted poly-(L)-lysine as a new synthetic gene carrier. J Biomed Mater Res 2003; 66A:854–859.

35. Kabanov AV, Lemieux P, Vinogradov S, Alakhov V. Pluronic block copolymers: novel functional molecules for gene therapy. Adv Drug Deliv Rev 2002; 54:223–233.

36. Gebhart CL, Sriadibhatla S, Vinogradov S, Kabanov AV. Pluronic-polyethyleneimine conjugates for gene delivery: cell transport and transgene expression. Polym Preprints 2001; 42:119–120.

37. Lemieux P, Guerin N, Paradis G, Proulx R, Chistyakova L, Kabanov A, Alakhov V. A combination of poloxamers increases

gene expression of plasmid DNA in skeletal muscle. Gene Ther 2000; 7:986–991.

38. Dautzenberg H, Zintchenko A. Polycationic graft copolymers as carriers for oligonucleotide delivery. Complexes of oligonucleotides with polycationic graft copolymers. Langmuir 2001; 17:3096–3102.

39. Lee H, Jeong, JH, Park TG. PEG grafted polylysine with fusogenic peptide for gene delivery: high transfection efficiency with low cytotoxicity. J Contr Rel 2002; 79:283–291.

40. Wyman TB, Nicol F, Zelphati O, Scaria PV, Plank C, Szoka Jr FC. Design, synthesis, and characterization of a cationic peptide that binds to nucleic acids and permeabilizes bilayers. Biochemistry 1997; 36:3008–3017.

41. Mannisto M, Vanderkerken S, Toncheva V, Elomaa M, Ruponen M, Schacht E, Urtti A. Structure-activity relationships of poly(L-lysines): effects of pegylation and molecular shape on physicochemical and biological properties in gene delivery. J Contr Rel 2002; 83:169–182.

42. Maruyama A, Watanabe H, Ferdous A, Katoh M, Ishihara T, Akaike T. Characterization of interpolyelectrolyte complexes between double-stranded DNA and polylysine comb-type copolymers having hydrophilic side chains. Bioconjug Chem 1998; 9:292–299.

43. Read ML, Dash PR, Clark A, Howard KA, Oupicky D, Toncheva V, Alpar HO, Schacht EH, Ulbrich K, Seymour LW. Physicochemical and biological characterisation of an antisense oligonucleotide targeted against the bcl-2 mRNA complexed with cationic-hydrophilic copolymers. Eur J Pharm Sci 2000; 10:169–177.

44. Toncheva V, Wolfert MA, Dash PR, Oupicky D, Ulbrich K, Seymour LW, Schacht EH. Novel vectors for gene delivery formed by self-assembly of DNA with poly(L-lysine) grafted with hydrophilic polymers. Biochim Biophys Acta 1998; 1380:354–368.

45. Kopecek J, Bazilova H. Poly[N-(hydroxypropyl) methacrylamide]. I. Radical Polymerization and copolymerization. Eur Polym J 1973; 9:7–14.

46. Sprincl L, Exner J, Sterba O, Kopecek J. New types of synthetic infusion solutions. III. Elimination and retention of poly-[N-(2-hydroxypropyl)methacrylamide] in a test organism. J Biomed Mater Res 1976; 10:953–963.

47. Jensen KD, Kopeckova P, Kopecek J. Antisense oligonucleotides delivered to the lysosome escape and actively inhibit the hepatitis B virus. Bioconjug Chem 2002; 13:975–984.

48. Oupicky D, Konak C, Ulbrich K. DNA complexes with block and graft copolymers of N-(2-hydroxypropyl)methacrylamide and 2-(trimethylammonio)ethyl methacrylate. J Biomater Sci Polym Ed 1999; 10:573–590.

49. Oupicky D, Konak C, Ulbrich K, Wolfert MA, Seymour LW. DNA delivery systems based on complexes of DNA with synthetic polycations and their copolymers. J Contr Rel 2000; 65:149–171.

50. Dick CR, Ham GE. Characterization of polyethyleneimine. J Macromol Sci Chem 1970; A4:1301–1314.

51. Lukovkin G, Pshezhetsky V, Murtazaeva G. NMR 13C study of the structure of polyethyleneimine. Eur Polym J 1973; 9: 559–565.

52. Coll JL, Chollet P, Brambilla E, Desplanques D, Behr JP, Favort M. In vivo delivery to tumors of DNA complexes with linear polyethyleneimine. Hum Gene Ther 10:1659–1666.

53. Suh J, Paik HJ, Hwang BK. Ionization of polyethyleneimine and polyallylamine at various pH's. Bioorganic Chem 1994; 22:318–327.

54. Behr JP. The proton sponge: A trick to enter cells the viruses did not exploit. Chimia 1997; 51:34–36.

55. Tseng WC, Tang CH, Fang TY. The role of dextran conjugation in transfection mediated by dextran-grafted polyethylenimine. J Gene Med 2004; 6:895–905.

56. Tseng WC, Jong CM. Improved stability of polycationic vector by dextran-grafted branched polyethylenimine. Biomacromolecules 2003; 4:1277–1284.

57. Shi L, Tang GP, Gao SJ, Ma YX, Liu BH, Li Y, Zeng JM, Ng YK, Leong KW, Wang S. Repeated intrathecal administration

of plasmid DNA complexed with polyethylene glycol-grafted polyethylenimine led to prolonged transgene expression in the spinal cord. Gene Ther 2003; 10:1179–1188.

58. Tang GP, Zeng JM, Gao SJ, Ma YX, Shi L, Li Y, Too HP, Wang S. Polyethylene glycol modified polyethylenimine for improved CNS gene transfer: effects of PEGylation extent. Biomaterials 2003; 24:2351–2362.

59. Sung SJ, Min SH, Cho KY, Lee S, Min YJ, Yeom YI, Park JK. Effect of polyethylene glycol on gene delivery of polyethylenimine. Biol Pharm Bull 2003; 26:492–500.

60. Petersen H, Fechner PM, Martin AL, Kunath K, Stolnik S, Roberts CJ, Fischer D, Davies MC, Kissel T. Polyethylenimine-graft-poly (ethylene glycol) copolymers: influence of copolymer block structure on DNA complexation and biological activities as gene delivery system. Bioconjug Chem 2002; 13:845–854.

61. Brus C, Petersen H, Aigner A, Czubayko F, Kissel T. Efficiency of polyethylenimines and polyethylenimine-graft-poly (ethylene glycol) block copolymers to protect oligonucleotides against enzymatic degradation. Eur J Pharm Biopharm 2004; 57:427–430.

62. Brus C, Petersen H, Aigner A, Czubayko F, Kissel T. Physicochemical and biological characterization of polyethylenimine-graft-poly(ethylene glycol) block copolymers as a delivery system for oligonucleotides and ribozymes. Bioconjug Chem 2004; 15:677–684.

63. Fischer D, Osburg B, Petersen H, Kissel T, Bickel U. Effect of poly(ethylene imine) molecular weight and pegylation on organ distribution and pharmacokinetics of polyplexes with oligodeoxynucleotides in mice. Drug Metab Dispos 2004; 32:983–992.

64. Lemieux P, Vinogradov SV, Gebhart CL, Guerin N, Paradis G, Nguyen HK, Ochietti B, Suzdaltseva YG, Bartakova EV, Bronich TK, St-Pierre Y, Alakhov VY, Kabanov AV. Block and graft copolymers and NanoGel copolymer networks for DNA delivery into cell. J Drug Target 2000; 8:91–105.

65. Nguyen HK, Lemieux P, Vinogradov SV, Gebhart CL, Guerin N, Paradis G, Bronich TK, Alakhov VY, Kabanov AV.

Evaluation of polyether-polyethyleneimine graft copolymers as gene transfer agents. Gene Ther 2000; 7:126–138.

66. Gebhart CL, Sriadibhatla S, Vinogradov S, Lemieux P, Alakhov V, Kabanov AV. Design and formulation of polyplexes based on pluronic-polyethyleneimine conjugates for gene transfer. Bioconjug Chem 2002; 13:937–944.

67. Gebhart CL, Kabanov AV. Evaluation of polyplexes as gene transfer agents. J Contr Rel 2001; 73:401–416.

68. Ochietti B, Lemieux P, Kabanov AV, Vinogradov S, St-Pierre Y, Alakhov V. Inducing neutrophil recruitment in the liver of ICAM-1-deficient mice using polyethyleneimine grafted with Pluronic P123 as an organ-specific carrier for transgenic ICAM-1. Gene Ther 2002; 9:939–945.

69. Ochietti B, Guerin N, Vinogradov SV, St-Pierre Y, Lemieux P, Kabanov AV, Alakhov VY. Altered organ accumulation of oligonucleotides using polyethyleneimine grafted with poly (ethylene oxide) or pluronic as carriers. J Drug Target 2002; 10:113–121.

70. Belenkov AI, Alakhov VY, Kabanov AV, Vinogradov SV, Panasci LC, Monia BP, Chow TY. Polyethyleneimine grafted with pluronic P85 enhances Ku86 antisense delivery and the ionizing radiation treatment efficacy in vivo. Gene Ther 2004 [Epub ahead of print].

71. van de Wetering P, Moret EE, Schuurmans-Nieuwenbroek NM, van Steenbergen MJ, Hennink WE. Structure-activity relationships of water-soluble cationic methacrylate/ methacrylamide polymers for nonviral gene delivery. Bioconjug Chem 1999; 10:589–597.

72. Jeong B, Choi YK, Bae YH, Zentner G, Kim SW. New biodegradable polymers for injectable drug delivery systems. J Contr Rel 1999; 62:109–114.

73. Jeong B, Bae YH, Lee DS, Kim SW. Biodegradable block copolymers as injectable drug-delivery systems. Nature 1997; 388:860–862.

74. Li Z, Ning W, Wang J, Choi A, Lee PY, Tyagi P, Huang L. Controlled gene delivery system based on thermosensitive biodegradable hydrogel. Pharm Res 2003; 20:884–888.

75. Ahn CH, Chae SY, Bae YH, Kim SW. Synthesis of biodegradable multi-block copolymers of poly(L-lysine) and poly(ethylene glycol) as a non-viral gene carrier. J Contr Rel 2004; 97:567–574.

76. Ahn CH, Chae SY, Bae YH, Kim SW. Biodegradable poly(ethylenimine) for plasmid DNA delivery. J Contr Rel 2002; 80: 273–282.

77. Maruyama A, Ishihara T, Kim JS, Kim SW, Akaike T. Nanoparticle DNA carrier with poly(L-lysine) grafted polysaccharide copolymer and poly(D,L-lactic acid). Bioconjug Chem 1997; 8:735–742.

78. Wu GY, Wu CH. Receptor-mediated in vitro gene transformation by a soluble DNA carrier system. J Biol Chem 1987; 262:4429–4432.

79. Kim JS, Kim BI, Maruyama A, Akaike T, Kim SW. A new non-viral DNA delivery vector: the terplex system. J Contr Rel 1998; 53:175–182.

80. Kim JS, Maruyama A, Akaike T, Kim SW. Terplex DNA delivery system as a gene carrier. Pharm Res 1998; 15:116–121.

81. Suh W, Han SO, Yu L, Kim SW. An angiogenic, endothelial-cell-targeted polymeric gene carrier. Mol Ther 2002; 6: 664–672.

82. Choi YH, Liu F, Park JS, Kim SW. Lactose-poly(ethylene glycol)-grafted poly-L-lysine as hepatoma cell-targeted gene carrier. Bioconjug Chem 1998; 9(6):708–718.

83. Benns JM, Maheshwari A, Furgeson DY, Mahato MI, Kim SW. Folate-PEG-folate-graft-polyethylenimine-based gene delivery. J Drug Target 2000; 9:123–139.

84. Lee ER, Marshall J, Siegel CS, Jiang C, Yew NS, Nichols MR, Nietupski JB, Ziegler RJ, Lane MB, Wang KX, Wan NC, Scheule RK, Harris DJ, Smith AE, Cheng SH. Detailed analysis of structures and formulations of cationic lipids for efficient gene transfer to the lung. Hum Gene Ther 1996; 7:1701–1717.